Fruit Phenolics

Authors

Jean-Jacques Macheix, D.Sc.
Department of Plant Physiology
Université des Sciences
et Techniques du Languedoc
Montpellier, France

Annie Fleuriet, D.Sc.
Department of Plant Physiology
Université des Sciences
et Techniques du Languedoc
Institut Universitaire de Technologie
Montpellier, France

Jean Billot, D.Sc.
Department of Plant Physiology
Université d'Orléans
Orléans, France

CRC Press
Taylor & Francis Group
Boca Raton London New York

CRC Press is an imprint of the
Taylor & Francis Group, an informa business

First published 1990 by CRC Press
Taylor & Francis Group
6000 Broken Sound Parkway NW, Suite 300
Boca Raton, FL 33487-2742

Reissued 2018 by CRC Press

Library of Congress Cataloging-in-Publication Data

Macheix, Jean-Jacques.
Fruit phenolics/authors, Jean-Jacques Macheix, Annie Fleuriet, Jean Billot.
p. cm.
ISBN 0-8493-4968-0 1. Fruit—Physiology. 2. Phenols. 3. Fruit—Composition.
I. Fleuriet, Annie. II. Billot, Jean III. Title.
SB357.28.M33 1990
582'.0464—dc20 89-17461

A Library of Congress record exists under LC control number: 89017461

Publisher's Note
The publisher has gone to great lengths to ensure the quality of this reprint but points out that some imperfections in the original copies may be apparent.

Disclaimer
The publisher has made every effort to trace copyright holders and welcomes correspondence from those they have been unable to contact.

ISBN 13: 978-1-315-89307-5 (hbk)
ISBN 13: 978-1-351-07217-5 (ebk)

Visit the Taylor & Francis Web site at http://www.taylorandfrancis.com and the
CRC Press Web site at http://www.crcpress.com

PREFACE

Why write a book today on phenolic compounds in fruits? The rapid evolution of knowledge on the subject seems to us to be the best justification. Indeed, considerable progress has been made over the past 20 years, thanks in particular to the development of analytical techniques which enable both the accurate determination of new molecules and their quantitative estimation. However, although much experimental data concerning analytical, metabolic, physiological, and technological aspects can now be found in the literature, there has been no recent overall study of phenolics of fruits. Likewise, few of the numerous works devoted to the phenolic compounds in plants cover fruits, even partially. Fruits are nevertheless very interesting plant material, partly because of their position in the biological cycle of the higher plants and partly because of their use as food, either as they are or in processed form. Phenolic compounds are always involved in the relations of a fruit with its biological and physicochemical environment and contribute to the numerous reactions which lead to the sensory properties characteristic of the fruit itself or of fruit products.

The format of the book follows a simple rule: rather than examine fruits separately, we have as far as possible grouped the data around a few major ideas which form the basis for the chapters (qualitative and quantitative aspects, metabolism and role of phenolic compounds in fruits, and technological implications). This system appears to have the merit of making common features and differences in phenolic patterns and their significance stand out better. It has also made it possible to stress the aspects which have received most study in fruits and, in contrast, to show the need for developing research in certain other fields. In addition, the phenolic patterns of the main fruits of economic interest have been assembled in Chapter 2 to facilitate the analytical approach of the work.

Faced with the many fields in which phenolics can be involved, we have limited the study to fleshy fruits of economic interest. With a few exceptions, neither ripe dried fruits nor seeds alone are covered. The data that we report and discuss, therefore, mainly concern fleshy fruits themselves, although in some cases it is not easy to draw a line between the fruit and the seeds that it contains.

Several thousand bibliographical references have been consulted during the writing of the book. We have reported the most recent as far as possible and have voluntarily chosen not to mention those earlier than 1969 to 1970 which can be found in two very complete reviews published then: first, the chapter written by Van Buren[593]* on fruit phenolics in a multi-author work on the biochemistry of fruits; and second, the book by Singleton and Esau[528] on phenolic compounds in grape and wine. However, older references are cited when the work in question made noteworthy progress at the time (e.g., the revelation of a new phenolic compound in a fruit) or when the subject investigated has not subsequently received closer attention.

We hope that this work will be useful for all plant physiologists, phytochemists, and technologists who take an interest in phenolic compounds and in fruits within the framework of fundamental research or for more applied purposes.

We thank all the colleagues who have helped us in the preparation of the book and we hope that they will not be disappointed.

* References for the Preface are to be found at the end of Chapter 2.

THE AUTHORS

Dr. Jean-Jacques Macheix is Professor of Plant Physiology at Université des Sciences et Techniques du Languedoc (Montpellier, France). After studying at Ecole Normale Supérieure de St. Cloud, France, he was awarded a Master of Science degree at Université Pierre et Marie Curie (Paris) in 1962 and then a Doctorat es Sciences in 1974. Dr. Macheix holds an Agrégation degree (Biology). After teaching for 12 years at Paris University first as assistant lecturer and then as lecturer, he was appointed Professor of Plant Physiology at Montpellier University in 1976. He is now co-director of the Applied Plant Physiology Laboratory. Dr. Macheix is a member of the French Plant Physiology Society, of the French section of the International Association of Plant Tissue Culture and of the Phytochemical Society of Europe. He has been President of the Groupe Polyphénols, an international scientific association whose special interest is phenolic compounds in plants, since 1986. Dr. Macheix is the author of over 50 papers on phenolic compounds in plants and in particular in fruits. The major lines of research followed concern metabolic aspects and determination of the role of phenolics in enzymatic browning phenomena.

Dr. Annie Fleuriet is Lecturer in Plant Physiology and Biochemistry at Institut Universitaire de Technologie (Montpellier, France). She obtained a Master of Science degree (Biology) at Université Pierre et Marie Curie (Paris) in 1968 and a Doctorat es Sciences in 1982 from Université des Sciences et Techniques du Languedoc (Montpellier). She taught as Assistant Lecturer in Paris from 1970 to 1980 and then at Institut Universitaire de Technologie (Montpellier) where she was appointed Lecturer in 1984. Dr. Fleuriet is a member of the French Plant Physiology Society and the Groupe Polyphénols, an international scientific association whose special interest is phenolic compounds in plants. She carries out research at the Applied Plant Physiology Laboratory (University of Montpellier) and is the author of some thirty papers on the metabolism of phenolic compounds, in particular during the development of fruits, and on variations in this metabolism following stress.

Dr. Jean Billot is Professor of Plant Physiology at Orléans University (France). He studied at Ecole Normale Supérieure de St. Cloud, France from 1956 to 1960 and was awarded a Master of Science degree by Paris University in 1958 and an Agrégation degree in Natural Science in 1960. He received a Doctorat es Sciences in Natural Science at Orsay University, France, for work on anthocyanic pigments. From 1961 to 1972 he was Assistant Professor and then Professor of Plant Physiology at the University of Antananarivo (Democratic Republic of Madagascar). He has been Professor at Orléans University since 1973. Dr. Billot is a member of the French Plant Physiology Society, affiliated to the FESPP, and of the Groupe Polyphénols. He has published some thirty papers on anthocyanins and other phenolic compounds. His current major research interest concerns analysis of phenolic compounds, especially in fruits, in relation to their possible involvement of the control of development.

ACKNOWLEDGMENT

The editors wish to express their deepest thanks to Simon Barnard who translated this book from French to English.

TABLE OF CONTENTS

Chapter 1

THE MAIN PHENOLICS OF FRUITS

I. INTRODUCTION

In addition to the primary metabolites common to all plants, analysis has led to distinguishing numerous secondary metabolites which belong to a variety of chemical groups. Among them, phenolics embrace a considerable range of substances which possess an aromatic ring bearing one or more hydroxyl substituents. These compounds are present in all plant tissues and frequently form the most abundant secondary metabolites in fruits, where they sometimes reach high concentrations.

The several thousand polyphenols which have been described in plants can be grouped in several classes,[220]* most of which are found in fruits (Table 1). Distinction between these classes is drawn first of all on the basis of the number of constitutive carbon atoms and then in the light of the structure of the basic skeleton. In addition, besides simple soluble forms found mainly in vacuoles, there are also polymerized forms of varying solubility (tannins) or which are completely insoluble (lignins).

We have mentioned as examples for each of these classes a fruit in which the constituents of this class are particularly abundant (Table 1), which does not exclude the presence of other phenolic compounds in the same fruit. As is often the case with all secondary metabolites, distribution of the constituents of the different classes is not homogeneous in all fruits. Some compounds, such as hydroxycinnamic acids and flavonoids are very widely represented whereas others are only found in the fruit of a particular genus and species.

The diversity observed in fruits with regard to the distribution of phenolics is also found in terms of quantity. We have grouped in Table 2 the main results obtained in different ripe fruits for the six principal classes found and for total phenolics when this was estimated. Analysis of the data in Table 2 should nevertheless be handled cautiously for various reasons: (1) the absence of figures in a column may mean either absence of the corresponding compounds in the fruit or absence of reliable quantitative data; (2) as is discussed later, estimations of total phenolics are often questionable; (3) results display great variability for different cultivars of the same species and also from one author to another: only extreme values have been reported; and (4) in certain fruits (e.g., citrus and olive), the major phenolic compounds do not appear in the table since they belong to none of the six classes used.

Rapid analysis of Table 2 reveals considerable differences in the total phenolics contents according to the fruit: whereas some have a fairly low phenolics content, others, e.g., plum and persimmon, contain amounts totaling 5 to 10% of their dry weight (DW); the maximum is in sapodilla (approximately 6.5 mg/100 g of fresh weight [FW]). Such variations are also found in the various classes of phenolics: whereas hydroxybenzoic acids and flavonols are generally not quantitatively very abundant, the opposite is true for hydroxycinnamic derivatives and for anthocyanins and tannins when they are present. Certain fruits display remarkable accumulations of compounds belonging only to one or two classes, e.g., hydroxycinnamic derivatives and anthocyanins in cherry, or tannins in persimmon. In these cases, the amounts of total phenols broadly reflect the presence of the major compounds. In contrast, distribution of the various classes is more balanced in other fruits, such as red varieties of grape. All these data are examined in detail below, but these preliminary observations already show how the qualitative and quantitative expression of the phenolic metabolism may differ between species and varieties. This notion is amply illustrated by studying the distribution of phenolics within each of the classes defined above. This study

* All References for this chapter can be found at the end of Chapter 2.

TABLE 1
The Major Classes of Phenolics in Fruits

Number of carbon atoms	Basic skeleton	Class	Example	Fruit (example)
7	C_6-C_1	Hydroxybenzoic acids	*p*-Hydroxybenzoic	Strawberry
9	C_6-C_3	Hydroxycinnamic acids	Caffeic	Apple
		Coumarins	Scopolin	Citrus
10	C_6-C_4	Naphthoquinones	Juglone	Walnut
13	C_6-C_1-C_6	Xanthones	Mangiferin	Mango
14	C_6-C_2-C_6	Stilbenes	Resveratrol	Grape
15	C_6-C_3-C_6	Flavonoids	Quercetin, cyanidin	Cherry
		Isoflavonoids	Daidzein	French bean
n		Lignins		Stone fruits
		Tannins		Persimmon

is preceded by a rapid summary of the techniques used in the analysis of phenolic compounds, with particular stress on applications to fruits.

II. TECHNIQUES OF ANALYSIS

The methods used for studying phenolic compounds are very varied and have been the subject of numerous reviews and updates.[82,262,263,288,332,333,358-360,411,469,595] The methods are applied to both fruits and other parts of plants. Before approaching the study of the various categories of phenolic compounds which have been identified in fruits, a brief reminder of the various techniques used to obtain both qualitative and quantitative data seems useful. The examples and references given are for soluble phenolic compounds in fruits. Indeed, we do not approach the study of insoluble residues — lignins in particular. In addition, tannins raise special analysis problems which are discussed in the chapter on these polymers. We examine successively techniques for the extraction, purification, separation, identification, and measurement of phenolic compounds, as summarized in Figure 1.

A. EXTRACTION AND PURIFICATION

The first problem to be solved is that of having an extraction method suited to the plant material chosen (a given fruit in this case) and to the types of phenolic compound to be studied. Such a method must enable complete extraction of phenolic compounds and must avoid chemical modification — which result in artifacts — as much as possible.

Soluble phenolic compounds are generally extracted by means of alcohol-water solutions. For example, an extraction technique using 80% ethanol was perfected for hydroxycinnamic derivatives and flavonols in apple[336,340] and then applied to tomato.[150,151] Extraction is performed by ethanol and water mixture (80/20, v/v) applied to the powder obtained by crushing frozen or freeze-dried material in the presence of liquid nitrogen. Oxidation is avoided by working at a low temperature. Methanol (80%) is also used to extract soluble phenolic compounds from fruits, for example from cherry[381,382] and pear.[56] It was checked in all these cases that hydroxycinnamic derivatives, flavonols, and catechins were fully extracted and that no soluble phenolic compounds remained in the plant material. A solvent mixture of methanol-ethyl acetate-H_2O (4:4:2, v/v/v) also gave excellent results.[56a]

Extraction of anthocyanins is conventionally carried out under cold conditions with methanol containing 1% concentrated hydrochloric acid[469] to obtain the flavylium cation form which is red and stable in a highly acid medium. Fuleki[180] used a methanol-water-concentrated HCl mixture (50:49:1) to extract anthocyanins from cranberries (*Vaccinium macrocarpon*). A detailed study of the conditions of extraction of pulp of elderberry (*Sambucus nigra*) was carried out recently.[78] Anthocyanin extraction from elderberry pomace

TABLE 2
Phenolic Contents in Ripe Fruits

Fruits	Hydroxybenzoic acid derivatives	Hydroxycinnamic acid derivatives	Anthocyanins	Flavonols	Flavan-3-ols	Tannins	Total phenolics
Apple		6.2—134[388,389,395,475,593]	10—2160 (peel)[593] 70 nmol/cm^2 (peel)[144]	17.8—40.5 (peel)[342,568] 154—285 (peel)[126]	0.2—16.3[395,475] 45.8 (peel)[342]	40—350 mg/100 ml (cider)[629]	50—1100[336,593] 78—120 (pulp)[273] 790—1160 (peel)[273] odiP[a]: 63—79[403]
Cell culture		8.3—27.9[388,389,475]		1.3—1.6[247]	9.3—24.6[395,475]		0.6 μmol/g DW[a309]
Apricot							odiP: 12—56[590]
Aronia Melanocarpa			0.56% DW[293]			7.38% DW[293]	4.52% DW[293]
Avocado		2.8—46[196,438]			Peel = 3300[438] Pulp = 9[438] Susceptible = 6—13[442] Resistant = 63.6—174[442]		2.8—28.7[196,594] 110-180[580] odiP: 1.4—1.6[196] peel: 11550 ?[438]
Banana						0.6[371]	150[23] 0.53% DW[593]
Boysenberry						90[205]	
Blackberry	0.9—3.0[512]	6.2—8.8[512]	82—180[579] 234—326[579]	0.85—2.15[b57] 20.7—31.4[246]	7.3—17.7[396]		
Blackcurrant	1.0—2.1[512]	12.1—14.6[512]	250[305]	7.6—31.5[523]	0.9—1.4[550]	0.21—0.37[c]%[205]	
Cherry							
Sweet	0.05—1.3[552] 4.1[304]	28.5—140[380,388,389,475,593]	350—450[h]	1.1—3.1[249]	2.0—7.0[395,475,552]	Bn[d]: 70[380]	Bn[d]: 360[380]
Sour	0.1—0.8[b552]	27.0—67.1[388,389,475]	28.8[120]	4.2—6.0[249]	15.2—20.2[395,475,552]		200[593]
Chinese quince							0.4[371]
Citrus[f]							
Grapefruit		12.0—21.7[551]					
Orange		13.6—16.3[551]					
Coffee pulp	0.09—0.35 mg/g DW[451]	3.87—40.8 mg/gDW[108,451]		0.09—0.70 mg/g DW[451]	1.59—4.94 mg/gDW[451]		62.9 mg/g DW[659]
Crowberry			300—460[296,326]				
Date	0.17 mg/100g DW[458]	0.42 mg/100 g DW[458]				114.5[349]	500[593] 133[e349]
Eggplant		59.8—66.0[639]					
Elderberry			448[78]	10.5—23.7[548]			
Gooseberry	1.1—2.1[512]	2.5—14.8[512]			1.7—3.3[550]	0.06—0.1%[c205]	
Grape							
White	0.4 mg/100 g DW[463] (skin)	1.33—86.55 (skin)[70] 16—346 mg/l[c533]		0.81—8.19[103]	1.4—52.7 (skin)[72,530]	17.2 (skin)[72]	350 (skin)[593] 100 (pulp)[593]
Red	24.73 mg/100 g DW[463] (skin)	10—109 (skin)[70] 0.25—4.0 μmol/g DW[486] 64—477 mg/l[c533]	8—388[114,313,432]	1.85—9.75[103]	22—37[c] ppm[404] 2.5—16.7 (skin)[72]	32.1—78.3 (skin)[72]	900—950[114]
Teinturier		29—147.5 (skin)[70] 127—274 mg/l[c533]	161—282[432]		23.4 (skin)[72]	64.2 (skin)[72]	
Huckleberry			100[167]				
Kiwi		60—100[178]				35—60[178]	
Loganberry						0.19%[c 205]	

TABLE 2 (continued)
Phenolic Contents in Ripe Fruits

Fruits	Hydroxybenzoic acid derivatives	Hydroxycinnamic acid derivatives	Anthocyanins	Flavonols	Flavan-3-ols	Tannins	Total phenolics
Mango						31—75[555]	2% peel[555]
							1% flesh[555]
Miracle fruit			14.3[88]	7.2[88]			
Olive[g]		2.5—5.1[8]		3.3[8]			
Passion *fruit*							1.4[593]
Peach		8.1—75.0[89,388,389,475,593]		0.4—1.0[247]	5.3—14.1[396,475]		28—180[593]
							odiP: 3—40[89]
Pear		170[593]	5—10 (peel)[185]	4[56];20—160 (peel)[185]	0.1—6.2[56,395,475]	53[56]	123—400[56,593]
		6.4—51.8[56,388,389,475]					odiP:1—47[54,589]
Persimmon					0.06% DW[657]	0—2000[281,282,371,372]	1—12% DW[593,657]
Pineapple		5.3—14.9[128,139]					
Plum	0.2—1.2[b,552]	12.1—94[388,389,444,475,593]	1.9—5.3[593]	2—5.2[258]	1.8—6.1[395,475,552]	76—150[307]	167—200[307]
	2.2[304]	2.76 mg/g DW (pulp)[457]	6.34 mg/g DW (peel)[457]	2.20 mg/g DW (peel)[457]	0.10 mg/g DW (pulp)[457]		2—5.7% DW[593]
		7.66 mg/g DW (peel)[457]			0.74 mg/g DW (peel)[457]		odiP: 20.4—101.8[591]
Raspberry							
Red	3.4—4.4[512]	2—4[512]	23—59[579]	7.2—10.2[246]	3.2—4.9[396]	0.10—0.14[c,205]	
Black			214—428[579]				
Red currant	0.9—1.3[512]	1—2.6[512]	11.9—18.6[425]	2—5.1[523]	0.4—3.6[550]		
Sapodilla					4170[312]	1360[312]	6470[312]
Strawberry	0.7[512]	1.4—3.1[273]	28—70[645,648]	2.1—17.4[246]	2.4—9.6[549]	110—150[205]	
Tamarillo fruit			58 (peel)[646]				
Tomato	0.15[507]	7—23.2[158,639]					45—57[255]
		Pericarp: 9.4[158]					22 skin[620]; 3.9 (peeled)[620]
		Pulp: 8.4[158]					
Vaccinium							
Bilberry		250[593]	460[326]	14.8—22.8[548]	2—2.8[550]		
Blueberry	0.5—1.7[512]	188.1—211.2[512]	160—503[28,30]	2.4—2.85[57]			
Cranberry			46—172[323,496]	13.9—33.4[57,323]			

Note: All values in mg/100 g fresh weight except where otherwise specified.

[a] odiP: *o*-diphenol; DW: dry weight.
[b] After hydrolysis.
[c] Juice values.
[d] Bn: cv Bigarreau Napoléon.
[e] Simple polyphenols and soluble tannins.
[f] Flavanones and coumarins are abundant in citrus.
[g] Oleuropein is abundant in olive fruit.
[h] Billot and Okombi: unpublished data.

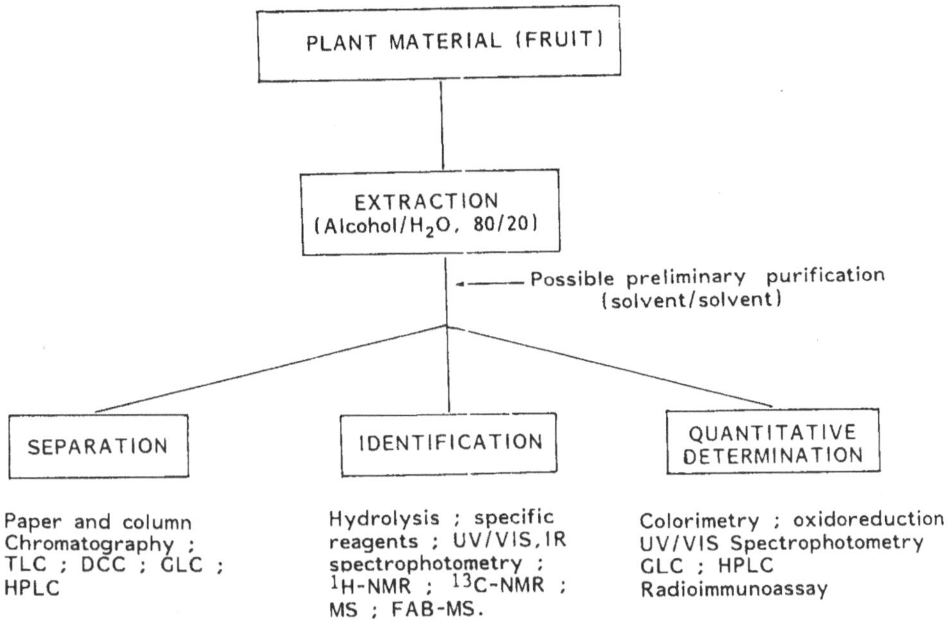

FIGURE 1. Analysis scheme of fruit phenolics.

was improved by mechanical disintegration and use of aqueous HCl instead of acidified ethanol as extractant, with pH being important in determining extractant effectiveness.

Acylated anthocyanins pose a problem since acyl groups — particularly those formed from dicarboxylic acids, such as malonic acid — are labile in solution containing mineral acid and are lost with methanol-HCl. Replacing HCl by acetic acid or formic acid makes it possible to isolate these pigments with the dicarboxylic substituents intact: e.g., recent study of malonated anthocyanins, especially in flowers.[200,222,223,560] Extraction of anthocyanins from fruits can be performed by maceration in a methanol solution containing 2% or 10% formic acid; this has been described for blueberry[31] and grape.[198,650] In the latter case, the anthocyanin-acetates present in grape skin were obtained. Bakker and Timberlake[26] showed that artifacts could be obtained from *Vitis* berries by using formic acid or acetic acid combined with HCl. It is, therefore, essential to ensure that the acetylated derivatives observed are in fact natural forms; this is the case, for example, of malvidin-3-acetylglucoside in grape. The authors have checked that this compound is obtained with satisfactory stability with a solution of 0.6% perchloric acid and methanol (80/120, v/v). Use of nonacidified solvents has been recommended to prevent the formation of artifacts and the degradation of the least stable anthocyanins.[596]

The water-alcohol extract obtained is a raw extract containing numerous nonphenolic substances (sugars, organic acids, proteins, pigments, etc.) which can interfere during chromatographic separation and assays. Purification is thus essential:

- The first stage is the removal of chlorophylls and carotenoids using the conventional depigmentation technique with petroleum ether applied to the aqueous extract obtained after the removal of organic solvents. This method can be improved by the addition of metaphosphoric acid (1 to 2%) and ethanol (10 to 20%) according to Poessel.[435] It may be necessary to check that this method does not cause a loss of certain lipophilic phenolic molecules which have been found in various species.[220]
- The second stage of purification of phenolic compounds consists of extracting them from the depigmented aqueous extract with ethyl acetate, using a method previously

TABLE 3
Preparative Separation Methods for Plant Phenolics

Paper chromatography		
Thin-layer chromatography	⟶	Centrifugal TLC
Open column chromatography (polyamide, cellulose, Sephadex, silica)		
Pressure column chromatography	⟶	Flash chromatography
(Polyamide)		Low-pressure liquid chromatography
Silica		Preparative liquid chromatography
Chemically bonded phases		(axial or radial compression of the column)
(RP-18,RP-8, RP-2, NH_2, CN)		Semi-prep. high performance liquid chromatography
Liquid-liquid chromatography	⟶	Craig distribution
		Droplet counter-current chromatography
		Rotation locular counter-current chromatography

From Hostettman, K., *Ann. Proc. Phytochem. Soc. Eur.*, Vol. 25, Van Sumere, C.F. and Lea, P. J., Eds., Clarendon Press, Oxford, 1985, 1. With permission.

described[153,340] in which ammonium sulfate (20%), metaphosphoric acid (2%), and ethanol (20%) are added. These conditions are necessary to enable quantitative extraction of the main phenolic compounds by ethyl acetate. It should be noted that anthocyanins, which are practically insoluble in ethyl acetate, remain in the aqueous phase. This method of purification has been used to study phenolic compounds in various fruits.[8,56,151,340,380]

B. SEPARATION AND IDENTIFICATION

1. Separation

Separation of the various phenolic compounds in an extract is an essential stage prior to the application of identification and measurement techniques. Given on the one hand the diversity and complexity of the phenolic structures found in an extract, and on the other the low contents of many of these compounds, only modern methods of analysis have made it possible to make rapid progress in knowledge of phenolic compounds in fruits. Reference should be made to Hostettmann and Hostettmann[263] and Hostettmann[262] for a general review of isolation techniques for flavonoids and plant phenolics. These techniques can be used for preparative separation (Table 3), but most can be applied for analytical purposes. We give here only a few examples of applications concerning fruits.

Column chromatography is a particularly useful technique for the isolation of large amounts of the compounds in a plant extract. It has been used with various supports (silica gel, magnesium silicate, alumina, cellulose and DEAE-cellulose, polyamide, PVP and Polyclar AT, Sephadex) and has been used in numerous applications for the separation of phenolics.[274,333,358] In fruits, examples include the separation of anthocyanins and chlorogenic acid on Polyclar AT,[15,597] of flavonol glycosides on Polyamide SC6,[102,127,249] and the utilization of Sephadex LH20 for grape.[322,416,532] Although separation of the compounds is not complete, running raw extract through the column can achieve both purification and preliminary fractionation. Thus, for example, an SC6 polyamide column has the advantage of separating the two groups of hydroxycinnamic acid derivatives: glucose and quinic, tartaric, malic, or galactaric derivatives.[387,388,389,477,512,638,639] Filtration on Sephadex gel is frequently the final stage of purification of fractions isolated using other methods. It can be noted that considerable hydrolysis of glycosides appears in the PVP column during separation of anthocyanins in elderberries.[79]

Paper and thin-layer chromatography (TLC) techniques have been used for some 30 years to separate the phenolic compounds in fruits. Numerous solvent systems have been recommended[166,218,333,340,469] and applied, in particular for the separation of anthocyanins, hydroxycinnamic derivatives, and flavonols in fruits.

Chromatography on cellulose, silica gel, or polyamide plates is still very widely used for the separation of various groups of phenolic compounds.[263,358] In fruits, much work on the analysis of phenolics has been carried out using TLC. We mention only the isolation of flavonol glucuronides of *Rubus idaeus*,[492] of anthocyanins in *Rubus*,[37] and flavonol glycosides and hydroxycinnamic esters in *Ribes nigrum*.[305] Analysis of hydroxybenzoic and hydroxycinnamic acids by TLC on silica gel[510] can be applied to fruit extracts after hydrolysis of esters and glycosides. A technique using high performance silica chromatoplates (HPTLC) has been developed recently.[263] It enables very fine separation of compounds and can be used to separate the substances produced by total or partial microhydrolysis of caffeic acid heterosidic esters.[10]

The separation of phenolics has greatly progressed over the past decade thanks to use of high performance liquid chromatography (HPLC) which enables both **separation** and **quantitative analysis.**[263,582] All phenolic groups have benefited from this effective technique which has the advantage of not requiring prior transformation. In particular, the development of reversed phase columns has greatly improved the separation performance of anthocyanins by HPLC. For example, 18 anthocyanins in a mixture of 20 in *Vitis lambrusca* berries can be separated in 2 h.[636] The use of various supports and mobile phases has enabled the separation and measurement of most phenolic compounds in fruits (Table 4 and Figure 2): flavonol glycosides,[102,103,127,246-249,523] polymethoxylated flavones,[51,188,211,245,578] flavanones and flavanonols,[148,290,430,489,530] monomeric flavan-3-ols and procyanidins,[318,319,530] hydroxycinnamic derivatives,[34,159,363,387-389,421,486,512,638,639] and anthocyanins.[11,25,26,31,79,180,198,297,404,484,498,544,603,632,636,650] A diode array detector was used for simultaneous recording of chromatographic analysis at several wavelengths, e.g., for anthocyanins of *Vitis vinifera*.[242]

A general procedure for the analysis of phenolic compounds by reversed phase HPLC with ternary elution gradients has been designed; this method was applied to the separation and estimation of phenolic acids, phenolic aldehydes, and coumarins in plant extracts, e.g., grape juices.[605]

Gas-liquid chromatography is relatively little-used today for the study of phenolic compounds since it requires prior transformation of the compounds and a high operating temperature. Nevertheless, there are some experimental results concerning the study of free phenolic acids[261,387,388,394,511] and hydroxycinnamic esters in various fruits.[153,512] Yonemori et al.[657] used gas chromatography to analyze tri-methyl silylated derivatives of catechin and gallic acid in persimmon. Gas-liquid chromatography was also used in combination with mass spectrometry for the separation and identification of anthocyanins,[62,63] particularly in bilberries.[25]

Droplet counter-current chromatography (DCCC) is a fairly recent technique and has been used for the separation of anthocyanins, particularly of blackcurrant, raspberry, and various *Vaccinium*.[11,13,169] This technique makes it possible to isolate anthocyanins on a preparative scale and would appear to complement column and plate chromatography (Figure 3).

2. Identification

Identification of phenolic compounds after isolation and purification is a delicate problem since standards for complex forms (esters and glycosides) are not always available for comparisons using cochromatography on paper, TLC, or HPLC.

Various conventional criteria of identification are first used to determine the precise nature of the various phenolic compounds isolated: UV-visible spectrophotometry, reagents for visualization on the chromatograms, hydrolysis releasing the various constituents, comparison with authentic standards using TLC, HPLC, etc.

TABLE 4
HPLC Separation of Phenolic Compounds in Fruits

Fruits	Compounds	Column	Mobile phase	λ (nm)	Ref.
Apple	Flavonol glycosides; phloridzin; catechins; phenolic acids	RP C_{18}	25—50% tetrahydrofuran in 0,1% aqueous trifluoroacetic acid over 20 min	270	126,127
Juice	Chlorogenic acid; (+)-catechin; (−)-epicatechin; phloridzin; phloretin xyloglucoside; procyanidins	Zorbax CN	A: tetrahydrofuran (THE)-hexane-acetic acid-formic acid-isopropyl alcohol (IPA)(32:60:4:2:2) B: THF-hexane-acetic acid-formic acid-IPA(64.5:27.5:4:2:2) 2 min with solvent A then linear gradient (10 min) to solvent B	280	637
		μ Bondapak C_{18}	Linear gradient from aqueous 0,1 *M* KH_2PO_4-methanol(9:1) to aqueous 0.1 *M* KH_2PO_4 methanol (6:4) in 15 min		637
	Procyanidins dimers, oligomers; phloridzin; epicatechin; chlorogenic acid	LiChrosorb RP8 (5 μm)	From 100% water pH 2 with 0.1% perchloric acid to 100% methanol in 20 min	280	318,319
Apricot and cherry, peach, plum	Flavonol glycosides	LiChrosorb Si60 (5 μm)	Benzene-acetonitrile (80:20)	300	247,248,249
	Caffeoyl, *p*-coumaroyl, feruloyl quinic (*cis*, *trans*, 3′, 5′, 4′)	LiChrosorb RP 18 (5 μm)	A: 2% acetic acid; B: methanol; from 7% B to 15% in 10 min; from 15% B to 35% in 50 min	320	387—389
Bilberry	Anthocyanins	Aquapore RP-300 10 μm	A: water-formic acid (90:10) B: methanol-acetonitrile-water-formic acid (22.5:22.5:45:10); from 0 to 30% B in 40 min	530	25
	Anthocyanins	LiChrosorb RP18 (10 μm)	A: water-acetic acid (85:15) B: methanol-water-acetic acid (50:35:15)-gradient nonlinear	525	603,604
Blackberry	Flavonol glycosides	LiChrosorb Si60 (5 μm)	Toluene-acetonitrile (100:20) (35°C)	310	246
	Acylated flavonol glycoside	ODS-Hypersil (5 μm)	A: 1% acetic acid B: 5% acetonitrile To 30% B in 45 min	360	613
	Anthocyanins	C_{18} 5 μm	A: phosphate buffer 0.1 *M* B: acetonitrile From 12% B to 20% in 25 min	546	498
Blackcurrant	Flavonol glycosides	Ultrasphere ODS (5 μm)	A: 1% acetic acid B: acetonitrile From 5% B to 30% in 45 min.	360	523
Blueberry	Anthocyanins	HS-5-C_{18} RP	A: acetonitrile B: 10% formic acid 7% A in B for 10 min followed by 17% to 15% A in B in 17 min and then 15% to 30% A in B in 5 min.	530	31
Citrus					
Orange, tangerine juices	Polymethoxylated flavones	Zorbax C-8	Tetrahydrofuran-water (25:75)	313	578

Orange, Tangerine Mandarin	Polymethoxylated flavones	LiChrosorb Si 60 (6 μm)	Heptane-ethanol (75:25) or heptane-isopropanol (60:40)	280	51,188
Orange peel and juice	Polymethoxylated flavones	μ Bondapak C_{18}	Water-acetonitrile (60:40)	λ max	211
	Polymethoxylated flavones	ODS-Hypersil (5 μm)	Water-acetonitrile (60:40)	330	245
Grapefruit juice	Naringin, narirutin (naringenin rutinoside)	μ Bondapak C_{18}	Water-acetonitrile (80:20)	280	148
Orange juice	Hesperidin	μ Bondapak C_{18}	Water acetonitrile (80:20)	285	147
Grapefruit, sweet orange, sour orange, tangerine	Naringin, narirutin, hesperidin, neohesperidin	Zorbax ODS (C_{18})	Water-acetonitrile-acetic acid (79.5:20:0.5)	280	489
Orange and lemon peel	Flavanones; polymethoxylated flavones; 6-8-di-C-glycosylflavones.	Ultrasphere C-8 RP	A: methanol-acetonitrile-acetic acid 0.5% (5:10:85) B: methanol-acetic acid 0.5% (95:5) From A to B linearly in 45 min	280 or 340	430
Orange juice	Narirutin, hesperidin.	Li Chrosorb Si 60 (5 μm)	Isooctane-ethanol-acetonitrile (70:16:5.5) (37°C)	270	184
Citrus Orange juice (pigmented varieties)	Anthocyanins, acylated anthocyanins with acetic and hydroxycinnamic acids	μ Bondapak C_{18}	A: water-acetic acid (85:15) B: water-acetic-acid-methanol (65:15:20) From 1% B to 100% in 40 min	520	334,335
Cowberry	Anthocyanins	ODS-hypersil (3 μm)	A: formic acid-water (10:90) B: formic acid-water methanol (10:40:50) 0—1 min,:10% B in A (isocratic) 1—2 min,:10—30% B in A 2—10 min,:30—70% B in A	535 and λ max	11
Cranberry	Anthocyanins	NOVA-PAK C_{18} + μ Bondapak C_{18} RP	A: 4,5% formic acid in water B: methanol-8.2% formic acid in water-acetonitrile (55:33:10) C: methanol-1.3% formic acid in water-acetonitrile (55-35-10) From 0% B to 100% B in 70 min according to a complex convex gradient	520	180
	Anthocyanins	Polymeric reversed phase styrene divinylbenzene	A: acetic acid-water (10:90) B: methanol, acetic acid, water (60:10:30) 0% B to 90% B in 30 min	530	213
Crowberry	Anthocyanins	μ Bondapak C_{18}	Water-methanol-formic acid (74:16:10)	520	297
Elderberry	Anthocyanins (flavylium cations and chalcone forms)	Nucleosil 5 C_{18}	Tetrahydrofurane (THF) in 0.05 *M* phosphoric acid (pH 1.8). From 1% to 40% THF in 15 min	510 340	79
Fig	C-Glycosylflavones	ODS-Hypersil (5 μm)	A: 1% acetic acid B: 5% acetonitrile 5% B to 30% B in 45 min	333	525
Grape	Anthocyanins	μ Bondapak C_{18}	Water-acetic acid-methanol (71:10:19)	530	632
	Anthocyanins	μ Bondapak C_{18}	A: acetic acid-water (15:85) B: water-acetic acid-methanol (65:15:20) From 1% B to 100% B in 40 min	520	636
	Anthocyanins	LiChrosorb ODS	A: 10% formic acid in water B: 10% formic acid—50% methanol in water From 25% B to 75% B in 40 min.	520	650
	Anthocyanins	Spherisorb-Hexyl (5 μm)	A: Water acidified with 0.6% $HClO_4$ B: methanol From 20% B to 23% B (6 min), 30% (8 min), 40% (5 min), 50% (8 min), 95% (5 min)	520	26

TABLE 4 (continued)
HPLC Separation of Phenolic Compounds in Fruits

Fruits	Compounds	Column	Mobile phase	λ (nm)	Ref.
Grape (suite)	Anthocyanins	μ Bondapak C_{18}	A: formic acid-water (10:90) B: acetone, formic acid, water (25:10:65) From 20% B to 70% B in 1 h	546	484
	Anthocyanins	μ Bondapak C_{18} (10 μm)	A: 4,5 % formic acid in water B: acetonitrile 10% to 15% B over 10 min, 15% to 20% B over 10 min, 20% to 30% B over 17.5 min	525, 330,370 280	242
	Anthocyanins, Hydroxycinnamic esters, catechin, epicatechin	Zorbax ODS or LiChrosorb ODS	Acetonitrile and 0,5% formic acid in water (9:91), or A: 10% formic acid in water B: 25% acetone-10% formic acid in water 0—5 min: 20% B then to 80% B in 30 min	320 or 530	404
	Flavonol glycosides	Spherisorb ODS (5 μm)	A: acetonitrile; B: 4% acetic acid in water From 10% A to 25% A (30 min), 40% (10 min), 100% (5 min) (26°C)	280 340 365	102,103
	Epicatechin, catechin, astilbin, engeletin, gallic, caftaric, coutaric acids	RP-8	A: 0.5% formic acid in water B: methanol (0.1% formic acid); 0—5 min: A then to 65% B in 65 min	292	530
	Flavonol glycosides	Zorbax ODS	From 20 to 100% tetrahydrofurane in water	360	397
	Hydroxycinnamic acid esters	Zorbax ODS	0—15 min: 5% acetonitrile in water, pH 2.6 then linear gradient from 6 to 16% aqueous acetonitrile (15 min)	320	397
	Catechin, epicatechin, procyanidins B_1, B_2, B_3, catechin-gallate, caffeoyl and coumaroyl tartaric acid.	Radial-Pak C_{18}	A: 5% acetic acid in water B: 40% acetonitrile in water From 0 to 100% B	320 280	322
	Caffeoyl, *p*-coumaroyl, feruloyl tartaric acids	RP C_{18} (5 μm)	A: water pH 2.6 (H_3PO_4) B: acetonitrile From 6% B to 40%	254 or 312	486
	Caffeoyl, *p*-coumaroyl, feruloyl tartaric acids	C_{18} (5 μm)	0.05 ammonium dihydrogen phosphate pH 2.6 (H_3PO_4) 16% methanol	320	535
	Catechins, procyanidins, flavonol glycosides	Zorbax ODS	A: water with 1% acetic acid B: Acetonitrile From 6% B to 26%	280	651
Wine	Procyanidin dimers, trimers, oligomers, catechin, epicatechin.	Spherisorb-Hexyl	A: water pH 2.0 with 0.1% $HClO_4$ B: methanol From 2% B to 25% in 25 min then 25% B to 98% in 10 min	280	318
	Caffeoyl, *p*-coumaroyl, feruloyl tartaric acids.	Zorbax ODS	0.5% formic acid (750 s) then 9% acetonitrile in 0.5 % formic acid	280	34
	Phenolic acids Catechin, epicatechin, procyanidins Flavonols	C_{18}	A: acetonitrile B: water (pH = 2,6) From 3% A to 100% in 45 min	313 280 365	424

Olive	Oleuropein and derivatives, verbascoside, luteoline 7-glucoside, rutin	Micropak (MCH-5) 5 μm	A: water pH 2.6 (H_3PO_4) B: acetonitrile From 15% B to 40% in 20 min	280 320 340	8,159
Raspberry	Flavonolglycosides	LiChrosorb Si60 (5 μm)	Toluene-acetonitrile (100:20) (35°C)	310	246
Redcurrant	Flavonol glycosides	Ultrasphere ODS (5 μm)	A: 1% acetic acid; B: acetonitrile From 5% B to 30% in 45 min	360	523
Strawberry	Flavonol glycosides	LiChrosorb Si60 (5 μm)	Toluene-acetonitrile (100:20) (35°C)	310	246
Tomato	Naringenin, quercetin 3-glucoside, kaempferol 3-rutinoside, rutin	LiChrosorb Si60 (5 μm)	Benzene-acetonitrile (85:20) (45°C)	300	183
Tomato and bell pepper, eggplant, bushbean, broad bean strawberry, raspberry, blackberry, redcurrant, blackcurrant, gooseberry, blueberry	5′,4′,3′ Caffeoyl quinic caffeoyl, *p*-coumaroyl, feruloyl-glucose and glucoside caffeoyl, *p*-coumaroyl feruloyl-malic, galloylquinic and glucose	Ultrasphere ODS	A: 10% methanol in 2% acetic acid B: methanol From 0% B to 24% in 50 min	320	638,639
		Lichrosorb RP 18 (5 μm)	A: 10 m*M* H_3PO_4 B: methanol From 10% B to 26% (35 min), 30% (1 min) then 10 min at 30%		512
Umbelliferae	Furanocoumarins	Varian Si-5	Cyclohexane-Isopropanol-n-amyl alcohol (14:4:0.5)	254	96

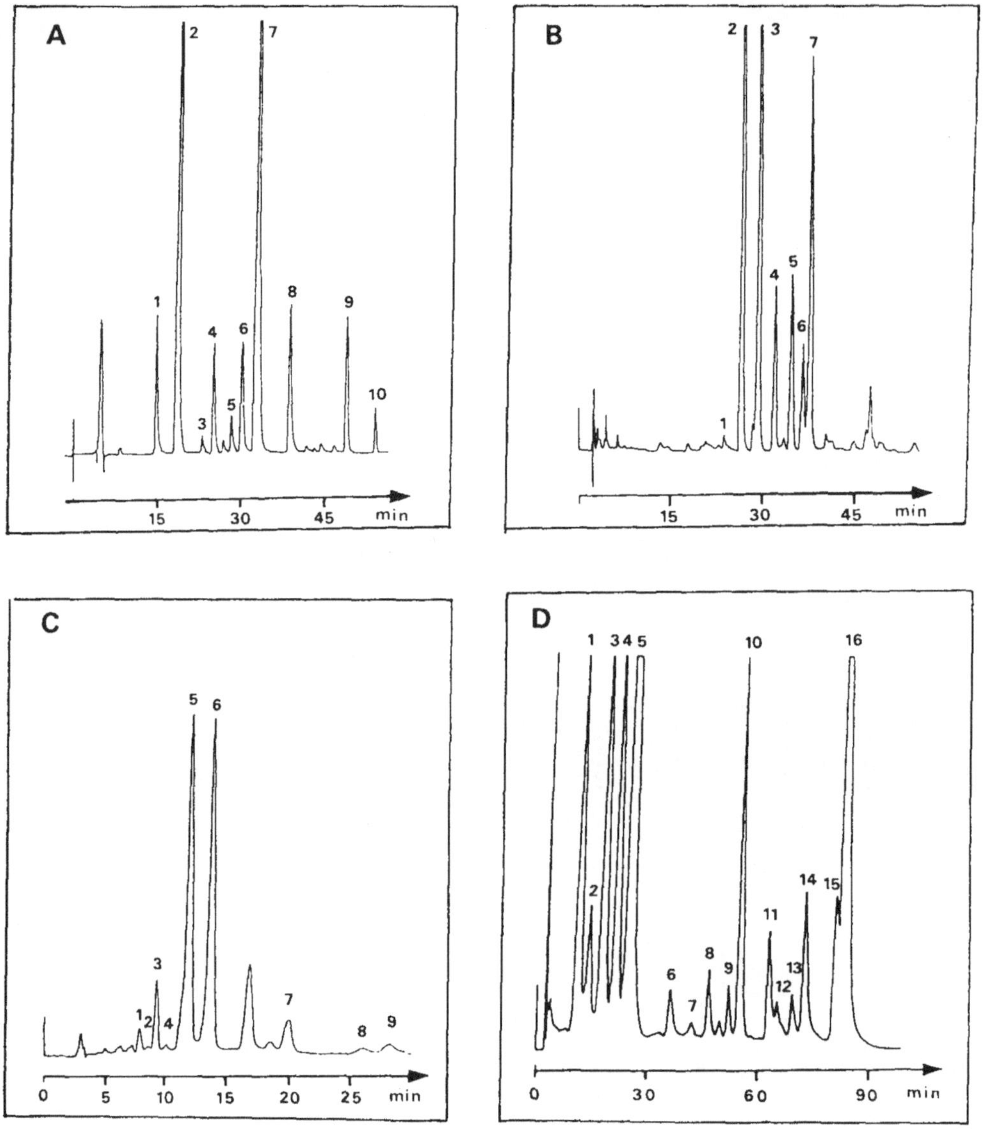

FIGURE 2. HPLC analysis of fruit phenolics. (A) Hydroxycinnamoylquinic esters in apricot: 1: *cis*-3-pCQ; 2: *trans*-3-pCQ; 3: *cis*-4-pCQ; 4: *trans*-4-pCQ; 5: *cis*-5-pCQ; 6: *trans*-5-pCQ; 7: *cis*-3-FQ; 8: *trans*-3-FQ; 9: *cis*-4-FQ; 10: *trans*-4-FQ. pCQ: *p*-coumaroylquinic acid; FQ: feruloylquinic acid. (B) Glucose esters, glucosides, and quinic esters of hydroxycinnamic acid in tomato: 1: C-glucose; 2: *p*C-glucoside; 3: C-glucoside; 4: *p*C-glucose; 5: F-glucoside; 6: 4-CQ; 7: 3-CQ; C: caffeic, *p*C: *p*-coumaric, F: ferulic, Q: quinic acids. (C) Flavonol glycosides of sour cherry: 1: K 3-glucoside; 2: K 3-galactoside; 3: Q 3-glucoside; 4: Q 3-galactoside: 5: K 3-rutinoside; 6: Q 3-rutinoside; 7: K 3-galactoside-7-diglucoside; 8: K 3-rutinosyl-4′-diglucoside; 9: Q 3-rutinosyl-4′-diglucoside (K = Kaempferol; Q = quercetin). (D) anthocyanins of Syrah grape skin: 1: Dp-Glu; 2: Cy-Glu; 3: Pt-Glu; 4: Pn-Glu; 5: Mv-Glu; 6: Dp-Glu-Ac; 7: Cy-Glu-Ac; 8: Pt-Glu-Ac; 9: Pn-Glu-Ac; 10: Mv-Glu-Ac; 11: Dp-Glu-pC; 12: Mv-Glu-C; 13: Cy-Glu-pC; 14: Pt-Glu-pC; 15: Pn-Glu-pC; 16: Mv-Glu-pC. The peak between 8 and 9 is very likely Mv. (Dp = delphinidin; Cy = cyanidin; Pt = petunidin; Pn = peonidin; Mv = malvidin; Glu = 3-glucoside; Ac = acetic acid ester; C = caffeic acid ester; pC = p-coumaric acid ester). ([A] Adapted from Möller, B. and Herrmann, K., *J. Chromatogr.*, 241, 371, 1982. [B] Adapted from Winter, M. and Herrmann, K., *J. Chromatogr.*, 315, 243, 1984. [C] Adapted from Henning, W. and Herrmann, K., *Z. Lebensm. Unters. Forsch.*, 170, 433, 1980. [D] Adapted from Roggero, J. P., Coen, S., and Ragonnet, B., *Am. J. Enol. Vitic.*, 37, 77, 1986.)

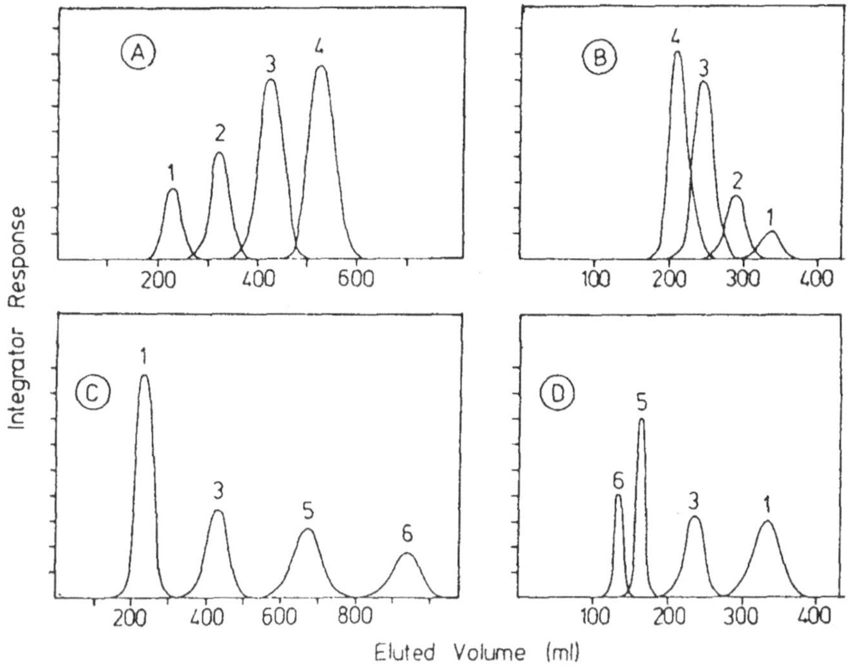

FIGURE 3. DCCC of anthocyanins using n-butanol-acetic acid-water (4:1:5) as solvent system. (A and B) Blackcurrant anthocyanins using the upper layer (A) or the lower layer (B) as mobile phase. (C and D) Raspberry anthocyanins, using the upper layer (C) or the lower layer (D) as mobile phase. 1: cyanidin 3-glucoside; 2: delphinidin 3-glucoside; 3: cyanidin 3-rutinoside; 4: delphinidin 3-rutinoside; 5: cyanidin 3-sophoroside; 6: cyanidin 3-(2^G-glucosyl)rutinoside. (From Francis, G. W. and Anderson, O. M., *J. Chromatogr.*, 283, 445, 1984. With permission.)

Each group of phenolic compounds is characterized by one or several UV or visible light absorption maxima (Table 5). The absorption spectra of the total extracts display peaks which reveal the presence of groups of dominant phenolic compounds, e.g., hydroxycinnamic derivatives and catechins in apple and peach.[89,340] Examination of the spectra of total extracts can thus be used to guide close analysis towards a particular group of phenolic compounds.

UV-visible spectroscopy identification of each compound isolated, flavonoids in particular, is carried out from spectra in methanol after addition of shift reagents.[333] The addition of sodium acetate, sodium hydroxide, and hydrochloric acid reveals the phenolic and free carboxylic groups present in the molecules.[291,340]

Visualization of chromatograms is carried out mainly by examining them under UV light at 254 and 350 nm with or without the presence of ammonia vapor. Liquid nitrogen (−196°C) applied to the surface of a thin layer plate observed under UV stimulates the fluorescent capacities of numerous phenols intensely and can make them phosphorescent. The spraying of chemical reagents completes UV observation. Among the numerous reagents available,[469] the most commonly used are

- Neu's reagent (1% sodium 2-aminoethyldiphenylborate in methanol) which visualizes most flavonoids and hydroxycinnamic derivatives.[410] The addition of PEG 400 (5%) improves even further this extremely sensitive reagent which stabilizes the fluorescence of the compounds determined.[75]
- Benedikt's reagent[461] which forms a complex with substances possessing a free *o*-diphenol group. The latter lose their fluorescence. In contrast, the fluorescence of monophenolic derivatives increases. This visualization technique is extremely sensitive.

TABLE 5
Spectral Properties of Various Phenolic Compounds (λ max)

	Class of compounds	UV B	UV A	Visible
1[a]	Benzoic acids	270—280		
	Hydroxycinnamic acids	(290—300)[b]	305—330	
	Anthocyanic pigments	270—280	(315—325)[c]	500—550
	Flavonols	250—270	(300)[b] 350—380	
	Flavan-3-ols	270—280		
2[a]	Coumarins	220—230	310—350	
	Flavones	250—270	330—350	
	Flavanones, Flavanonols	270—295	(300—330)[b]	
	Chalcones	220—270	(300—320)[b] 340—390	
	Aurones	240—270		370—430
	Isoflavones	245—270	300—340	

Note: Solvent: methanol, except for anthocyanic pigments where the solvent was methanol HCl 0.01%.

[a] 1: Classes of common compounds in fruits; 2: classes of less widespread compounds in fruits, except in particular cases.
[b] shoulder.
[c] In the case of acylation by hydroxycinnamic acids.

- Hydrochloric vanillin[469] which gives specific red visualization of flavans, flavanols, and flavan-diols and of their polymers (condensed tannins) to a lesser extent.
- Diazotized *p*-nitraniline[469] which gives specific visualization of most phenolic compounds in different colors. Colors are modified by spraying with 15% sodium carbonate.
- Aluminum chloride, which can be used to visualize flavonols.[333]

Hydrolysis of the purified compound makes it possible to identify the constitutive molecules and the type of bond connecting them.[469] Various types of hydrolysis are used (acid with HCl 2 *N*, 10 to 30 min in a waterbath at 100°C, basic with NaOH 2 *N* for 3 h under nitrogen, and enzymatic with β-glucosidase). These methods provide information about the type of linkage in complex compounds (esters or glycosides). Hydrolysis of amino compounds of the phenolamide type is much longer.

The set of characteristics obtained for each compound generally leads to correct identification but does not make an absolute conclusion possible. Only physicochemical techniques, such as nuclear magnetic resonance (NMR) spectra or mass spectrometry (MS) techniques can give a decisive answer. These modern analysis techniques (NMR, MS) have not been used in much work — particularly older research — on phenolic compounds in fruits. Nevertheless, all the techniques listed lead to a set of presumptions which enable identification of a number of compounds with a fair degree of certitude; identifications are compared continuously with data in the literature.

The development of modern techniques of analysis over the past 10 years has enabled great progress to be made in accurate knowledge of the structure of natural phenolic molecules. Several reviews have been published on ^{1}H-NMR,[333,359] ^{13}C-NMR,[360] and on MS.[132,332,509] Analysis of fruit phenols has recently benefited from utilization of these methods: for example, ^{1}H-NMR and ^{13}C-NMR have been used to determine the structure of flavonol glycosides in apple[127] and grape[102,103] and the structure of flavonoid glycosides and hydroxycinnamic derivatives of blackcurrant.[305] These two techniques, combined with fast atom bombardment mass spectroscopy (FAB-MS) make it possible to determine complex structures such as that of an acylated flavonol glycoside in blackberry.[613] In *Citrus*, complex flavone glycosides have been determined in the fruits of *C. sudachi*[257] as have flavonoid glycosides in *C. unshui*.[368] The obtaining of mass spectra for anthocyanins involves difficulties of low volatility and poor stability. These problems were solved recently by using FAB-MS.[576] Thus, six anthocyanins were identified by FAB-MS and the structures of malvidin derivatives (3-glucoside, 3-acetylglucoside, and 3-*p*-coumarylglucoside) were confirmed by ^{1}H-NMR.[26] NMR spectroscopy study of malvidin 3-glucoside obtained from *Vitis vinifera* berries revealed seven different structures in aqueous solution according to pH and temperature.[101]

Finally, it is mentioned that resonance Raman spectroscopy has been applied to anthocyanins and that *in vivo* and *in vitro* spectra have been obtained, e.g., in grape skin.[81,383]

C. QUANTITATIVE ANALYSIS OF PHENOLIC COMPOUNDS

Quantitative analysis of phenolic compounds has evolved considerably over the past 20 years or so. Analytical approaches can be classed in three trends which are often developed in a complementary manner by the different authors.

A preliminary approach consists of determining in a more or less purified total extract either all the phenolic compounds or one category of these compounds. The numerous determination methods available use the general properties of phenols, e.g., the oxidation capacity and the chelation capacity with certain metal ions or absorption in UV and visible light. More specific information can be obtained when the property in question only applies to one category of phenolic compounds: determination of tannins and proanthocyanidins (described in Chapter 1 Section V), determination of anthocyanins, *o*-diphenols, etc. Techniques include the following:

- Conventional measurement of total phenols using Folin-Ciocalteu's reagent.[356,529,536,558] This reagent is very sensitive but not very specific. Utilization of Polyclar AT enables differential determination as it removes interferences caused by nonphenolic substances.[356] Utilization of formaldehyde[310] or 1% gelatin in an aqueous solution containing 10% sodium chloride[68] enables separation of the tannin and flavonoid fractions and measurement of the various groups of phenolic compounds.[356] This procedure has been applied to cherry.[380]
- Direct spectrophotometric determination of phenolics in more or less purified total extracts.[53,56,121] These rapid approaches which are easy to use give fairly accurate results when applied to purified extracts. For example, the absorption spectrum of a purified total extract of pear (Figure 4A) makes it possible to estimate hydroxycinnamic derivatives (chlorogenic acid) from absorption measured at 325 nm.[56] A spectrophotometric method was perfected to estimate groups of compounds in extracts of Passe-Crassane pear,[53] with caffeic acid derivatives estimated as chlorogenic acid, *p*-coumaric acid derivatives as *p*-coumaroyl-5′-quinic acid, and flavonols as rutin. In apple and pear, for example, catechins can be measured from the spectrum of the difference between total extract and the standard chlorogenic acid solution at the same concentration as that determined in the extract (Figure 4A).[56,121]
- Determination in nonaqueous medium after oxidation by Cu^{2+} ions.[54,342] This method can be used to determine total *o*-diphenols and also various groups of phenolic compounds with *o*-diphenol functions (chlorogenic acid, catechin, flavonols derived from quercetin). It is suitable for application to certain fruits, such as apple, peach, and pear (Figure 4B).[54,89,342]

Total anthocyanins are frequently measured in total extracts from absorption measured at about 520 nm using an average extinction coefficient (e.g., $\epsilon = 33{,}000$ l·mol^{-1}·cm^{-1}),[397] but anthocyanin estimation methods are very varied.[166,469]

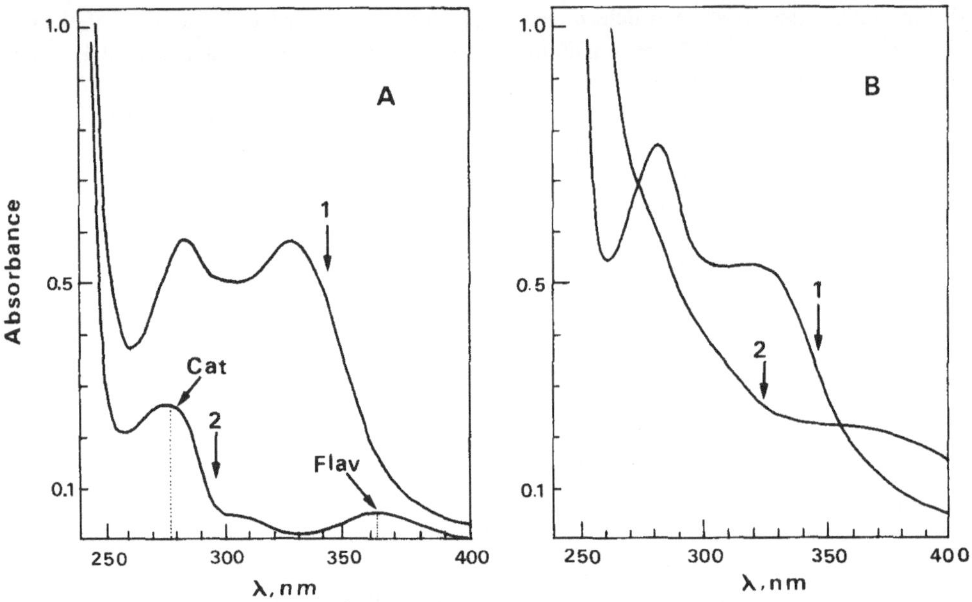

FIGURE 4. UV spectra of total extracts of cv. Passe-Crassane pear fruit. (A) Young entire fruit (1 1/2 month after full bloom); 1: total extract in methanol after purification with ethylacetate; 2: spectrum of difference (1) — ACQ; ACQ = chlorogenic acid reference solution of the same concentration of hydroxycinnamic derivatives present in the extract; Cat = catechins; Flav = Flavonols. (B) Peel fruit at harvest; 1: nonoxidized total extract; 2: extract after oxidation by Cu^{+2} ions in acetonitrile.

This type of overall approach has often been used in analysis of fruits, but it is only applied occasionally today and generally as a preliminary before other techniques. Indeed, in spite of the many improvements made in these methods, the information that they give is often affected by considerable inaccuracy. This is caused first, by the possibility of interference from nonphenolic substances with similar properties (although this can be remedied to some extent by prior purification of extracts), and second, by the very variable reactivity of the various molecules present, even within a single class of phenolic compounds. As a result, the same overall value obtained may represent extracts whose phenolic compositions are different.

These overall methods were soon joined by determination and measurement after the separation of the various compounds of a phenolic extract. Nevertheless, given the diversity and complexity of the combined forms naturally present in plant material, it has frequently been easier to determine molecules released after hydrolysis of the extract. For example, an estimation of hydroxycinnamic acids reflects the three or four commonest acids, whereas several tens of derivatives may be present in the initial material. This approach, which is widely used for fruits,[57,395,549,550,552,620] has given interesting results, in particular when it is combined with different types of hydrolysis. Nevertheless, it remains doubtful from the quantitative point of view (variable hydrolysis yields, degradation of the products of hydrolysis, etc.) and in addition completely masks the diversity of native molecules and their relative proportions.

Determination and individual measurement of the components of a phenolic extract, in fact, form an ideal situation on which efforts have been concentrated for many years. Final success should make it possible to measure each of the compounds present in a fruit and to monitor each one during the technological processes to which it may be subjected, e.g., during winemaking and the aging of wine in the case of grape.[527]

Chromatographic techniques form the major analytical basis of this approach.

Paper, thin-layer, column and gas chromatography have given interesting results in the separation and measurement of phenolic compounds in fruits,[150,153,250,251,340,543] but HPLC techniques appear to be the most suitable and have been widely developed today.[103,126,249,397,404,489,498,523,530] The use of columns with different characteristics[74,637] makes possible complementary separations and the quantitative study of native forms.[512] One of the limiting factors nevertheless remains the absence of standards which would make it possible to define UV detection and absorption characteristics for each compound. The use of a family of compounds is, therefore, frequent; the characteristics correspond to the most abundant of them.[404,486,533] Preparative HPLC combined with analytical chromatography[74] should lead to improvement of this aspect in years to come.

Finally, a radioimmunoassay method has been perfected for the flavanone glycosides naringin in *Citrus paradisi* and hesperidin in *C. sinensis*.[40,289,290] The sensitivity of this method (0.3 pmol/0.1 ml) makes it possible to study the distribution of these compounds in the different parts of fruit (Chapter 1 Section VII). Given the very high specificity and sensitivity of this radioimmunological technique, application to other phenolic molecules should develop rapidly in the years to come.

III. PHENOLIC ACIDS AND COUMARINS

A. HYDROXYBENZOIC ACIDS

Hydroxybenzoic acids (HBA) have a general structure of the C_6–C_1 type derived directly from benzoic acid (Figure 5). Variations in structure lie in the hydroxylations and methoxylations of the aromatic cycle. Four acids are apparently universal in the Angiosperms: *p*-hydroxybenzoic, vanillic, syringic, and protocatechuic acids. The first three are constituents of lignin, from which they are released by alkaline hydrolysis.[499] It can be stated as a fair approximation that plants which do not contain lignin do not contain these acids either. Gentisic acid has a particular pattern of hydroxylation which can be related to that of salicyclic acid but it is much less common however. Gallic acid is another well-known plant acid which participates in the formation of hydrolysable gallotannins. Its dimeric condensation product (hexahydroxydiphenic acid) and the related dilactone, ellagic acid, are also common plant constituents, the former usually occurring bound as ellagitannin.[210,220,238]

Research on benzoic acids in fruits has been mainly concentrated on the study of free acids determined after hydrolysis of extracts. It has made it possible to obtain quantitative estimation in the case of the most common fruits (Table 6). Nevertheless, benzoic acids are frequently present in bound form and constitute either complex structures like hydrolyzable tannins or simple molecules by combining with sugars or organic acids. The observation of these simple combined forms in fruits is fairly recent.[129,304,512] Glucosides of *p*-hydroxybenzoic, salicylic, protocatechuic, vanillic, and syringic acids were shown recently in cherry and plum by HPLC;[304] contents varied from 1 to 14 mg/kg.

Very small concentrations of gallic acid are found in the form of esters with quinic acid (theogallin) or glucose (glucogallin) and in the form of glucosides (Table 7). Glucogallin has also been identified in persimmon and isolated only from astringent immature fruit,[406] whereas free gallic acid was found in immature fruit of both astringent and nonastringent varieties. Glucogallin was thus proposed as being a good index for distinguishing between astringent and nonastringent varieties. Gallic acid is also found combined with naringenin in fruits of *Acacia farnesiana*[147] (Figure 5) or with (−)-epicatechin to form epicatechin-gallate, which is a constituent of unripe grapes.[161,162,554,624] Nevertheless, this compound is probably not present in berry pulp but only in seeds.[554] It also contributes to the formation of condensed tannins in grape and may appear as a product of thiolysis or acid degradation from grape polymers[161] or in persimmon.[369,371] As gallic and hexahydroxydiphenic acids are constituents of hydrolyzable tannins, they may appear during any natural or forced degra-

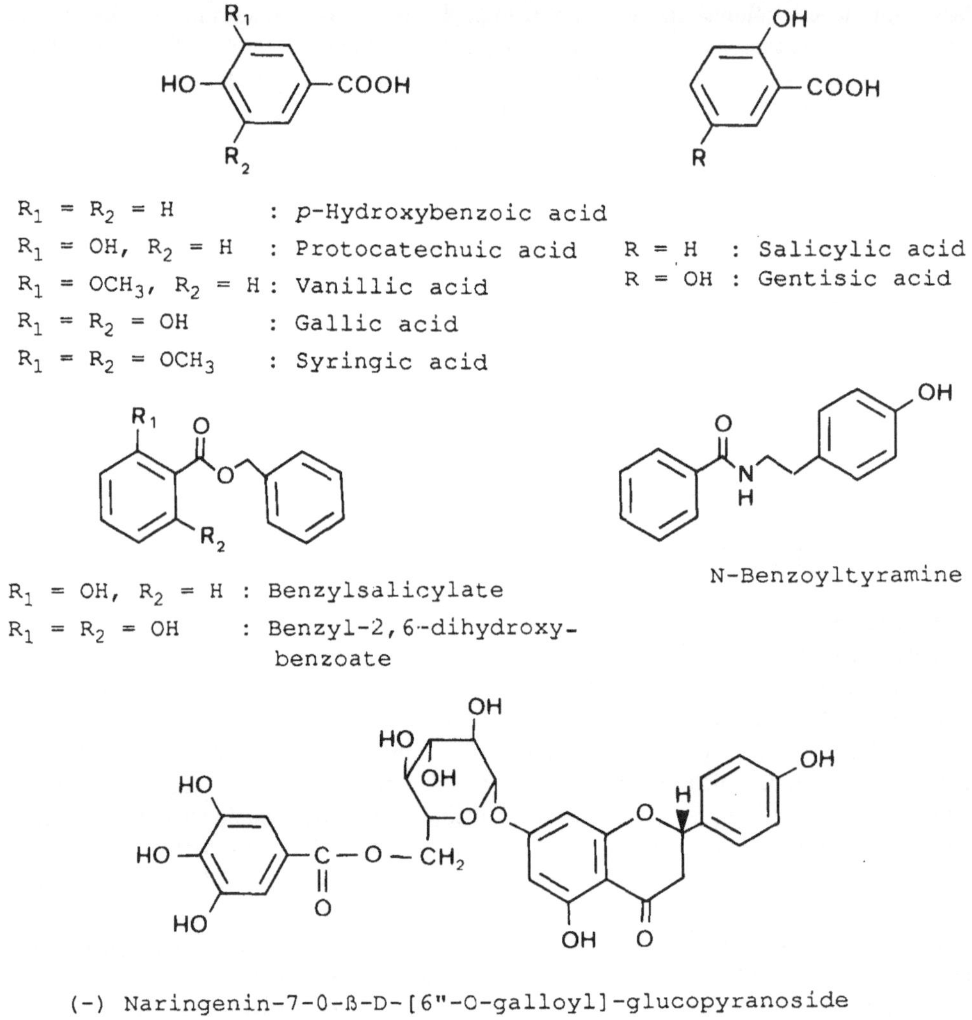

FIGURE 5. Chemical structures of benzoic acids and derivatives.

dation of these condensed forms. More exceptionally, forms of gallic acid esterified with methanol or prunin have been reported in strawberry and the fruit of *A. farnesiana*.[142,252]

p-Hydroxybenzoic and vanillic acids are also present in numerous fruits (Table 6).[395,396,507,549,550,552] In soft fruits, the native forms of *p*-hydroxybenzoic acid are simple combinations with glucose (Table 7).[512] More exceptionally, conjugation may occur with certain glucosides such as glucopyranoxylosyl in the fruits of Apiaceae,[129] catalpol in fruits of *Catalpa bignonioides* (6-*p*-hydroxybenzoyl-catalpol), or aucubin in those of *Vitex agnus castus* (0-*p*-hydroxylbenzoyl-aucubin).[252] Other derivatives of hydroxybenzoic acid have been detected in certain fruits: the methylester of *p*-hydroxybenzoic acid in passion fruit,[252] 3-4-dihydroxybenzoic aldehyde in banana,[401] 4-methoxy-3,5-bis (3′-methyl-2′-butenyl)-benzoic acid in fruits of Jamaican *Piper* sp.,[90] and benzyl-2,6-dihydroxybenzoate and other derivatives in the fruits of *Aniba riparia*[36] (Figure 5). Syringic acid has only been reported in grape;[463] this compound thus appears to be of very limited distribution in fruits. The same applies to sinapic acid, its equivalent in the cinnamic series. It is not impossible that *p*-hydroxybenzoic, vanillic, and syringic acids derive at least partially from the degradation of certain lignified zones of the fruit when these exist (stone, seed teguments, etc.).

TABLE 6
Contents[a] (mg/kg Fresh Weight) of Hydroxybenzoic Acids[b] in Ripe Fruits

Fruits	*p*-Hydroxy-benzoic	Protocatechuic	Vanillic	Salicylic	Gentisic	Gallic	Ref.
Blackberry	6—16	68—189				8—67	396
Blackcurrant	0—6	<10—52				30—62	550
Blueberry	0	0				1—2	550
Cherry							
Sour			1—8				552
Sweet	<0.5—5		2—11				552
Cucumber			<0.5	<0.5—3	<0.5		507
Eggplant			<0.5	<0.5	<0.5		507
Gooseberry	0	<0.5—8				0	550
Grape[c]	0—0.07	t	0.07—2.75	0.04	—	t—0.46	463, 481
Grapefruit				0.1			482
Kiwi				0.02			482
Mango						t	555
Melon (skin)	2—6		2—9	<0.5	<0.5		507
Pepper			11	<0.5—1	<0.5		507
Persimmon						0—28	657
Plum			2—12				552
Pumpkin (skin)			<0.5				507
Raspberry	15—27	25—37				19—38	396
Redcurrant	10—23	3—8				<1	550
Strawberry	10—36	<0.5—6	<0.5—4			11—44	549
Tomato			<0.5—1	<0.5—1			507
White currant	5—19	<1—7				3—38	550

Note: t, traces.

[a] Extreme values for several cultivars.
[b] After hydrolysis.
[c] Syringic acid: t-11 (mg/kg FW)

TABLE 7
Contents[a] (mg/kg Fresh Weight) of Hydroxybenzoic Acid Derivatives in Soft Fruits

Fruits	Gal G	Gal Q	Gal Gluc	pHB Gluc	Pro Gluc
Blackberry	2—3	—	t	5—21	2—6
Blackcurrant	4—7	t—1	0—3	4—13	0—2
Blueberry	—	—	3—7	5—6	4
Gooseberry	—	—	t	9—14	2—7
Raspberry	t	t	t	34—59	t
Redcurrant	—	t	—	9—13	t—2
Strawberry	—	t—1	—	6	—

Note: t, traces; —, not detectable; Gal G: galloylglucose; Gal Q: 5′-galloylquinic acid; Gal Gluc: gallic acid glucoside; pHB Gluc: *p*-hydroxybenzoic acid glucoside; Pro Gluc: protocatechuic acid glucoside.

[a] Extreme values for several cultivars.

From Schuster, B. and Herrmann, K., *Phytochemistry*, 24, 2761, 1985. With permission.

Protocatechuic acid is found in a number of soft fruits in the form of glucosides (Table 7); the latter are generally much less abundant than those of *p*-hydroxybenzoic acid. Salicylic and gentisic acids have been reported in very small quantities in the fruits of certain Solanaceae (tomato, eggplant, pepper) and Cucurbitaceae (melon, cucumber) (Table 6)[507] and in kiwi fruit, grapefruit, and grape.[481,482] Salicylic acid was used for a long time as a preservative in wine but is now prohibited and the amounts in grape, as in other fruits, are very small (0.04 mg/kg).[482]

From the quantitative point of view, the overall HBA content is generally low (Tables 6 and 7) except in certain fruits of the Rosaceae family and in particular blackberry in which protocatechuic and gallic acid contents may be very high (Table 6). This observation should be seen together with the very high contents in the same fruit of hydrolyzable tannins derived from gallic acid.[162] *p*-Hydroxybenzoic, protocatechuic, and gallic acids are generally predominant among HBA (Table 6). When it has been possible to assay native derivatives,[512] glucosides dominate distinctly (Table 7) in particular in raspberry with regard to *p*-hydroxybenzoic glucoside.

There are thus great interspecific differences in HBA in fruits with regard to both quality and quantity. Such differences are also found between the varieties of the same species. The figures in Tables 6 and 7 show the extreme values obtained for different varieties of each fruit. Variations involve, first, HBA contents and, second, the presence or absence of compounds. In fact, qualitative and quantitative investigation of the native molecules of HBA derivatives is only just beginning and the results are too fragmentary today for it to be possible to draw general, definitive conclusions. In addition, HBA can play a role in the organoleptic qualities of fruits through their more complex forms of combination which lead to different types of hydrolyzable tannins (gallotannins, ellagitannins) or, for example their integration in condensed tannins in the form of epicatechin-gallate. This sensorial aspect is discussed in Chapter 4.

B. HYDROXYCINNAMIC ACIDS (HCA)

Among fruit phenolics, hydroxycinnamic derivatives (HCD) play an important role because of both their abundance and diversity.[252,343,346] They all derive from cinnamic acid and are essentially present as combined forms of the four basic molecules *p*-coumaric, caffeic, ferulic, and sinapic acids (Figure 6). Coumarins, also derived from hydroxycinnamic

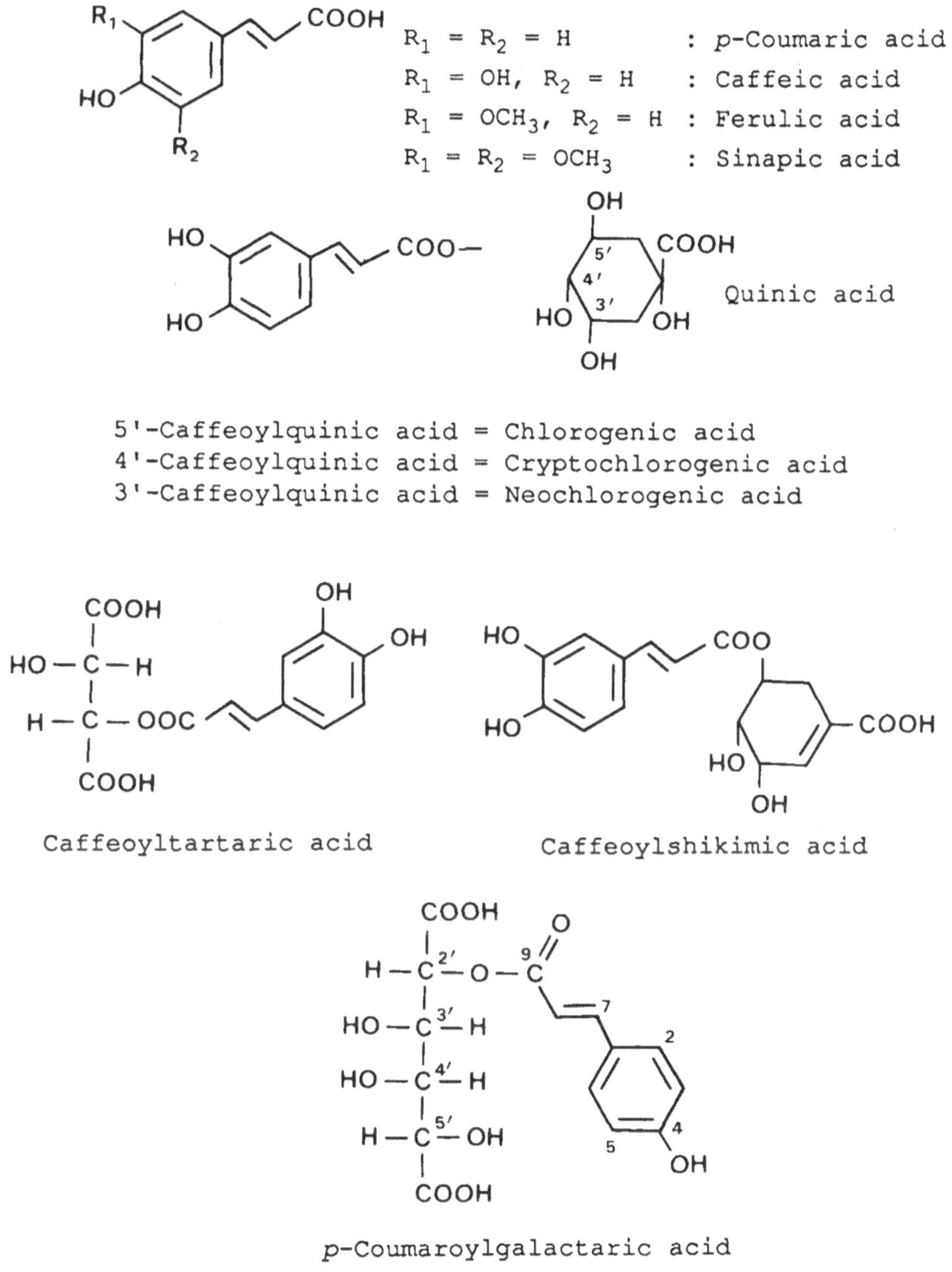

FIGURE 6. Chemical structures of hydroxycinnamic acids and their hydroxyacid esters.

acids, are examined separately. As is discussed below, the free forms of these acids are very rarely present in fruits, and two main types of soluble derivatives have been identified: (1) first, those involving an ester bond between the carboxylic function of phenolic acid and one of the alcoholic groups of an organic compound (i.e., quinic acid, glucose, etc.), for example chlorogenic acid which has been identified in numerous fruits; (2) second, those which involve a bond with one of the phenolic groups of the molecule, i.e., *p*-coumaric acid *O*-glucoside in tomato fruit. The diversity of HCD encountered in plants[386] and particularly in fruits thus results from the nature of the bonds and that of the molecule(s) involved.

In addition, for each of these compounds, the presence of a double bond in the lateral chain leads to the possible existence of two isomeric forms *cis* (Z) and *trans* (E). Although native compounds are mainly of the *trans* form,[469] interconversion of the two forms might occur *in situ* and be responsible for certain physiological responses in plants.[581] In any case, it takes place very easily under the effect of light,[292] in particular during extraction and purification operations prior to analysis of HCD. The *cis* form then appears next to the *trans* form which nevertheless remains predominant. They can be separated and determined quantitatively by HPLC,[533] but the physiological significance of these data remains very much a subject for discussion.

1. The Problem of Free Forms of HCA

HCA are present in fruits in combined forms, and only a few exceptional situations can cause them to accumulate in the free form. Distinction is made between five main possibilities:

1. Too brutal extraction conditions (e.g., temperature too high or medium too acid) can lead to partial hydrolysis of the combined forms. Chromatographic monitoring revealing the absence of free HCA can ensure that no chemical degradation has occurred during extraction.
2. Free acids can accumulate when the fruits display physiological disturbances as in "cloud" tomato[614] or when they are contaminated by microorganisms;[616] in this case, it is probable that the esterified forms are degraded by certain fungal hydrolases.
3. Anaerobiosis is also responsible for the hydrolysis of the combined forms. Thus, under nitrogen or carbon dioxide, caffeic *p*-coumaric, and ferulic acids appear in tomato[154] and the phenomenon is particularly marked in young fruits. Simultaneously, chlorogenic acid disappears although there is no quantitative relationship between the two phenomena. These evolutions suggest the degradation of the esterified forms under the effect of endogeneous esterases in the fruit.[155] Identical phenomena have been reported during the anaerobic fermentation of grape and are likely to be partly responsible for the appearance of certain free acids in wines.[528]

It remains to be determined whether these observations can be generalized; indeed, the appearance of free HCA during the postharvest storage of fruits under atmospheres with a very low O_2 content might be an early marker of the triggering of an abnormal metabolism which might then lead to browning. In addition, the much greater toxicity of free acids compared to the combined forms[220] may be the cause of an acceleration in metabolic deviation in fruits.

4. Free acids may also appear during the technological processing (see Chapter 5) to which fruits are sometimes subjected. Thus, free *p*-coumaric and caffeic acids are found in several fruits when they are frozen and stored at low temperature: e.g., avocado pears,[452] cherries,[506] the berries of mountain ash.[446] Likewise, free benzoic and cinnamic acids are reported in the juice obtained from bilberries after freezing and thawing.[22] Sterilization can also cause HCA to appear in canned peaches and apricots (see Reference 344) and canned tomato pastes.[478] The winemaking process also leads to fairly high HCA contents.[527,528] In all these cases, the factors mentioned above (and in particular chemical or enzymatic degradation and the effect of anaerobiosis) and their combining with subcellular decompartmentation are certainly responsible for the appearance of free acids.
5. The supplying of a metabolic precursor, such as phenylalanine sometimes leads to the accumulation of free HCA in fruits.[224] Nevertheless, combined forms of HCA frequently accumulate under these conditions.[341]

Free HCA thus only appear in fruits in special situations, and it is likely that the small quantities reported by certain authors (see References 344, 450, and 490) are only traces resulting from partial hydrolysis. Only the date seems to display high free ferulic and *p*-coumaric acid contents;[458] this should probably be associated with the original maturation of this fruit in which the browning which occurs is certainly the result of subcellular de-compartmentation, as is the hydrolysis of the combined forms of HCA.

2. Bound Forms of HCA

a. Hydroxyacid Esters

i. Quinic Esters

These have been reported for a long time in fruits; the first were chlorogenic acid (5′-caffeoylquinic acid*)[73] and *p*-coumaroylquinic acid[95] in apple (Figure 6). Chlorogenic acid was subsequently found in many other fruits (Table 8), often accompanied by other caffeoylquinic isomers,[543] such as neochlorogenic acid (3′-caffeoylquinic acid) and cryptochlorogenic acid (band 510, 4′-caffeoylquinic acid) in various other fruits,[593] (Table 8). The structure of each of these compounds, which were frequently first isolated from fruit extracts, was proposed in 1964[503,611] and they can now be separated in fruits by HPLC.[389] Transesterification between these various isomers appears to be possible, at least *in vitro*.[505]

Somewhat more complex forms of caffeoylquinic derivatives have frequently been demonstrated under the name of isochlorogenic acid;[543] they are abundant in coffee bean, which made it possible to identify different dicaffeoylquinic esters.[112,504] Fractions which were probably identical or at least very similar have been found in apple,[340] pineapple,[128] cherry,[381] and peach.[89]

Quinic derivatives of other HCA have also been identified in numerous fruits (Table 8), and the same series of isomers may be present although they have often been investigated with less fine precision. Cartwright et al.[95] and then Williams[633] reported the presence of several isomers of *p*-coumaroylquinic acid in apple; the main isomer was probably 4′-*p*-coumaroylquinic acid in the variety Yarlington Mill.[630]

5′-Feruloylquinic acid, which is very abundant in various plant organs[252] was also present in tomato but in small quantities;[151] it has also been reported in cell suspensions from apple fruit[309] although it was not present in the original fruit. In this case, it accumulates mainly after the exponential growth phase, throughout the stationary phase, especially when the aging of the culture has been extended by several days.

Quinic esters of sinapic acid have not been determined in fruit to date although they are present in various other plant organs.[252]

Although quinic derivatives are generally abundant in fruits, some contain none at all, e.g., those of the genus *Vitis*[420,468] or cranberries[106,364] and strawberry.[512]

ii. Shikimic Esters

5′-Caffeoylshikimic acid (Figure 6) is not very widespread in plants, but is very abundant in Palmeae[231] and in particular in date, together with other caffeoylshikimic isomers.[352] These compounds contribute to the browning reactions which take place during maturation of the fruit.[350] No derivatives of the feruloyshikimic type were identified, although a recent study reported a high ferulic acid content in a free and an unspecified combined form.[458]

iii. Tartaric Esters

These form an interesting example insofar as their distribution is limited to certain fruits of *Vitis* species. Their presence in *Vitis vinifera* was clearly demonstrated by Ribéreau-Gayon in 1965,[468] who also showed the absence in these fruits of HCA derivatives with

* The nomenclature of HCA quinic esters is in conformity with IUPAC recommendations.[284] Chlorogenic acid is thus 5′-caffeoylquinic acid and not 3-caffeylquinic acid as it was originally called.

TABLE 8
Contents[a] (mg/kg Fresh Weight) of Hydroxycinnamic Acid Derivatives in Ripe Fruits

	Quinic Esters									Glucose Derivatives							
	Caffeoylquinic			*p*-Coumaroylquinic			Feruloylquinic			Caffeic		*p*-Coumaric		Ferulic		Sinapic	
Fruits	5′	4′	3′	5′	4′	3′	5′	4′	3′	E	G	E	G	E	G	E	Ref.
Apple	49—506	t—12	—	t—12	3—46	2—4	2—4	—	—	t—6		t—19		t—9		—	340,388,389,403,475
Apricot	51—123	+	26—132	t—2	t—2	2—9	2—7	—	3—15	—		+		+		—	388,389,475
Banana										—		+		+		—	459
Blackberry	t—3	1	41—52	—	—	2—5	+	—	2—4	3—6	—	4—11	2—5	2—6	—		512
Blackcurrant	1—2	3—5	38—48	—	1—2	13—21	1—2	2—3	t—1	19—30	2	10—14	3—10	11—15	2—4	—	512
Blueberry	1851—2075	2—5	5—7	2—5	—	—	7.8	—	—	+	3	+	3—15	+	5—10	+	512
Cherry																	
Sour	53—140	1—10	79—536	t—2	11—25	40—226	2—5	—	t—9	—		+		—		—	388,389,475
Sweet	11—40	3—21	73—620	1	8—11	81—450	+	—	3—13	—		+		+		—	381,388,389,475
Cranberry												+		+		+	459
Eggplant	575—632	8—11	+	+			15—17			+	—	—	+	+	—	+	460,639
Elderberry										+		+		+		—	459
Gooseberry	—	—	3—26	—	—	t—14	—	—	1—2	5—35	2	7—15	4—11	t—32	2—7	—	512
Grapefruit										—		—		+		+	459
Kiwi	600—1000																178
Lemon										—		+		+		+	459
Orange										—		+		+		+	459
Peach	30—282	—	29—135	+	—	2—3	2—9	—	t—2	—		+		—		—	89,388,389,475
Pear	10—516	—	—	2—60	—	—	—	—	—	t		t—8		—		—	56,388,389,475,589
Pepper	—	—	—	—			—			+	—	+	—	t—11	—	5	460,639
Pineapple[b]	diCQ	7.8	ΣpCQ	34.5			—	—	—	—		4.1		15.7		9.5	128,459
Plum	18—142	6—100	98—771	t—2	t—4	4—40	1—7	—	9—34	2—7		3—34		2—12		—	388,389,444,475
Raspberry	t—1	—	—	1—2	—	—	+	—	—	3—7	—	6—14	4—10	4—7	t—2	—	512
Redcurrant	1	—	+	—	—	t—1	2	—	—	2—5	t—2	t—2	5—16	+	+	—	512
Strawberry	—	—	—	—	—	—	—	—	—	t—2	—	14—27	+	t—2	—	—	512
Tomato	12—71	5—11	+							+	15—48	6—19	19—68	+	8—15	—	460,639
Pericarp	38		—	2.7			2.2			—	20	1.8	5.6	21	—	2.6	158
Pulp	38		11.2	0.8			—			8	13.6	0.8	12	—	—		158

Note: t = traces; + presence but not quantitative value; — not detectable; di CQ = dicaffeoylquinic acid; Σ = total *p*-coumaroylquinic acids; E = glucose esters; G = glucosides.

[a] Extreme values for several cultivars.
[b] diCQ: 7.8; ΣpCQ: 34.5

quinic acid. Nevertheless, there was a certain amount of uncertitude on the subject for about a decade, in particular with regard to the presence or absence of chlorogenic acid in grape and derived products including wine (see References 527 and 528). HPLC separations[34,416,421,486,531] fully confirmed Ribéreau-Gayon's initial data by showing that the only combined form of caffeic acid in grape was in fact caffeoyltartaric acid or caftaric acid (Figure 6). In addition, *p*-coumaroyl and feruloyl-tartaric derivatives were found in varying proportions according to species and physiological stage.[72,486,531,533] As in any extraction of phenolic compounds, considerable precautions must be taken to prevent oxidation of HCD and particularly of caffeoyltartaric acid;[535] when protective substances are not present there is considerable degradation and the appearance of a product of oxidation, *S*-glutathionyl caftaric acid.[534]

iv. Other Derivatives with Hydroxyacids

Numerous HCA derivatives with other hydroxyacids have been reported in plants,[252] but only those with malic acid are present in small concentrations in the pods of Fabaceae (pea, broad bean)[639] and as traces in pear.[475] Recently, 2′-*O*-*p*-coumaroyl- and 2′-*O*-feruloylgalactaric acids and 2′-*O*-*p*-coumaroyl-, 2′-*O*-feruloyl-, and 2′,4′-*O*-diferuloylglucaric acids (Figure 6), which were all previously unknown in nature, were isolated and identified from orange peel.[476,477]

b. Sugar Derivatives

The first revelation of HCA derivatives with sugars in fruits was by Corner and Harborne[111] and then Harborne and Corner[224] in the case of the potato berry on which 1-*O*-*p*-coumaroylglucose (Figure 7) and caffeic acid 3-*O*-glucoside were identified. Ten years later, it was shown that 1-*O*-*p*-coumaroylglucose was one of the major *p*-coumaric derivatives in apple (cv. Calville blanc) together with several *p*-coumaroylquinic isomers.[338]

Since then, numerous derivatives of HCA with sugars were identified in various fruits.[460,512,639] It can be seen in particular that the presence of combined forms of HCA with glucose is being reported more and more often in fruits (Table 8) and it is probable that these compounds have long been underestimated or confused with other derivatives.

Two main types of HCA derivatives with glucose can be present in the same fruit: on the one hand, esters, such as *p*-coumaroylglucose and on the other, glycosides in which one of the phenolic groups of the molecule is involved in bonding with glucose (Figure 7). Although HCA glycosides have often been encountered in plants,[252] they have still only rarely been reported in fruits, e.g., in the potato berry,[111,224] in cranberries,[106,363,364] and tomato fruit.[150,156,639] In the latter case, *p*-coumaric and ferulic acids were present both as glucosides and as glucose esters whereas caffeic acid was only represented by caffeoylglucose.[156]

HCA derivatives with sugars and hydroxyacids are present simultaneously in numerous fruits (Table 8), e.g., in apple, tomato, cherry, etc. Several remarkable exceptions should nevertheless be reported. Glucose derivatives of HCA are not present[56] or only present as traces[459] in pear. Likewise, they appear to be absent in grape, where the assumed presence of glucosidic esters of *p*-coumaric or caffeic acids[421,541,542] has been invalidated[34] or requires confirmation. Strawberry on the one hand[512] and cranberry[106,363,364] contrast with the examples above since HCA are only present in the form of derivatives with sugars.

Verbascoside (Figure 7) is an example of a rather more complex chemical combination; this was identified fairly recently in olive.[159] The presence of caffeic acid in the olive had already been reported either after hydrolytic degradation of extracts[390] or in the form of combinations which had only been partly identified[600] or again in the form of chlorogenic acid.[315] A laboratory study carried out on six cultivars showed that chlorogenic acid was not present and that the major caffeic derivative was verbascoside[159] an heterosidic ester of

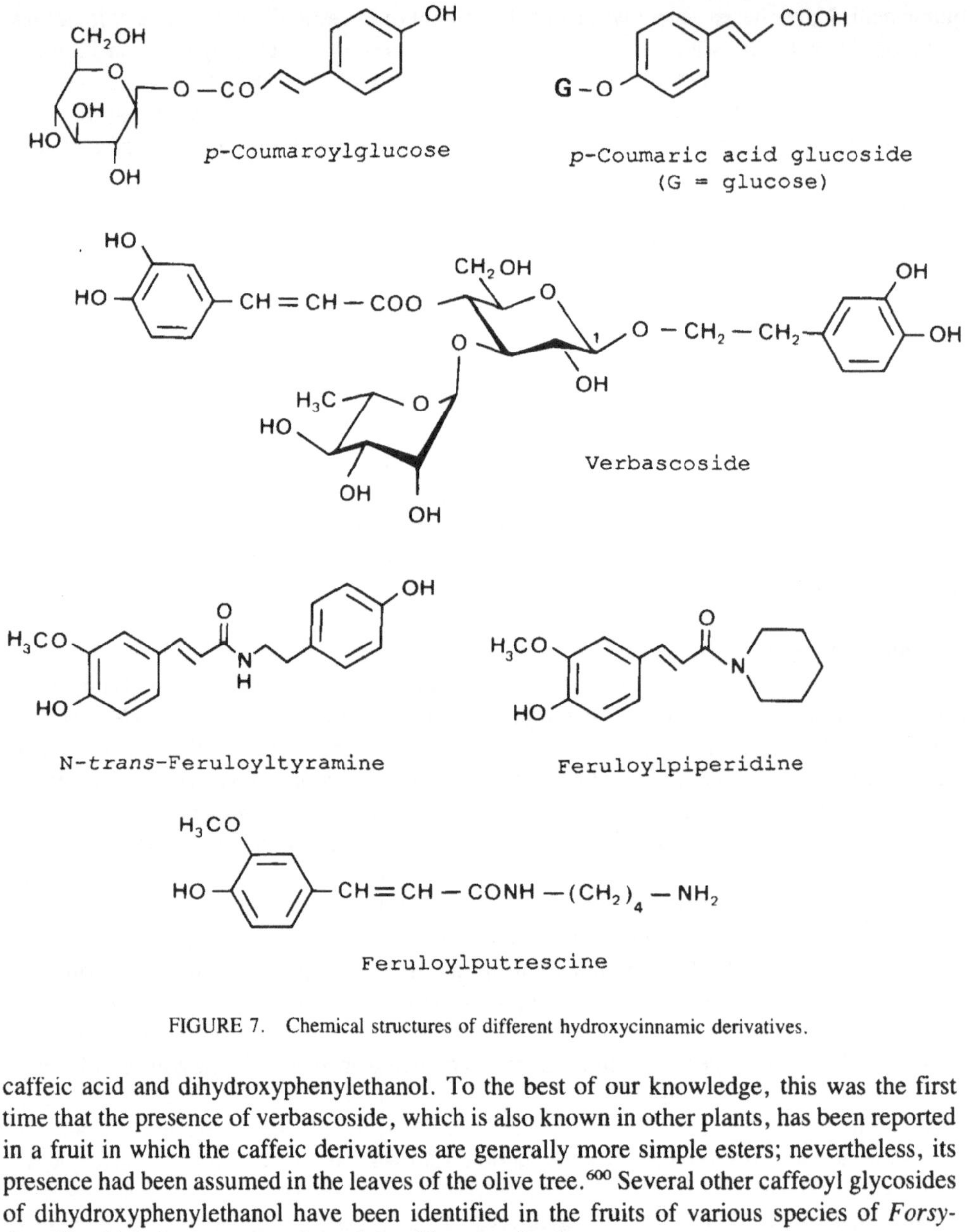

FIGURE 7. Chemical structures of different hydroxycinnamic derivatives.

caffeic acid and dihydroxyphenylethanol. To the best of our knowledge, this was the first time that the presence of verbascoside, which is also known in other plants, has been reported in a fruit in which the caffeic derivatives are generally more simple esters; nevertheless, its presence had been assumed in the leaves of the olive tree.[600] Several other caffeoyl glycosides of dihydroxyphenylethanol have been identified in the fruits of various species of *Forsythia*,[303] which is also a member of the Oleaceae family.

Many other derivatives of HCA and sugars have been identified in plants.[252] The variations involve the nature of the sugar, the number of molecules of HCA bound to a glucose molecule or the bonds that may be formed by the sugar with other molecules, e.g., with amines (asparagine, glutamine, etc.). However, not all these variations seem to have been reported in fruits, perhaps because they have not been sought sufficiently.

HCA, and ferulic acid in particular, can also be found in small quantities in the various glucidic fractions of the cell wall,[174] but this does not appear to have been studied in fruits to date. It would nevertheless appear to be an interesting subject to develop insofar as the maturation of fruit is characterized by deep biochemical modifications in the cell wall in which pectin-HCA bonds may play a regulatory role.[426] Ferulic acid may also be linked by ester bonds to the glucid complex of certain glycoproteins. This is the case of bromelain, a protease preparation extracted from pineapple stem juice,[325] and it is likely that bromelain from pineapple fruit displays the same characteristics. However, the main question is to specify whether this type of bond really exists *in vivo* or whether it only occurs during extraction of the protein.

c. Hydroxycinnamoyl Amides

Although these are frequently encountered in plants,[362,539] their presence in fruits has only been reported fairly infrequently. Feruloyputrescine (Figure 7) occurs in grapefruit and orange juice[627] but has not been found in tangerine and lemon. Likewise, two new phenolic amides were isolated from the fruit of white pepper (*Piper nigrum* L.): *N-trans*-feruloyltyramine and *N-trans*-feruloylpiperidine (Figure 7), together with some other derivatives of piperidine and phenolics[277,408] which appear to have a role in the biosynthetic pathway of piperine, the major pungent component of pepper.

Caffeoylputrescine and caffeoylspermidine accumulate massively in the ovaries of *Nicotinia tabacum*,[91,92,362] whereas these amides are not found in anthers. The concentration of these compounds decreased rapidly and drastically after fertilization while feruloyltyramine and diferuloylputrescine became predominant.[361] Their evolution after fertilization during the development of the ovary into fruit has not been examined, except in a small number of special cases. It would also be interesting to investigate whether these compounds accumulate during chilling injury of fruits, as has been shown for free putrescine.[377]

d. HCA Linked with Flavonoids

Acylation of anthocyanins by various organic acids, and by certain phenolic acids in particular, has been known for a long time[215] and is discussed in the study of flavonoids (see below). The case of grape has been studied extensively[404,593,636,650] and has shown that *p*-coumaric acid plays a major role in the acylation of all the anthocyanins present, whereas caffeic acid only combines with malvidin 3-glucoside.[26,404,484] This situation is in fact common in fruits and vegetables.[573] Caffeic acid has nevertheless been reported in the acylated anthocyanins of olive[345] and eggplant.[573,621]

Flavonoid glycosides other than anthocyanins can also be acylated by HCA,[252] but have only been reported in fruits fairly rarely, e.g., in the form of kaempferol *p*-coumaroylglycoside in *Tribulus terrestris*[49] or 7-*O*-*p*-coumaroylglycoside naringenin in *Mabea caudata*.[39] These compounds probably occur more frequently but have not been identified to date.

e. HCA Linked with Lignin, Suberin, and Cutin

HCA and ferulic and *p*-coumaric acids in particular, form part of the structure of the various polymers which make up cell walls either as direct precursors of lignin monomers[491] or esterified with the same lignin[392,393] or with cell-wall polysaccharide.[173] They may also be covalenty attached to aliphatic components of cutin and suberin[306,472] (Figure 8).

Lignin is not generally formed in the pulp of fleshy fruits, with the exception of sclereids or "stone cells" in pear fruit which have been identified as being ligno-cellulosic in nature.[453] In contrast, lignin is abundant in the teguments of dry fruits and in certain seeds and forms the major part of the stone of drupes; however, no study seems to have been made of these materials. Deposition of polymeric material containing phenolic substances occurs as a result of injury caused by fungal attack or mechanical wounding. In such cases, the polymeric material has been variously designated as lignin, induced lignin, lignin-like, wound lignin, ligno-suberin, suberin, etc. because of lack of information concerning the chemical composition of such materials.[306] Lignin-like compounds are thus synthesized by young tomato fruits after infection by *Botrytis cinerea*[194] or mechanical injury.[152] The composition of these lignins has not been studied but they are probably wound-lignins[462] which have been described elsewhere. Their biosynthesis, together with that of aliphatic compounds typical of suberin

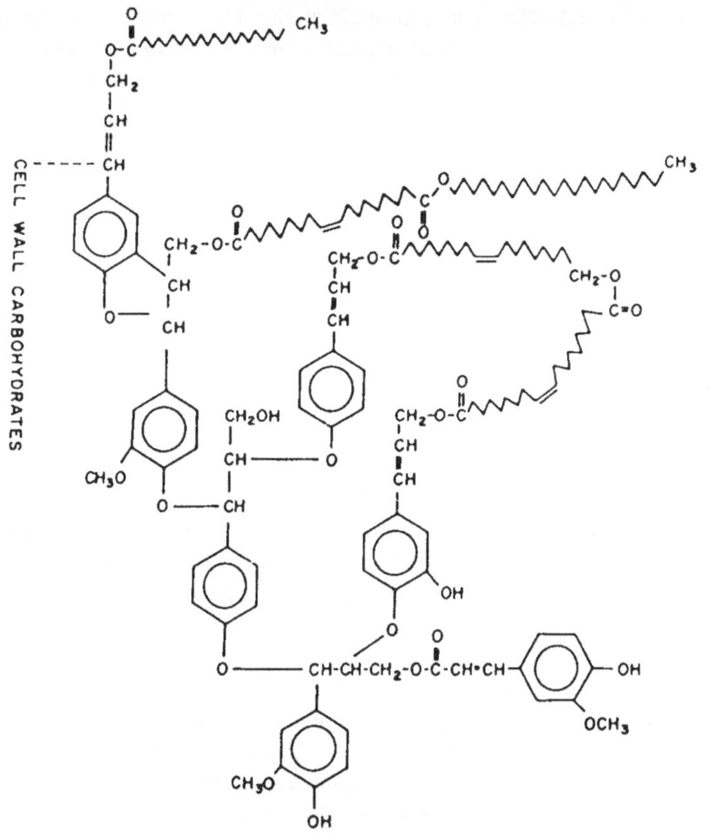

FIGURE 8. A tentative model for suberin structure. (From Kolattukudy, P. E., *Annu. Rev. Plant Physiol.*, 32, 539, 1981. With permission.)

which form in the wound periderm of tomato fruit,[118] provides information about the possible role of the phenols in the defence mechanism. Indeed, they form a mechanical barrier which protects the fruit from too great a loss of water and from the propagation of mycelium. Variations in the phenolic metabolism as a result of stress are discussed in Chapters 3 and 4.

The amount of covalently bound phenolic compounds (*m*, *p*-coumaric acids and flavonoids) in tomato fruit cutin increased during fruit development and climacteric and accounted for as much as 6% of cutin membranes.[272]

The relative amounts of phenolic fraction (esterified ferulic acid) present in suberin preparations depend upon the tissues examined and the methods used to isolate the preparation.[306] However, no analysis has been carried out on this fraction except in protoplasts isolated from immature tomato fruit; they secrete a wall which has been shown to contain suberin[454] and phenolic compounds formed 25% of total monomers.[455]

3. Occurrence of Hydroxycinnamic Derivatives in Fruits: Comparative Study

The HCD content in fruits varies according to the physiological stage. This is examined elsewhere (Chapter 3) and only the variations connected with specific or varietal differences in healthy and ripe fruit are discussed here. However, comparisons are sometimes difficult to establish since it is not always easy to define the stage of maturity.

The variations observed between fruits may concern the qualitative aspects (presence or absence of certain HCD) and also quantitative features (levels of each HCD and relative proportions of each of the compounds present). These two points are examined successively.

TABLE 9
Variability in Hydroxycinnamoyltartaric Ester Levels (10^{-2} μmol/g FW) and Percentages in Skins of *Vitis vinifera* Berries (cv. Carignane)

	pCT	CT	FT	ΣDHC	%pCT	%CT
Mean	74.4	42.1	0.8	117.3	62.8	36.4
Standard error	33.0	15.8	0.1	48.6	2.5	2.4
Average variation %	29.9	25.4	9.2	27.9	2.7	4.5

Note: Mean values were obtained from 17 different vineyards, pCT, CT, FT: *p*-coumaroyl, caffeoyl, and feruloyltartaric esters; ΣDHC: sum of hydroxycinnamic esters.

Data from Boursiquot, J. M., Thèse Doctorat Ing., ENSA, Montpellier, 1987.

a. Qualitative Aspects

Each cultivar of a fruit species displays an HCD pattern which depends little on the set of environmental factors; it generally remains constant while amounts of individual HCD and their relative proportions can vary considerably. In most cases, the qualitative differences between varieties are not sufficiently well marked to enable them to be distinguished unequivocally. Thus, in the case of grape, for example, caffeoyltartaric acid is present not only in all varieties of *Vitis vinifera* but also in all the Euvitis;[70,71,533] however, it is not present in *V. rotundifolia*, a species which occupies a special position and is frequently classified in the genus *Muscadinia* (muscadine grape).

Generally speaking, true biochemical characterization of a species or a cultivar should be sought through the whole phenolic pattern rather than through a particular compound. Models have been developed previously with anthocyanins for different cultivars of *V. vinifera*[198,514] or for azalea flowers for which computerized fingerprints have been proposed.[596] This approach should also make it possible to use the various HCD (as a complement to other categories of phenolics) as a criterion of adulteration in analysis of fruit-based food products.[639] It is probable that in most cases qualitative analysis should be completed by quantitative approaches based in particular on the relative proportions of the various HCD.

b. Quantitative Aspects

Quantitative analysis of fruit HCD generally displays considerable variability. This is the result of the different techniques of analysis used by authors and also because of difficulties in sampling; this is a delicate problem in perennial plants and in particular in fruit trees and grape vine.

i. Influence of Fruit Environment

Numerous factors can be at the origin of considerable qualitative and quantitative modifications in phenolic patterns in fruits and particularly in HCD. Most of the factors (physiological evolution, attack by microorganisms, effect of wounding and irradiation, etc.) are examined in a separate study (Chapters 3 and 4) and only a few variations which depend on agrometeorological conditions are discussed here. An in-depth example was developed in the Carignane variety of *V. vinifera*;[70] samples from 17 vineyards with very different geographical, climatic, and edaphic features displayed great variability in levels of tartaric esters of HCA in the berries (Table 9). Variability was nevertheless considerably reduced when the relative proportions of each of the esters were taken into account. This type of expression can thus be used for much more accurate characterization of the cultivar in question than is possible with the absolute values of these levels. These results were confirmed when comparison was made of data for three consecutive years: although caffeoyltartaric acid contents varied very strongly (from 9 to 45 mg/100 g FW), this compound formed approximately 45% of all HCD in cv. Carignane grape in each case.[344a]

The position of the fruit on the mother-plant can also form a factor in variation on the one hand because of competition that can occur between fruits and vegetative organs and on the other because of the microenvironment of the fruit. However, in the case of apple, the chlorogenic acid contents in the sides of the fruit exposed to the sun or shaded are not significantly different.[342]

ii. Comparison between Different Species

Comparison of HCD contents in numerous fruits (Tables 2 and 8) reveals the enormous variations between species, e.g., from a few hundred mg/100 g FW (blueberries) to only traces in Cucurbitaceae.[460,640] The physiological, sensorial, and technological consequences of these variations are examined in subsequent chapters.

The relative proportions of each HCD essentially represent the best characteristic of a fruit in the mature stage. Thus, the main compounds accumulated are, respectively, 5′-caffeoylquinic type derivatives in apple, pear, peach, tomato, and blueberry and 3′-caffeoylquinic acid type derivatives in cherry, plum, blackberry, blackcurrant, etc. (Table 8).

The preferential accumulation of quinic derivatives is frequently homogeneous between the different constituents of the cinnamic series, involving either 5′-quinic derivatives (e.g., in tomato which contains 5′-*p*-coumaroyl, caffeoyl, and feruloylquinic acids) or 3′-quinic derivatives (e.g., in plum). Apple is probably an exception in that the major *p*-coumaric derivative present is 4′-*p*-coumaroylquinic acid[388,630] whereas the predominant caffeic derivative is chlorogenic acid (5′-caffeoylquinic acid). Nevertheless, there may be varietal differences insofar as cv. Calville blanc is rich in 5′-*p*-coumaroylquinic acid.[338,340]

In fact, only the most recent studies have provided quantitative data on the various types of HCD[157,512] and make it possible to outline a few general ideas (Table 8):

- When both types of derivative (quinic or glycosylated) are present simultaneously, their relative proportions may vary according to the fruit. In most cases (e.g., apple and tomato), glycosylated derivatives are distinctly less abundant than quinic esters (70 mg/100 g FW compared to 290 in young Calville blanc apple).[340] In rarer cases, certain fruits, such as raspberry and redcurrant contain more glycosylated that quinic derivatives.[512]
- When HCA glycosides are present, these are frequently more abundant than the corresponding glucose esters, as is seen in particular in the case of tomato.[157,639]
- One of the most remarkable facts is that according to the terms of the HCA series, accumulation of esters can take place very irregularly either with quinic acid or with glucose. The example of cv. Calville blanc apple is very clear in this respect:[340,344] whereas *p*-coumaric acid is present in the form of two esters (with quinic acid and glucose) in practically identical proportions, caffeic acid exists almost only in the form of quinic derivatives (mainly as chlorogenic acid) whereas ferulic acid is only present as 1-*O*-feruloylglucose.[344] A special effort should be made to understand the regulation of such situations and their meaning from both the metabolic and physiological angles.

iii. Intraspecies Variation: Comparison between Cultivars

Variations between cultivars can be considerable in terms of both absolute levels and relative percentages of the various HCD. Old data, which frequently used global assay methods, made it possible to stress certains aspects in the case of peaches, apples, and grapes (see References 340, 528, and 593). More recent and more accurate data are shown in Table 8 in the form of extreme variations of each HCD between cultivars. Only a few of the most significant examples are discussed here.

TABLE 10
Hydroxycinnamoyltartaric Ester Contents of *Vitis vinifera* Berries or Juice (pCT: *p*-coumaroyltartaric acid; CT: caffeoyltartaric acid)

Type	Number of cultivars	pCT	CT	% pCT[a]	% CT[a]
Red cultivars	47[b] (Skin)	55.0	71.7	43.4	56.6
	17[c] (Press juice)	(8.2—163.4)	(12.9—199.1)	(14.3—64.8)	(35.2—85.7)
		20.7	165.4	15.7	84.3
		(7.0—42.0)	(50.0—435.0)	(2.9—30.8)	(69.2—97.1)
White cultivars	46[b] (Skin)	17.9	30.5	37.0	63.0
	20[c] (Press juice)	(0.6—79.1)	(2.0—176.0)	(16.5—56.7)	(43.3—83.5)
		16.0	129.0	12.8	87.2
		(tr[d]—53.0)	(16.0—299.0)	(3.6—25.4)	(74.6—96.4)

Note: Each value is the mean of various cultivars. Extreme values are given in brackets.

[a] Percentages of the sum (pCT + CT).
[b] From Boursiquot.[70] Contents are expressed as 10^{-2} μmol/g FW berry skin.
[c] From Singleton et al.[533] Contents are expressed as mg/l of press juice.
[d] Traces.

Very complete studies have been carried out on a large number of cultivars of *V. vinifera*.[70,71,487,533] Variations are very great in all the cases studied (Table 10) and partly account for the very different browning capacities which characterize each cultivar or derived products.[487,534] This aspect is discussed in Chapter 5.

As in the case of interspecies comparison, the most reliable criteria in comparison of cultivars of the same species appear to be the percentage of each HCD. Thus, in the case of *V. vinifera*,[70,71] the percentages of *p*-coumaroyl and caffeoyltartaric esters can be used to discriminate between varieties and can be used for taxonomic purposes. Principal component analysis[70] of 47 red varieties showed that it was possible to group the latter in several categories, e.g., by differentiating those with a high percentage of caffeoyltartaric acid (Jurançon, Gamay Freaux) from those with a high percentage of *p*-coumaroyltartaric ester (Carignane, etc.). Less dispersion seems to be observed in white varieties, with the most marked feature being nevertheless the high density of cultivars with somewhat low ester levels, balanced caffeoyl and *p*-coumaroyltartaric ester levels and a non-negligible percentage of feruloyltartaric ester.

A particularly interesting problem is the search for relationships which may exist between the capacity to accumulate HCD and that of synthesizing anthocyanins. Globally, in *V. vinifera* a loss of anthocyanin synthesis capacity is accompanied by a fall in total phenolics content.[537] The red *vinifera* cultivars averaged slightly higher in both caffeoyl and *p*-coumaroyltartaric esters content than did the white ones, but they overlapped broadly (Table 10). In varieties with both a colored (red or pink) and a white form,[70] the HCD contents were always higher in the skins of colored berries than in those of white berries of the same cultivar (Figure 9). Nevertheless, this general reduction associated with the absence of anthocyanin synthesis was not distributed evenly between the three main tartaric HCD (Table 10 and Figure 9). The relative proportion of *p*-coumaroyltartaric ester increased whereas those of caffeoyl and feruloyltartaric esters decreased with the change from the white to the colored form. Thus, in the species *V. vinifera*, the evolution from the original form with colored or teinturier grapes to white forms[58] appears to be accompanied by a fall in their tartaric HCD contents and an increase in the percentages of caffeoyl and feruloyltartaric esters. In the present state of knowledge, these data cannot, however, be extended to other fruit species since there have been no studies as complete as those on grape.

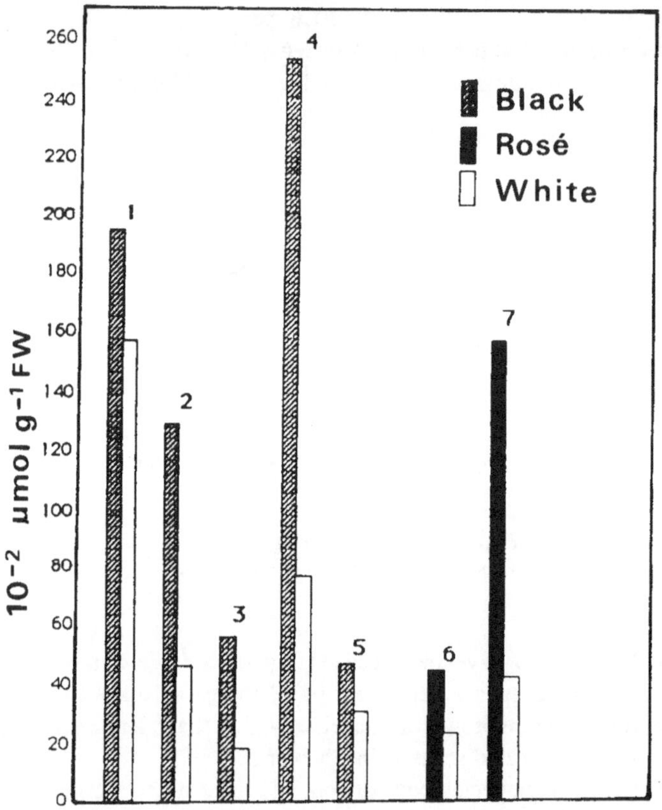

FIGURE 9. Comparison of total hydroxycinnamoyltartaric acid concentration in the skin of seven grape cultivars. 1: Aramon; 2: Grenache; 3: Picpoul; 4: Terret; 5: Pinot; 6: Clairette; 7: Savagnin. (Adapted from Boursiquot, J. M., Thèse Doct. Ing., ENSA, Montpellier, France, 1987.)

Considerable differences in HCD contents were also observed according to the varieties of peach[89] and olive.[159] In the latter case, a negative correlation was found[8] between levels of verbascoside, the main HCD in this fruit, and oleuropein, a compound with a partially common structure and which is the main cause of bitterness.

iv. Overall Balance: Relative Proportions of Each HCA in Fruits

As a complement to the results described above, it is possible to define a few main trends with regard to the presence of each HCA in fruits without taking into account the combined form in which they occur. Depending on the authors and the techniques used, the quantitative data (Table 11) can thus be from two sources: either the direct result of analyses when measurements were made after hydrolytic degradation, or the sum calculated from the different combined forms previously isolated and assayed. In the latter case, which generally applies to the more recent analyses, the results can generally be considered to be more accurate. Only extreme variations are given when the data concern several cultivars of the same species (Table 11).

Caffeic acid is frequently the most abundant (Table 11). It commonly exceeds 75% of total HCA in numerous fruits (e.g., apple, plum, tomato, etc.) and may even form almost the entire HCA content in extreme cases, such as eggplant or certain blueberries. It is still the major representative of the cinnamic series in several other fruits (*Citrus*, numerous small red fruits, etc.) in which the proportion is nevertheless smaller (in the order of 50%). In other more rare cases (pineapple, white currants, etc.) there is a smaller proportion of

TABLE 11
Percentages[a] of Hydroxycinnamic Acids[b] in Ripe Fruits

FRUITS	*p*-Coumaric	Caffeic	Ferulic	Sinapic	Ref.
Apple	5—20	84—94	1—5	—	340,344,388
Apricot	3—5	87—91	6—9		388
Blackberry	23—35	38—77	12—27		396
Blackcurrant	53—67	17—41	6—16		550
Blueberry	0—8	90—100	0—2		550
Cherry					
Sour	20—65	35—79	0.2—1.2	—	388,552
Sweet	10—65	35—87	0.4—3.5	t	344,381,388,552
Coffee pulp		67	1		451
Eggplant		99	0.5	0.5	507
Gooseberry	34—48	47—60	5—6		550
Grape					70,344,533
White	25	50	25	—	
Red	66	33	1	—	
Grapefruit	21—37	19—42	29—47	0.6—6	551
Lemon	41—55	18—28	20—31	2—13	551
Melon (skin)	27—73	0.03—32	0.2—42		507
Orange	13.5—14.7	53—23	18—43	16—19	551
Peach	1—3	93—97	2—4		344,388
Pear	1—8	99	t		344,388
Pepper	2—27	14—67	3—74	0—28	507
Pineapple	53.8	10.8	22.1	13.2	128
Plum	2—25	72—95	0—8		388,552
Pumpkin (skin)	91			9	507
Raspberry	27—73	0—36	22—50		344,396
Redcurrant	33—37	42—48	19—21		550
Strawberry	56—92	8—44			549
Tomato	2—21	69—91	t—9	0.3—7	344,507
White currant	48—85	10—31	6—21		550

Note: t = traces.

[a] Extreme values for several cultivars.
[b] After hydrolysis or calculated from derivatives.

caffeic acid, and in exceptional cases — in a few varieties of raspberry for example[396] — only traces are found whereas the other HCA are dominant.

In *V. vinifera* (Table 10), the mean proportion of caffeic acid in press juice was found to be around 84% for red cultivars and 87% for white.[533] These values fell to 58 and 64%, respectively, when only the skins of the berries were analyzed.[70] The difference found between the two groups of data result from the sampling methods: press juice mainly contains berry pulp in which the proportion of caffeic derivatives is higher than in the skin.[70] The mean values obtained cover considerable intervarietal variability (Table 10). For example, the percentage of caffeic acid varied from 36.4% in the skins of Carignane berries (Table 9) to 82.9% in Alphonse Lavallée.

***p*-Coumaric acid,** although present in most fruits, is frequently less abundant than caffeic acid (Table 11). Nevertheless, it may be predominant in certain *Citrus* and certain varieties of raspberry and in pineapple. Particularly high percentages of *p*-coumaric acid have been found in strawberries and in some varieties of white currants. In *V. vinifera* (Table 10), this proportion varies in pulp between 3 and 25% according to variety but can exceed 60% in the skins of certain red cultivars (Carignane and Terret Noir).

Ferulic acid usually only forms a small percentage of total HCA in fruits. However, in certain cases it can reach and even exceed 50% of HCA as in peppers, certain *Citrus*, and in the white grape cultivar Clairette at maturity.[486]

Sinapic acid has more rarely been reported in fruits and is generally only observed as traces (Table 11). However, it forms several percent of the HCA in several *Citrus*,[51] in pineapple,[128] and in tomato[150,620] where it is, nevertheless, located only at the pericarp.[158] It also appears in *in vitro* cell suspension cultures from apple parenchyma[309] although it was not found in the original fruit.

The other HCA do not appear to have been reported in fruits. Likewise, the presence of cinnamic acid s.s. derivatives, which have been reported only in cranberries,[106] requires confirmation.

C. COUMARINS

Coumarins are lactones derived from *o*-hydroxycinnamic acids by cyclization and ring closure between the *o*-hydroxy and carboxy groups. Today, nearly a thousand coumarins are known in nature, and there have been several reviews of their distribution, role, and biosynthesis in plants.[85,86,220,402] Distinction can be made between several groups according to their chemical structure; the most important are the simple coumarins (including coumarin itself (**1**) and above all the 7-oxygenated derivatives; see Table 12) and the more complex coumarins, including furano and pyranocoumarins. Although some of these coumarins are present in plants as their glycosides,[220,378,402] most of them exist in the free form in the cell.[86]

Considerable study of fruit coumarins has been carried out in the Umbelliferae and the Rutaceae, and in particular in citrus fruits because of their economic importance. Only the main points are examined here, reference is made to studies or reviews which have been published on the subject.[203,204,402]

1. Coumarins in the Rutaceae (essentially in Citrus)

There are over 350 coumarins in the Rutaceae.[203] They have been reported in a wide variety of plant parts, leaf and bark (stem or root) being the main sources although, as observed by Gray and Waterman,[204] this may simply be a reflection of the greater frequency of studies of these parts. According to Gray,[203] simple coumarins account for about 55% of total coumarins in Rutaceae, furanocoumarins for 28%, and pyranocoumarins for 11%. The first two types are plentiful in citrus fruits and seeds and in the oil extracted from them (Table 12). Coumarins can also be used to complement methoxyflavones and flavanones as useful taxonomic markers[203,378,567] in the nomenclature of *Citrus*, which is still complex and confused.[515] Aurapten (**4**) appears to be the major component of grapefruit peel oil,[564] the approximate order of decreasing concentration being **4** = **29**>**7**>**5**>**26**>**23**>**28** = **11.** Some of these compounds, and in particular merazin hydrate (**11**) and isomeranzin (**10**) may be extraction or chromatographic artifacts,[378] as is the case of several other coumarins "isolated" from plant material.[203,402]

Approximate concentrations of coumarin in flavedo are given in Table 13. In comparison, concentrations of coumarins observed in the albedo were very low: these substances are essentially flavedo constituents.[378] In addition to several well-known coumarins, two coumarin glycosides were recently reported in aqueous extracts of bitter orange flavedo. They were shown to be meranzin hydrate glucopyranoside (**12**) and columbianetin-*O*-β-D glucopyranoside (**35**). Concentrations of these two coumarin glycosides are of possible taxonomic significance and are consistent with the close relationship between pummelo, grapefruit, and bitter orange,[378] as shown by examination of certain enzymes and substrates.[348,515] As expected, aqueous extracts of sweet orange flavedo contained no coumarins or coumarin glycosides.[378]

Scoparone (**20**) was extracted from grapefruit peel 7 d after gamma irradiation at 30 krad[473] and was recently shown to be present in bark of trunks and branches and in fruit

TABLE 12
Main Coumarins Identified in Citrus Fruits

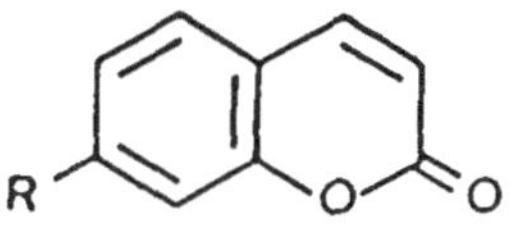

Number	Trivial name		Citrus[a]
1	Coumarin	R = H	
2	Umbelliferone	R = OH	B,I
3	Herniarin	R = OMe	I
4	Aurapten	R = $O{-}CH_2{-}CH{=}C{-}CH_2{-}CH_2{-}CH{=}C(Me)_2$ (Me on C)	B,D,I,J,K
5	(+)-Epoxyaurapten	R = $O{-}CH_2{-}CH{=}C{-}CH_2{-}CH_2{-}CH{-}C(Me)_2$ (epoxide O bridging CH—C(Me)$_2$) (Me on C)	I
6	Marmin	R = $O{-}CH_2{-}CH{=}C{-}CH_2{-}CH_2{-}CH{-}C(Me)_2$ Me OH OH	I

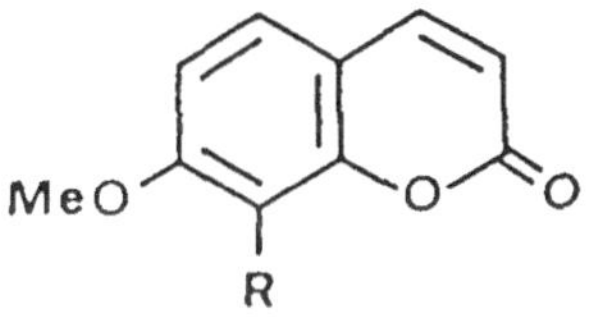

Number	Trivial name		Citrus[a]
7	Meranzin (= Aurapten)	R = $CH_2{-}CH{-}C(Me)_2$ (epoxide O bridging CH—C(Me)$_2$)	B,I,J
8	Auraptenol (artifact ?)	R = $CH_2{-}CH{-}C{=}CH_2$ OH Me	B,C
9	Citrusal (artifact ?)	R = $CH_2{-}C(Me_2){-}CHO$	I
10	Isomeranzin (artifact ?)	R = $CH_2{-}C({=}O){-}CH(Me)_2$	B,I
11	Meranzin hydrate (artifact ?)	R = $CH_2{-}CH{-}C(Me)_2$ OH OH	B,I
12		R = $CH_2{-}CH{-}C(Me)_2$ OH O-β-D glucopyranosyl	B,I

TABLE 12 (continued)
Main Coumarins Identified in Citrus Fruits

Number	Trivial name		Citrus[a]
	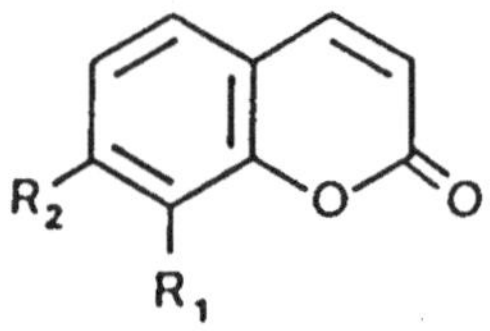		
13	Poncimarin	$R_1 = CH_2–CH—C(Me)_2$ \ / O $R_2 = O–CH_2–CH—C(Me)_2$ \ / O	K
14	Isoponcimarin	O ‖ $R_1 = CH_2–C–CH(Me)_2$ $R_2 = O–CH_2–CH—C(Me)_2$ \ / O	K
	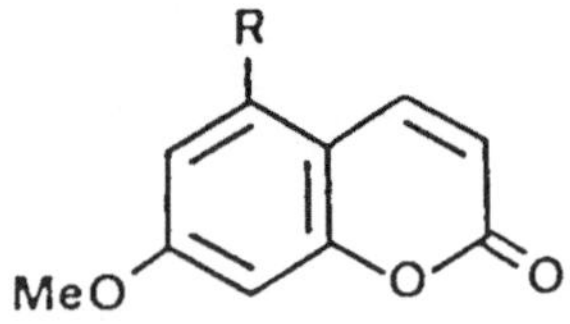		
15	Limettin (= Citropten)	R = OMe	A,B,C,D, F,G,I,J
16		$R = O–CH_2–CH{=}C–CH_2–CH_2–CH{=}C(Me)_2$ | Me	A,C,D,E
17		$R = O–CH_2–CH{=}C(Me)_2$	E
	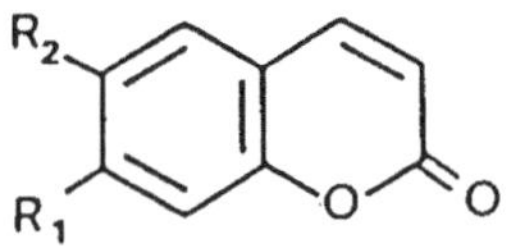		
18	Aesculetin	$R_1 = R_2 = OH$	I
19	Scopoletin	$R_1 = OH$ $R_2 = OMe$	I
20	Scoparone	$R_1 = OMe$ $R_2 = OMe$	A(?),B,H, I,J
21	6-Methoxyaurapten	$R_1 = O–CH_2–CH{=}C–CH_2–CH_2–CH{=}C(Me)_2$ | Me	K

TABLE 12 (continued)
Main Coumarins Identified in Citrus Fruits

Number	Trivial name		Citrus[a]
		R_2 = OMe	
22	Scopolin	R_1 = O-β-D-glucosyl R_2 = OMe	I

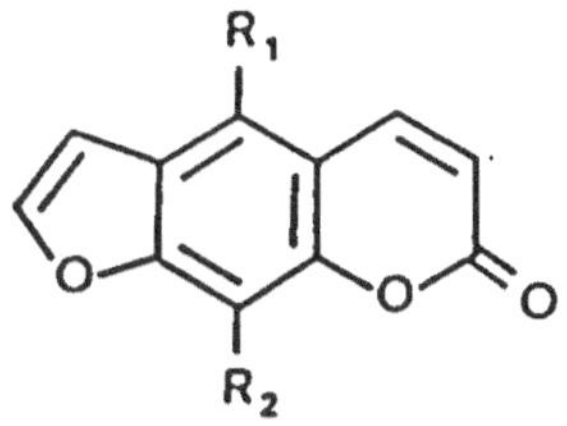

Number	Trivial name		Citrus[a]
23	Bergaptol	R_1 = OH R_2 = H	A,B,C,E,I
24	Bergapten	R_1 = OMe R_2 = H	A,B,C,K
25	Isoimperatorin	R_1 = O–CH_2–CH=C$(Me)_2$ R_2 = H	A,B,E,J
26	Bergamottin	R_1 = O–CH_2–CH=C(Me)–CH_2–CH_2–CH=C$(Me)_2$; R_2 = H	A,B,C,D E,F,I
27	Aurantiumal (artifact ?)	R_1 = O–CH_2–CH=C(Me)–CH_2–CH_2–C$(Me)_2$–CHO; R_2 = H	I
28		R_1 = O–CH_2–CH=C(Me)–CH_2–CH_2–C(OH)–C(OH)$(Me)_2$; R_2 = H	F,I
29		R_1 = O–CH_2–CH=C(Me)–CH_2–CH_2–CH–(O)–C$(Me)_2$ (epoxide); R_2 = H	I
30	Imperatorin	R_1 = H; R_2 = O–CH_2–CH=C$(Me)_2$	A,K
31		R_1 = H; R_2 = O–CH_2–CH=C(Me)–CH_2–CH_2–CH=C$(Me)_2$	A,E
32	Isopimpinellin	R_1 = OMe; R_2 = Me	A,D
33	Phellopterin	R_1 = OMe; R_2 = O–CH_2–CH=C$(Me)_2$	A
34		R_1 = O–CH_2–CH=C(Me)–CH_2–CH_2–CH=C$(Me)_2$; R_2 = OMe	A

TABLE 12 (continued)
Main Coumarins Identified in Citrus Fruits

Number	Trivial name	Citrus[a]

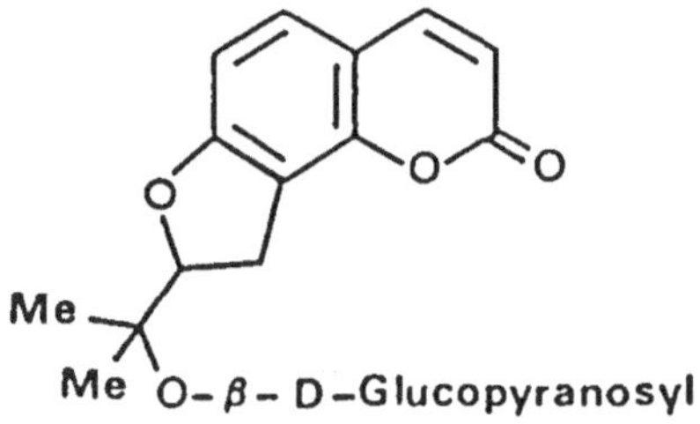

35	Columbianetin-O-β-D-glucopyranosyl	

[a] Identification of *Citrus* species: A: *C. aurantifolia;* B: *C. aurantium;* C: *C. bergamia;* D: *C. limetto;* E: *C. limon;* F: *C. macrocarpa* and *C. macroptera;* G: *C. medica;* H: *C. mitis;* I: *C. paradisi* and *hybrids;* J: *C. sinensis;* K: *Poncirus trifoliata.*

Adapted from References 203, 204, 378, 402, 474 and 564.

TABLE 13
Approximate Concentrations of Coumarins in Flavedo of Citrus Fruits (g/100 g FW)

Compounds (see Table 12)	11	10	7	35	12	Total
Bitter orange	0.01	0.08	0.25	0.06	0.31	0.71
Grapefruit	0.01	0.04	0.16	—	0.04	0.25
Pummelo	0.02	0.04	0.28	—	0.17	0.51

Data from McHale, D., Khopkar, P. P., and Sheridan, J. B., *Phytochemistry,* 26, 2547, 1987.

peel in sweet orange, bitter orange, lemon, trifoliate orange *(Poncirus trifoliata)*, and Troyer citrange *(C. macrophylla)* following inoculation with *Phytophtora citrophtora.*[3] Bioassays with scoparone inhibited the growth of various phytopathogenic fungi *in vivo*. This compound is thus similar to a phytoalexin and is responsible for the defense mechanism against *P. citrophtora* in citrus; it accumulates after inoculation in the bark, with the concentration increasing more rapidly in resistant species than in sensitive ones.

Ionizing radiation also stimulates the biosynthesis of two coumarins: scopoletin **(19)** and its 7-glucoside, scopolin **(28)** in grapefruit peel[474] (see Chapter 3).

2. Coumarins in Umbelliferous Fruits

Plants of the Umbellifereae family are very rich in coumarins, particularly those of the subfamily Apoideae. Nevertheless, simple coumarins are not very plentiful whereas furanocoumarins occur frequently. A full review of these compounds and of their occurrence in the different species of Umbellifereae has already been published.[402] Some of them are potent photosensitizers which cause severe dermatitis after contact with plants in the presence of sunlight. Furthermore, furanocoumarins can be phototoxic, mutagenic, and photocarcinogenic.

The various parts of umbelliferous plants contain furanocoumarins, but the highest concentrations are found in the fruits.[96,402] It should, nevertheless, be noted that the term fruit is used here to designate achenes, which at maturity consist esentially of seeds and dehydrated teguments. It would seem, however, that embryos do not contain coumarins,[93,402] which probably accumulate principally in seed coats, including fruit teguments. Umbelliferous fruits or the oils extracted from them are traditionally used by man for flavoring foods and beverages and for use in perfumery, medicine, and in the making of alcoholic beverages. Quantitative studies were carried out recently[48,96] in which HPLC separation and an ultrasensitive photobiological assay were used[96] to determine the concentration of photoactive furanocoumarins in the umbelliferous fruits most frequently used by man (Table 14). The total concentration of photoactive furanocoumarins is extremely high in the fruits of Angelica (*Angelica archangelica* L.), where it is over 12 mg/g in ripe fruits; it has even reached 2.59% in ripening fruits.[402] The fruits of parsnip (*Pastinaca sativa* L.) are also rich in furanocoumarins (0.06 to 2.63% DW according to variety).[48,96] In this case, fruits of wild plants contain almost three times the quantity of furanocoumarins as the fruits of cultivars; this is probably a result of the selection pressure imposed by cultivation or by herbivores.[48] All the other umbelliferous fruits have much lower furanocoumarin contents: less than 0.2 mg/g (hogweed, parsley, celeriac, celery, lovage, fennel) or mere traces (carrot, cumin, coriander, anise, etc.).

Although the most common furanocoumarins found in umbelliferous fruits are bergapten **(24)** and xanthotoxin (8-methoxypsoralen), imperatorin **(30)** is often the most plentiful, for example in angelica fruits and in certain varieties of parsnip.[48,96] In fact, according to the cultivar studied, there may be considerable variation in the proportions of the various constituent furanocoumarins. New qualitative and quantitative data[96] have made it possible to discuss and re-evaluate earlier observations on the chemosystematics of Umbelliferae.

The fruits of *Amni visnaga* and *in vitro* cultures derived from them contain furochromones (kellin, visnagin) which are used in the treatment of heart diseases.[300]

3. Other Cases

Apart from the two examples above, coumarins — generally simple coumarins — are more rarely reported in fruits than other plant organs.[402] Thus, in Rosaceae, fruits of *Prunus mahaleb* contain several coumarins,[402,658] including umbelliferone **(2)**, herniarin **(3)**, aesculetin **(18)**, scopoletin **(19)**, and mahaleboside. Aesculetin is also present in the fruit of *P. spinosa.* Scopoletin has been reported in very small quantities (0.05 to 0.2 mg/100 g FW) in plums *(P. domestica)* whereas none was found in sour and sweet cherries.[552] In contrast, aesculetin was present in sour cherries (0.1 to 0.4 mg/100 g FW) although not in sweet cherries or plums.[552] However, these identifications have not been confirmed by other authors, and the possibility of their being artifacts formed during the extraction of phenolic compounds cannot be excluded. Aesculin is also plentiful in horse chestnut.[220] Among Solanaceae, the eggplant contains scopoletin **(19)** and scoparone **(20)**.[17] The fruit of *Kielmeyera pumila* has recently been shown to contain two phenylcoumarins, mammeigin and isomammeigin.[405]

IV. FLAVONOIDS

Flavonoids are compounds which all possess the same C_{15} (C_6–C_3–C_6) basic skeleton. The various classes differ in the oxidation level of the central pyran ring except in chalcones where it is open. These compounds are extremely numerous and varied in plants and they have been studied in detail in very well-documented works or reviews, such as those of Harborne et al.,[230] Harborne,[220] and Harborne and Mabry[226] which comprises complete references, including the oldest.

Three of the numerous classes of flavonoids are very widespread in fruits and are

TABLE 14
Photoactive Furanocoumarins in Various Umbelliferous Fruits Used by Man (μg/g fruit)

	Psoralen	24[a]	Xanthotoxin	32[a]	25[a]	30[a]	Oxypeucedanin	Sphondin	Σ
Angelica	—	3477	427	128	3852	5030	—	—	12914
Parsnip	—	213—430[b]	170—682	91—205	—	161—355	—	3—8	639—1956
Hogweed	—	136	25.5	8	—	36	—	t	205.5
Parsley	0—9.7	t—12.5	t—5.3	—	t—5.6	0—5.9	0—26	—	0—46.6

Note: t, traces.

[a] For identification, see Table 12.
[b] Extreme values for several cultivars.

Data from Ceska, O., Chaudhary, S. K., Warrington, P. J., and Ashwood-Smith, M. J., *Phytochemistry*, 26, 165, 1987.

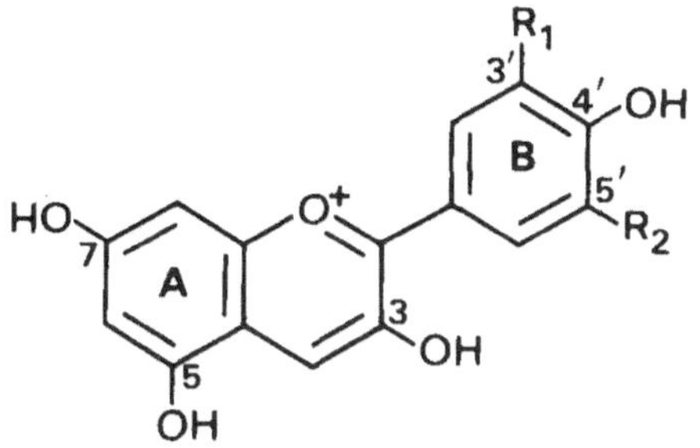

R_1 = H	R_2 = H	: Pelargonidin (Pg)
R_1 = OH	R_2 = H	: Cyanidin (Cy)
R_1 = OH	R_2 = OH	: Delphinidin (Dp)
R_1 = OCH_3	R_2 = H	: Peonidin (Pn)
R_1 = OCH_3	R_2 = OH	: Petunidin (Pt)
R_1 = OCH_3	R_2 = OCH_3	: Malvidin (Mv)

FIGURE 10. Chemical structures of anthocyanidins present in fruits.

quantitatively dominant: **anthocyanins, flavonols,** and **flavan-3-ols,** the latter being present as monomers and in condensed forms (tannins). The other classes (in particular **flavones, flavanones, flavanonols** or **dihydroflavonols, chalcones,** and **dihydrochalcones**) are less important except in certain special cases, such as *Citrus*. Some of these compounds, e.g., flavanones and flavanonols, are intermediate in the biosynthesis pathway of other flavonoids which leads from hydroxycinnamic acids and malonylCoA to anthocyanins, flavones, and flavonols.[207,212] This may explain why only traces are found.

Finally, there are certain classes of flavonoids (e.g., isoflavonoids and C-glycosylflavonoids) which are only found in particular groups of plants. These compounds have rarely been reported in fruits.

The diversity and complexity of the flavonoids of fruits depends on at least two factors: first, the variety of aglycones and the high number of glycosides possible, sometimes in acylated form, and second, the possibility of condensation into complex molecules, e.g., condensed tannins which are studied later (Chapter 1, Section V).

A. ANTHOCYANIC PIGMENTS

Anthocyanins are pigments which give most fruits their red, violet, and blue color. They are, nevertheless, not present in certain fruits (tomato, red pepper) in which the color is caused by carotenoid pigments. Much work has been carried out on anthocyanins, particularly in flowers, and reviews have been published at various dates.[217,241,265,274,573,575] Natural anthocyanic pigments (anthocyanins) are glycosides which release aglycone forms (anthocyanidins) by hydrolysis.

1. Anthocyanidins

These are hydroxylated and methoxylated derivatives of phenyl-2-benzopyrylium (flavylium salt structure, Figure 10). A feature of anthocyanins among flavonoids is that they are present in cation form in acid media. They are, thus, charged molecules whose structure displays stabilization by resonance with numerous mesomeric forms. The special properties of anthocyanins (color, stability, formation of metal complexes) are linked with this structure. They have been the subject of much investigation of the past 10 years in relation with the natural expression of the color of parts of plants and the utilization of anthocyanins as coloring additives in foodstuffs (cf. Chapters 4 and 5). Some 15 anthocyanidins are known today,[575] but only 6 are widespread and commonly contribute to the pigmentation of plant organs. Only these six common anthocyanidins are found in fruits; they are all 3,5,7 and 4′ hydroxylated derivatives (Figure 10). The only exception is the banana: the presence of a red pigment has been described after UV irradiation of this fruit.[526] It was probably a 3-deoxyanthocyanidin whose structure may require confirmation. 7 and 5-methoxylated anthocyanidins found in *Primula* and *Plumbago* flowers were not found in the fruits. The six aglycones in the fruits differed in 3′ and 5′ hydroxylation and methoxylation on ring B (Figure 10).

A study of some 40 fleshy fruits (Table 15) showed that cyanidin is the most common anthocyanidin (90% of fruits examined). It is followed in decreasing order by delphinidin (35%), peonidin (30%), pelargonidin (20%), and finally petunidin and malvidin (15%). Very few red fruits do not contain cyanidin. This is the case, however, of mango, eggplant, and the garden huckleberry *(Solanum guineese)*. A study on anthocyanic pigments in the fruits of 44 species belonging to 25 families of Angiosperms reached the same conclusion:[280] cyanidin was present in 82% of the fruits.

Among the fruits examined (Table 15), a dozen common fruits contained only cyanidin: apple, pear, peach, blackberry, black raspberry, redcurrant, gooseberry, mulberry, mangosteen, and elderberry. The same applies to privet and viburnum fruits[209,265] and to those of less common species.[280,575] All these fruits seem to be incapable of the methylation reactions which lead from cyanidin to the other anthocyanidins. The same situation is observed in most leaves which synthesize anthocyanins.[52,217]

Delphinidin alone is present in eggplant.[571,621] There is no common fruit which contains either petunidin, malvidin, or pelargonidin alone. In contrast, the fruit of *Polygonum perfoliatum* contains malvidin alone, and the fruits of three Rubiaceae (genera *Damnacanthus* and *Mitchella*) contain only pelargonidin.[280]

The most frequent case is that of fruits which contain two aglycones (15 out of 40 cases examined). Cyanidin is generally combined with peonidin, delphinidin, or pelargonidin (Table 15). The fruit of *S. guineese* contains two aglycones different to cyanidin: petunidin and malvidin.[167]

A very small number of fruits possess three aglycones; these are cyanidin and delphinidin accompanied by either pelargonidin (e.g., pomegranate[136] and tamarillo fruit, *Cyphomandra betaceae*[646]) or peonidin (e.g., cranberry, *Vaccinium macrocarpon*). In the latter fruit, the two major anthocyanidins are cyanidin and peonidin[181,182] and it has recently been shown that small quantities of delphinidin are present.[180]

The most complex case is that of certain fruits which contain five aglycones with only pelargonidin absent and sometimes only very small amounts of cyanidin. The fruits concerned are blueberry, whortleberry, crowberry, and grape. Fruits of *Berberis fortunei* also probably contain these five aglycones.[280]

2. Anthocyanins

Anthocyanins have been analyzed in a fairly large number of fruits. Detailed anthocyanin compositions of common fruits of economic interest are given in Chapter 2. For other fruits, the results reported by Timberlake and Bridle,[575] Ishikura[278,279] Lowry,[329] Ishikura and Sugahara,[280] and Hrazdina[265] are assembled in Table 16.

a. Monoglycosides

In fruits as in other parts of plants, glycosylation of anthocyanidins almost always occurs in the 3-position. (See Figure 11.) One exception may be the 7-arabinoside cyanidin described in the skin of Red Delicious apples.[557] However, this minor pigment has not been found by other authors and its presence has been questioned.[573]

TABLE 15
Anthocyanidins in Fruits

	Pg[a]	Cy	Pn	Dp	Pt	Mv	Ref.
Apple		7[b]					265,557,573,574
Blackberry		5					225,287,498,579
Blackcurrant		2		2			169,305
Blueberry		3	2	3	3	3	28,29,30,353
Cherry							
Sour		5	1				119,120,519,573
Sweet		2	2				417,418,573,575
Chokeberry		4					423
Cowberry		3		1			11,180
Cranberry		3	3	3			180-182,213,323
Crowberry		2	2	3	3	2	296,297
Eggplant				8			571,573,621
Elderberry		4					79
Fig	1	3					265,443,573
Garden huckleberry					3	1	167,573
Gooseberry		2					217,225
Grape		3	3	3	3	4	26,198,404,465,471,483,484,650
Litchi	2	2					439
Mangosteen		2					134
Mango fruit			1				441
Miracle fruit		3		2			88
Mulberry		2					265,278
Olive		6		1[c]			345,355
Peach		2					278,592,593,575
Pear		2					165,574,575
Plum		2	2				575,593
Pomegranate	2	2		2			136,265
Raspberry							
Red	4	5					38,169,225,287,385
Black		4					225,287,573,579
Redcurrant		6					225,425
Strawberry	2[d]	1					265,575,647
Sweet orange	2	6	2	2	2		334,335
Tamarillo fruit	2	2		2			646
Whortleberry		3	3	3	3	3	25,603,604

[a] Pg: pelargonidin: Cy: cyanidin; Pn: peonidin; Dp: delphinidin; Pt: petunidin; Mv: malvidin.
[b] Number of glycosides.
[c] Only in some varieties (From Marekov[355]).
[d] Five derivatives of pelargonidin have been reported by Fuleki.[179]

Five oses are involved in the formation of the 3-monoglycosides. Three of them (glucose, arabinose, and galactose) can combine with the 6 anthocyanidins and lead to the 18 anthocyanins identified in fruits. On the contrary, xylose and rhamnose are only linked with one aglycone in fruits: cyanidin 3-xyloside is known in apple[574] and in blackberry, where identification should be confirmed.[498] Pelargonidin 3-rhamnoside is one of the major pigments in litchi pericarp.[439]

The 3-glucosides of the six aglycones, cyanidin 3-arabinoside and 3-galactoside, are fairly widespread. In contrast, the 3-arabinosides and 3-galactosides of malvidin, petunidin, delphinidin, and pelargonidin are found almost only in Ericaceae[12,25,28,29,31,32,168,180,353,603] and Empetraceae.[296,297] Exceptions include peonidin 3-galactoside in mango[441] and delphinidin 3-galactoside and 3-arabinoside in miracle fruit (*Synsepalum dulcificum*, Sapotaceae).[88]

Harborne's publications should be consulted for a detailed study of monoglycosides and their constituent oses.[215,217,218] Glucose is generally in the form of β-pyranose, but in straw-

TABLE 16
Anthocyanins of Noncommon Fruits

1. Monocotyledones

Liliaceae:	
Smilacina japonica	Cy 3-rutinoside, Pn glycoside
Disporum sessile	Cy 3-glucoside, Cy glycoside
Liriope platyphylla	Dp,Pt,Mv 3-glucosides Dp,Pt,Mv 3-rutinosides
Ophiopogon japonica	Dp 3-rutinoside, Pt glycoside
Smilax riparia	Cy 3-glucoside, Cy 3-rutinoside
Smilax china	Cy 3-glucoside, Cy 3-rhamnoglucoside
Majanthemum bifolium	Cy 3-glucoside, 3-rhamnoglucoside
Amaryllidaceae	
Clivia nobilis	Cy,Pg 3-rutinosides Cy,Pg 3-glucosides
Zingiberaceae	
Renealmia regmelliana	Pt 3-glucoside Dp,Pt,Mv 3,5-diglucosides

2. Dicotyledones

Anacardiaceae	
Schinus molle	Cy 3-galactoside, 3 rutinoside Pn 3-glucoside
Aquifoliaceae	
Ilex buergeri	Pg,Cy 3-xylosylglucosides
I. crenata	Cy 3-glucoside, 3-sambubioside
I. geniculata	Pg,Cy 3-xylosylglucosides
I. kiusiana	Cy 3-glucoside; Pg,Cy 3-xylosylglucosides
I. macropoda	Cy 3-xylosylglucoside
I. micrococca	Cy 3-glucoside
I. nipponica	Pg,Cy 3-xylosylglucosides
I. sugeroki	Cy 3-glucoside, 3-xylosylglucoside
Araliaceae	
Aralia elata	Cy 3-xylosylgalactoside
A. canescens	Cy 3-xylosylgalactoside
Acanthopanax divaricatus	Dp 3-xylosylgalactoside
Dendropanax trifidus	Cy 3-xylosylgalactoside
Berberidaceae	
Berberis fortunei	Cy,Dp,Pt, Mv 3-glucosides; Pn glycoside; Cy,Dp,Pt,Mv 3-rutinosides
B. buxifolia	Pn, Pt 3-glucosides Pn, Pt 3-rutinoside-5-glucosides Pt 3-rutinoside,-3-gentiobioside
Buxaceae	
Buxus microphylla	Cy 3-glucoside,-3-rhamnoglucoside
Caprifoliaceae	
Lonicera hypoglauca	Cy 3-glucoside,-3,5-diglucoside
L. japonica	Cy,Pn 3,5-diglucosides
Viburnum lantana	Cy 3-glucoside
V. urceolatum *V. furcatum* *V. sargentii* *V. erosum* *V. japonicum* *V. wrightii*	Cy 3-xyloxylglucoside
V. awabuki *V. phlebotrichum* *V. dilatatum*	Cy 3-xylosylglucoside, Cy 3-glucoside

TABLE 16 (continued)
Anthocyanins of Noncommon Fruits

Celastraceae	
Euonymus fortunei	Cy 3-glucoside
E. oxyphylla	Cy 3-glucoside
E. sieboldianus	Cy 3-glucoside
Chloranthaceae	
Chloranthus glaber	Cy, Pg 3-rhamnosylglucosides
Combretaceae	
Terminalia catappa	Cy 3-glucoside
Cornaceae	
Cornus alba	Cy 3-glucoside, 3-rutinoside; Pt 3-rutinoside Cy 3-galactoside; Cy 3-arabinoside
C. alternifolia	Cy 3-glucoside, 3-rutinoside; Pt 3-rutinoside Cy 3-galactoside; Cy 3-arabinoside
C. brachypoda	Cy glycoside; Dp glycoside
C. canadensis	Pg,Cy 3-robinobiosides
C. controversa	Cy,Dp,Mv glycosides
C. florida	Cy 3-glucoside
C. kousa	Cy 3-glucoside
C. mas	Cy,Pg,Dp 3-galactosides Cy,Pg,Dp 3-rhamnosylgalactosides
Aucuba japonica	Cy,Pg 3-xylosylglucosides
Lauraceae	
Cinnamomum camphora	Cy 3-rhamnoglucoside; Cy 3-glucoside
C. japonica	Cy 3-rhamnoglucoside; Cy 3-glucoside
Lindera glauca	Cy 3-glucoside, 3-rutinoside
Litsea japonica	Cy 3-glucoside, 3-glucosylrutinoside
Machilus thumbergii	Cy,Pn 3-glucosides, 3-rutinosides
Neolitsea sericea	Cy 3-rhamnoglucoside
Leguminosae	
Dumosia truncata	Cy 3-glucoside; Mv 3,5-diglucoside Mv 3-rutinoside-5-glucoside; Mv glycoside
Magnoliaceae	
Michelia compressa	Cy 3-glucoside, 3-rutinoside
Melastomaceae	
Clidemia hirta	Acylated Dp 3,5-diglucoside
Dissochaeta bractata	Dp 3-(p-coumarylglucoside)-5-glucoside
D. celebica	Dp,Mv 3-acylglucoside-5-glucosides
Memecylon amplexicaule	Cy 3,5-diglucoside
Melastoma malabrathricum	Cy 3-glucoside, 3,5-diglucoside
Menispermaceae	
Cocculus trilobus	Cy 3-glucoside
Myricaceae	
Myrica rubra	Cy 3-glucoside
Myrsinaceae	
Ardisia crispa	Cy,Pn 3-galactosides
Myrtaceae	
Eugenia jambolana	Dp,Pt-3 gentiobioside, Mv 3-laminaribioside
E. aquaea	Cy 3-glucoside
Neomyrtus pedunculata	Cy 3-glucoside
Psidium guajavica	Cy 3-glucoside
Myrciara jaboticaba	Pn 3-glucoside
Oleaceae	
Ligustrum ovalifolium	Cy 3-caffeylglucoside
L. obtusifolium	Cy 3-caffeylrutinoside
L. ibota	Cy glycosides
L. vulgare	Cy 3-glucoside
L. amurense	Cy 3-rutinoside
L. lucidum	Mv 3-rutinoside-5-glucoside

TABLE 16 (continued)
Anthocyanins of Noncommon Fruits

L. sinense	Mv 3-rutinoside-5-glucoside
Osmanthus fragrans	Cy 3-rhamnoglucoside,-3-glucoside
O. ilicifolius	Cy 3-glucosylrhamnoside
Phillyrea latifolia[a]	Mv glucoside
Passifloraceae	
Passiflora edulis	Cy 3-glucoside
Polygonaceae	
Polygonum perfoliatum	Mv 3,5-diglucoside
Renonuculaceae	
Clematis ternifolia	Cy 3-glucoside
Paeonia japonica	Pn 3-glucoside, 3,5-diglucoside
Rhamnaceae	
Rhamnus crenata	Cy 3-rhamnoglucoside, 3-glucoside
Rhizophoraceae	
Gynotroches axillaris	Cy 3-glucoside
Rosaceae	
Cotoneaster melanocarpus	Cy 3-glucoside
C. uniflora	Cy 3-glucoside
Duchesnea chrysantha	Cy,Pn 3-glucosides; Pn 3-rutinoside
Laurocerasus officinalis	Cy,Pn 3-arabinosides
Photinia glabra	Cy 3-glucoside, 3-rhamnoglucoside
Pourthiaea villosa	Cy 3-glucoside
Spiraea cantoniensis	Cy 3-glucoside
Sorbus commixta	Cy 3-glucoside
Rubiaceae	
Damnacanthus indicus	Pg 3-glycoside
D. macrophyllus	Pg 3-glycoside, 3-rutinoside
Mitchella undulata	Pg 3-rutinoside; Pg 3-glycosides
Rubia akane	Cy 3-rutinoside
Lasianthus japonicus	Cy 3-rutinoside, 3-glucoside
Rutaceae	
Skimmia japonica	Pg 3-glucoside, 3-rhamnoglucoside
Zanthoxylum piperitum	Cy 3-rhamnoglucoside
Sabiaceae	
Meliosma tenuis	Cy,Dp 3-rutinosides; Dp 3-glucoside
Schizandraceae	
Kadsuea japonica	Cy 3-(2^G-xylosylrutinoside)
Solanaceae	
Atropa belladonna	Pt 3-(*p*-coumarylglucoside)-5-glucoside
Staphyleaceae	
Euscaphis japonica	Cy 3-xylosylglucoside
Symplocaceae	
Symplocos chinensis	Dp 3-xylosylglucoside
S. coreana	Dp 3-xylosylglucoside
	Dp 3-diglucoside-5-glucoside
Symplocos lucida	Dp 3-glucoside,-3 sambubioside
S. myrtacea	Dp 3-xylosylglucoside
Theaceae	
Camellia japonica	Cy 3-glucoside
Cleyera japonica	Cy 3-glucoside, 3-rhamnoglucoside
Eurya emarginata	Cy 3-glucoside, 3-rutinoside, Cy glycosides
Ternstroemia japonica	Cy 3-glucoside
Thymeleaceae	
Daphne kiusiana	Cy 3-glucoside
Umbelliferae	
Angelica decursiva	Cy sambubioside, 3-rutinoside
Verbenaceae	
Callicarpa spp	Cy glucoside; Pn glycoside

TABLE 16 (continued)
Anthocyanins of Noncommon Fruits

Vitaceae	
Parthenocissus tricuspidata	Cy,Pn,Dp,Pt,Mv 3-glucosides Cy,Mv 3-rutinosides Cy,Mv 3-rutinosides-5-glucosides Cy glycosides

Note: Abbreviations: Cy = cyanidin; Pn = peonidin; Pt = petunidin; Pg = pelargonidin; Mv = malvidin; Dp = delphinidin.
[a] From Reference 125.

Adapted from References 265, 278—280, 329, 575.

berry, apart from the main pigment (Pg 3-β-D-glucopyranoside), there is a minor pigment which is thought to be the furanose form of the α-anomer of pelargonidin 3-glucoside.[647]

b. Diglycosides

The two ose molecules may be linked either to two different hydroxyls of the aglycone or only to position 3. (See Figure 11.) In the first case, only glucose is involved in fruits and positions 3 and 5 are the two main glycosylation sites. 3,5-Diglucosides are known for each of the six aglycones and are found as minor anthocyanins in a small number of common fruits: fig and elderberry with cyanidin 3,5-diglucoside,[79,443] litchi with pelargonidin 3,5-diglucoside,[439] pomegranate with 3,5-diglucosides of cyanidin, pelargonidin, and delphinidin.[136] Eggplant may contain delphinidin 3,5-diglucoside.[575] Berries of species of the genus *Vitis*, other than *Vitis vinifera*, possess 3,5-diglucosides corresponding to malvidin, petunidin, peonidin, delphinidin, and cyanidin.[198,471] The interest of the presence or absence of 3,5-diglucosides in *Vitis* fruits is discussed later in the study of interspecific variations. Hrazdina[265] reported 3,5-glucosides in several less common fruits including two species of *Lonicera*. In 3-diglycosides, the two ose molecules which form the diholoside are attached to the anthocyanidin in position 3. The following diholosides occur more frequently in fruits:

- **Rutinose or 6-*O*-α-L-rhamnosyl-D-glucose:** cyanidin 3-rutinoside is very common in fruits, most commonly associated with cyanidin 3-glucoside. The other 3-rutinosides are less common.
- **Sambubiose or 2-*O*-β-D-xylosyl-D-glucose:** cyanidin 3-sambubioside is present in elderberry,[79] red-currant,[425] and black raspberry.[573] Pelargonidin and cyanidin 3-xylosylglucosides have also been reported in the fruits of several species of *Ilex* and *Viburnum*.[265]
- **Sophorose or 2-*O*-β-D-glucosyl-D-glucose:** cyanidin 3-sophoroside is the main pigment in fruits of *Rubus idaeus* and hybrids, loganberry and boysenberry,[38,169] and of mangosteen, *Garcinia mangostana*.[134] It is one of the minor pigments in redcurrant and sour cherry.[119,425]
- **Gentiobiose or 6-*O*-β-D-glucosyl-D-glucose:** 3-gentiobiosides are found in the fruits of *Berberis buxifolia*[575] and of *Eugenia jambolana*.[265]

3-Xylosyl-galactosides of delphinidin and cyanidin have been reported in fruits of Araliaceae and 3-rhamnosylgalactosides in those of *Cornus*.[265,575]

c. Triglycosides

These are not very numerous and not widely distributed in fruits. (See Figure 11.) Distinction can be made between several types. There are, first, 3-diglycosides to which is bound an extra glucose molecule in position 5. Delphinidin 3-diglucoside-5-glucoside and delphinidin 3-rutinoside-5-glucoside are present in eggplant.[575] Petunidin 3-rutinoside-5-

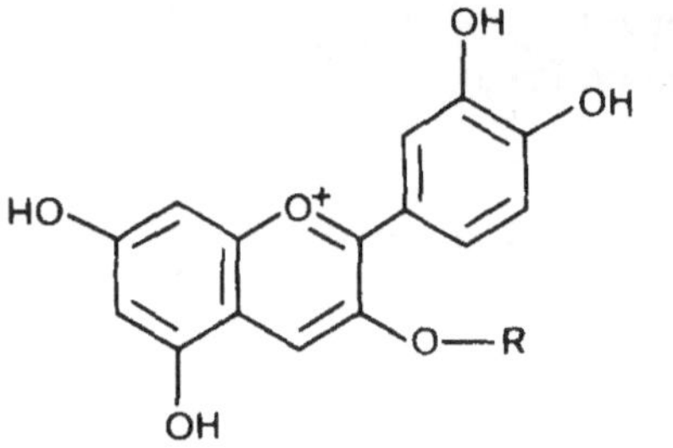

R = Glu : Cy 3-glucoside
R = Gal : Cy 3-galactoside
R = Ara : Cy 3-arabinoside

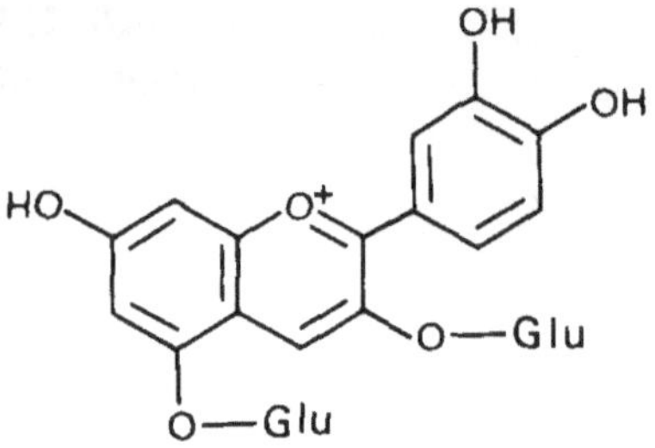

Cy 3,5-diglucoside

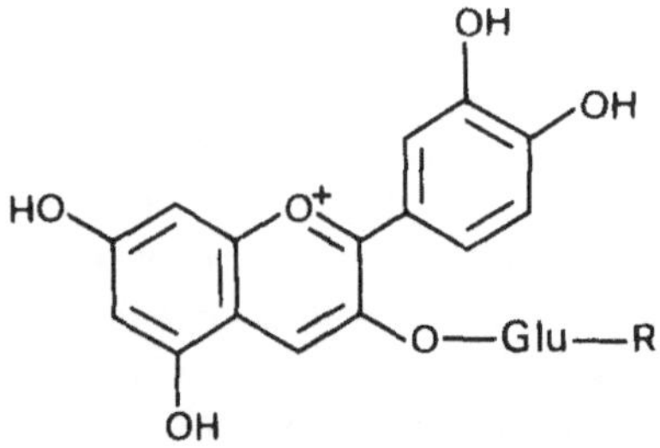

2"R = Glu : Cy 3-sophoroside
6"R = Rha : Cy 3-rutinoside
2"R = Xyl : Cy 3-sambubioside

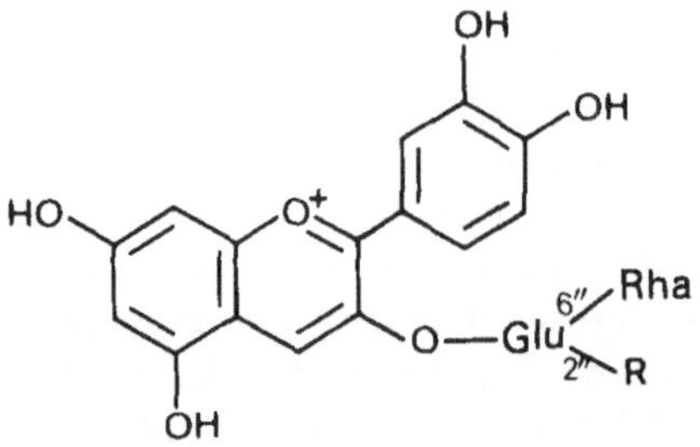

R = Glu : Cy 3-glucosylrutinoside
R = Xyl : Cy 3-xylosylrutinoside

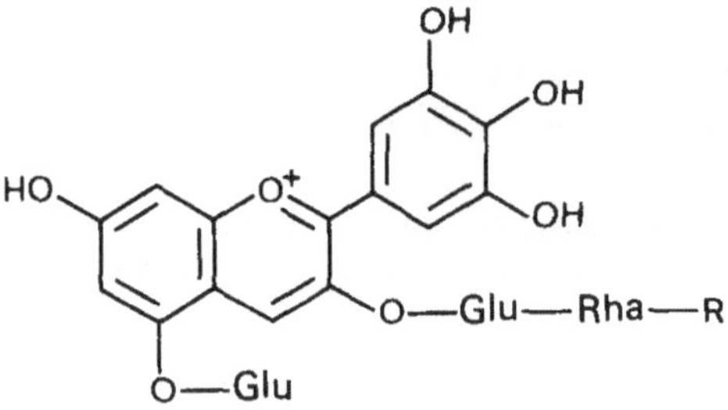

R = H : Dp 3-rutinoside-5-glucoside
R = *p*-coumaryl : Dp 3-(p-coumarylrutinoside)-5-glucoside (=nasunin).

FIGURE 11. Examples of anthocyanins found in fruits. Cy = cyanidin; Dp = delphinidin; Glu = glucosyl; Gal = galactosyl; Ara = arabinosyl; Rha = rhamnosyl; Xyl = xylosyl.

glucoside is a minor constituent in fruits of *Solanum guineese.*[167] In these two Solanaceae, these triglycosides are principally in the form of acylated derivatives. Elderberry contains a very small amount (<1%) of the cyanidin 3-sambubioside-5-glucoside.[79] Certain varieties of olive *(Olea europea)* are thought to contain delphinidin 3-rhamnoglucoside-7-xyloside, a new anthocyanin identified by Tanchev and Codounis (quoted by Marekov[355]). This would appear to be the only example of a triglycoside with glycosylation in the 7-position. Malvidin 3-rutinoside-5-glucoside was reported by Hrazdina[265] as being a constituent of fruits of several species of *Ligustrum* (Oleaceae) and of *Dumosia truncata* (Leguminosae).

In the other triglycosides, the three ose molecules are linked to C-3, thus forming a triholoside. Peonidin 3-rhamnosyldiglucoside and linear and branched cyanidin 3-glucosylrutinosides have been reported to be constituents of black olive although the nature of the bonds between ose molecules has not been specified.[573,575] Anthocyanins with branched triglycosides are worthy of consideration. Two pigments, cyanidin 3-*O*-β-(2^G-glucosylrutinoside) and cyanidin 3-*O*-β-(2^G-xylosylrutinoside), have been identified in fruits of the genera *Prunus, Rubus,* and *Ribes.*[225] The first (glucosylrutinoside) is found in red raspberry,[573] redcurrant,[425] and sour cherry.[119] The second (xylosylrutinoside) is a constituent of black raspberry, blackcurrant, and perhaps sour cherry.[519] These two pigments appear to be more characteristic of fruits than of flowers.[225] According to Timberlake and Bridle,[575] the fruits of *Kadsura japonica* (Magnoliales) also contain cyanidin 3-(2^G-xylosylrutinoside).

d. Acylated Anthocyanins

Acylated anthocyanins are found fairly often in fruits. (See Figure 11.) Acylation by phenolic acids has been known for a long time, but other types of organic acids can also be involved.[422] The significance of this acylation appears to be related to the stabilization of anthocyanins in the acid environment of the cell sap.[80,81,222] Three acylated delphinidin anthocyanins have been identified in eggplant:[571,573,621]

1. Delphinidin 3-(*p*-coumarylrutinoside)-5-glucoside, a major pigment in purple-skinned varieties
2. Delphinidin 3-(*p*-coumarylglucoside)-5-glucoside
3. Delphinidin 3-(caffeyldiglucoside)-5-glucoside

A Bulgarian cultivar of eggplant probably differs in that acylation was absent.[355]

In the fruit of *Solanum guineese* (garden huckleberry), petunidin 3-(*p*-coumaryl-rutinoside)-5-glucoside forms over 90% of total anthocyanins; it is accompanied by very small quantities of two other acylated derivatives, malvidin 3-(*p*-coumarylrutinoside)-5-glucoside and petunidin 3-(di-*p*-coumarylrutinoside)-5-glucoside.[167]

The case of grape *(Vitis vinifera)* is more complex since 11 acylated derivatives have been found in 16 pigments separated and identified by HPLC. The 3-monoglucosides corresponding to the five aglycones present in grape (see above) can all be acylated by *p*-coumaric acid and by acetic acid, giving 3-*p*-coumarylglucosides and 3-acetylglucosides. Malvidin 3-glucoside is also acylated by caffeic acid. The presence of these acylated anthocyanins in species of the genus *Vitis* is of great chimiotaxonomic interest.[14,198,242,404,471,636,650]

Acylated derivatives — in particular with caffeic acid — have been reported in olive.[345,573] Acylated derivatives of four cyanidin 3-monoglycosides are thought to be among the minor pigments of apple, but the nature of the acids involved has not been specified.[574] The fruits of two species of *Ligustrum* contain cyanidin 3-caffeyl glucoside and rutinoside.[265]

It has recently been seen that anthocyanins are probably in zwitterion form because of acylation with dicarboxylic aliphatic acids, malic, succinic, and malonic acids; the latter appears to be more frequent.[222,223,560] According to Teusch and Formann,[569] these rather labile acid groups were often overlooked during earlier studies due to their hydrolysis under usual extraction procedures based on methanolic HCl. These acylated derivatives have mainly been found in flower petals, and very little information is available about their presence in fruits, with the exception of those of *Actea* sp. (Ranunculaceae).[222] However, data on acylated anthocyanins, especially with aliphatic acids, should be enlarged in the coming years, in particular as regards purple and blue fruits. It can be noted that delphinidin 3-(*p*-coumarylglucoside)-5-glucoside or awobanin is found in eggplant and has been recognized as a derivative acylated with malonic acid in various blue flowers.[561]

3. Comparisons between Fruits

a. *Qualitative Aspects*

Each fruit is characterized by its anthocyanidin glycosides. Some 70 different anthocyanins have been found in fruits, whereas at least 200 are known in various parts of plants.[265,575] The most numerous glycosides are those of cyanidin and delphinidin and then, in decreasing order, the glycosides of malvidin, petunidin, peonidin, and pelargonidin (Figure 12A). An important difference can be seen when comparison is made with all parts of plants. The fruits are characterized by a distinctly higher number of delphinidin glycosides and a small number of pelargonidin derivatives.

Fruits also stand out from other plant organs in that they contain a relatively large quantity of monoglycosides compared to diglycosides (Figure 12B). The importance of acylated anthocyanins should also be noted.

The six anthocyanins most commonly found in fruits (Figure 13) consist of five 3-monoglycosides and a single diglycoside, cyanidin 3-rutinoside. Among them, cyanidin 3-glucoside is the most common; the same applies to plant organs as a whole where this pigment is the commonest anthocyanin.[217]

The number of anthocyanins present in a fruit is also a criterion which makes it possible to distinguish between fruits. It is fairly rare to find only one glycoside. This is the case of mango with peonidin 3-glucoside[441] and of certain red-skinned peach cultivars[592] and the black berries of *Viburnum lantana*[209] with cyanidin 3-glucoside. In fact, in addition to the anthocyanin identified, minor pigments are often present in very small quantities, as has been shown in peach.[278]

In contrast, very great complexity, with 12 to 16 glycosides or more, is found in the berries of Ericaceae (*Vaccinium* sp.), Empetraceae *(Empetrum nigrum)* and Vitaceae. For example, 16 anthocyanins were separated by HPLC in the berries of *Vitis vinifera,* var. Cabernet Sauvignon.[404,650] The very high resolution capacity and the sensitivity of the detection system in reversed phase HPLC make it possible to detect a higher number of pigments than with classic chromatographic techniques. Fuleki[180] obtained 22 clearly separated peaks for *Vaccinium macrocarpon* (cranberry), whereas only 10 strips were observed when paper chromatography was used (Figure 14). A total of 31 peaks were separated in fruits of *Vitis rotundifolia* (Muscadin grape).[198] This would appear to be the highest number of anthocyanins in a plant organ. Such a complex situation with at least ten anthocyanins is very rarely observed outside fruits. Only the flowers of *Pisum sativum* and *Petunia hybrida* appear to possess as complex a set of anthocyanins.[575]

Most of the fruits examined possess two to six anthocyanins, with the highest frequency being for four pigments. In this case, it can be four cyanidin glycosides (e.g., in black raspberry and elderberry) or two aglycones, each in the form of two glycosides (e.g., in blackcurrant).[79,305,579]

b. *Quantitative Aspects*

In most fruits of commercial interest, numerous cultivars have been obtained whose anthocyanin contents may display considerable differences: e.g., 3 to 80 nmol/cm^2 of peel for seven red apple varieties.[440] However, variations between different cultivars are often relatively small and cannot be used as a recognition criterion. For example, the anthocyanin contents of five redcurrant cultivars varied from 11.9 to 18.6 mg/100 g, with a mean of 16.2 ± 2.7.[425] Likewise, the total anthocyanin content for eight cultivars of red raspberry varied from 23 to 59 mg/100 g, with a mean of 39 ± 13.[579] In 45 cranberry cultivars, anthocyanins totalled 46 to 172 mg/100 g, i.e., a variation of 1 to 3.7.[496] Furthermore, variability in anthocyanin content of individual fruits was greater than variation between cultivars.[497]

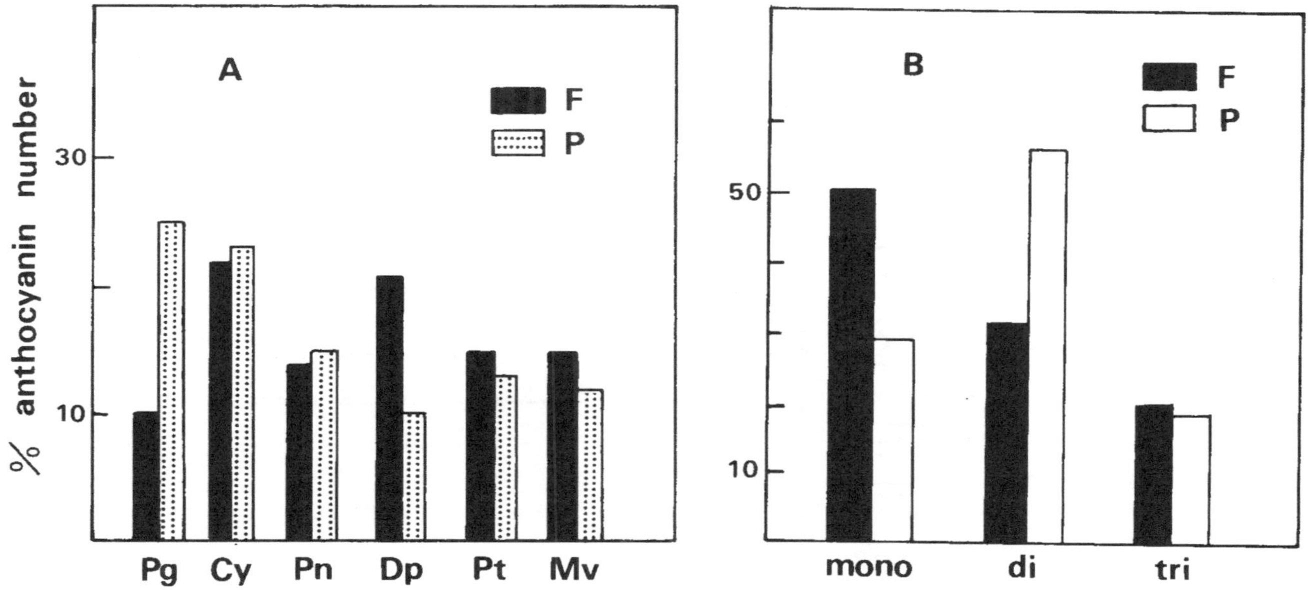

FIGURE 12. Relative importance of the different anthocyanins in fruits (F) and whole plant organs (P). A: Percentages of each aglycone; Pg = pelargonidin; Cy = cyanidin; Pn = peonidin; Dp = delphinidin; Pt = petunidin; Mv = malvidin. B: Percentages of the different types of glycosides; mono = monoglycosides; di = diglycosides; tri = triglycosides. The acylated anthocyanins have not been taken into account. All values are expressed in per cent of the total anthocyanin number.

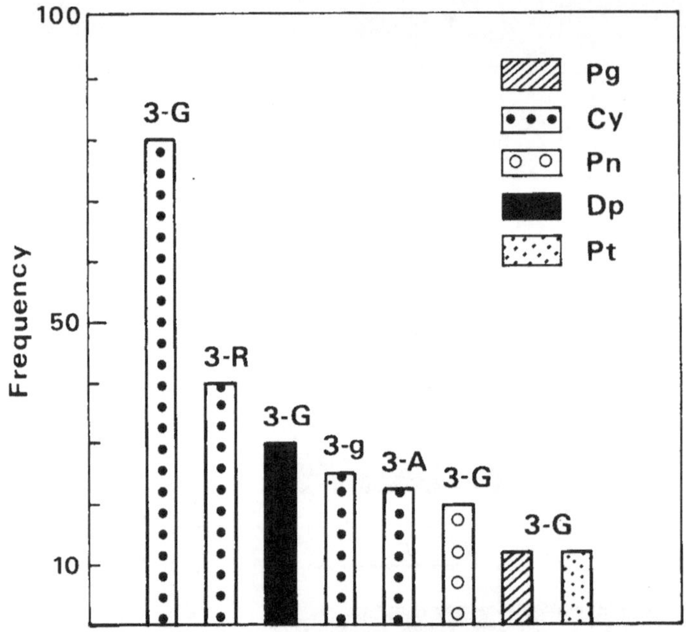

FIGURE 13. The anthocyanins most commonly found in fruits. Frequency is expressed in percent of the total fruit number. (Cy = cyanidin; Pg = pelargonidin; Pn = peonidin; Dp = delphinidin; Pt = petunidin; G = glucoside; R = rutinoside; g = galactoside; A = arabinoside).

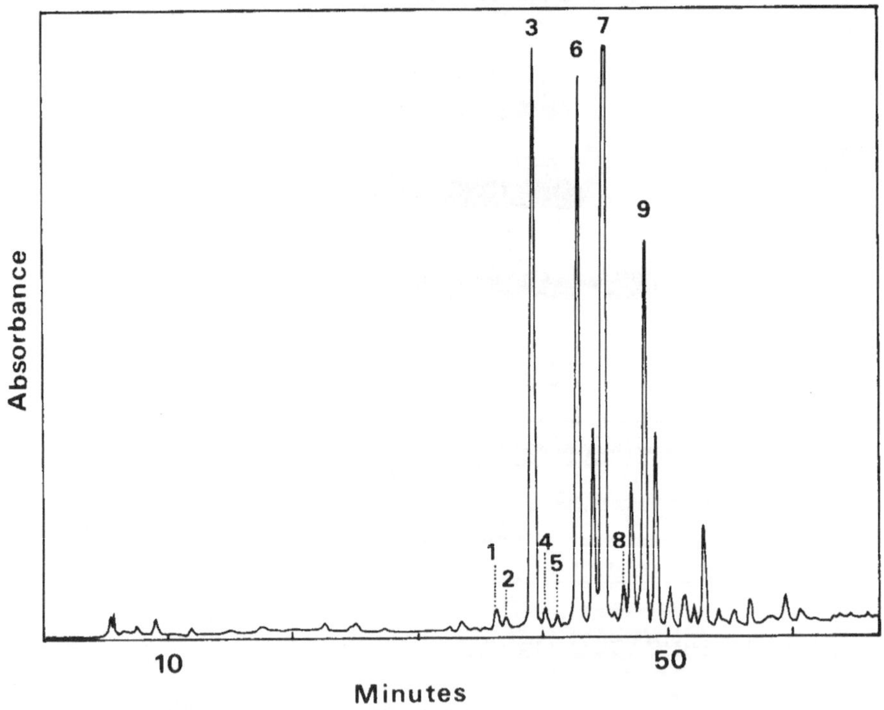

FIGURE 14. Separation of cranberry anthocyanins by HPLC. Major pigments: 3: Cy 3-galactoside; 6: Cy 3-arabinoside; 7: Pn 3-galactoside; 9: Pn 3-arabinoside; minor pigments: 1: Dp 3-galactoside; 2: Dp 3-glucoside; 4: Dp 3-arabinoside; 5: Cy 3-glucoside; 8: Pn 3-glucoside. (Cy = cyanidin; Pn = peonidin; Dp = delphinidin). (Adapted from Fuleki, T., *Bull. Liaison Groupe Polyphénols*, 13, 374, 1986.)

The anthocyanin content of ripe fruits varies considerably in relation to climatic factors, light and temperature in particular.[16,143,413] In cv. Cabernet Sauvignon of *V. vinifera*, significant differences were found between sun-exposed and shaded fruits for both anthocyanin concentration and total content per berry at preharvest. However, there was no significant difference at harvest.[114] This sensitivity to environmental factors means that very different values are observed from one year to the next, which sometimes makes it difficult to compare different cultivars of the same fruit and even more difficult to compare different fruits. These comparisons are made even more delicate by the diversity of global determination methods for anthocyanins: densitometry, polarography, colorimetric measurement at fixed pH values, spectrophotometry, bleaching with sulfur dioxide or sodium bisulfite.[166,182,469] Fortunately, HPLC techniques now make it possible to obtain much more accurate results for each constituent.[25,404,483,484]

Levels expressed in mg/100 g FW range from 0.25 (pear) to 500 (highbush blueberry) and even 760 for *Vaccinium elliotti*[28] (Table 2). In certain varieties of pear and apple with red skin, only 10 to 50% of epidermal or hypodermal cells contain anthocyanic pigments.[165] Even when concentrations are related to unit area of fruit surface or to unit weight of peel, the values remain distinctly lower that those observed in other fruits: 5 to 10 mg/100 g FW of peel in pear. The example of cherry is a good illustration of the three cases which can exist in fruits:

- Pigments located in the skin which is only pale pink; anthocyanins are very scarce at 4 mg/100 g peel, as for example in *Prunus avium* cv. Bigarreau Napoléon
- Pigment also located in the skin only, but the latter is bright red; there is a high concentration of anthocyanins: 30 mg/100 g of peel, e.g., *P. cerasus* cv. Montmorency
- Pigment located throughout fruit strongly colored in red or black and very rich in anthocyanins: 350 to 450 mg/100 g FW e.g., *P. avium*, cvs. Bigarreau Burlat, Guignes, etc.,

The fruits which are richest in anthocyanins (>200 mg/100 g FW) are very strongly colored berries (deep purple, black) belonging to the genera *Rubus, Ribes, Vaccinium, Empetrum, Vitis,* and *Sambucus* (see Table 2). These fruits, together with certain varieties of sweet cherry, are the most important source of natural coloring agents for the food industry.[264,357,572] Numerous works have been devoted to the modifications undergone by anthocyanic coloring agents during technological processes. These aspects are discussed in a later chapter.

The anthocyanins of a fruit can frequently be separated into two groups according to quantitative importance. These are the major anthocyanins which as a whole form over 90% of total pigments, and the minor anthocyanins which are each only present as traces or at levels of only a few percent. In general, one or two anthocyanins stand out among the major anthocyanins. For example, in 11 apple cultivars, cyanidin 3-galactoside is the main pigment, forming 83 to 94% of total anthocyanins.[574] Cyanidin 3-glucoside is the major pigment in blackberry, peach, elderberry, and mulberry.[79,278,498] Other predominant anthocyanidins include pelargonidin 3-glucoside in strawberry, cyanidin 3-rutinoside in fig[443] and sweet cherry, cyanidin 3-xylosyl-rutinoside in *Ribes rubrum* var. red Lake[425] and malvidin 3-glucoside in *Vitis vinifera* berries.[198,404,465,484,650] Acylated anthocyanins are predominant in the fruits of *Solanum guineese* and in certain eggplant cultivars.[167,571,621]

In blackcurrant and cranberry (Table 17), no anthocyanin is strongly predominant and the relative proportions of four pigments characterize the fruit. The possibility of determining the qualitative and quantitative "anthocyanin pattern" of each fruit made it possible to approach chemotaxonomic aspects.

TABLE 17
Anthocyanin Composition of Two Ripe Fruits: Blackcurrant and Cranberry

Variety		Content mg/100 g FW	% of total anthocyanins
Blackcurrant[a]	Cy[c] 3-rutinoside	90	36
(*Ribes nigrum*)	Cy 3-glucoside	37.5	15
	Dp 3-rutinoside	85	34
	Dp 3-glucoside	37.5	15
Cranberry[b]	Cy 3-galactoside	30.5	30
(*Vaccinium macrocarpon*)	Cy 3-arabinoside	20.4	20
	Pn 3-galactoside	40.7	40
	Pn 3-arabinoside	10.2	10

[a] cv. Silvergieters Schwarze. (From Koeppen and Herrmann).[305]
[b] cv. Early Black (From Fuleki and Francis[182]); minor anthocyanins are present undertraces (see Figure 14).
[c] Cy = cyanidin; Dp = delphinidin; Pn = peonidin.

c. Genetic and Chemotaxonomic Aspects

A number of examples can be given in which there are differences in the nature and levels of the anthocyanins in several cultivars of the same fruit. Sun and Francis[557] examined 74 apple cultivars and observed that over 60 of them contained the three pigments (1) Cy 3-galactoside, (2) Cy 3-arabinoside, and (3) Cy 7-arabinoside. Six cultivars contained pigments 1 and 2 and two contained only pigment 1. The main pigment in *P. avium,* var. Bigarreau Napoléon is the peonidin 3-rutinoside which replaces cyanidin 3-rutinoside, the major pigment in the other cultivars of *P. avium* analyzed.[382,417] Crossing or mutation can give cultivars which have a low anthocyanin content or even in which these pigments are lacking, such as the pink fruited hybrid of highbush blueberry[30] and varieties of eggplant whose fruits lack anthocyanin.[20,571]

The genetic determinism of these qualitative and quantitative determinations has been studied closely in eggplant and fruits of the genus *Rubus*. The coloring of eggplant depends on nine genes which are involved in the development of anthocyanins and their distribution in the fruit. A dominant gene Ac determines the nature of the major pigment by ensuring conversion of Dp 3-rutinoside into Dp 3-(*p*-coumarylrutinoside)-5-glucoside (Table 18).[571] However, it is not certain that the same gene controls both acylation by *p*-coumaric acid and the fixation of glucose at 5.[19] In parallel with this transformation, the color of the fruit changes from vinaceous to purple, and the various cultivars can be classified in two main groups according to whether the fruits are mauve-violet or purple.[19,571]

After the work of Torre and Barrit,[579] in which numerous species of *Rubus* were examined, Jennings and Carmichael[287] published a fundamental work on "anthocyanin variation in the g. *Rubus*." The presence or absence of the diglycosides rutinose, sophorose, and sambubiose is determined by the segregation of the genes R (rhamnose), So (sophorose), and Xy (xylose). Synthesis of triglycosides occurs when the genes So and R or Xy and R are present simultaneously. Thus, black raspberry *(R. occidentalis)* does not possess the gene So and does not, therefore, form sophorose but two pigments which contain xylose, Cy 3-sambubioside and Cy 3-(2^G-xylosylrutinoside), a branched triglycoside which is characteristic of this species. In contrast, the red raspberry, *R. idaeus,* is characterized by absence of xylose (no gene Xy) and the presence of a branched triglycoside, Cy 3-(2^G-glucosylrutinoside) which accompanies the Cy 3-sophoroside. In certain cultivars, the absence of gene R means that the only pigments are Cy 3-glucoside and 3-sophoroside. In addition, the quantitative variations depend on three major genes, T, Y, and B1 which determine the quantity of pigments and the color of the fruits.

TABLE 18
Anthocyanin Coloration in Eggplant[a]

Variety of cross	Major pigment: Nasunin	Major pigment: Dp 3-rutinoside	Expected ratio[b]
Rosita (R)	—	9	...
Sinkuro (S)	9	—	...
(R × S) F1	10	—	...
(R × S) × R	71	83	1:1
(R × S) F2	105	44	3:1
Puerto Rican Beauty (PRB)	—	99	...
(PRB × S) F1	8	—	...
(PRB × S) × PRB	68	76	1:1
(PRB × S) F2	100	39	3:1

[a] Segregation for the gene Ac (acylated anthocyanin) in crosses and backcrosses of Rosita × Sinkuro and Puerto Rican Beauty × Sinkuro.
[b] Following genotypes assumed: R = acac; S = AcAc; PRB = acac; reaction was:

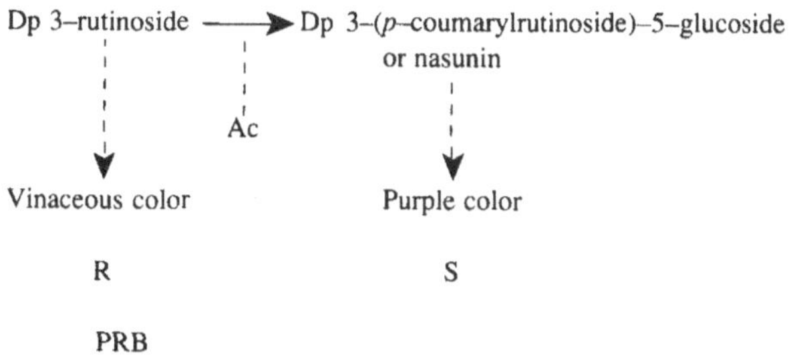

From Tigchelaar, E. C., Janick, J., and Ericson, H. T., *Genetics*, 60, 475, 1968. With permission.

Chemotaxonomic studies have been carried out principally on the genera *Ribes*,[225] *Rubus*,[37,38,287,385,498,579] *Vaccinium*,[12,28,32] and *Vitis*.[198,465,483,485]

Genus *Ribes* — Harborne and Hall[225] examined 29 species and 4 cultivars of the genus *Ribes;* they distinguished between 5 groups:

1. The majority of the species (14) possessed only two pigments, Cy 3-glucoside and Cy 3-rutinoside, e.g., *Ribes grossularia* (gooseberry).
2. Six species with black-colored fruits possessed four anthocyanins: Cy and Dp 3-glucosides and 3-rutinosides, e.g., *Ribes nigrum* (blackcurrant), a species in which the four pigments were separated by DCCC (see Figure 3).[169]
3. The three other groups possess a single aglycone, Cy, and 11 species possess branched triglycosides. Cy 3-(2^G-xylosylrutinoside) is very common and almost all the species with triglycosides are part of the same section and subsection of the genus *Ribes*. These three groups differ from each other in the number of pigments (6, 4, or 3). The various species of currants belong to these three groups.

Genus *Rubus* — Sapers et al.[498] distinguished between five groups of cultivars and clones of thornless blackberry in which Cy 3-glucoside is the major pigment (72 to 94%) but in which the proportions of minor pigments are different (Table 19). Pelargonidin, which is not present in blackberry and in *Rubus occidentalis,* is present as traces in numerous *R.*

TABLE 19
Anthocyanin Distribution Patterns in Ripe Thornless Blackberries

			Mean HPLC peak area (%)				
			Peak no.[a]				
Pattern	Representative clone	Year	1	2	3	4	5
A	Black Satin	1982	81.3	0.9	8.5	2.4	5.6
		1983	87.5	0.2	5.8	1.7	4.0
B	SIUS 50	1982	94.2	0.2	0.1	1.0	3.8
C	Hull Thornless	1982	72.2	1.0	12.3	3.1	9.4
		1983	72.1	1.0	16.9	2.8	6.4
D	C-57	1982	84.8	0.5	4.2	0.9	8.7
		1983	86.1	0.3	3.3	0.9	7.0
E	Comanche[b]	1983	88.9	7.6	0.2	1.7	—

[a] Pigments detected at 546 nm and designated as peaks 1 to 5 on basis of HPLC retention time: 1 = Cy 3-glucoside; 2 = Cy 3-rutinoside; 3 = Cy 3-xyloside? 4 and 5 = dicarboxylic acylated derivatives of Cy 3-glucoside.

[b] Not classified as thornless cultivar.

Data from Sapers, G. M., Hicks, K. B., Burgher, A. M., Hargrave, D. L., Sondey, S. M., and Bilyk, A., *J. Am. Soc. Hortic. Sci.*, 111, 945, 1986.

idaeus cultivars.[38,544] Only two species of *Rubus*, including *R. parviflorus*, are characterized by the predominance of pelargonidin glycosides; an Asian species, *R. pileatus*, contains only two pelargonidin anthocyanins.[287,579]

From the quantitative point of view, the fruits of *R. idaeus* contain less than 60 mg/100 g of anthocyanins. Those of *R. occidentalis* have a very high anthocyanin content (200 to 400 mg/100 g). The blackberry group, of which 16 species were studied, had an intermediate level (80 to 180 mg/100 g), with the exception of 3 species which displayed high anthocyanin contents similar to that of *R. occidentalis*. Boysenberries and loganberries, which are raspberry/blackberry hybrids, have levels which are typical of blackberries (161 mg/100 g in boysenberry and 77 mg/100 g in loganberry), whereas their pigment composition is similar to that of red raspberry *(R. idaeus)*. The total anthocyanin content is, therefore, an interesting criterion which makes it possible to characterize species of the genus *Rubus*.

Genus *Vaccinium* — In addition to cultivated species of *Vaccinium*, numerous wild species in North America were examined by Ballinger et al.[28,32] Their general characteristic is the presence of 3-monoglycosides in which five aglycones are combined with three oses, galactose, arabinose, and glucose. This corresponds to 15 anthocyanins, but the content varies in relation with classification in subgenera (see Chapter 2). Thus, only four pigments were identified in the fruits of the *V. vitis idaea* (cowberry).[11] HPLC analysis of seven blueberry species (including *V. corymbosum*) revealed differences in relative percentages of individual anthocyanins and made it possible both to distinguish between species and to observe the rate of ploidy.[31] Derivatives of pelargonidin have recently been found in an Asian species, *V. japonicum*.[12]

Genus *Vitis* — The situation is complex here because of the large number of cultivars, the large number of species of *Vitis*, and the possibility of obtaining hybrids. Ribereau-Gayon[471] described the distribution of anthocyanins in 15 species of *Vitis*. Distinction is made between two groups:

- Euvitis, including *V. vinifera* and many other species, American species in particular, including *V. labrusca*, *V. riparia*, and *V. rupestris*. These species are characterized by acylated anthocyanins. *V. vinifera* is distinct from American species (e.g., *V. riparia* and *V. rupestris*) since it lacks 3,5-diglucosides.

- Muscadine grapes, including *V. rotundifolia* which, in contrast, does not possess acylated anthocyanins.

These characteristics made it possible to recognize hybrids.[198,465,471,636] For example, the anthocyanins of 12 *Euvitis* × *Vitis rotundifolia* hybrids were analyzed recently using HPLC.[198] The "diglucoside" character of *V. riparia* and *V. rupestris* (the two species of *Vitis* most commonly used for cross-breeding) was dominant in relation to the recessive "monoglucoside" character of *V. vinifera*[465,471] and the hybrids contain anthocyanidin 3,5-diglucosides. However, although the presence of diglucosides proves the hybrid origin of a grape or a wine, their absence is not final proof of *V. vinifera* origin. Indeed, as an example, crossing on F1 hybrid of *V. vinifera* and *V. riparia* with *V. vinifera* gives 50% of F2 hybrids without the "diglucoside" character.

HPLC techniques have been used for a number of years for the accurate determination of anthocyanosides in grape and wine extracts, making it possible to compare cultivars. For example, Table 20 shows the distribution of anthocyanins in the skin extracts of 16 grape cultivars used for Port wine production.[26] The percentages of acylated derivatives of malvidin vary considerably from one cultivar to another: 2 to 51% of anthocyanins identified for Mv 3-*p*-coumarylglucoside and 1 to 15% for Mv 3-acetylglucoside. The ratio of Mv 3-acetylglucoside to total Mv glucosides seems to be characteristic of a cultivar and independent of the site of production. Likewise, the distribution of anthocyanins was studied in Cabernet Sauvignon and Merlot, two *V. vinifera* cultivars.[404,650] According to Ribereau-Gayon,[471] absence of acylated anthocyanins is demonstrated in Pinot varieties *(V. vinifera)* where it would seem to be a specific character.

A systematic effort is being made today to determine the criteria to be retained to enable characterization of a cultivar, based on analyses of the anthocyanins in different grape varieties. Experimental data are processed by computer using factorial analysis.[483] This work has led to concluding that the following parameters can be selected from the chemotaxonomic point of view:

- Malvidin monoglucoside contents
- Anthocyanin acetic ester contents
- Anthocyanin cinnamic ester contents
- Ratio of malvidin and peonidin glucosides contents

B. FLAVONOLS

1. Flavonol Aglycones

Flavonols are flavonoids characterized by an unsaturated 3-C chain with a double bond between C-2 and C-3 and by the presence of an hydroxyl group in the 3-position (Figure 15). They can be considered to be 3-hydroxyflavones. Extremely complete reviews have been drawn up of these compounds which are very widespread in the higher plants.[202,641] Nearly 200 flavonol aglycones have been identified in plants and many of them are methylated derivatives. Approximately 90% of these flavonols are hydroxylated in 5- and 7-positions in addition to the hydroxyl in 3 (3,5,7-tri-OH). The most common do not possess other hydroxyl groups in the A-ring and differ in hydroxylation in the B-ring.

In fruits, almost only the glycosides of four aglycones are found: kaempferol (4′OH), quercetin (3′,4′diOH), myricetin (3′,4′,5′triOH) and isorhamnetin (3′-methylether of quercetin) (Figure 15). Other methylated derivatives of kaempferol and quercetin have been identified in the berries of *Rhamnus cathartica* and *Rhamnus* sp. These are rhamnetin (= quercetin-7-methylether), rhamnazin (= quercetin-7,3′-dimethylether), and rhamnocitrin (= kaempferol-7-methylether).[232] Galangin (3,5,7-trihydroxyflavone), without hydroxyl in the B-ring, has been reported in the fruits of *Citrus sinensis*, but this presence suggested by

TABLE 20
Area Percentages of Known Anthocyanins (A) in Grape Skin Extracts of 1982 Port Wine Grape Cultivars

Cultivar	Site[a]	Mv[b] (%)	Mv-ac (%)	Mv-cou (%)	Pn (%)	Dp (%)	Pt (%)	Cy (%)	Tot Mv / Known A. (%)	Mv-ac / Tot Mv (%)	Mv-ac / Mv + Mv-ac (%)	Known A / Tot A. (%)
Tinta Cão	Tua	33	11	51	1	2	3	tr	94	11	24	88
Tinta Barroca	BC	35	5	51	4	2	3	tr	91	6	13	92
Tinta Barroca	Tua	37	5	47	6	2	3	tr	89	6	12	92
Touriga Nacional	P	36	14	38	6	2	3	tr	89	16	28	95
Touriga Nacional	V	38	15	36	6	2	3	tr	89	17	28	93
Touriga Nacional	Tua	41	15	31	6	3	4	tr	87	17	26	93
Tinta da Barca	Tua	37	13	39	4	3	4	tr	88	15	26	93
Touriga Francesa	BC	41	11	36	3	4	5	tr	87	15	23	93
Touriga Francesa	SB	43	13	30	4	4	6	tr	86	12	21	93
Tinta Francisca	Tua	46	5	35	4	4	5	1	86	5	9	94
Tinta Francisca	SB	48	1	32	5	6	7	1	82	2	3	92
Rufete	Tua	45	5	30	5	7	7	1	80	6	10	91
Malvesia Preta	SB	53	4	20	6	7	9	1	77	5	7	95
Souzão	Tua	57	4	15	6	8	8	2	76	5	7	96
Tinta Roriz	Tua	41	4	29	3	11	11	1	74	6	10	88
Tinta Santarem	Tua	49	3	19	23	2	3	1	72	5	7	96
Viera da Natividade	V	52	4	14	27	1	2	1	69	5	7	98
Tinta Amarela	Tua	49	4	18	10	8	9	2	71	6	8	94
Tinta Amarela	BC	47	4	18	7	11	11	2	69	5	7	93
Mourisco	Tua	48	2	16	16	7	9	3	66	3	4	94
Cornifesto	P	60	1	2	22	6	7	2	63	2	2	99
Donzelinho	Tua	44	1	6	25	9	9	6	50	1	1	98

[a] BC = Baixo Corgo, P = Pinhão, V = Vilariça, SB = Santa Barbara.

[b] Mv = malvidin 3-glucoside; Mv-ac = malvidin 3-acetylglucoside; Mv-cou = malvidin 3-*p*-coumarylglucoside; Pn = peonidin 3-glucoside; Dp = delphinidin 3-glucoside; Pt = petunidin 3-glucoside; Cy = cyanidin 3-glucoside.

Data from Bakker, J. and Timberlake, C. F., *J. Sci. Food Agric.*, 36, 1315, 1985.

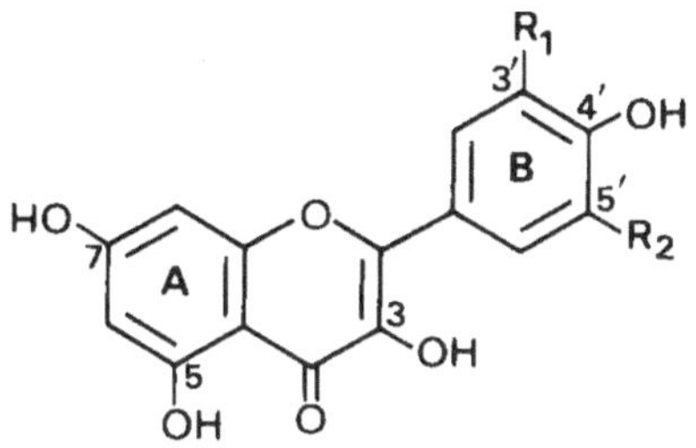

R_1 = H	R_2 = H	: Kaempferol (K)
R_1 = OH	R_2 = H	: Quercetin (Q)
R_1 = OH	R_2 = OH	: Myricetin (M)
R_1 = OCH_3	R_2 = H	: *iso*-Rhamnetin (iR)

FIGURE 15. Chemical structures of flavonol aglycones present in fruits.

TLC remains to be confirmed by isolation of the substance.[202] Polymethoxylated flavonols are found in the fruits of *Citrus, Murraya omphalocarpa* (Rutaceae), and *Vitex agnus castus* (Verbenaceae). They are generally 6-OH or 8-OH flavonols.

Table 21 shows that quercetin is present in all and kaempferol in approximately 80% of some 30 fleshy fruits. The two other aglycones are much less common and were only found in five fruits. The distribution of aglycones is different to that observed for anthocyanidins (see above) since although the 3′ and 4′ hydroxylated derivative in the B-ring is predominant in both cases, flavonols are characterized by distinctly greater importance of the monohydroxylated derivative (4′-OH, kaempferol) and by the relative unimportance of trisubstituted derivates in the B-ring. Myricetin is less frequent than delphinidin, the corresponding anthocyanin, and only one methylated flavonol (isorhamnetin) is found whereas there are three methylated anthocyanidins.

Most fruits contain two aglycones (Table 21), and the quercetin and kaempferol combination is by far the most frequent. Only two fruits are characterized by isorhamnetin accompanying quercetin: pear[138,176,253,412] and lemon.[217] The latter fruit also contains polymethoxylated flavonols. Approximately 30% of the fruits considered contain three aglycones; the most common combination is kaempferol + quercetin + myricetin. There are only small quantities of myricetin in redcurrant and only in certain varieties.[523,548] Only quercetin and kaempferol glycosides have been reported in bilberry *(Vaccinium myrtillus)*,[21,22,172] with the exception of two varieties where the presence of myricetin has been reported after hydrolysis.[548] Kaempferol is present as traces in cranberry *(V. macrocarpon)*.[57] The combination kaempferol + quercetin + isorhamnetin is rare; it has been reported in *Prunus avium* cv. Bigarreau Napoleon[382] and in grapefruit.[351]

Quercetin was the sole aglycone found in four of the fruits examined, including apple where the diversity of flavonols is at glycoside level since there are six or seven quercetin derivatives.[127,568] The simultaneous presence of four aglycones (kaempferol, quercetin, myricetin, and isorhamnetin) has been found in the skin of red varieties of *Vitis vinifera*, but depending on cultivar, isorhamnetin may be present as traces. The white varieties analyzed contain only quercetin and kaempferol[102,103] (Table 22).

The predominant combination in flavonols (two aglycones) is, therefore, the same as that found in anthocyanidins, but the overall pattern is less complex. Nevertheless, analytical data for fruit flavonols are less plentiful and complete than for anthocyanins and the difficulties encountered frequently stem from the small quantities found.

Some fruits, including certain *Citrus*, contain polymethoxylated flavonols substituted on carbons 6 and 8 (Figure 16). They can be classified as follows:

TABLE 21
Repartition of Flavonol Aglycones in Fruits

Number of aglycones	Fruits	K[a]	Q	M	iR	Ref.
1	Apple		6[b]			127,146,568
	Elderberry		+[c]			205,548
	Olive		+			600
	Privet		+			548
2	Apricot	2	2			247,253
	Blackberry	4	6			246,613
	Fig	2	2			525
	Litchi	1	1			439
	Peach	3	4			247
	Plum	4	5			248,548
	Raspberry	5	4			246,492
	Sour cherry	7	7			249,520
	Sweet cherry	6	4			249
	Strawberry	7	5			107,246,491
	Tomato	1	2			150,254,620
	Viburnum	+	+			209
	Lemon[d]		+		+	217
	Pear		3		4	138,176,253 412
3	Bilberry	1	4	+?		22,172
	Blackcurrant	2	2	2		305,523,548
	Cranberry	+	3	2		57,323,445
	Grapefruit	+	+		+	351
	Redcurrant	1	3	1		523,548
	Miracle fruit	1	1	1		88
	Sweet cherry Bn[e]	5	2		1	380,382
4	Grape[f]	4	5	2	1	102,103,253 465

[a] K: kaempferol; Q: quercetin; M: myricetin; iR: isorhamnetin.
[b] Number of glycosides.
[c] Undetermined or no precisely known number.
[d] With, in addition, polymethoxylated flavonols.
[e] Bn: cv. Bigarreau Napolèon.
[f] Myricetin and isorhamnetin are absent in white varieties.

- **6-OH-flavonols:** these have been identified in the form of methylated derivatives in the fruits of *Vitex agnus-castus* (Verbenaceae).[643] The main component is casticin, a tetramethylated derivative of quercetagetin (or 6-hydroxyquercetin), which is accompanied by three other flavonols (Figure 16). Murrayanol and the corresponding entirely methylated derivative have been isolated from *Murraya omphalocarpa* (Rutaceae) (Figure 16).[649]
- **8-OH-flavonols:** lemon peel *(Citrus limon)* contains the well-known di- and trimethylated derivatives limocitrin, limocitrol, and isolimocitrol (Figure 16).[202,217] Limocitrin is the 8,3′-dimethylether derivative of gossypetin or 8-hydroxyquercetin. Two limocitrin glycosides have recently been characterized in *C. unshiu.*[368] The entirely methylated derivative, 3,5,6,7,8,3′,4′-heptamethoxyflavone, is found in the peel and essential oil of orange *(C. sinensis)*, tangerine *(C. tangerina)*, mandarin *(C. reticulata)*, and clementine *(C. clementina)*. *C. aurautium* contains auranetin or 3,6,7,8,4′-pentamethoxyflavone, which is a 5-deoxyflavonol.[202]

TABLE 22
Flavonol Glycosides in Grape (*Vitis vinifera*)

Flavonol Content (mg/kg of berries)

	Varieties	K 3-Glucoside	Q 3-Glucoside	Q 3-Glucuronide	M 3-glucoside	iR 3-Glucoside	Total
A	Syrah	2.3	38.1	23.4	22.8	10.9	97.5
	Carignane	4.6	25.0	31.0	33.1	1.7	95.4
	Merlot	3.9	35.0	34.0	15.5	4.8	93.2
	Cinsault	7.2	15.9	18.6	10.5	3.4	55.6
	Negrette	1.0	16.6	23.1	11.7	Traces	53.0
	Gros verdot	Traces	1.7	10.2	26.1	Traces	37.9
	Malbec	Traces	1.3	10.9	14.7	0.9	27.7
	Tannat	Traces	1.9	7.4	9.2	Traces	18.5
B	Terret	19.1	33.3	29.5	0	0	81.9
	Muscat de F,[a]	10.1	23.0	35.0	0	0	68.0
	Maccabeu	1.1	3.3	3.7	0	0	8.1

Note: A: red varieties; B: white varieties.

[a] Muscat of Frontignan.

Data from Cheynier, V. and Rigaud, J., *Bull. Liaison Groupe Polyphénols,* 13, 442, 1986.

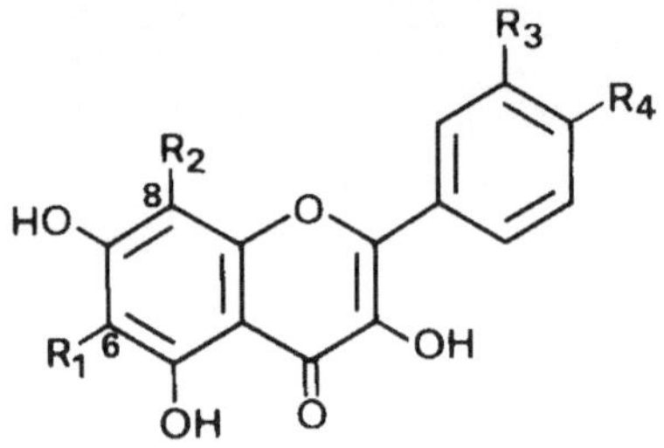

$R_1 = R_2 = R_3 = OCH_3$, $R_4 = OH$: Limocitrol
$R_1 = H$; $R_2 = R_3 = OCH_3$, $R_4 = OH$: Limocitrin
$R_1 = R_2 = R_4 = OCH_3$; $R_3 = OH$: Isolimocitrol

(Citrus limon)

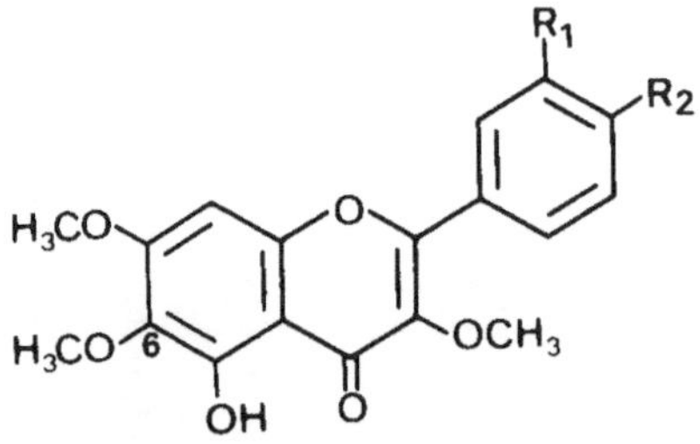

$R_1 = OH$	$R_2 = OCH_3$	:	Casticin
$R_1 = OH$	$R_2 = OH$	:	Chrysosplenol-D
$R_1 = H$	$R_2 = OH$	:	Penduletin
$R_1 = H$	$R_2 = OCH_3$	:	3,6,7,4' - tetramethylether of 6-hydroxykaempferol

(Vitex agnus castus)

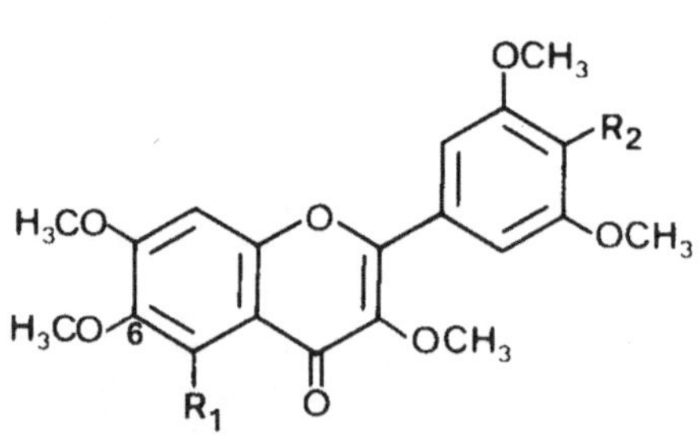

$R_1 = R_2 = OH$: Murrayanol
$R_1 = R_2 = OCH_3$: 3,5,6,7,3',4',5'-heptamethoxyflavone

(Murraya omphalocarpa)

FIGURE 16. Polymethoxylated flavonols of fruits.

2. Flavonol Glycosides

a. Monoglycosides

Glycosylation occurs preferentially on the hydroxyl group in the 3-position. In fruits as in other parts of plants, 3-*O*-monoglycosides are very predominant. Examples of 7 or 4′ glycosylation are rare in fruit flavonols: kaempferol-7-*O*-glucoside in strawberry[491] and sweet cherry var. Bigarreau Napoléon,[382] quercetin 7-xyloside in pear,[138] kaempferol and quercetin-4′-*O*-glucoside in sour cherry.[520] Characterization is still incomplete in all these cases and the structures proposed remain to be confirmed.

Five oses and one ose derivative occur in glycosylation at position 3. These are glucose, galactose, rhamnose, arabinose, xylose, and glucuronic acid. Although kaempferol and quercetin 3-glucosides and 3-galactosides are common, myricetin 3-glucoside has been found only in blackcurrant[305,523] and red grape varieties.[102,103] Isorhamnetin 3-glucoside is found in grape and in pear.[176] Concerning the three other oses (xylose, rhamnose, and arabinose), only quercetin 3-monoglycosides are found in fruits, an exception being myricetin-3-arabinoside in cranberry.[323,445] Quercetin and kaempferol 3-glucuronides have been identified in strawberry, raspberry, and blackberry[246,492] and in grape, which also contains myricetin 3-glucuronide.[102] It should be remembered that glucuronic acid derivatives are not found in anthocyanins.

Determination of the complete structure of a flavonol monoglycoside requires knowledge of whether the bond between ose and hydroxyl in the aglycone is α or β type, and whether the ose is in pyranose or furanose form. Glucose, galactose, xylose, and glucuronic acid, which all possess a D configuration, are usually linked to aglycone by β bonds whereas L-arabinose and L-rhamnose have α bonds. Oses are generally in pyranose form, but furanose form has sometimes been reported, e.g., in quercetin 3-*O*-α-L-arabinofuranoside or avicularin in apple.[127] Structures have not usually been fully described, and flavonol glycosides have only been separated and obtained in crystallized form in a few cases to enable full analysis. In the case of apple,[127] glycosides have been characterized by ^1H and ^{13}C NMR. The coupling constants in the ^1H NMR spectra were used to establish anomeric configuration of all glycosides. The authors concluded that the following were present: quercetin 3-*O*-α-L-arabinofuranoside, Q 3-*O*-β-D-galactopyranoside, Q 3-*O*-β-D-glucopyranoside (isoquercitrin), Q 3-*O*-α-L-rhamnopyranoside (quercitrin), and Q 3-*O*-β-D-xylopyranoside. The structure of Q 3-galactoside has also been proposed (in apple) as being that of Q 3-*O*-α-D-galactoside.[568]

b. Diglycosides

These are of two types, depending on whether the two ose molecules are attached to the same carbon (always in position 3) or to two different carbons. 3-Diglycosides are more numerous, forming approximately 80% of fruit diglycosides. They differ in the nature of the two oses and in the type of osidic linkage. Cases of two glucose molecules are rare in fruits, being found only in apple and in pear[146,176] in the form of Q 3-diglucoside and K 3-diglucoside; the type of linkage was not specified. It can be noted that although they occur fairly frequently in leaves, petals and pollen, flavonol sophorosides, in which the two glucoses are bound 1 → 2, have not been reported in fruits, with the exception of pods of *Sophora japonica*.[228] Cranberries contain myricetin 3-digalactoside.[445] The most widespread flavonol diglycosides are the 3-rutinosides in which a rhamnose molecule and a glucose molecule are joined by a 1 → 6 bond (e.g., rutin, Figure 17). The combination of xylose and glucuronic acid is found in strawberry, raspberry, and blackberry.[246] Other types of diholosides linked in the 3-position and involving xylose, galactose, arabinose, and glucose, are also found in fruits, e.g., quercetin 3-glucosyl-galactoside and 3-glucosyl-xyloside in grape[102] and quercetin and kaempferol 3-glucosyl(1 → 4) rhamnoside in fruits of *Rosa multiflora*.[229]

3,7-Diglycosides are distinctly less frequent in fruits than in plant organs as a whole.

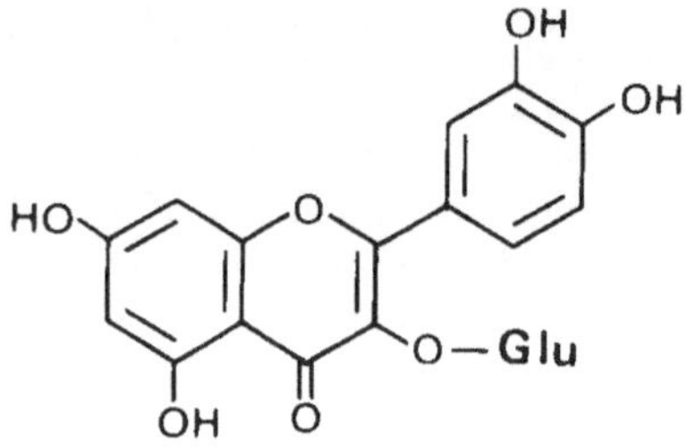

Q 3-glucoside (isoquercitrin)

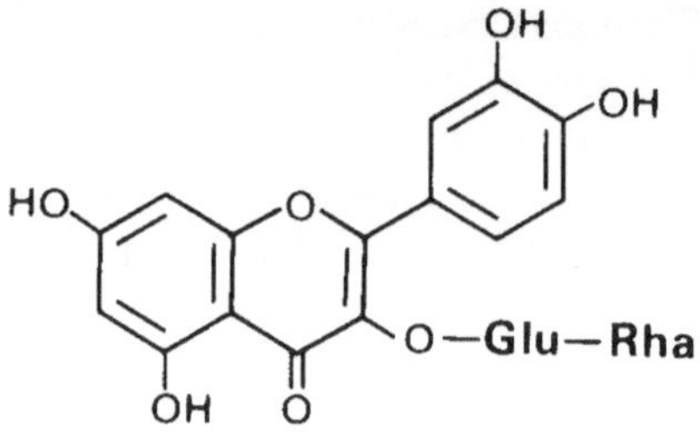

Q 3-rutinoside (rutin)

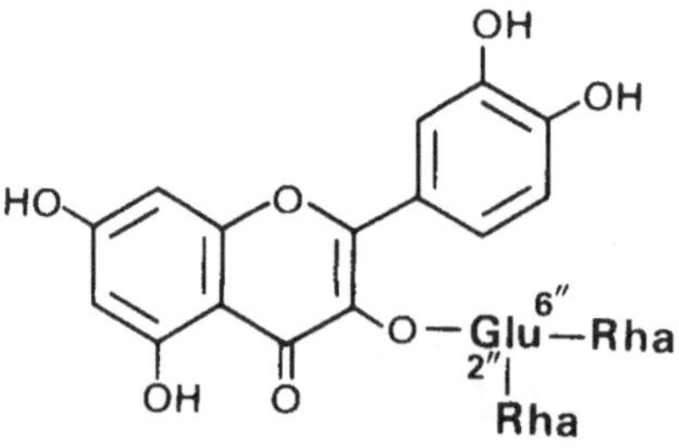

Q 3-dirhamnosyl-glucoside

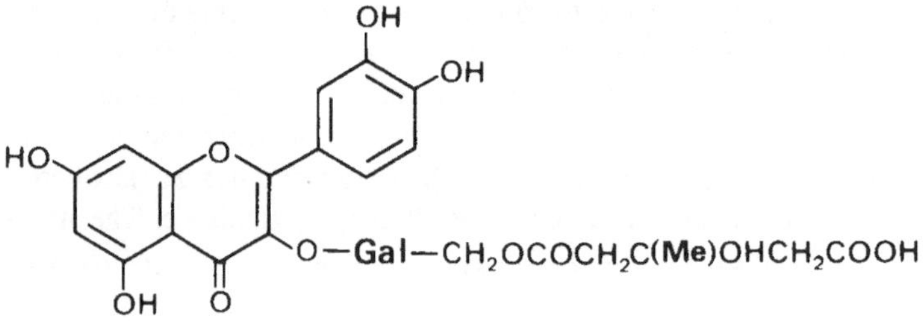

Acylated derivative of Q 3-galactoside

FIGURE 17. Examples of quercetin glycosides found in fruits (Q = quercetin; Glu = glucosyl; Rha = rhamnosyl; Gal = galactosyl). Q 3-dirhamnosyl-glucoside in redcurrant. (Adapted from Siewek, F., Galensa, R., and Herrmann, K., *Z. Lebensm. Unters. Forsch.*, 79, 315, 1984.) Acylated derivative of Q 3-galactoside in blackberry. (Adapted from Wald, B., Galensa, R., Herrmann, K., Grotjahn, L., and Wray, V., *Phytochemistry*, 25, 2904, 1986.)

Although about 20 of them are known, only 5 or 6 have been reported in a few fruits, for example the kaempferol 3,7-dirhamnoside in fruits of *Prunus salicina,*[248] kaempferol and quercetin 3-glucoside-7-xylosides and the quercetin 3,7-diglucoside in strawberry.[246] A single 3,4′-diglycoside, the kaempferol 3-rhamnoside-4′-galactoside has been reported in sour cherry.[520]

c. *Triglycosides*

There are not many triglycosides and they have only been identified in a few fruits. The identification of quercetin and kaempferol 3-glucoside-7-diglucosides and 3-galactoside-7-

TABLE 23
Relative Importance of Kaempferol, Quercetin, Myricetin, and Isorhamnetin Glycosides in Fruits and in Whole Plant Organs

	Fruits	Whole plant organs[a]
Kaempferol glycosides	37.5	35
Quercetin glycosides	41.7	41
Myricetin glycosides	12.5	7
Isorhamnetin glycosides	8.3	16.5

Note: Values are expressed in per cent of total identified glycosides.

[a] Data from Reference 229.

diglucosides in sour and sweet cherries[249] probably requires confirmation. The same fruits are thought to contain tetraglycosides, kaempferol, and quercetin 3-rutinoside-4′-diglucosides, which is exceptional.

In the present state of research, branched triglycosides appear to be less important for flavonols than for anthocyanins. One of them (quercetin 3-dirhamnosyl-glucoside, Figure 17) has been identified accurately in redcurrant *(Ribes rubrum)*[523,524] and is not found in blackcurrant *(R. nigrum)*. One of the flavonols in Bigarreau Napoleon cherry may be the kaempferol 3-(2^{G}-glucosyl-rutinoside).[382] Linear triglycosides are known in plants but have only very rarely been found in fruits, with the exception of *Rhamnus cathartica, R. alaternus,* and *R. leptophylla* which contain kaempferol, kaempferol 7-methylether, and quercetin 7-methylether glycosylated with the rhamninose.[228,618] The fruits of *Rosa multiflora* contain the quercetin 3-glucosyl(1 → 6)-glucosyl(1 → 4) rhamnoside.[229]

d. Acylated Flavonols

Flavonol glycosides have been found acylated with *p*-coumaric, ferulic, caffeic, *p*-hydroxybenzoic, gallic, acetic, and malonic acid,[228] but they have rarely been reported in fruits for which data is still fragmentary. An acylated derivative of quercetin was isolated and identified (Figure 17) in a recent work on blackberry.[613] After alkaline hydrolysis, this compound releases the quercetin 3-*O*-β-galactoside and 3-hydroxy-3-methylglutaric acid. Acylation takes place in position 6 on the galactose. An acylated derivative of isorhamnetin 3-glucoside was found in pear.[176,412] It is probably the isorhamnetin 3-(*O*-malonyl)-glucoside.[253] Several malonated flavonol glycosides have been recently identified in this fruit.[612] The most widespread acylated glycosides in plant organs are the kaempferol 3-(*p*-coumaroyl)glucosides, including tiliroside.[228] Their presence in fruits has only been reported in *Rosa canina*[217] and *Tribulus terrestris*.[49] A acetylated derivative of kaempferol is present in the fruits of *R. multiflora*.[229] Data on acylated flavonol glycosides are hence even more fragmentary.

The presence of bisulfate flavonols has not been reported in fruits to date, although these compounds are fairly broadly distributed.[228]

3. Comparisons between Fruits

a. Qualitative Aspects

In contrast with the high number of flavonol glycosides known in the plant kingdom (approximately 250),[228,229] only about 50 have been identified in fruits and most of these are quercetin glycosides. In comparison with plant organs as a whole, fruits stand out by their relatively greater abundance of myricetin glycosides and the small number of isorhamnetin glycosides (Table 23). In contrast, there is no difference in quercetin and kaempferol glycosides.

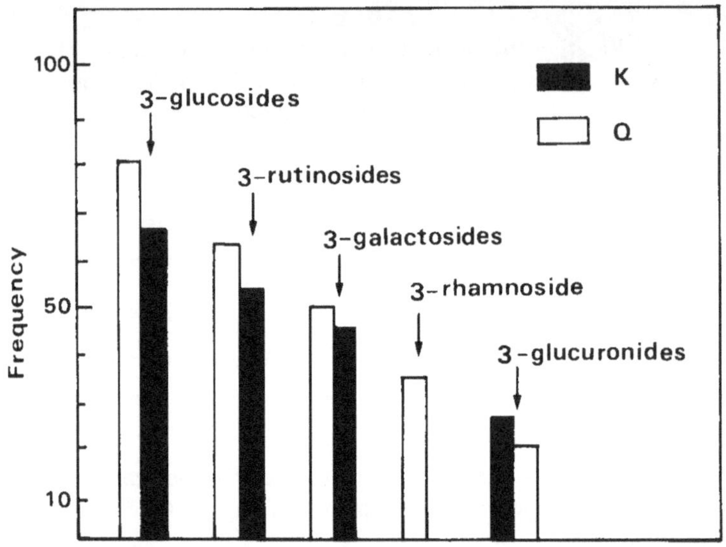

FIGURE 18. Relative importance of the 9 most common flavonol glycosides found in fruits. Frequency is expressed in per cent of the total fruit number. (K = kaempferol; Q = quercetin).

The distribution of the different types of glycosides in fruits corresponds to that of plants as a whole, with diglycosides being always the most numerous (about 40%). Nevertheless, fruits are characterized by a relatively greater number of monoglycosides (35% compared to 25% in plants as a whole), and above all by a smaller number of triglycosides and acylated glycosides. The situation thus contrasts with that observed for anthocyanins where acylated derivatives are numerous in fruits.

The most frequent flavonol glycoside is the quercetin 3-glucoside (isoquercitrin), which was found in 80% of the fruits examined, and whose equivalent is the cyanidin 3-glucoside which is also the most common anthocyanin in fruits. This corresponds to the situation observed for all plant organs.[215,216] It can also be pointed out (Figure 18) that:

- The most common glycosides are 3-monoglycosides in the following order: 3-glucosides>3-galactosides>3-rhamnoside>3-glucuronides
- Quercetin and kaempferol 3-rutinosides are the only diglycosides to occur very frequently
- Frequencies of the various quercetin and kaempferol glycosides are generally very close, except for rhamnose which is represented only by quercetin 3-rhamnoside (quercitrin)

Fruits contains from 2 to 14 flavonol glycosides (Table 21) although a maximum cannot be given, unlike the case of anthocyanins. The most complex pattern, with over ten glycosides, has been found in cherry,[249] strawberry,[246,491] and grape.[102,103] In the first two cases, this complexity is the result of a high number of quercetin and kaempferol glycosides (5 to 7 and 7, respectively), whereas the diversity in grape is caused partly by the presence of four aglycones.

Many fruits contain five to ten flavonol glycosides (Table 21), but here again the number may indicate very different situations, e.g., six to seven quercetin glycosides in apple,[127] four quercetin glycosides and three kaempferol glycosides in peach,[247] two glycosides of each of the three aglycones kaempferol, quercetin, and myricetin in blackcurrant.[523] Raspberry and blackberry, which are species of the genus *Rubus*, display the same complexity

TABLE 24
Comparison of Flavonol Contents in Fruits and Leaves for Some Species

	Flavonols in mg/100 g FW		Ratio
	Fruit	Leaf	leaf/fruit
Strawberry	7.0	227.5	33
Raspberry	8.5	361.3	42.5
Mirabelle plum	2.6	268.0	103
Apricot	1.5	300.0	200
Peach	1.0	736.5	736.5

Data from References 246 to 248.

in the glycosides present.[246] Pear is characterized by the predominance of isorhamnetin glycosides, four or five of which are present.[138,176]

Very few fruits contain a small number of flavonol glycosides, unlike the case observed for anthocyanins. However, apricot stands out from the other *Prunus* species by the presence of only three or four glycosides, depending on variety.[247] Flavonols — except for rutin — are very scarce in ripe tomato[150,620] and detection and identification pose problems.[254] Identification of three flavonol glycosides in miracle fruit *(Synsepalum dulcificum)*[88] and of only two in litchi *(Litchi chinensis)*[439] are certainly not complete as minor compounds may be present.

b. Quantitative Aspects

There are far fewer detailed quantitative studies on fruit flavonols than on anthocyanins, particularly with regard to variability in relation to genetic and environmental factors. Flavonol contents are sometimes very low in mature fruits, e.g., in cherry tomato,[150] whence difficulties in analysis, separation, and quantitative determination. Flavonol glycoside concentrations vary according to fruit from 0.2 mg/100 g FW to 30 mg/100 g FW, i.e., a variation of 1 to 150 (Table 2). This range of concentrations is comparable to that observed for monomeric flavan-3-ols. However, there does not seem to be a correlation between the flavonol and flavan-3-ol contents of fruits. The flavonols/flavan-3-ols ratio is over 1 in some fruits (e.g., 6 to 8 in blackcurrant and bilberry) and distinctly less than 1 in grape, peach, and apricot.

Although the fruits with the highest flavonol contents are red- or blackberries which are also rich in anthocyanins, a recent study[57] on varietal differences in the quercetin, kaempferol, and myricetin contents of highbush blueberry, cranberry, and thornless blackberry concludes that no correlation exists in any of the cultivars between total anthocyanin content and the amounts of flavonols.

Flavonols have been analyzed and determined in the fruits and leaves of a small number of plant species.[246-249] Fruits contained less flavonols than leaves in all cases; the leaf content/fruit content ratio varied from 30 to 700 according to species (Table 24).

The blackberry *(Rubus fructicosus)* appears to possess the highest flavonol content; this varies from 20.7 to 31.4 mg/100 g FW in five cultivars.[246] The figures for raspberry *(R. idaeus)* are only about a third of this. However, several cultivars or clones of thornless blackberry[57] display distinctly lower flavonol contents (determined after hydrolysis of glycosides) than those given above.[246] Elderberry fruits *(Sambucus nigra)* have a high quercetin content.[548] The fruits of *Vaccinium* are also rich in flavonols.[57,548] Thus the amounts in six cultivars of cranberry *(V. macrocarpon)* vary from 13.9 to 26.8 mg/100 g FW with predominance of quercetin, which represents over 90% of total aglycones.[57]

The flavonol content of ripe *Prunus* fruits is generally lower than 5 mg/100 g FW (Table

TABLE 25
Quercetin Glycoside Concentrations in Apple Peel (μg/g)

Variety	Quercetin 3-glycosides					
	Rutinose	Glucose + Galactose	Xylose	Rhamnose	Arabinose	Total
Jersey Mac	185	555	120	125	555	1540
Gravenstein	190	783	160	446	548	2127
Mc Intosh	115	696	254	361	777	2203
Cortland	160	734	192	80	590	1756
Spartan	55	783	244	71	663	1756
Golden Delicious	110	780	180	265	502	1837
Red Delicious	117	604	210	104	546	1581
Northern Spy	170	924	297	660	800	2851

Data from Dick, A. J., *Bull. Liaison Groupe Polyphénols,* 13, 358, 1986.

2). The smallest amounts are observed in peach *(P. persica)* in which no glycoside dominates distinctly.[247] In contrast, in certain apricot *(P. armeniaca)*[247] and plum cultivars *(P. domestica),*[248] the major flavonol glycosides are, respectively, quercetin 3-rutinoside and quercetin and kaempferol rutinosides.

Redcurrant *(Ribes rubrum),* with levels ranging from 2.0 to 5.1 mg/100 g FW in 14 cultivars analyzed, contains 5 to 6 times less flavonols than blackcurrant *(R. nigrum)*; the latter is also richer in anthocyanins.[523] The example of the strawberry *(Fragaria ananassa)* shows the variability which may exist between cultivars. In 20 cultivars, contents vary from 2.1 to 17.4 mg/100 g FW with an average of 7 mg.[246] Three glycosides are dominant in all cases: kaempferol 3-glucuronide, kaempferol 3-glucoside, and quercetin 3-glucuronide. Such variability is also found for concentrations of quercetin glycosides in the skin of eight varieties of apple (Table 25).[126] The main glycoside is generally quercetin 3-galactoside which can form up to 54% of total flavonols in Golden Delicious.[568]

In grape, where flavonols have been studied much less than other phenolic compounds, great differences can be observed in both red and white varieties[103] (Table 22). Skin contents range from 0.81 to 9.75 mg/100 g whereas the pulp contains no flavonols.

Considerable intraspecific variability is also found for total flavonols in pear peel.[185] The flavonol content in the cultivar Passe-crassane is eight times greater in peel than in pulp.[55]

C. MONOMERIC FLAVAN-3-OLS

Flavans are flavonoids characterized by a saturated 3-C chain. They are of two types, flavan-3,4-diols and flavan-3-ols. The monomeric flavan-3,4-diols, now referred to as leucoanthocyanidins[232,625] are found frequently in the wood and bark of trees, but they are not major compounds in fruits where they appear to be present only exceptionally.[321,623] In fact, the compounds previously identified in fruits by different authors as leucoanthocyanins[593] probably do not have the structure of flavan-3,4-diols. According to Haslam,[233] flavan-3,4-diols do not seem to participate in the structure of condensed tannins, although certain authors have postulated that they are involved as intermediate substances in the biosynthesis of flavan-3-ols and proanthocyanidins.[546,547] We examine only flavan-3-ols, which are important constituents of fruits.

1. Structure and Nomenclature

Flavan-3-ols participate in the structure of proanthocyanidins (condensed tannins) as their monomers. Detailed reviews of these compounds have been published.[232,237] The oligomeric and polymeric forms are discussed in the chapter on tannins below. A dozen flavan-

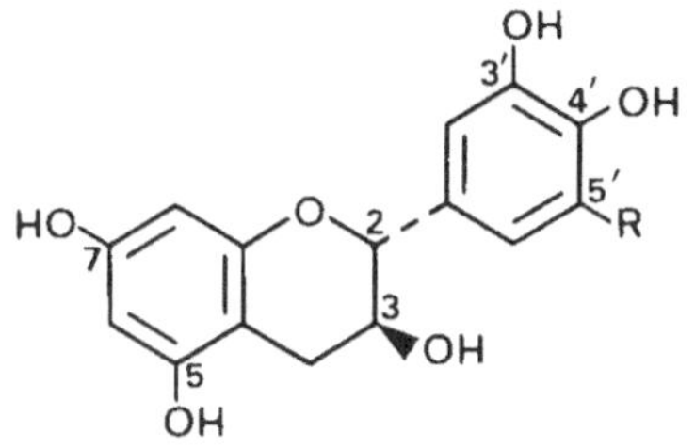

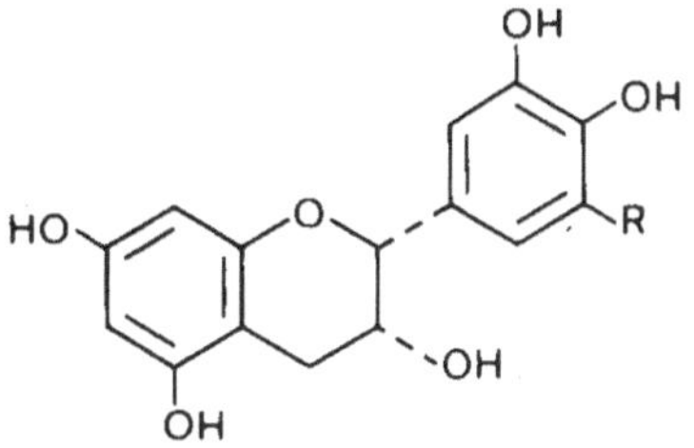

R = H : (+)-Catechin

R = OH : (+)-Gallocatechin

R = H : (-)-Epicatechin

R = OH : (-)-Epigallocatechin

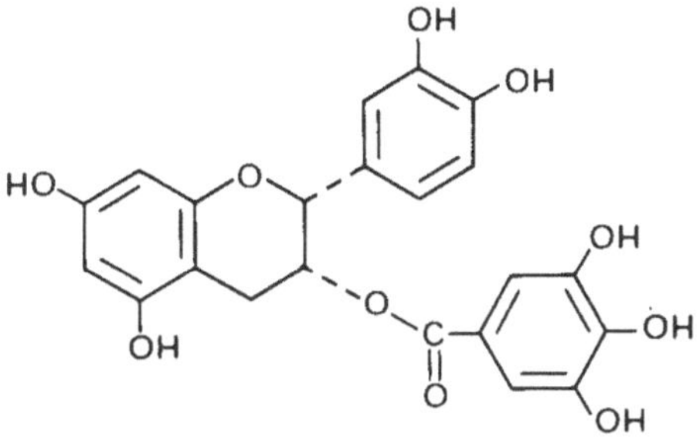

(-)-Epicatechin 3-gallate

FIGURE 19. Chemical structure of monomeric flavan-3-ols found in fruits. 2R:3S configuration for (+)-catechin and (+)-gallocatechin; 2R:3R configuration for (−)-epicatechin and (−)-epigallocatechin.

3-ols have been obtained, essentially from woody species (*Quercus, Castanea, Acacia, Eucalyptus,* etc.), but these compounds are more limited in fruits. The absence of flavan-3-ols which have lost the hydroxyl in the 5-position (robinetinidol, fisetinidol) and those monohydroxylated in the B-ring (4′OH, afzelechin, epiafzelechin) can be noted. Four main flavan-3-ols are found in fruits. Two are orthodiphenols hydroxylated in the 3′- and 4′-positions in the B-ring ([+]-catechin and [−]-epicatechin) and two are trihydroxylated in the 3′-, 4′-, and 5′-positions ([+]-gallocatechin and [−]-epigallocatechin) (Figure 19).

C-2 and C-3 represent two centers of asymmetry in the molecule, in such a way that the four flavan-3-ols in fruits are grouped in two pairs of diastereoisomers whose absolute configurations have been determined: 2*R* in all cases and either 3*S* for (+)-catechin and (+)-gallocatechin or 3*R* for (−)-epicatechin and (−)-epigallocatechin (Figure 19). *In vitro* epimerization is possible in the 2-position in hot solution, and so it is possible that the observation of small amounts of (+)-epicatechin and (−)-catechin in natural extracts may be caused by this phenomenon during extraction.[232] However, the large amount of (+)-epicatechin found in the drupes of a palm, *Chamaerops humilis,* appears to show that in this case it is a natural compound,[122] but this remains an exception.

Unlike the other classes of flavonoids, monomeric flavan-3-ols are generally found in free rather than glycosylated or esterified form in fruits. One exception is, nevertheless, the case of the (−)-epicatechin-3-*O*-gallate identified in grape.[554,624] This ester has been found in several red grape cultivars,[161,554,624] but only traces were found in the white cv. Müller-Thurgau;[321] this is probably a simple difference between cultivars since the compound was initially revealed in unripe berries of the white cv. Emerald Riesling.[554]

2. Comparisons between Fruits

a. Qualitative Aspects

Practically all the fruits examined contain monomeric flavan-3-ols at maturity, although

TABLE 26
Repartition of Flavan-3-ol Monomers in Ripe Fruits (Average Contents in mg/100 g FW)

	(+)-Catechin	(−)-Epicatechin	(+)-Gallo catechin	(−)-Epigallo catechin	Total	Ref.
Apricot	4.40	20.20	—	—	24.60	395
Sweet cherry, cv, Bn[a]	6.80	15.00	—	—	21.80	380
Grape (skin extract)	5.18	13.35	+	+	18.53	530
Sour cherry	2.08	13.06	1.90	0.53	17.57	395,552
Blackberry	1.40	11.20	—	—	12.60	396
Apple	1.24	6.14	0.42[b]	0.10[b]	7.90	395
Peach	5.40	0.50	0.40	0.30	6.60	395
Plum, cv. Mirabelle	4.40	1.70	—	—	6.10	395,552
Sweet cherry	1.60	3.65	—	—	5.25	395,552
Raspberry	0.83	3.55	—	—	4.38	396
Strawberry	3.71	0.35[b]	0.13[b]	—	4.19	549
Gooseberry (red)	2.50	0.40	—	—	2.90	550
Bilberry	1.17	1.27	—	—	2.44	550
Plum	1.17	1.18	—	—	2.35	395,552
Redcurrant	0.65	0.52	0.47	0.24	1.88	550
Gooseberry (yellow)	1.80	—	—	—	1.80	550
Pear	0.23	1.03	—	—	1.26	395
Blackcurrant	0.40	0.17	0.20	0.35	1.12	550
Redcurrant var. white	0.50	—	—	—	0.50	550

[a] Bn,: cultivar Bigarreau Napolèon.
[b] Occurring only in some varieties.

the tomato appears to be an exception.[153] There may, nevertheless, be great differences between the cultivars of the same species. Thus in the case of Japanese persimmon *(Diospyros kaki)* only the pollination constant nonastringent type cultivars still contain catechin at maturity whereas it disappears at an early stage in the other types.[657] In addition, in certain cases, such as that of mango,[175] flavanols have been measured in fruits but without any indication of the nature of the compounds present.

(+)-Catechin and (−)-epicatechin were always present in some 20 fruits analyzed accurately (Table 26). Several situations were observed among the dozen fruits which contained only these two flavan-3-ols.[395,396,549,550,552]

- **(−)-Epicatechin** dominates very distinctly in six fruits which are all members of the Rosaceae family: apple (where it forms 74 to 98% of flavan-3-ols according to cultivar), pear (54 to 80% according to year), apricot (82%), sweet cherry (52 to 83% according to cultivar), raspberry and blackberry (62 to 98% and 85 to 93%, respectively, depending on the cultivar). It can be noted that (+)-gallocatechin and (−)-epigallocatechin, which are normally absent in apple, may occur in two cultivars (Jonathan and Golden Delicious) in relation to climatic factors and conditions of growth.[395]
- **(+)-Catechin** is dominant in only three fruits: mirabelle plum (72% of flavan-3-ols, whereas it only reaches 44 to 55% in the other *Prunus domestica* cultivars), gooseberry and strawberry. The latter fruit is a good example of the variability of flavan-3-ol composition related to the nature of the cultivars.[549] Of the 14 cultivars studied, 6 contain only (+)-catechin, which remains predominant in the 7 other cultivars (78 to 94%), and (+)-gallocatechin is only present in small amounts (5 and 8%) in 2 of them. A single cultivar displayed approximately equivalent quantities of the three flavan-3-ols (+)-catechin, (−)-epicatechin, and (+)-gallocatechin. Thus in this case,

in spite of the variations related to environmental factors, the qualitative composition of flavan-3-ols may be a criterion for the identification of the different cultivars.

- **(+)-Catechin and (−)-epicatechin** are found in approximately equal proportions in two types of fruits: plum (three cultivars of *P. domestica*) and bilberry. In bilberry, (+)-gallocatechin and (−)-epigallocatechin, absent in the mature fruits, are found in unripe fruit where they form approximately 10% of the flavan-3-ols.

The presence of (+)-catechin and (−)-epicatechin has been reported in a number of fruits without their relative proportions being specified: cranberry,[619] sapodilla *(Achras sapota)*,[312] *P. amygdalus, Cola acuminata,* and *Wisteria sinensis* pods.[622] It would seem that (−)-epicatechin alone is present in the fruits of three species of *Crataegus* and the fruits of *Pyracantha coccinea*.[622]

In several other fruits (peach, sour cherry, blackcurrant, redcurrant, grape), (+)-catechin and (−)-epicatechin are always accompanied by (+)-gallocatechin and (−)-epigallocatechin, generally in fairly small proportions (less than 20%) (Table 26).[251,395,465,550,552] In comparison with sweet cherries *(Prunus avium)*, the presence of two 5′-OH flavan-3-ols appears to be characteristic of sour cherries *(P. cerasus)*. Likewise, unlike redcurrants, white currants contain only (+)-catechin.

In grape, (+)-catechin and (−)-epicatechin are always predominant in comparison with the other monomer flavan-3-ols, but proportions vary according to cultivar. Thus, (+)-catechin is generally more plentiful than (−)-epicatechin,[235,251,404,624] with the mean reaching 79% in berry skins whereas it is only 50% in the seeds.[72] There is considerable variation in percentage from one cultivar to another, ranging from 64 to 98% in skins. Lower values, of the order of 50% have been reported for certain cultivars;[115] these are probably for the whole of the berry including seeds. Surprisingly, (−)-epicatechin is reported as being the most abundant in skin extracts of several white cultivars.[530] It is possible that these differences are associated with the cultivars analyzed or the extraction conditions, but a certain amount of contamination of peaks in HPLC separations cannot be excluded.

b. Quantitative Aspects

It is important to consider the catechin content of fruits insofar as these compounds are, together with chlorogenic acid, the natural substrates of polyphenoloxidases.[89,337,366,373,374] They are involved in the browning phenomena which occur during the conservation or technological processing of fruits. These aspects are discussed in Chapter 5.

In the fruits studied, the amounts of monomeric flavan-3-ols (that is to say essentially (+)-catechin and (−)-epicatechin as mentioned above) vary considerably with species and cultivar and can attain 30 mg/100 g FW (Tables 2 and 26).

Ripe fruits can be classified in three groups in a fairly schematic manner as follows:

1. Low or very low flavan-3-ol contents. This is the case of fruits belonging to the genus *Ribes* (redcurrant, blackcurrant, gooseberry), raspberry, bilberry, strawberry, plum, and pear. This situation could be interpreted either by orientation of the phenolic metabolism towards the accumulation of other classes of flavonoids like anthocyanins (the case of the "small red fruits" above), or because these monomers are used for condensation into catechic tannins (e.g., astringent cultivars of Japanese persimmon). However, several cultivars already stand out in this category: the flavan-3-ol content of mirabelle plum is 2 to 3 times as high as that of other plums and thus appears to be different with regard to both the concentration and relative proportions of the two constituent flavan-3-ols. Unlike other cultivars of pear which have very low flavan-3-ol contents,[395] the variety Passe-Crassane contains much greater amounts.[54,56] Catechins are located above all in the skin of the fruit; the skin/pulp concentration ratio is 16.[55]

TABLE 27
Monomeric Flavan-3-ols in Skin Extracts of Grapes

Type		(+)-Catechin		(−)-Epicatechin		Total
		mg/100 g FW	%	mg/100 g FW	%	mg/100 g FW
Chenin Blanc	1980	—		4.09	100	4.09
	1981	6.68	100	—		6.68
French Colombard	1980	1.38	3.5	38.46	96.5	39.84
	1981	0.09	6.3	1.33	93.7	1.42
Semillon	1980	21.40	40.6	31.30	59.4	52.70
	1981	5.68	25.6	16.46	74.4	22.14
Thompson Seedless	1980	2.26	100	—		2.26
	1981	3.92	26.8	10.70	73.2	14.62

Data from Singleton, V. L. and Trousdale, E., *Am. J. Enol. Vitic.*, 34, 27, 1983.

2. Medium flavan-3-ol contents. In spite of strong variability, blackberries have a distinctly higher content than raspberries which also belong to the genus *Rubus*. The ratio of contents between the two species is about 3; this is similar to the ratios of flavonol (3.1) and anthocyanin (3.5) contents. In apple, there is also variability between cultivars (3.45 to 16.3 mg/100 g FW) and in a given cultivar depending on the location of the orchard.[395]
3. High flavan-3-ol contents. This is the case of sour cherries, apricot, grape, and persimmon (nonastringent constant pollination cultivars). In a general manner, sour cherries *(Prunus cerasus)* are clearly distinct from sweet cherries *(P. avium)* since they have higher flavan-3-ol contents. However, the cultivar Bigarreau Napoleon of *P. avium* has a higher flavan-3-ol content than *P. cerasus,* confirming the very special character of this sweet cherry.[381,382] In grape, there is both considerable variation in content according to variety (Chapter IV, Tannins, Table 32)[72] and even greater variability within the same variety depending on the year (Table 27).[530] The values given in Table 27 are for the white varieties which seem to be richer in flavan-3-ols than the red varieties (Table 32). Nevertheless, there is no distinct difference between the white variety (Grenache blanc) and red varieties in Languedoc-Roussillon.[72]

D. OTHER FLAVONOIDS

1. Flavones

Flavones are flavonoids characterized by a nonsaturated 3-C chain and have a double bond between C-2 and C-3, like flavonols, with which they differ by the absence of hydroxyl in the 3-position. It appears that this simple difference in structure between flavones and flavonols has very important consequences in the biogenesis, physiological and pharmacological roles, and the phylogenetic and chemotaxonomic signification of these compounds. Flavones are widely distributed among the higher plants in the form of aglycones or glycosides.[602,641] However, although over 100 natural flavones have been identified in plants, these compounds are not common in fruits and are never predominant. We will first examine the special case of *Citrus*.

a. Polymethoxylated Flavones of Citrus and Other Rutaceae

Citrus fruits contain several polymethoxylated flavones, some of which were isolated and identified a long time ago: nobiletin and sinensetin in orange peel *(Citrus sinensis)* and tangeretin in tangerine oil *(C. tangerina)*.[258,301,577] These flavones are located in the flavedo and are found in the essential oil of citrus fruits. They are extremely interesting since they have a physiological effect on animals and man. For example, they have an effect on the

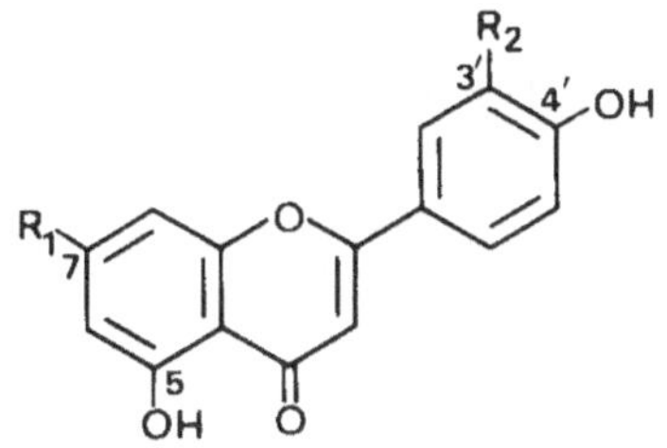

A

R1 = OH, R2 = H : Apigenin
R1 = O-Glucosyl, R2 = OH : Luteolin 7-glucoside

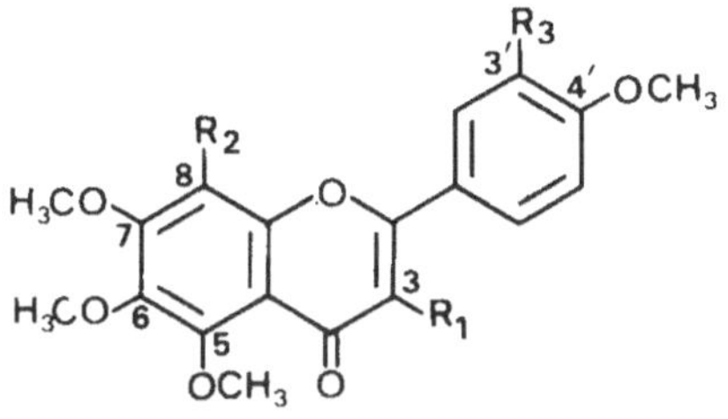

B

R1 = R2 = R3 = H : Tetramethylscutellarein
R1 = R2 = H, R3 = OCH3 : Sinensetin
R1 = H, R2 = OCH3, R3 = H : Tangeretin
R1 = H, R2 = R3 = OCH3 : Nobiletin
R1 = OCH3, R2 = H, R3 = OCH3 : Hexamethoxyflavone
R1 = R2 = R3 = OCH3 : Heptamethoxyflavone

FIGURE 20. Examples of flavone structures present in fruits. A: common structures. B: polymethoxylated flavones of Citrus.

high blood viscosity syndrome.[480] These compounds form part of the citroflavonoids which are defined as being "flavonoids with a vitamin P effect, extracted from the peel of various citrus fruits."[268] In addition to their pharmacodynamic effect, polymethoxylated flavones in *Citrus* possess an antimicrobial and antiviral activity. Together with the other components of the essential oil, they form part of the resistance mechanisms in *Citrus*.[268]

HPLC techniques can be used to separate and determine quantitatively polymethoxylated flavones from peel and essential oil of orange and tangerine.[51,187,188,578] The six polymethoxylated flavones isolated from essential oil of orange[187,188] have no free hydroxyl (Figure 20) and differ in their degree of methoxylation; heptamethoxyflavone displays the greatest methylation. Hexamethoxyflavone and heptamethoxyflavone, which have a methoxyl group in the 3-position, can be considered as methylated flavonols on C-3. Several other compounds are also present in very small amounts, with 11 peaks being distinctly separated in orange and tangerine juice.[578] Several of these minor components have been identified as being methoxyflavones with hydroxyl free in the 5-position and which could be nonmethoxylated in the 6-position.[562]

In hesperidin complex, a light-tan powder obtained from deoiled orange peel extracts, the overall flavone content is 317 μg/g on average with variations running from 155 to 420 μg/g. This is very little in comparison with flavanones, whose concentration is 790 mg/g.[430] Polymethoxylated flavones have been measured after separation by HPLC in peel oils of orange, tangerine, mandarin and clementine (Table 28). Standardized principal component

TABLE 28
Quantitative Determination of Polymethoxylated Flavones in Peel Oils of Various Citrus Samples

	Flavone content in g/l						
Code	**Tangeretin**	**Tetramethyl scutellarein**	**Hepta methoxy flavone**	**Nobiletin**	**Hexa methoxy flavone**	**Sinensetin**	**Total**
01	0.5	0.30	0.56	0.75	0.02	0.10	2.22
02	0.8	0.40	0.40	0.60	0.00	0.20	2.40
03	0.6	0.45	0.70	1.10	0.00	0.15	3.00
04	0.5	0.45	1.00	0.90	0.04	0.10	2.99
05	0.7	0.20	0.30	0.50	0.00	0.30	1.88
06	0.7	0.50	2.00	0.70	0.00	0.10	4.00
W	1.0	0.60	0.70	1.00	0.03	0.14	3.47
M1	2.8	0.50	1.10	2.00	0.02	0.07	6.49
M2	1.9	0.00	1.60	0.60	0.00	0.20	4.30
M3	2.0	0.20	0.30	0.80	0.00	0.20	3.50
M4	2.1	0.00	0.20	0.90	0.00	0.20	3.40
M5	2.2	0.05	0.70	0.70	0.00	0.30	3.95
K	2.5	0.00	2.70	0.60	0.00	0.20	6.00
C	1.1	0.25	0.75	0.40	0.03	0.07	3.60
T	1.5	0.35	0.22	1.50	0.00	0.07	3.64

Note: Code:

01 to 06	*Citrus sinensis*	oranges
W	*C. oblonga*	Washington navel
M1	*C. deliciosa*	common mandarin
M2 to M5	*C. reticulata*	mandarins
K	*C. nobilis*	king mandarin
C	*C. clementina*	clementine
T	*C. tangerina*	tangerine

Classification according to Tanaka, in Gaydou et al.[188]

From Gaydou, E. M., Bianchini, J. P., and Randriamiharisoa, R. P., *J. Agric. Food Chem.*, 35, 525, 1987. With permission.

analysis was carried out using seven parameters: the concentration of each of the six flavones measured separately and the sum of these concentrations.[187,188] Taxonomic significance of the content of these flavones was shown by factorial discriminant analysis for the differentiation of the orange and mandarin groups (Table 28). The latter group was characterized by a higher concentration in total flavones than oranges. The two major components were nobiletin and tangeretin, whose concentration varied from 0.5 to 0.8 g/l for oranges and 1.9 to 2.8 g/l for mandarins.

Tangeretin and heptamethoxyflavone have been reported in the grapefruit *(C. paradisi)*[217] as have 5,7,8,4′-tetramethoxyflavone and 5,7,8,3′,4′-pentamethoxyflavone.[602] Several polymethoxylated flavones are present in *C. aurantium:* nobiletin and its 5-demethyl derivative, sinensetin, tetramethylscutellarein, and a 5-deoxyflavonol, auranetin (= 3,6,7,8,4′-pentahydroxyflavone).[202,217,602] Hypolaetin 7,8,3′4′-tetramethylether (= 5-hydroxy-7,8,3′4′-tetramethoxyflavone) is found in bergamot oil *(C. bergamia)*.[602] In contrast, as mentioned in discussion of flavonols (Section B above), lemon (*C.* lemon) contains polymethoxylated flavonols and not flavones.[202]

Polymethoxylated flavones of *C. sudachi* fruits have been studied in detail. On the basis of UV, ^{1}H-NMR, and ^{13}C-NMR spectral data, six flavones have been identified including sudachitin or 5,7,4′-trihydroxy-6,8,3′-trimethoxyflavone, which is present in free form and as four glycosides called sudachin, A, B, C, and D (Figure 21).[256,257] These are 4′-*O*-β-D-glucoside or 7-*O*-β-D-glucoside which can be esterified by 3-hydroxy-4-methyl-glutaric acid.

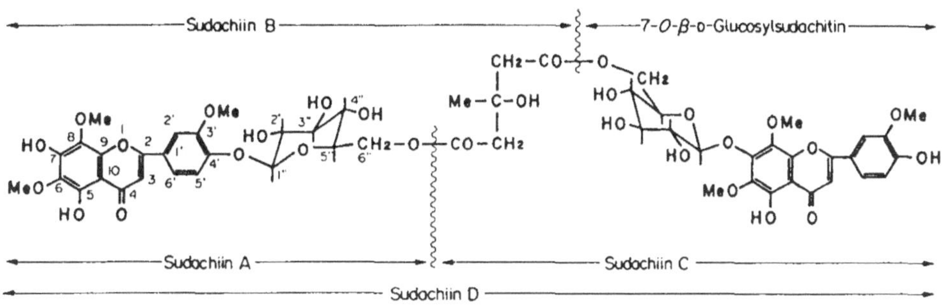

FIGURE 21. Flavone glycosides from *Citrus sudachi*. (From Horie, T., Tsukayama, M., Yamada, T., Miura, I., and Nakayama, M., *Phytochemistry*, 25, 2621, 1986. With permission.)

Polymethoxylated flavones have also been found in the fruits of other Rutaceae. Thus, in addition to three chalcones, the fruits of *Merrillia caloxylon* contain several tetra and pentamethoxylated flavones.[170,171] Expressed juice of the same fruit shows the presence, after mild acid hydrolysis, of the cytotoxic flavone eupatorin in 0.1% yield.[2] The fruits of *Casimiroa edulis* and *Sargentia greggi* also contain methoxylated flavones.[218,602]

Flavonoid production by callus cultures originating from orange and lemon flavedo was evident in the light grown tissue, with sinensetin and nobiletin as major components.[87] However, the amounts of flavonoids formed in the callus were relatively low as compared to those found in the peel extracts.

b. Other Flavones

With the exception of *Citrus*, flavones are not common chemical constituents of fruits. Thus the glucosides of apigenin and luteolin which are, nevertheless, widely distributed in leaves and flowers[220] are only infrequently represented in fruits: apigenin-4′-methylether-7-neohesperidoside in *Fortunella japonica* (Rutaceae)[228] and various glycosides in dry fruits of Umbelliferae where they are of distinct chemotaxonomic interest.[227] In Rosaceae, the 7-glucosides of apigenin and luteolin have been identified in the stems and leaves of several species but not in the fruits.[295] Apigenin-7-glucoside, luteolin-5-glucoside, and luteolin-7-glucoside are present in olive.[600] Although the latter substance is more abundant, it forms only 2% of all the phenolic compounds in the ripe fruits of the cultivar Picholine (13.3 mg/100 g FW).[8]

In citrus fruits, in addition to the polymethoxylated flavones described above, apigenin has been identified in grapefruit after hydrolysis[351] and in combined form (apigenin-7-rutinoside) in the fruits of *C. aurantium*.[217] Luteoline-7-rutinoside and diosmetin 7-rutinoside (diosmin) are also present in *C. limon*.[217,221,598] A particularly interesting case is that of the C-glycosylflavones present in certain *Citrus*[368,430] and in fig.[525] These compounds are examined in Section 4 below.

2. Flavanones and Flavanonols

These are two closely related classes of flavonoids characterized by a C_6-C_3-C_6 structure in which the 3-C chain is saturated and with an oxygen atom in the 4-position. Flavanones correspond to flavones whose double bond between carbons 2 and 3 is saturated, whence the name dihydroflavones which is sometimes used. Flavanonols have an hydroxyl in the 3-position and are also referred to as 3-hydroxyflavanones or dihydroflavonols. Full reviews of these compounds have been published.[59,61]

Because of the saturation of the bond between C-2 and C-3, flavanones possess an asymmetrical carbon (in the 2-position) and flavanonols, with their hydroxyl in the 3-position, have two centers of asymmetry, whence the existence of stereoisomeric forms. The most widespread configuration of natural forms is 2R:3R.

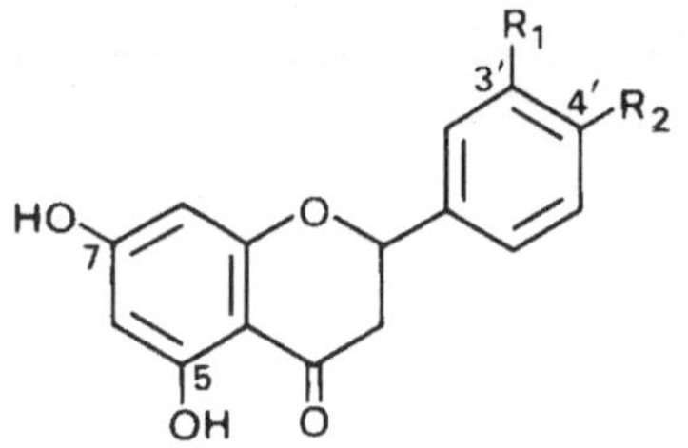

R_1 = H	R_2 = OH	: Naringenin
R_1 = H	R_2 = OCH_3	: Isosakuranetin
R_1 = OH	R_2 = OH	: Eriodictyol
R_1 = OH	R_2 = OCH_3	: Hesperetin

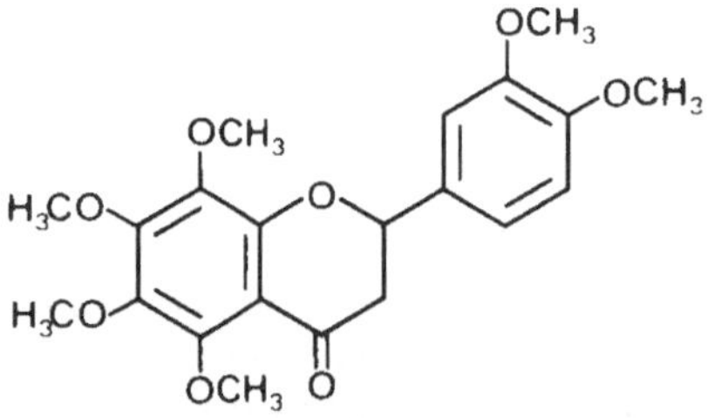

Citromitin

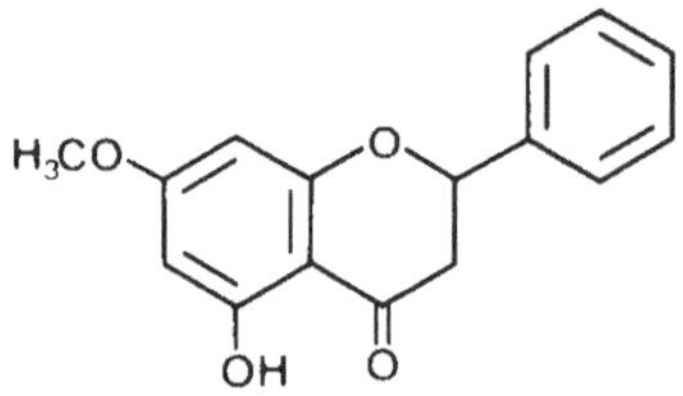

Pinostrobin

FIGURE 22. Chemical structures of fruits flavanones.

Flavanones and flavanonols are widely distributed in the plant kingdom but are very rarely present in fruits[59,61] except in *Citrus*.[258,301] We first examine flavanones in *Citrus* followed by flavanones in other fruits and finally flavanonols.

a. Flavanones in Citrus

The genus *Citrus* is characterized by substantial accumulation of flavanone glycosides at the expense of accumulation of more common flavonoids, flavonols, and anthocyanins.[59] Four aglycones are involved, all of them being 5,7-diOH: naringenin, eriodictyol, and their 4′-methylated derivatives isosakuranetin and hesperetin (Figure 22). These four flavanones, together with naringenin 4′-*O*-glucoside, are glycosylated in position 7 either by rutinose or by neohesperidose, which are diholosides formed by a rhamnose molecule and a glucose molecule and which only differ respectively in the type of bond: 1→6 or 1→2.

Flavanones in citrus fruits are of chemotaxonomic interest[5,294,489,566] and of considerable practical interest, in particular because of their relation to taste and bitterness[259,260,348] which are discussed in Chapter 4. In addition, the fungistatic and fungitoxic properties of citrus flavanones are well known.[301]

Each species of *Citrus* is characterized (especially in the fruits) by a particular flavanone glycoside pattern which can be separated by HPLC[290,430] (Figure 23). Concentrations of flavanone glycosides in juices from 52 *Citrus* cultivars were determined using HPLC[489] (Table 29).

In grapefruit *(C. paradisi)*, naringin (naringenin 7-neohesperidoside) is distinctly dominant, accompanied by narirutin (naringenin 7-rutinoside).[148,290] The naringin concentration in whole fruit varies from 170 to 280 mg/100 g FW.[184] A quantitative determination method using radioimmunoassay gives the naringin concentrations in flavedo and albedo; these are, respectively, 488 and 1165 mg/100 g FW.[40,289,290] Hesperidin (hesperetin 7-rutinoside) and neohesperidin (hesperetin 7-neohesperidoside) are also present in grapefruit juice, but their respective concentrations reach a maximum of 2% of total flavanones (Table 29).[489] In grapefruit juice, the average naringin content is 585 ppm for 3 years and 3 cultivars[6] with high variability (363 to 801 ppm). These values are comparable to those given by the statistical analysis of 6685 grapefruit juice samples[354] and to those given in Table 29.[489] After hy-

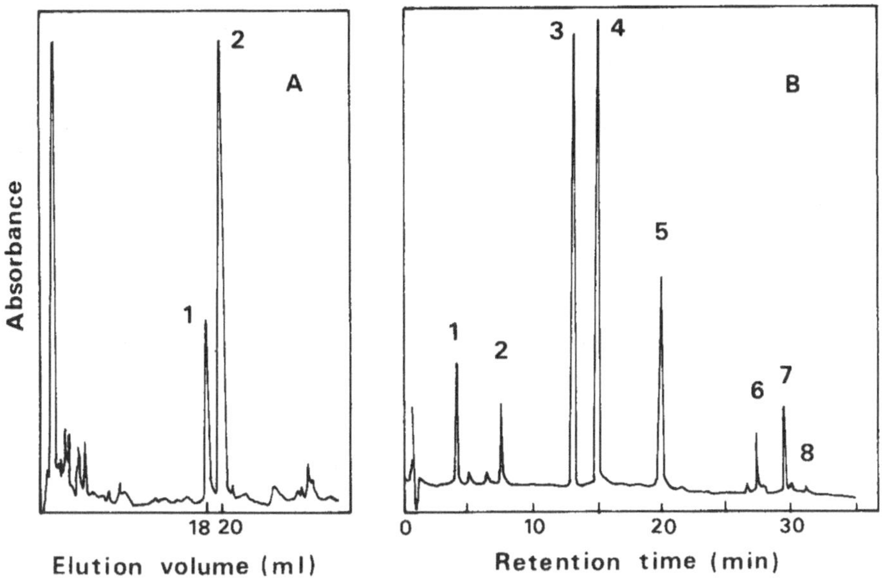

FIGURE 23. HPLC chromatogramms of Citrus flavanones. A: grapefruit extract; 1: narirutin; 2: naringin. B: Orange peel deoiled extract; 1: narirutin-4′-glucoside; 2: eriocitrin; 3: narirutin; 4: hesperidin; 5: isosakuranetin-7β-rutinoside; 6, 7, and 8: polymethoxylated flavones (sinensetin, nobiletin and tangretin). (Adapted from Jourdan, P. S., McIntosh, C. A., and Mansell, R. L., *Plant Physiol.*, 77, 903, 1985 and Park, G. L., Avery, S. M., Byers, J. L., and Nelson, D. B., *Food Technol.*, 37, 98, 1983.)

TABLE 29
Flavanone Glycosides in Juice of *Citrus* Species (Concentrations in ppm)

	Common names	Narirutin	Naringin	Hesperidin	Neohesperidin
Citrus aurantium	Sour oranges	PT[a]	133—362	PT	97—209
C. grandis	Pummelos	ND[b]	40—144	PT	ND
C. paradisi	Grapefruits	23—124	73—419	4—16	4—10
C. sinensis	Sweet oranges	18—65	ND	122—254	ND
	Tangor[c]	150—249	ND	80—191	ND
Hybrids	Tangelo[d]	17—100	ND	58—155	ND
	K-Early Tangelo	41	174	89	720

[a] PT = possible trace (small peaks 1—3 ppm at approximatively the correct retention time).
[b] ND = not detected (detection limit: 1 ppm).
[c] Tangor are orange × tangerine hybrids.
[d] Tangelos are tangerine × grapefruit or tangerine × pummelos hybrids.

Data from Rouseff, R. L., Martin, S. F., and Youtsey, C. D., *J. Agric. Food Chem.*, 35, 1027, 1987.

drolysis, in addition to naringenin which predominates very distinctly (88% of flavanones), very small amounts of isosakuranetin, eriodictyol and hesperetin have been revealed.[351]

Only 7-β-rutinosides are present in sweet orange *(C. sinensis)*; hesperidin is predominant, with a concentration of 270 to 600 mg/100 g FW in the whole fruit.[184] In orange juice, it is accompanied by narirutin (Table 29).[489] In "hesperidin complex" prepared by extracting ground, deoiled orange peel with water and alkaline pH, five 7-β-rutinosides were separated by HPLC[430] (Figure 23). Neither naringin nor neohesperidin could be detected in the juices of sweet orange cultivars, mandarin, and tangerine *(C. reticulata)*.[489] In contrast, sour orange juice *(C. aurantium)* contains neohesperidosides, naringin, neohesperidin, and neoeriocitrin (eriodictyol 7-β-neohesperidoside).[489] Naringin predominates in sour orange at 56 to 71% (Table 29). The presence of hesperidin has been reported in certain cultivars of *C. aurantium*.[59]

Pummelos *(C. grandis)*, close to grapefruits, are characterized by the almost exclusive presence of naringin in juice, whereas the juices of the hybrids tangor and tangelo have the same composition as sweet orange juice, with the exception of K-Early tangelo which contains four flavanone glycosides (Table 29).[489] Thus, naringin cannot be used as the sole indicator of the presence of grape fruit juice in orange juice. However, the neohesperidin concentration can be used to differentiate between grapefruit, sour orange and K-early tangelo juices (Table 29).

Lemon peel *(C. limon)* is characterized by two flavanone glycosides: hesperidin and eriocitrin (eriodictyol 7-β-rutinoside). 7-neohesperidosides are absent except in *C. limon f. ponderosa* where, however, hesperetin-7-neohesperidoside has been reported.[59] 7-Neohesperidosides of hesperetin, naringenin, and isosakuranetin (poncirin) have been revealed in *Poncirus trifoliata*.[59,217] Narirutin has been identified in *C. unshiu*.[368] Finally, two highly methoxylated derivatives, citromitin (5,6,7,8,3′,4′-hexamethoxyflavanone) and 5-demethylcitromitin are present in *C. mitis* and *C. reticulata* peel[59] (Figure 22).

b. Flavanones in Other Fruits

Data on flavanones in fruits other than *Citrus* are very fragmentary. In tomato fruit, naringenin is present in a free state[620] or in the form of naringenin 7-rutinoside.[254] It is abundant in the cuticle in both the free form and as naringenin 7-glucoside (prunin) accompanied by the corresponding chalcone, chalconaringenin.[272] At the postclimacteric stage, the naringenin content of the skin is 6.4 mg/100 g FW, i.e., 80 times higher than that of the pulp.[620]

Two naringenin derivatives acylated by *p*-coumaric acid were revealed in the fruits of *Mabea caudata* (Euphorbiaceae):[39] naringenin 7-*O*-β-(3-*p*-coumaroylglucoside) and naringenin 7-*O*-β-(3,6-di-*p*-coumaroylglucoside). Prunin-*O*-6″-gallat is present in *Acacia farnesiana* seed pods.[142] *Prunus cerasoides* seeds contain naringenin 4′-methylether-7-xyloside[521] but no information was given on fruits. Pinostrobin (5-hydroxy-7-methoxyflavanone) has recently been identified in the fruits of two species of *Piper*[90] (Figure 22).

c. Flavanonols (or Dihydroflavonols)

Precise analytical data on fruit flavanonols is very scarce. Dihydrokaempferol has been reported in the peel and endocarp of grapefruit *(C. paradisi)* after hydrolysis[351] but in much smaller amounts than naringenin. The dihydrokaempferol 4′-methylether-7-*O*-rhamnoside is present in pulp of mature fruit of *C. sinensis*.[100] Engeletin (dihydrokaempferol-3-*O*-rhamnoside) and astilbin (dihydroquercetin-3-*O*-rhamnoside) have recently been isolated and identified in the skin of white grape cultivars. The structures (Figure 24) were confirmed by ^{1}H-NMR and the absolute configuration was determined.[587] These two compounds, whose mean concentrations in skin extracts were 0.92 and 0.06 mg/100 g FW for astilbin and engeletin respectively, formed 4.74% of total phenolic compounds in grape skin.[530] They are also found in white wines.

3. Chalcones and Dihydrochalcones

Chalcones are characterized by a C_{15} skeleton with an open 3-C chain (Figure 25). Natural chalcones are all hydroxylated at nuclei A and B and there are numerous methoxylated derivatives. They play an important role in the biosynthesis of flavonoids (Chapter 3). Stereospecific isomerization of chalcones into flavanones is a reversible reaction[131,140,207] and the chalcone-flavanone-isomerase activity may be responsible for the absence of chalcones or the very small amounts detected in fruits except in special cases.

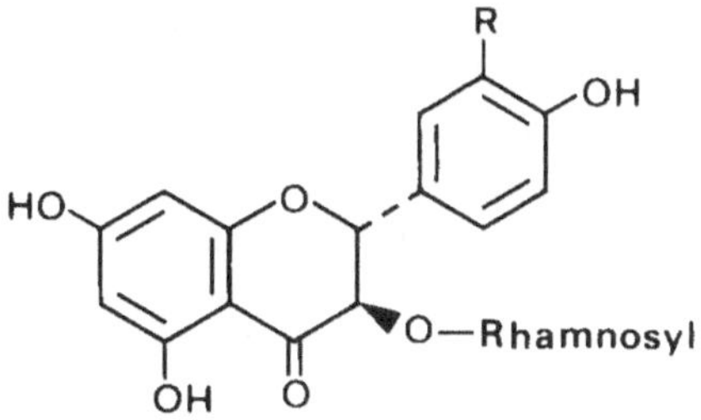

R = H : Engeletin
R = OH : Astilbin

FIGURE 24. Flavanonols present in *Vitis vinifera* berries. Engeletin = dihydrokaempferol-3-*O*-rhamnoside. Astilbin = dihydroquercetin-3-*O*-rhamnoside; configuration: 2R:3R.

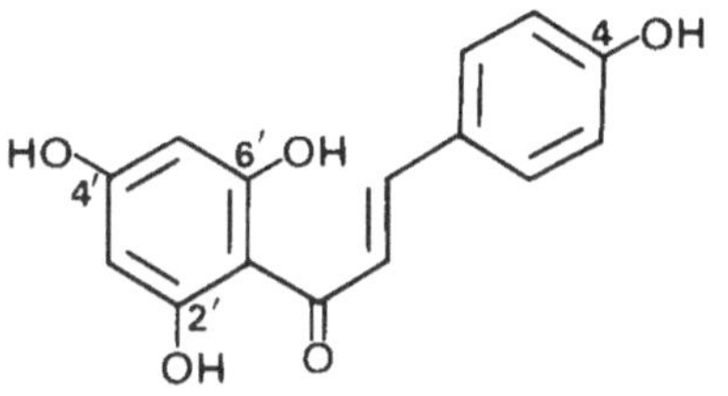

2',4,4',6'-tetrahydroxychalcone
= chalconaringenin

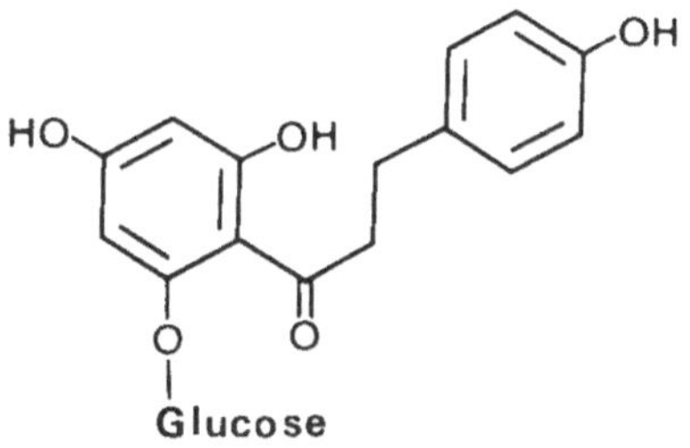

Phloretin-2'-glucoside
= phloridzin

FIGURE 25. Examples of chalcones (A) and dihydrochalcones (B) identified in fruits.

Although a large number of chalcones have been identified in flowers, wood, and bark, data on fruit chalcones are scarce.[60,61] Chalconaringenin (2',4,4',6'-tetrahydroxychalcone) has been identified in tomato cuticle by gas chromatography combined with MS.[272] It forms 29 to 37% of the major phenolic compounds in the cuticle of three tomato cultivars; the two other compounds were naringenin and coumaric acid.

Three chalcones are found in *Merrillia caloxylon* fruits (Rutaceae);[170,171] one of them (6'-hydroxy-2',4',3,4,5-pentamethoxychalcone) is one of the very rare chalcones to be trisubstituted at nucleus B. According to Bohm,[60] the essential oil of *C. limon* contains 4'-methoxy-chalcone but the result is presented as being tentative since the structure proposed was not discussed. The major flavonoid in pods of *Tephrosia pumila* (Leguminosae) has recently been identified as being praecansone-A, a β-oxygenate chalcone whose structure has been reconsidered.[117] The yellow chalcone okanin has been identified in the fruit of the Cyperaceous plant *Kyllingia brevifolia.*[219]

Transformation of the anthocyanin flavylium cation into the chalcone form has been studied on the basis of measurement of kinetic, thermodynamic, and spectrometric characters.[83] The chalcone forms are important in the question of the stability of anthocyanic extracts and in particular in the case of thermal degradation of anthocyanins, since the chalcone form is the result of an endergonic reaction which is considerably enhanced when the temperature rises.[84] Fruit extracts rich in anthocyanins can thus lead to the formation of chalcones. For example, after heat treatment of extracts of elderberry *(Sambucus nigra)* at pH 4.5, the chalcone forms which correspond to the flavylium cations of the four anthocyanins were separated by HPLC.[79] Likewise, six additional pigments were separated in *Vitis vinifera* berries; these are the chalcone forms of the anthocyanins.[102,103]

Dihydrochalcones are derived from chalcones by reduction of the double bond of the 3-C chain. All species of the genus *Malus* (Rosaceae) contain dihydrochalcones: phloridzin or phloretin 2′-*O*-glucoside in most species (Figure 25). As this compound is not present in neighboring genera, including *Pyrus*, this is of chemotaxonomic interest for the genus *Malus*.[97,634,635] It is classically stated that phloridzin is located in leaves, stems, and seeds but appears to be absent in fruits. For example, phloridzin is the most abundant phenolic compound in seeds of apple var. Calville blanc[109] whereas the fruits contain none. However, the presence of phloridzin was reported in methanol extracts of the skin of Democrat apple in 1966.[146] HPLC analysis of the press juice from mature apple (cv. Granny Smith) revealed the presence of a small amount of phloridzin and of another phloretin glycoside, perhaps phloretin-xyloside.[637] Likewise, phloridzin has been characterized by HPLC in apple juice of the cider variety Dabinett.[319] Contamination from seeds cannot be excluded in any of these cases, but recent data now appear to confirm the presence of phloridzin in apple skin, where it has been separated by HPLC and characterized by ^{1}H and ^{13}C-NMR.[127] It is accompanied by phloretin xyloglucoside.[419] Study was broadened to include eight apple cultivars grown in Canada and phloridzin contents in fruit skins varied from 87 to 330 μg/g at commercial harvest.[126]

A dihydrochalcone was revealed recently in the fruits of two species of Jamaican *Piper*.[90] It corresponds to pinostrobin, a flavanone which is also found in these two fruits. The fruits of *Iryanthera laevis* (Myristicaceae) were also found to contain new dihydrochalcones.[186]

It can be noted that the hydrogenation of the natural flavanones naringin and neohesperidin, the predominant bitter constituents of grapefruit and bitter orange, gave the corresponding intensely sweet dihydrochalcones. Taste-structure relationships were studied for some simple dihydrochalcones[137,348] and these aspects are discussed elsewhere (Chapter 4).

4. Minor Flavonoids

C-glycosylflavonoids are characterized by the fact that one or several sugar molecules are directly bonded to the flavonoid skeleton (generally a flavone) by a carbon-carbon bond which withstands conventional enzymatic or acid hydrolysis conditions. These compounds are fairly broadly distributed in the plant kingdom[104,105] and are found in Rosaceae, Vitaceae, Ericaceae, and Rutaceae, to mention only families of Angiosperms whose fruits are of economic importance. However, they have only been revealed with certitude in the fruits of *Citrus*[104,221,368,430] and in figs.[525]

HPLC was used to separate and identify four 6,8-diC-glycosylflavones in lemon peel extracts. They differed in the nature of the flavone: apigenin, luteolin, chrysoeriol, and diosmetin (Figure 26). These four compounds form only 7% of flavonoids, the remainder being made up of two flavanones.[430] 3,6-diC-glucosylapigenin was identified with five other flavonoids in the peel of *C. unshiu*.[368] Two apigenin isomeric C-glycosides called schaftoside and isoschaftoside were isolated from leaves and fruits of *Ficus carica* and identified by UV, ^{1}H-NMR, ^{13}C-NMR spectroscopy (Figure 26). They are much more abundant in leaves (42.3 ± 11.7 mg/100 g FW) than in fruits (0.53 ± 0.09 mg/100 g FW).[525]

Aurones are pigments which were first isolated in the yellow flowers of various species, especially of Compositae and Leguminosae, and then in other parts of plants. However, they have not been reported in fruits[60,61] except for aureusidin in *Gahnia clarkei*.[219] In addition, these compounds have generally not been identified in Angiosperm families with economically important fleshy fruits, such as the Rosaceae. In the present state of knowledge, it would, therefore, seem that with certain exceptions, aurones are not present in fleshy fruits, even though some publications have reported the presence of yellow pigments whose characteristics appeared to be those of aurones but which were not identified with accuracy. This is the case of the litchi for example.[439]

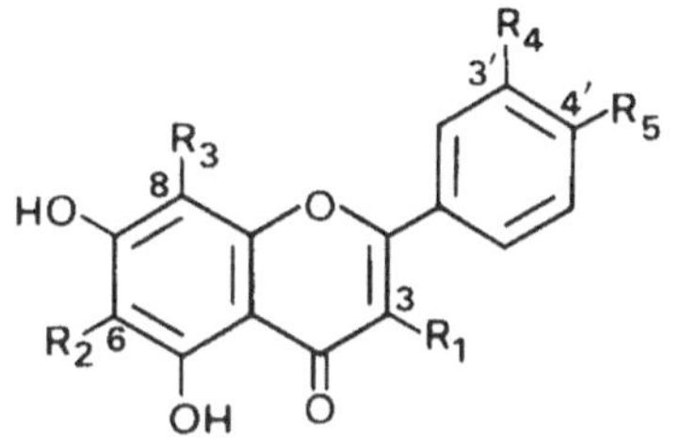

Citrus limon

$R_1 = R_4 = H$; $R_5 = OH$; $R_2 = R_3 = Glu$: 6,8-di-C-glucosylapigenin
$R_1 = H$; $R_4 = R_5 = OH$; $R_2 = R_3 = Glu$: 6,8-di-C-glucosylluteolin
$R_1 = H$; $R_4 = OCH_3$; $R_5 = OH$; $R_2 = R_3 = Glu?$: 6,8-di-C-glycosylchrysoeriol
$R_1 = H$; $R_4 = OH$; $R_5 = OCH_3$; $R_2 = R_3 = Glu?$: 6,8-di-C-glycosyldiosmetin

Citrus unshiu

$R_1 = R_2 = Glu$; $R_3 = R_4 = H$; $R_5 = OH$: 3,6-di-C-glucosylapigenin

Ficus carica

$R_1 = R_4 = H$; $R_2 = Glu$; $R_3 = Ara$; $R_5 = OH$: 6-C-glucosyl-8-C-arabinosylapigenin
$R_1 = R_4 = H$; $R_2 = Ara$; $R_3 = Glu$; $R_5 = OH$: 6-C-arabinosyl-8-C-glucosylapigenin

FIGURE 26. Examples of C-glycosylflavones identified in fruits. (Glu = glucosyl; Ara = arabinosyl).

Isoflavonoids are fairly numerous; they are represented above all by isoflavones but also by isoflavanones.[124,644] Taxonomic distribution is limited mainly to the Leguminosae Lotoideae where these molecules frequently appear as phytoalexins,[24] and it is probable that they are present in pods after infection (see Chapter 4). Two minor compounds in pods of *Tephrosia pumila* were recently identified as novel isoflavones and called pumilaisoflavone A and B.[117] Isoflavonoids are very rarely reported in fruits apart from in legume pods. Osajin and pomiferin, two isoflavones which contain two isoprenoid residues, contribute to the color of the Osaje orange *(Maclura pomifera)*.[219] Prunetin (isoflavone) and the corresponding isoflavanone have been reported in *Prunus puddum* but without reference to the fruit.[644]

The term **biflavonyl** was proposed by Harborne[217] to designate the dimer of apigenin, which is clearly distinct from other dimeric flavonoids such as proanthocyanidins. Since then, other types of **biflavonoid** have been discovered.[190,191] Their systematic distribution is limited (Gymnosperms and about ten families of Angiosperms) and they have not been reported in fruits, with the exception of several species of Clusiaceae (Guttiferae).[191]

Neoflavanoids are compounds whose taxonomic distribution appears to be limited to Guttiferae, Rubiaceae, and Leguminosae.[133] They are essentially derivatives with a 4-phenylcoumarin structure. They have been reported in the fruits of two species of Guttiferae: *Mammea americana,* where mammeisin was identified in peel,[133] and *Kielmeyera pumila* where mammeigin and isomammeigin have recently been identified.[405]

V. TANNINS

As discussed by Haslam,[234] it is always difficult to give a concise definition of the word tannin. It can, nevertheless, be stated that tannins are plant polyphenols which possess the

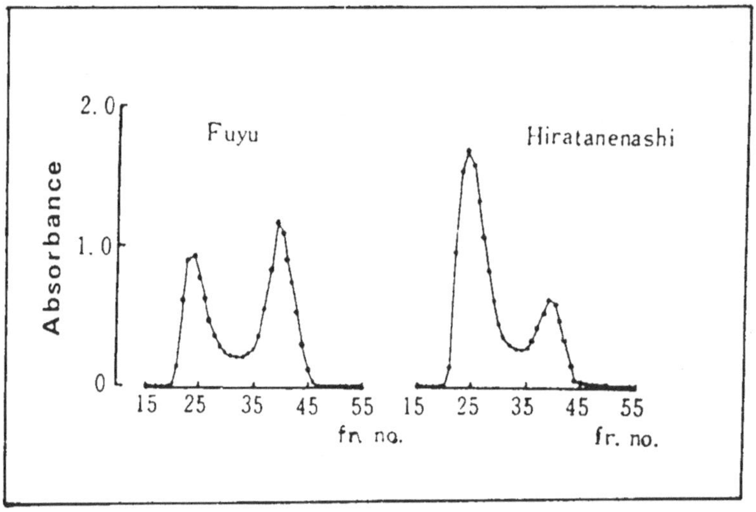

FIGURE 27. Separation pattern of aqueous acetone extracted tannins from "Fuyu" (PCNA cultivar) and "Hiratanenashi" (PVNA cultivar) by size exclusion chromatography on CPG-10 2000 A. (Adapted from Yonemori, K. and Matsushima, J., *J. Jpn. Soc. Hortic. Sci.*, 53, 121, 1984.)

property of precipitating proteins from aqueous media. The most acceptable major division of tannins is based on structural types, making it possible to separate the tannins into two classes:[234] hydrolyzable tannins and nonhydrolyzable or condensed tannins.

A number of remarks concerning the extraction, analysis, and assaying of tannins are made before examination of the distribution of the two groups in fruits.

A. PRELIMINARY COMMENTS ON THE ANALYSIS OF TANNINS

Tannins possess a great number of hydroxyl groups which give them the ability to bond reversibly with other natural macromolecules (polysaccharides, proteins, etc.) and with various metabolites, e.g., alkaloids. This bonding either occurs naturally during the physiological development of the fruit or during the decompartmentation which occurs when the fruit is crushed for technological processing or for the extraction of tannins themselves. The type of extraction solvent is, therefore, determined by three major objectives: (1) solubilization of tannins, (2) prevention or breaking of bonds with other compounds, and (3) avoiding of any condensation of native forms and in particular their transformation into phlobaphenes or anthocyanidins.

Little-polymerized forms of tannins are generally extracted using absolute methanol or methanol/water or ethanol/water mixtures containing up to 50% water.[44,233,370,488] However, the polymerized forms are not very soluble and extraction requires an acetone/water mixture (70:30 or 50:50 v/v) which breaks the phenol-proteins links. Polymerized forms of tannins in many fruits have been extracted and analyzed using this technique.[161,162,654,657] Nevertheless, plant tissues contain polymeric forms which are resistant to solubilization, and proanthocyanidins of this class often predominate over soluble forms.[239]

Extraction is generally followed by a purification phase in which a combination of more or less specific techniques may be used: affinity or adsorption chromatography in grapes,[414,415] gel filtration in the apple and other fruits,[116,235,316,637] exclusion chromatography (Figure 27) combined with ultracentrifugation in the Japanese persimmon,[654,657] successive precipitations combined with column chromatography in young fruits;[370] the latter approach resulted in kaki tannins whose degree of purity was evaluated at 95% by three complementary methods of analysis.[370]

Qualitative analysis of fruit tannins has progressed considerably over the past 20 years or so. Traditional techniques (various types of chromatography, colorimetric reactions, etc.) have been combined with chemical and physicochemical methods (chemical degradation, MS, ^{1}H and ^{13}C-NMR; IR, etc.) used more recently for the purpose. These approaches have made it possible to determine the structural characteristics not only of oligomer forms[233,234] but also of highly condensed polymers: oxidation of the B ring, stereochemistry of the heterocyclic ring, nature of the terminal unit of chains, number-average molecular weight.[45,116,149,161,162] These data also resulted in better understanding of the relationship which can occur between these polymers and other macromolecules like proteins and polysaccharides.[238,379,426,545]

A large number of detailed studies have been devoted to the problems raised by the quantitative estimation of tannins, in particular in food products; a recent review by Despandhe et al.[123] provides a critical analysis of the main methods used and their bibliographical references. Brief mention is made here of the principles of the major quantitative approaches which have been applied to fruits. They can be classified according to two main trends:

- First, global approaches for simultaneous evaluation of all the tannins or one of the two main categories of tannins: hydrolyzable or condensed.[44] The methods are generally rapid and sensitive but have two major disadvantages: they are not always specific to tannins alone, and the global result obtained masks the diversity of the chemical composition of the tannins present in the extract. In addition, the reference used for calibration curves is unavoidably arbitrary. Several of these methods have been widely used for assaying tannins in fruits or derived food products,[123] and in particular the transformation of colorless proanthocyanidins into red anthocyanidins in hot acid medium or their condensation with vanillin.[77,197,437,558] The precipitation of certain proteins, such as hemoglobin,[41,43] the reducing of enzyme activity,[426] or the application of reactions concerning total phenolics to tannins, and in particular the oxidation of hydroxyl groups by Folin-Denis' or Folin-Ciocalteu's reagents or even potassium permanganate[197,470,528,556,558] are also used. Use of the UV spectra extracts has also enabled rapid but global quantitative estimation of the tannins in persimmon.[189] Improvements for the process initially described have frequently been proposed;[77,356,437] they have made it possible for example to correct interference with the anthocyanins in red fruits in the use of the vanillin reagent.[77]
- Second, techniques which are based on the purification and separation of tannins and other phenolic constituents before any assaying. Among the different forms of chromatography,[262] HPLC analyses are now widely used for the separation and quantitative assessment of oligomeric forms, with either normal stationary phase[637] or reversed phases.[596] This technique has been successfully applied to tannins in apple and derived products,[317,319,637] in grape or wine,[72,488] and in various other fruits (see Table 4). Separation and purification techniques using column chromatography can also be used prior to assaying.[72,370]

B. HYDROLYZABLE TANNINS

Hydrolyzable tannins are complex polyphenols which may be degraded under hydrolytic conditions (either acid, alkaline, or hydrolytic enzymes) into simpler fragments, mainly sugars and some phenolic aicds. Hydrolyzable tannins are polyesters based on gallic acid and/or hexahydroxydiphenic acid and their derivatives (Figure 28); the two acids are usually esterified with D-glucose, leading to molecules with molecular weights of 500 to 2800 or more.[210,238] In rarer cases, esterification can take place with polyols, such as quinic acid in Tara gallotannin (Figure 28), glycosides, and other phenols. Depending on the products of hydrolysis, distinction is generally made[233] between gallotannins (which release gallic acid)

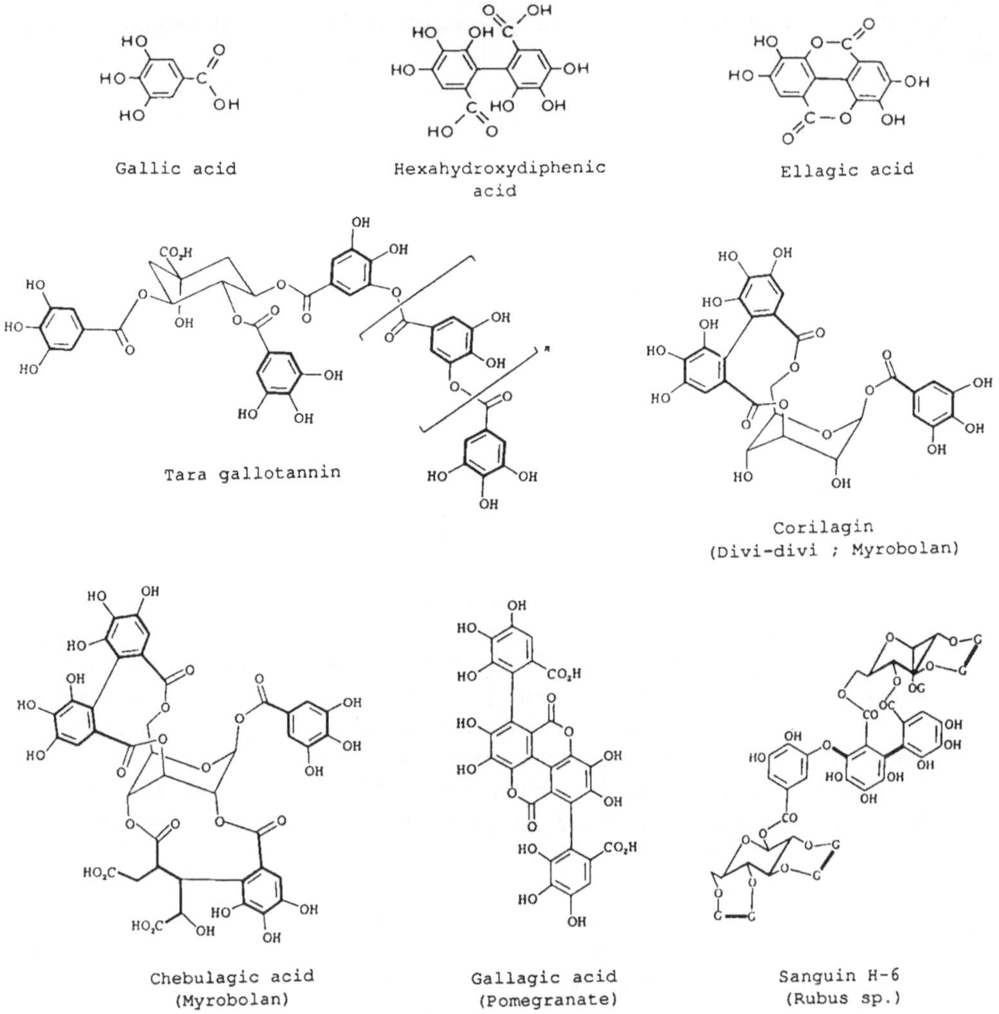

FIGURE 28. Components of hydrolyzable tannins found in fruits. (From Haslam, E., in *The Biochemistry of Plants,* Vol. 7, Academic Press, New York, 1981, 527; and Ozawa, T., Lilley, T. H., and Haslam, E., *Phytochemistry,* 26, 2937, 1987. With permission.)

and ellagitannins (which release ellagic acid from the lactonization of the hexahydroxydiphenyl groups initially present in the molecules). Some fruits are major commercial sources of gallo and ellagitannins, found in particular in the pods of Leguminosae (Table 30). The structure of numerous ellagitannins was reported by Haslam,[236] and several examples concerning fruits are shown in Figure 28. In pomegranate, gallagic acid is found esterified to glucose in punicalin and punicalagin;[375] the latter compound has also been reported in the fruit of *Rubus* sp. combined with the dimer sanguin H-6.[26]

C. PROANTHOCYANIDINS OR CONDENSED TANNINS

According to Haslam,[233,236] the term *proanthocyanidins* now means the condensed or nonhydrolyzable tannins, whereas the term *leucoanthocyanidins* is reserved for the monomeric flavan-3,4-diols.[625] They possess the general structure of more or less polymerized flavan-3-ols (Figure 29) in which the flavan bonds are most commonly C-4 to C-8 (or C-6). Fruit bearing plants have proved to be particularly rich sources of oligomeric procyanidins which generally occur unglycosylated and with one or both of the flavan-3-ols, (+)-catechin

TABLE 30
Fruits as Source of Hydrolysable Tannins of Commercial Importance

Ellagitannin	Gallotannin
Leguminosae	
Pods of *Caesalpinia coriaria* (Divi-divi) (corilagin)	Leguminosae
Pods of *Caesalpinia brevifolia* (Algarobilla) (brevilagin)	Pods of *Caesalpinia spinosa* (Tara)
Combretaceae	
Fruit of *Terminalia chebula* (Myrobolan) (chebulagic acid, terchebin; corilagin)	

Data from Haslam, E., *The Biochemistry of Plants*, Vol. 7, Conn, E. E., Ed., Academic Press, New York, 1981, 527.

or (−)-epicatechin (Figure 19). These monomer forms, which have no tanning properties, have been examined previously. Monomers with tri-hydroxylation in the B-ring, (+)-gallocatechin and (−)-epigallocatechin (Figure 19) can also contribute to the formation of condensed tannins in numerous fruits.[161] In addition, in *Vitis vinifera*, gallic acid can esterify (−)-epicatechin in position 3 to form (−)-epicatechin 3-*O*-gallate (Figure 19).[115,161,554,624] Exceptionally, the epicatechin unit in the polymeric fraction of ripe quince proanthocyanidin may be glycosylated.[436]

Both chemical and biochemical investigations first concentrated on simpler flavans, such as the monomers reported above or the four major dimeric proanthocyanidins B-1, B-2, B-3, and B-4 (Figure 29), and made it possible to identify their structure and stereochemistry, each procyanidin dimer having a preferred conformation.[236] Dimeric procyanidins have been obtained from fruits, fruit pods, and seeds or other organs in different kinds of plant, and simultaneous chromatographic analysis of mono, di, and oligomeric forms has made it possible to record "procyanidin fingerprints" which can be divided into four main categories.[233,236]

Oligomeric and polymeric forms of procyanidins are built up by the addition of further flavan-3-ol units, giving two helical structures[233,236] in which the central core of these polymers is composed of rings A and C of the flavan repeat unit; ring B projects laterally from this core, making possible bonds with proteins.[233] In fact, until 10 years ago, knowledge of the nature of fruit proanthocyanidins was limited to the isolation and structural characterization of the simpler flavans and the procyanidin dimers that co-occur with the polymer. However, the predominant stereochemistry of the proanthocyanidin units in the polymer is not always an accurate reflection of the structure of the dimers.[162] For example, the procyanidin polymer in strawberry consists mainly of *cis*-procyanidin units (the monomeric units being mainly of the epicatechin type) whereas the major procyanidin dimer in strawberry is B-3,[162,570] i.e., of the *trans* configuration (catechin type). In recent years, physicochemical techniques have made it possible for numerous researchers to obtain accurate data on the composition of tannin polymers.[115,116,160-163] According to these authors, condensed tannin polymers in fruits can be characterized by five main features (Table 31):

1. The stereochemistry of the constitutive units; data obtained by Foo and Porter[162] from 17 fruits clearly indicate that all polymers are based largely on 2,3-*cis* units (analogous in stereochemistry to [−]-epicatechin). The predominant occurrence of *trans* units is apparently somewhat rare, and although it has been reported in leaves of different *Ribes*, no polymer in which this stereochemistry is predominant has been isolated from the various fruits studied by Foo and Porter.[162] However, data obtained using acid and toluene-α-thiol degradations of tannins led to opposite results in banana, Japanese persimmon, carob bean, Chinese quince and loquat,[371] and in guava fruit and leaves.[283]

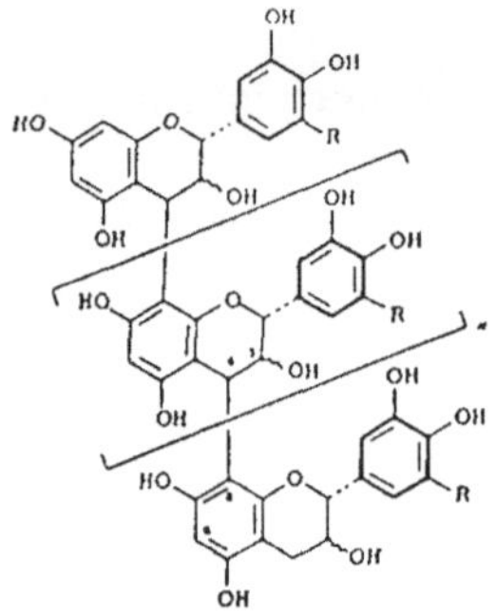

R = H : procyanidin units
R = OH : prodelphinidin units

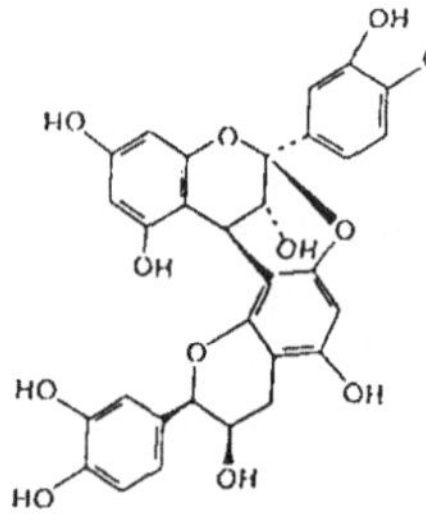

Proanthocyanidin A-2

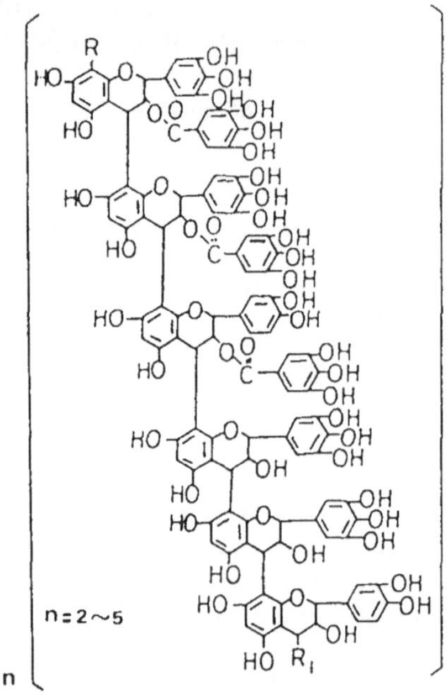

Kaki-tannin[a]

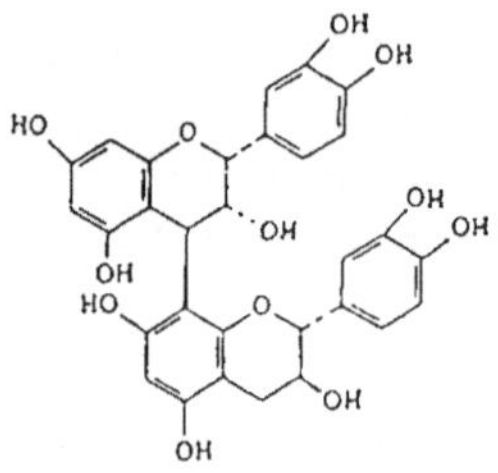

Procyanidin B-1

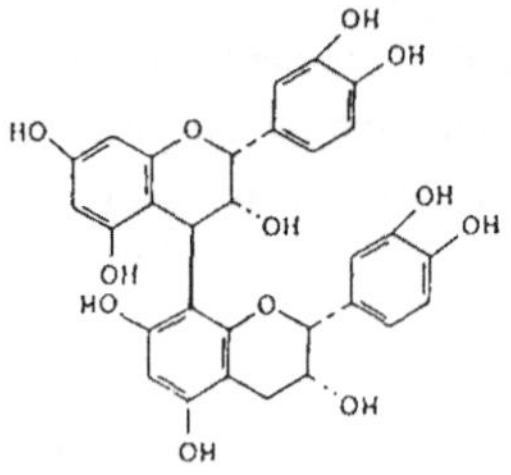

Procyanidin B-2

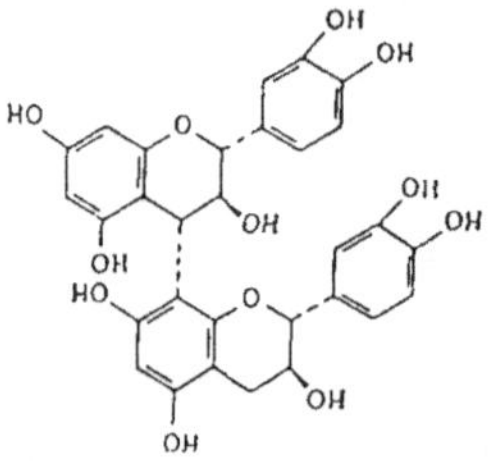

Procyanidin B-3

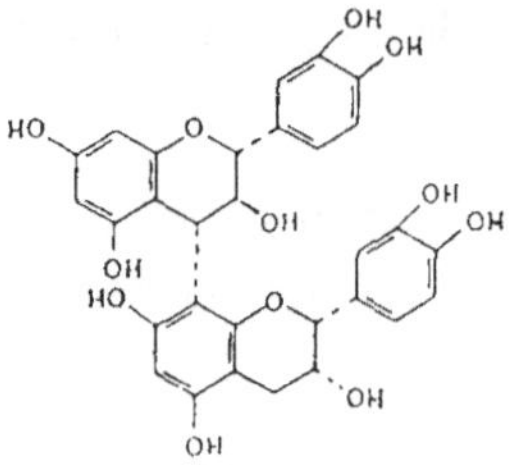

Procyanidin B-4

FIGURE 29. Chemical structures of condensed tannins. (From Foo, L. Y. and Porter, L. J., *Phytochemistry*, 19, 1747, 1980; and Haslam, E., in *The Biochemistry of Plants*, Vol. 7, Academic Press, New York, 1981, 527. With permission.) [a]Presumed structure of kaki-tannin. (Adapted from Itoo, S., in *Handbook of Fruit Set and Development*, CRC Press, Boca Raton, FL, 1986, 355.)

Furthermore, 2,3-*trans* configuration is always present in small amounts (approximately 5 to 25%) in condensed fruit tannins.[162]

2. The PC/PD ratio (procyanidin/prodelphinidin units) in the polymer (i.e., the ratio of monomeric units which are mainly of the (−)-epicatechin or (+)-catechin types to units which are mainly of the (−)-epigallocatechin or (+)-gallocatechin types). PC

TABLE 31
Structure of Fruit Tannins (Polymeric Forms)

Species	Family	Common name	Tannin[a] type	% of[b] hydrolysable tannins	Terminal[c,d] group (with %)	$\overline{Mn}$[c,e]	PC: PD[c,f]	*cis: trans*[c,g]	Class[c,h]	Comments
Actinidia chinesis	Actinidiaceae	Kiwi	C	—	e:c (78:22) (leaf)	2200	90:10	95:5	A	Ripe fruit
Aesculus x *carnea*	Hippocastanaceae	Horse chestnut	C	—	e:c (80:20)	2200	100:0	93:7	A	Presence of proanthocyanidin A-2[286]
Aesculus hippocastanum	Hippocastanaceae	Horse chestnut	C	—	e:c (90:10)	1750	100:0	97:3	A	Presence of proanthocyanidin A-2[286]
Chaenomeles sinensis	Rosaceae	Chinese quince	C	—	e:c (70:30)	1900	100:0	94:6	A	Unripe fruit
Cotoneaster serotina	Rosaceae		C	—	n.d.	n.d.	93:7	90:10	A	Unripe fruit
Crataequs oxycantha	Rosaceae	Hawthorn	C	—	n.d.	n.d.	100:0	94:6	A	Ripe fruit
Cydonia oblonga	Rosaceae	Quince	C	—	e:c (60:40)	3000	100:0	95:5	A	Unripe fruit[(i)]
Feijoa sellowiana	Myrtaceae	Feijoa	C	—	n.d.	n.d.	70:30	85:15	A	Ripe fruit
Fragaria ananassa	Rosaceae	Strawberry	C + H	37 (gt + et)	n.d.	n.d.	100:0	74:26	A	Ripe akenes cv. Redgauntlet
Malus pumila	Rosaceae	Apple	C	—	n.d	n.d	100:0	93:7	A	Ripe fruit cv. Granny Smith
Musa sapientum	Musaceae	Banana	C	—	n.d.	1900	78:22	95:5[j]	A	Unripe fruit
Ribes grossularia	Saxifragaceae	Gooseberry	C	—	n.d.	2700	63:37	77:23	C	Unripe fruit
Ribes nigrum	Saxifragaceae	Blackcurrant	C	—	e:c:egc (19:26:55)	large	40:60	74:26	C	Unripe fruit
Ribes rubrum	Saxifragaceae	Redcurrant	C	—	n.d.	2700	24:76	78:22	D	Unripe fruit
Ribes sanguineum	Saxifragaceae		C	—	c:egc (9:81)	2700	23:77	72:28	D	Ripe fruit
Rosa centifolia	Rosaceae		C + H	33	C(100)	n.d.	90:10	76:24	A	Ripe hips
Rubus fructicosus	Rosaceae	Blackberry	H (+C)	#100 (gt + et)	—	—	—	—	—	Slight amount of condensed tannins (B-4)

TABLE 31 (continued)
Structure of Fruit Tannins (Polymeric Forms)

Species	Family	Common name	Tannin[a] type	% of[b] hydrolysable tannins	Terminal[c,d] group (with %)	$\overline{Mn}$[c,e]	PC: PD[c,f]	*cis: trans*[c,g]	Class[c,h]	Comments
Rubus idaeus	Rosaceae	Raspberry	H (+C)	#100 (gt+et)	—	—	—	—	—	Slight amount of condensed tannins (B-4); unripe fruit
Vaccinium corymbosum	Ericaceae	Blueberry	C	—	e:c(30:70)	3500	100:0	94:6	A	Unripe fruit
Vaccinium oxycoccus	Ericaceae	Cranberry	C	—	n.d.	1750	63:37	77:23	C	Unripe fruit
Vitis vinifera	Vitaceae	Grape	C	—	e:c:egc:ecg (37:37:21:5)	3200	85:15	92:8	A	Unripe and ripe fruits cv. Gamay Beaujolais

[a] C: condensed tannins (proanthocyanidins); H: hydrolysable tannins.
[b] gt: gallotannins; et: ellagitannins.
[c] Concerns proanthocyanidins only.
[d] e: (−)-epicatechin; c: (+)-catechin; egc: (−)-epigallocatechin; ecg: (−)epicatechin 3-*O*-gallate.
[e] $\overline{Mn}$: number-average MW.
[f] PC: procyanidin units; PD: prodelphinidin units.
[g] *cis:* (+) epicatechin type unit; *trans:* (+)-catechin type unit.
[h] Classes as defined by Foo.[160]
[i] Three *O*-β-glucosides of procyanidin polymers have been reported in ripe fruit.[436]
[j] According to Matsuo and Itoo[371] banana tannins would only consist of catechin and gallocatechin.
n.d.: not determined.

Data from References 116, 160 to 162, and 436.

units are very common in fruit tannins and frequently predominate. In some cases, they are the only form of proanthocyanidin[371] (Table 31). PD units are more rarely predominant as in blackcurrant (60%) and redcurrant (76%). Nevertheless, relatively high proportions of PD units occur in the tannins in banana, cranberry, feijoa, kiwi, and grape.

3. The number-average molecular weight ($\overline{Mn}$), readily calculated from the ^{13}C-NMR spectrum of polymers.[115,116] There are no important differences between fruits; $\overline{Mn}$ values vary throughout the 2000 to 4000 range.
4. The nature of terminal units. It is comparatively rare to observe a single flavan-3-ol as a chain terminating units, and it may occur, as in the fruits of *Vaccinium corymbosum* for example, that the predominant terminal unit of a polymer is of opposite stereochemistry to that of the constitutive monomers (Table 31).
5. The nature of the linkage between constitutive units; interflavan bonds are broadly C-4 to C-8 or C-6 (type B proanthocyanidins), but the unusual double-linked dimer proanthocyanidin A-2 (Figure 29) may also be encountered, and trimers of similar structure in the fruit of *Aesculus x carnea* or *hippocastaneum*.[161,286] Nevertheless, proanthocyanidin A-2 does not participate in the formation of polymers in *Aesculus*[161] but has been used as a taxonomic marker[42] because of its scattered distribution.

Use of these data has made it possible to separate condensed proanthocyanidins into five main classes,[160] three of which are found in fruits (Table 31):

- Class A proanthocyanidins which contain mainly procyanidins whose monomers have a high proportion of *cis* configuration (epicatechin type). They are found in many fruits.
- Class C proanthocyanidins; these are less common and consist of approximately equal proportions of procyanidin and prodelphinidin units and have both *cis* and *trans* configurations
- Class D proanthocyanidins which contain mainly prodelphinidins (over 70%) with mainly the *cis* configuration. They are only found in fruits in *Ribes* fruits.

This classification calls for a number of remarks:

- Designed initially from the study of 26 proanthocyanidin polymers from 24 different species,[161] it will need to be extended to cover many other cases and, in particular, fruits or seeds in which the tannins are subject of prior, well-documented research but in which different analytical techniques were used (e.g., persimmon, sorghum, etc.).
- The boundaries between the different classes should only be considered as guide values; some polymers may obviously belong to several classes.
- It is probable that other classes in addition to the five defined by Foo[160] can be described, either in function of newly obtained characteristics of polymers or in function of different or complementary analytical approaches.
- Although it is interesting *a priori* from the chemotaxonomic point of view, this classification should be used with considerable caution since proanthocyanidin polymers in leaves or fruits (as in the case of *Ribes nigra* for example) may sometimes belong to two different classes.[160]

D. OCCURRENCE OF TANNINS IN FRUITS

The distribution of hydrolyzable tannins is relatively limited; they are found in highly evolved plant families,[42] such as the Leguminosae and the Rosaceae; in the latter family (Table 31) they are present in fruits of the genera *Fragaria* and *Rubus* which belong to the

Rosoideae, a developed subfamily of the Rosaceae.[162] Certain fruits may contain both hydrolyzable and condensed tannins, but hydrolyzable tannins predominate in *Rubus* in the form of mixed gallic and ellagic tannins with glucose as the sugar moiety.[162] In fact there are only small quantities of proanthocyanidins or condensed tannins in blackberry *(Rubus fructicosus)* and raspberry *(R. idaeus)* and none at all in boysenberry and bramble fruit *(Rubus cv. Aurora)*. Gallotannins are also the only type found in pistachio fruit.[67] Some fruits like the strawberry may contain almost equal proportions of the two groups of tannins, but the situation varies with the degree of maturation.[160] Numerous fruits (Table 31) in fact display complete absence of hydrolyzable tannins whereas condensed tannins are plentiful: *Actinidia, Crataegus, Musa, Malus, Ribes, Vaccinium, Vitis*, etc.

In the light of the remarks made concerning the diversity of extraction methods, the quantitative results found by different authors are fairly difficult to compare. In addition, many data only apply to qualitative aspects and reliable quantitative data are comparatively scarce, in particular for ripe fruits. We have grouped some such data in Table 2 and they are sufficient to make two general conclusions. First, as for the other groups of phenolic compounds, there is very great variation between species and between cultivars (e.g., comparison of the astringent and nonastringent varieties of Japanese persimmon). In addition, tannin contents are sometimes very high and can reach, for example, values of 2% or more of fresh matter in persimmon.[370] The organoleptic and technological consequences of these high concentrations are discussed in Chapters 4 and 5.

E. TWO SPECIAL CASES: GRAPE AND JAPANESE PERSIMMON

Numerous studies have been carried out on the tannins in certain fruits. A number of data are reported here on their structure in grape and Japanese persimmon.

1. Grape Tannins

These are condensed tannins[161,233,321,527,528] and have been studied either in the berry itself (including the seeds, which are particularly rich in tannin) or in wine. In the latter case (studied in detail in Chapter 5), the tannins are of mixed origin (berry, seeds, stalks). Data are given here, as far as possible, on the berry itself, but it is not always clear from the work of authors whether the seeds were analyzed with the fruit.

The lower mol wt flavans of grape[115,233,235,321,432,624] are formed of monomer units: (+)-catechin, (−)-epicatechin, (+)-gallocatechin, and (−)-epigallocatechin to which is added a gallate ester from (−)-epicatechin ([−]-epicatechin-3-*O*-gallate).[115,161,624] Traces of free gallic acid have also been reported,[115] but this was probably a product of the degradation of the gallate ester. Oligomeric procyanidins are represented by the four dimers B-1, B-2, B-3, and B-4, the first two being the most plentiful. The trimeric forms are probably C-1 and C-2[72,321,488] and tetrameric forms have not been characterized; they are certainly formed by a mixture of the possible combinations of dimers.[527] The presence of a 3-*O*-gallate ester of procyanidin B-2[115] does not appear to have been confirmed.

The flavan-3-ol and procyanidin oligomer contents in grapes can vary with the variety, the degree of maturity and the part of the berry studied.[72,115] Table 32 gives an idea of this variability for the skin of berries from nine varieties of *Vitis vinifera*;[72] procyanidins (separated and assayed by HPLC) are always present in greater quantities than monomers. The values obtained are also much higher in seeds, whereas there are only traces of flavan-3-ol derivatives in pulp, except in two varieties (Alexandre Bouchet and the hybrid Seibel 8357).

Variations may also occur in the relative proportions of polymeric forms. Procyanidin B-1 is generally the most abundant (Table 33); nevertheless, certain varieties display predominance of B-2[115] or equivalent B-1 and B-4 contents.[321]

The polymeric fraction of grape tannins is always substantial, and can form up to 25% of all polyphenols.[235,624] In addition, the proportion of polymeric forms is always high

TABLE 32
Monomeric Flavan-3-ols and Procyanidins (B-1,B-2,B-3,B-4, Trimers, and Tetramers) Contents (mg/100 g FW) in the Berry Skin from *Vitis vinifera* Cultivars

	(+)-Catechin and (−)-epicatechin	Procyanidins	Total
Alicante-Bouschet	2.34	6.42	8.76
Cabernet-Sauvignon	2.5	44.5	47.0
Carignane	9.7	55.2	64.9
Cinsaut	16.7	78.3	95.0
Grenache blanc	22.6	17.2	39.8
Grenache noir	9.5	55.0	64.5
Merlot	11.2	32.1	43.3
Mourvedre	14.0	64.9	78.9
Pinot noir	10.3	45.0	55.3
Average	13.3[a]	50.5[b]	64.0

[a] (+)-Catechin: 79%
[b] Percentages of different procyanidins are given in Table 33.

Data from Bourzeix, M., Weyland, D., and Heredia, N., *Bull. O.I.V.*, 669—670, 1171, 1986.

TABLE 33
Procyanidin Percentages in Grape Skin

Procyanidins	Percentages
B_1	34 (20—49)
B_2	8 (20—5)
B_3	6 (20—3)
B_4	5 (17—2)
Trimers	24 (41—9)
Tetramers	23 (40—14)

Note: Extreme values are indicated in brackets.

Data from Bourzeix, M., Weyland, D., and Heredia, N., *Bull. O.I.V.*, 669—670, 1171, 1986.

compared to simple flavan-3 ols and dimeric procyanidins.[235] The degree of condensation also differs, depending on whether tannins are extracted from skin (highly condensed forms) or from seeds (lightly condensed forms).[194a] The polymeric fraction (Table 31) is formed of a mixture of procyanidin (PC) and prodelphinidin (PD) units, with the proportion of PD being always relatively high — in the order of 15 to 20%.[161,162,235] Units with *cis* stereochemistry predominate in polymers (approximately 92%); the terminal groups of the chains are (−)-epicatechin, (+)-catechin, (−)-epigallocatechin, or (−)-epicatechin-3-O-gallate in proportions of about 37:37:21:5.[162] The solubility of proanthocyanidin polymers depends on their degree of polymerization. Mol wt polymers of 2300 to 2500 (n = 6, Figure 29) appear to be the most important in the winemaking process.[235,624]

2. Kaki Tannins

Tannins in Japanese persimmon *(Diospyros kaki)* are found in tannin cells whose size and number increase as the fruit ripens.[281,282,656] The structure of these tannins has been studied for a long time, but often using insufficiently purified fractions. Nevertheless, a partially purified tannin prepared by a K_2HPO_4 precipitation procedure from immature fruit made it possible to specify the chemical structure.[369,371] The results of acid and toluene-α-thiol degradation have shown that the tannin from an astringent type of persimmon consists of catechin, catechin-3-gallate, gallocatechin, and gallocatechin-3-gallate in the ratio 1:1:2:2 and an unknown terminal residue. Kaki tannin thus appears to be an heteropolymer belonging to the proanthocyanidin B group with a carbon-carbon interflavan linkage between the C-4 of one unit and the C-6 or the C-8 of another unit.[369] The four constituents may form a repeating chemical unit (Figure 29). The molecular weight of intact kaki tannin was estimated to be 1.12×10^4 (mol wt) and 0.52×10^4 ($\overline{Mn}$).

There are differences in the chemical characteristics of tannin between astringent and nonastringent type fruits.[654] The tannin substances from the second type lie mainly in a lower molecular weight range than those of the astringent type. In addition, they coagulate more slowly in the presence of acetaldehyde. Variation between the high and low molecular fractions have been observed during the growth and maturation of the fruit[657] and is related to changes in astringency (Chapter 3).

Free gallic acid and β-D-glucogallin are also present in young fruits.[282,406] The latter component is only found in astringent fruits and might be a good index for the differentiation of astringent and nonastringent varieties.[282]

VI. MISCELLANEOUS PHENOLS

In addition to the groups studied above, special phenolic compounds are found in certain fruits. A few examples of these are discussed.

A. PHENYLETHANOL DERIVATIVES

These are compounds whose molecules contain the C_6-C_2 structure corresponding either to 3,4-dihydroxyphenylethanol or to 4-hydroxyphenylethanol (tyrosol). They have been revealed mainly in fruits of the Oleaceae family and include the following substances:

1. Oleuropein, an heterosidic ester of elenolic acid with dihydroxyphenylethanol (Figure 30) which is clearly predominant among phenolic compounds in olive.[7-9] Oleuropein is responsible for the characteristic bitterness of this fruit (Chapter 4). Demethyloleuropein, a demethylated derivative of oleuropein, is also found in the pulp of ripe olive fruit.[447,448] The corresponding aglycones are released by acid hydrolysis whereas dihydroxyphenylethanol is obtained by basic hydrolysis.
2. Ligstroside, which is also an heterosidic ester of elenolic acid, but with hydroxyphenylethanol, has been identified in olive.[311] Oleuropein and ligstroside are phenolic secoiridoids which are thought to have molluscicide properties.[311]
3. Caffeoylglycosides of dihydroxyphenylethanol have been identified in olive (verbascoside, Figure 7)[159] and in the fruits of various species of *Forsythia*.[303]
4. 3,4-dihydroxyphenylethanol glycosides in fruits of *Ligustrum obtusifolium*,[302] of *Piper nigrum*,[599] and in olive juice.[601] The pulp of *Carica papaya* fruit contains 4-hydroxyphenyl-2-ethyl β-D-glucoside.[513]

B. AMINES AND AROMATIC AMINO ACIDS

Phenolic amides have been reported in numerous fruits and fruit juices and have been studied particularly closely in *Citrus*, as can be seen in the review by Maga.[347] In addition

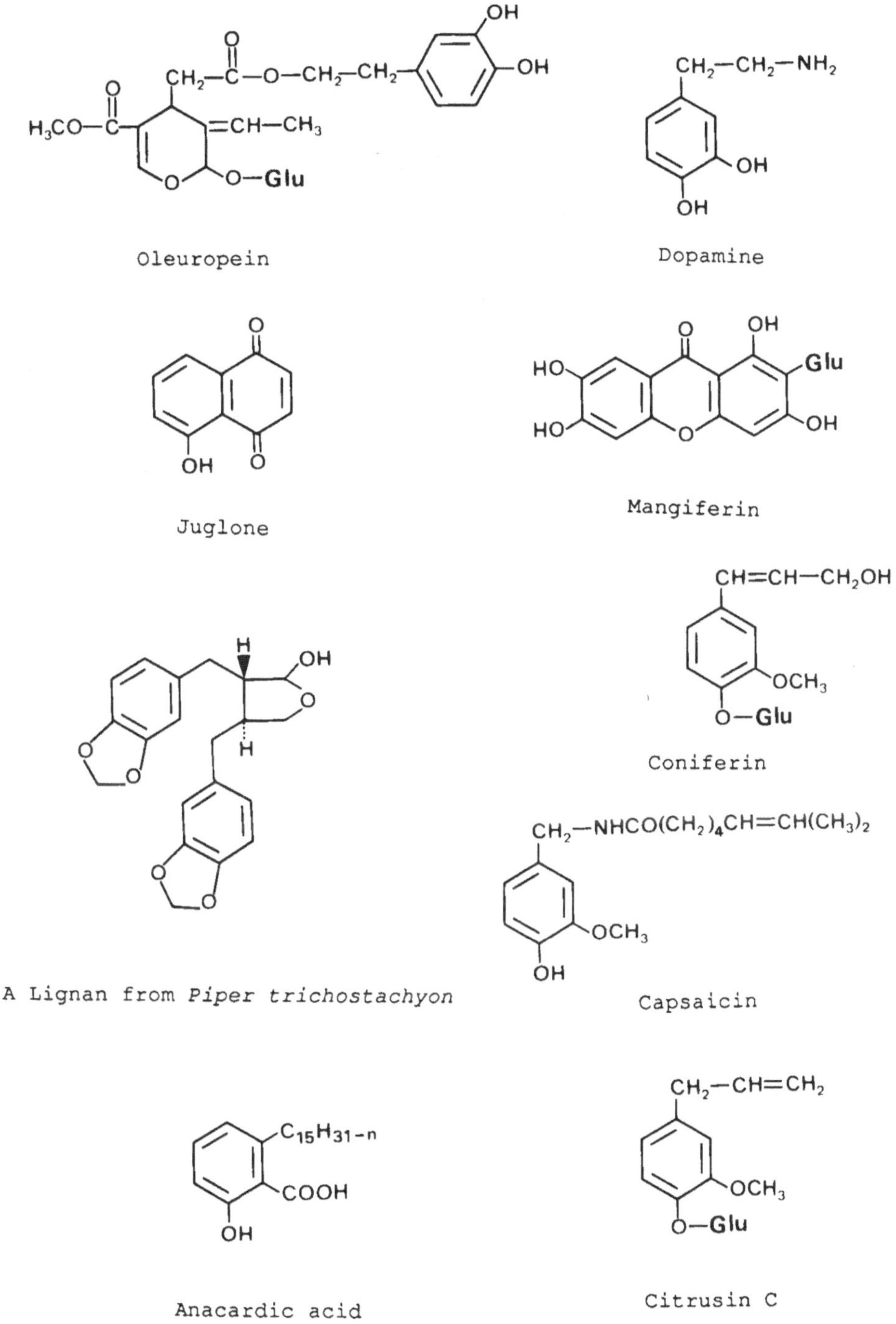

FIGURE 30. Chemical structures of miscellaneous phenolic compounds found in fruits.

to free phenolic amines (tyramine, methyltyramine, serotonin, dopamine, octopamine, hordenine, synephrine), forms combined with hydroxycinnamic derivatives are also found (see above). Contents vary with the variety and the part of the fruit studied.[347] Synephrine is abundant in orange but absent in grapefruit where, in contrast, feruloylputrescine is well represented.[627,628]

Dopamine or 3,4 dihydroxyphenylethylamine (Figure 30) is present in banana where concentration is high in the skin at 70 mg/100 g FW but only 0.8 mg in pulp.[427] It is the only major phenolic compound in peel in 16 banana clones studied. It is the substrate of polyphenoloxidases during enzymatic browning. Peel discoloration is a problem in the banana industry and the dopamine content is determined by genotype, being highest in *Musa acuminata* cultivars, intermediate in hybrids and lowest in *M. Balbisiana* itself.[427]

Dopamine and tyramine (4-hydroxyphenylethylamine) are aromatic amines metabolically related to the corresponding aromatic amino acids L-tyrosine and L-DOPA (3,4-dihydroxyphenylalanine). DOPA has not been detected in banana, and the normal pathway of dopamine biosynthesis is via tyramine. DOPA is found in broad bean, *Vicia faba*,[220] and in the form of glucosides in plant beans.[123]

C. QUINONES

Quinones form an important group of natural compounds because of their reactivity, their involvement in electron transport and their biological and pharmacological properties.[324] Most naturally occurring quinones are *p*-quinones whose structures are based on the benzo, naphto, or anthraquinone ring systems. It is fairly rare for them to be found in large quantities in fruits. We mention only juglone, 5-hydroxy-1,4-naphtoquinone (Figure 30), isolated from walnut shells *(Juglans regia)*. In addition to juglone, several compounds have been identified in the fruits of different species of *Juglans:* 1,4-naphtoquinone and two β-D-glucopyranosides, 4-hydroxy- and 4,8-dihydroxy-1-naphthalenyl-glucosides.[400] The metabolic relationships between the naphthalene derivatives and in particular the role of 1,4-naphthoquinone as an intermediary in the biosynthesis of juglone have been studied in *Juglans regia*.[398,399] Juglone is also found in other Juglandaceae: pecan nuts *(Carya illinoensis)* and hickories *(Carya sp.)*.[65,66,243,244]

Quinones may occur in a reduced state as hydroquinones. An example of this is arbutin, or hydroquinone-β-D-glucopyranoside, found in leaves and seeds of pear *(Pyrus communis)*, but which does not seem to be present in the fruit.[97,98]

Anthraquinones, which are anthracenic derivatives, are known mainly in the Rubiaceae. They have been isolated from leaves, stems, and roots but have only rarely been reported in fruits,[324] with the exception of pods.[4] Like other parts of the plants, extracts of berries of alder buckthorn *(Rhamnus frangula)* and buckthorn *(R. catharticus)* contain anthracenic heterosides which, together with tannins, are responsible for their laxative properties.

D. XANTHONES

Xanthones are compounds with a C_6-C_1-C_6 structure that is encountered in several families of Angiosperms and mainly in the Guttiferae and Gentianaceae.[94] Little work has been carried out on their presence in fruits:

1. Mangiferin, the 2-C-β-glucoside of 1,3,6,7-tetrahydroxyxanthone (Figure 30) has been identified in the pulp of three varieties of mango *(Mangifera indica)*.[494] This compound is abundant in leaves and bark, but its presence in mango fruit is controversial.[270] No trace of mangiferin was detected in either healthy or infected fruits.[193] Recent work on florets of *M. indica* shows that mangiferin, which is present as traces in healthy florets, increases considerably after infection with *Fusarium monoliforme*.[192] The presence or absence of mangiferin in mango may thus be related either to the variety analyzed or to the state of health of the fruit.
2. Several xanthones with isoprenoid substituents have been isolated from the fruit hulls of mangosteen *(Garcinia mangostana)*: mangostin, γ-mangostin, gartanins, and garcinones A, B, and C.[516,517] Two new xanthones, a *bis*-pyrano xanthone and 1-methoxy-2,4,5,-trihydroxyxanthone, have also recently been identified in this fruit.[27] Xanthones are known in other species of Guttiferae, particularly those of the genus *Kielmeyera*,

but analysis of the stems and fruits of *K. pumila* shows that these compounds are only located in the stem.[405]

E. LIGNANS AND NEOLIGNANS

Lignans are dimers, $(C_6\text{-}C_3)_2$, formed by the tail-to-tail linkage of two coniferyl or sinapyl alcohol units (Figure 30).[201,220] These compounds were revealed recently in the fruits of *Piper trichostachyon*,[308] *Schizandra chinensis*,[276,407] *Virola elongata*,[298] and *Forsythia* sp.[303] Related dimers called neolignans can be formed by other condensation between two $C_6\text{-}C_3$ units, for example by head-to-tail linkage. Numerous neolignans, with complex structures have been identified recently in the fruits of *Aniba riparia*,[36] *Licaria armeniaca*,[35] and *Virola elongata*.[298] The species for which lignans or neolignans have been revealed in fruits belong to a small number of families of Dicotyledoneae: Lauraceae, Myristicaceae, Oleaceae, Piperaceae, and Schizandraceae. Work on these natural molecules which are frequently of pharmacological interest should develop in years to come. Finally, it can be mentioned that in fruits of milk thistle *(Silybum marianum)*, several compounds with an antihepatotoxic action form a new class of compounds called flavonolignans since they are formed by a combination of a coniferyl alcohol unit and a flavonoid.[559,608-610]

F. OTHER MISCELLANEOUS PHENOLICS

1. Coniferin and Other Phenylpropanoids

Conferin has been identified from the peel of *Citrus sinensis* cv. Valencia, *C. paradisi* cv. Duncan, and *C. paradisi x C. reticulata* cv. Murcott.[565] It is a β-D-glucopyranoside of coniferyl alcohol (Figure 30) with astringent properties. Both syringin (glucoside of syringyl alcohol) and coniferin are found in peel of *C. hassaku* and *C. sinensis*.[502]

Fruits may also contain phenylpropenes, such as eugenol which is present in the form of a glucoside, (1-[4-glucosyl-3-methoxyphenyl]-propane-2-ene or citrusin C), in *Citrus* peel (Figure 30).[502] A new phenylpropene derivative was recently isolated from the fruits of Jamaican *Piper* species (*P. aduncum* and *P. hispidum*).[90] Like myristicin in nutmeg *(Myristica fragrans)*,[220] this molecule possesses a methylenedioxy heterocycle.

The three styryl-α-pyrones isolated from green fruits of *Aniba riparia*, Lauraceae,[36] can be included in the phenylpropanoid group. This was the first report of the co-occurrence of representatives of styrylpyrones and neolignans in the same species.

2. Volatile Compounds

A small number of phenolic compounds are volatile and contribute to the aroma of certain fruits. This is the case for example of vanillin (an aldehyde corresponding to vanillic acid at $C_6\text{-}C_1$) in vanilla pods *(Vanilla planifolia)* and eugenol, which is found with vanillin in banana. These compounds and their role in fruit aroma are discussed in Chapter 4.

3. Capsaicin

Capsaicin is the irritant (pungent) element in fruits of various species of *Capsaicin* (Solanaceae), including *Capsicum annuum*.[113,493] The molecule possesses an amide bond between an unsaturated fatty acid at C_9 and an aromatic amine at $C_8\text{-}C_1$ (Figure 30). It is also present in the vacuole.[177] Dihydrocapsaicin and other alkaloids of this group are also common in *Capsicum*.[113,493]

4. Phenolic Compounds in the Anacardiaceae

The Anacardiaceae contain special phenolic compounds whose substituent on the phenyl nucleus is a 13-C, 15-C, or 17-C carbon chain which is a saturated or unsaturated fatty acid. The fruits shell of cashew nut *(Anacardium occidentale)* contains principally anacardic acid, a derivative of salicylic acid with a 15-C side chain (Figure 30), with n = 0, 2, 4, 6). In addition, small quantities of cardol (resorcinol with a 15-C side chain) and traces of C_{17} component are also found.[588] C_{13} anacardic acids predominate in outer green shell of pistachio nut *(Pistacci vera)*: saturated (13:0, 46% of total anacardic acids) and *cis*-monoenoic (13:1, 17.4%).[652] These phenolic molecules with an unsaturated lateral chain are irritant when in contact with skin. For example, urushiol is the vesicant principle of poison ivy *(Rhus toxicodendron)*, which is well known in North America.[220] The poison is distributed throughout the plant, with the leaves being the most dangerous part.

In conclusion, it should be remembered that this review of miscellaneous fruit phenols is not exhaustive since other phenolic compounds may be present in fruits. For example, phlorin, the glucoside of phloroglucinol, has been reported in *Citrus* peel[220] and several acylresorcinols were identified in the fruits of *Virola*.[229] Likewise, pulverin, a new chromone, was reported from the fruits of *Neochamaelea pulverulenta*.[199] Other examples of phenolic molecules, such as resveratrol, a stilbene in C_6-C_2-C_6, are phytoalexins and are discussed in Chapter 4. Very closely related molecules may, nevertheless, be present in healthy grape bunches.[433]

VII. COMPARTMENTATION OF PHENOLIC COMPOUNDS IN FRUITS

Spatial compartmentation is one of the most important regulatory features in the secondary metabolism.[330] The distribution of phenolic compounds in fruits at *tissue* level can be revealed qualitatively in a very simple manner by using staining reagents such as vanillin-HCl or ferric chloride on slices of fruits. This demonstrates for example that the bulk of the polyphenolic compounds are found in the skins, seeds, and conduction tissue of grape berries.[240]

Cellular localization is based on microscopic observations which can be made without any particular treatment of the fruit either by direct observation of vacuoles colored red by anthocyanins or by epifluorescence after excitation under UV light,[606] as was used in studies of cell walls in wounded tomatos.[152] However, a number of preliminary treatments improve observation. In the Japanese persimmon,[655,656] the natural coagulation of tannins in cells was observed by staining with ferric chloride. Likewise, the size of the tannin cells can be observed on a light micrograph owing to the browning which results from reaction of coagulated tannin with EDTA. Furthermore, the morphology of the tannin cell can be observed by epifluorescent illumination after staining with a fluorochrome and scanning electron microscopy after coagulation of tannin by fixatives. A novel approach is the use of immunological techniques for *in situ* flavonoid localization,[275] but it has only rarely been applied to fruits.[40,289,290]

There are few quantitative data on the sites of accumulation in cells or tissues specific to phenolic compounds, mainly because of the difficulties encountered in isolating individual tissues free of cross-contamination.

A. SUBCELLULAR LEVEL

Phenolic compounds accumulate at two major sites in plant cells: the *cell walls*, where lignin is deposited[206] and also simpler molecules, such as flavonoids[99] and esterified ferulic acid,[173] and the *vacuole* where various soluble phenolic compounds and derivatives accumulate.[69,208,367,456,606,631] As noted above, in fleshy fruits, lignin is found only in stones and in certain cells of pear pulp or is formed as a reaction to wounding.[152,194] However, as in all parts of plants, the vacuole in fruit cells forms the main compartment in which phenolic compounds accumulate.[456] The destruction of this compartmentation sometimes causes deterioration of the fruit during transportation or storage. Vacuoles have been obtained from protoplasts of fruit cells only recently; this has confirmed that practically all the soluble

TABLE 34
Subcellular Content and Distribution of Phenolic Compounds from Apple Fruit Flesh

Subcellular Content (μg/10⁵)		Distribution		
			%	m*M*
Protoplasts (P)	891	Free space	3	5.5
Vacuoles (V)	965	Cytoplasm	—	—
P/V	1.08	Vacuole	97	108

Data from Yamaki, S., *Plant Cell Physiol.*, 25, 151, 1984.

TABLE 35
Anthocyanins, Flavonol Glycosides, and Hydroxycinnamic Acid Esters Composition of DeChaunac Grape Berry Subepidermal Tissue and Vacuoles

	Concentration		
	Subepidermal tissue	Vacuoles	
	μmol · g^{-1} fresh wt	pmol/vacuole	*M*
Anthocyanins			
Delphinidin 3,5-diglucoside	0.352	1.36×10^{-1}	8.00×10^{-3}
Cyanidin 3,5-diglucoside	0.225	8.15×10^{-2}	4.79×10^{-3}
Petunidin 3,5-diglucoside	0.733	3.81×10^{-1}	2.24×10^{-2}
Malvidin 3,5-diglucoside + cyanidin 3-glucoside	1.23	6.37×10^{-1}	3.75×10^{-2}
Petunidin 3-glucoside	0.254	4.29×10^{-2}	2.52×10^{-3}
Malvidin 3-glucoside	0.0898	1.64×10^{-2}	9.65×10^{-4}
Flavonol glycosides			
Quercetin 3-rhamno glucoside	0.0503	1.22×10^{-3}	7.18×10^{-5}
Myricetin 3-glucoside	0.0132	6.35×10^{-4}	3.73×10^{-5}
Quercetin 3-glucoside	0.0240	4.39×10^{-3}	2.58×10^{-4}
Quercetin 3-glucuronide	0.0994	1.80×10^{-3}	1.06×10^{-4}
Hydroxycinnamic acid esters			
p-Coumaroyl tartrate	0.222	6.88×10^{-2}	4.05×10^{-3}
Caffeoyl tartrate	0.757	2.81×10^{-1}	1.65×10^{-2}

From Moskowitz, A. H. and Hrazdina, G., *Plant Physiol.*, 68, 686, 1981. With permission.

phenolic compounds in fruits accumulate there. Thus 97% of the phenolic compounds in apple cells accumulate in vacuoles, where the concentration is of the order of 108 m*M*[653] (Table 34). The same applies to the vacuoles of grape berry subepidermal cells in which the anthocyanin concentration is approximately 97 m*M*.[397]

Although it is very clear that anthocyanins are accumulated in fruit vacuoles, some authors have shown the presence of anthocyanoplasts inside and outside vacuoles[431,538] and it is suggested that these structures may be the subcellular site of anthocyanin biosynthesis. In particular, dark grains were identified as anthocyanoplasts in the vacuoles of the skin of mature 'Pinot noir' grape berries; a Raman microspectrometer was used to obtain vibrational spectra of *in situ* anthocyanin pigments.[384] Thus, in the dark grains of the inner cells of skin, anthocyanins are mainly in the neutral quinonoidal base form whereas in the outer cell vacuoles anthocyanins are found essentially in the flavylium cationic form.

In grape berry, the vacuolar concentration of hydroxycinnamic esters approaches that of the anthocyanins and that of flavonol glycosides is about two orders of magnitude (Table 35). The vacuole is also the site of intracellular accumulation of capsaicin in *Capsicum* fruit.[177]

Isolated vacuoles do not contain the main enzymes of the flavonoid pathway.[267,607] However, a transferase responsible for the biosynthesis of sinapoyl-L-malate has recently been revealed in vacuoles from cotyledons of *Raphanus sativus*.[518,553] It is possible that biosynthesis might be located at an endoplasmic reticulum bound multienzyme complex; in this case, phenolic compounds would be transported into the central vacuoles as ER-derived vesicles.[266,607] The presence of phenylpropanoids and flavonoids in plastids, not reported in fruits but found in other parts of plants,[376,501,626] is contradictory and the localization in chloroplasts could have been caused by contamination during extraction.[266]

Although subcellular distribution of soluble phenolic compounds thus appears to be essentially vacuolar, distribution in the various fruit tissues is far from uniform.

B. TISSUE LEVEL

In most cases, soluble phenolic compound levels are higher in the external tissues of fleshy fruits (epidermal and subepidermal layers) than in internal tissue (mesocarp, pulp). Apple peel thus contains 10 times as much total phenols as flesh[273] and pear peel contains 27 times as much.[438,453] The figures are 15 and 50 for avocado[438] and mango,[1] respectively. However, three types of distribution can be observed (1) certain phenolic compounds are strictly located in a particular tissue and can thus be considered as being specific to this part of the fruit; (2) compounds are identical, but the concentrations differ according to the part of the fruit; (3) finally, certain compounds may be specific to a tissue or to a certain part at a precise physiological stage of fruit life.

It can thus be considered that there are two types of compartmentation, either in terms of space between the different parts of a fruit or in time between the various physiological stages. For reasons of clarity, only the first type is discussed here.

No flavonol glycosides are detected in many fruits after removal of the skin[583-586,593] as these compounds are mainly or even solely located in the outer part of fruits or in the epicarp. In apple cv. Calville blanc for example, the quercetin glycosides content is 17.8 mg/100 g FW in the external zone comprising the epiderm and several subepidermal layers, whereas it is less than 4 mg/100 g in the other parts of the fruit.[342] Levels of quercetin derivatives (apple, pear, tomato, and grape), isorhammetin (cherry, grape), and kaempferol (grape) are always low or nil in the internal tissues.[103,150,342,381,382] The disappearance of these compounds from external tissues following various types of traumatism (bruising, wounding, etc.) can result in the deterioration of the pulp of fleshy fruits;[584] it would seem that flavonols serve as an epidermal filter, in particular with regard to UV light.[479] In addition, flavonol contents may be totally different depending on the direct exposure of the fruit to sunlight. Thus, quercetin glycosides are twice as plentiful in the side of Calville blanc apple exposed to sunlight as on the shaded side.[342]

Anthocyanins may be distributed throughout the whole fruit as in strawberries, currants, and raspberries, but they are frequently located in the external tissue (epiderm and a few subepidermal cell layers). Both cases exist in grape: anthocyanins are located in the skin of most cultivars with the exception of teinturier grapes which have colored pulp.[471,650] In cv. De Chaunac, which is a complex interspecies *Vitis* hybrid, anthocyanins are present in subepidermal layers and not in the epiderm.[397] The first subepidermal layer contains dark red-colored vacuoles whereas the next two cell layers contain significantly smaller amounts of the pigments. Anthocyanins are very frequently distributed in different proportions between skin and pulp, as has been shown in different teinturier grape cultivars for malvidin or peonidin derivatives (Table 36). Similar results were obtained in the tamarillo fruit, where the major pigments are delphinidin 3-rutinoside in the jelly surrounding the seeds and cyanidin 3-rutinoside in the skin.[646] In the pomegranate *(Punica granatum)*, cyanidin 3 and 3,5-glucosides predominate in the seed coats and pelargonidin 3 and 3,5-glucosides predominate in peel whereas delphinidin 3 and 3,5-glucosides were only detected in seed coats.[135] Un-

TABLE 36
Comparison of Anthocyanin Percentages in Skin, Pulp, and Juices of Teinturier Grapes

	Peonidin 3-glucoside (%)			Malvidin 3 *p*-coumaroylglucoside (%)		
Cultivars	Skin	Pulp	Juice	Skin	Pulp	Juice
Tinta Santarem	23—36[a]		41—52	16—19		2
Viera Nativitade	27		41—76	8—14		1—3
Alicante Bouschet	12—23	65.5	45	18—42.7	8.1	17
Souzào	6		10	13		4
Gamay Freaux	14.6	63.2		61.5	23.3	
Gamay de Bouze	17.5	76.5		63.5	9.6	

[a] Values for 2 different years.

Data from References 26 and 483.

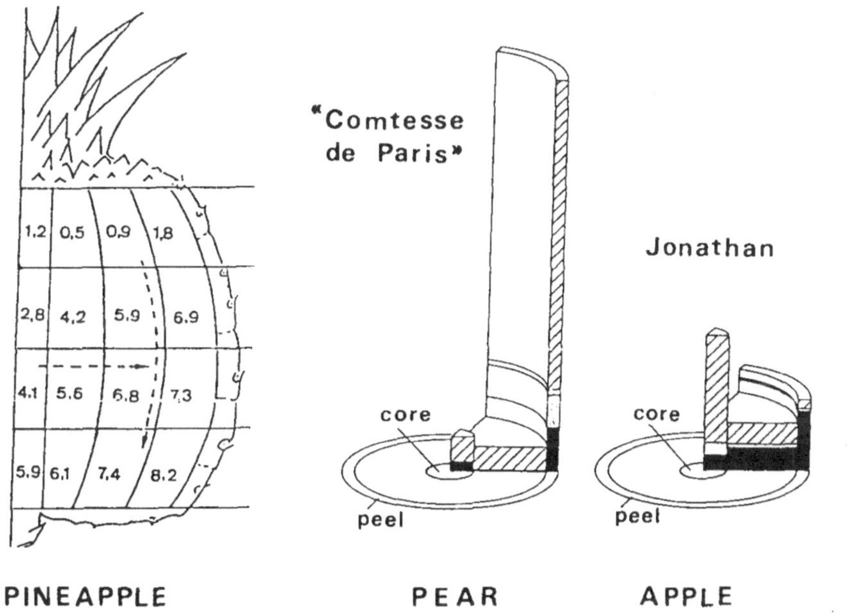

FIGURE 31. Tissue compartmentation of phenolic compounds in fruits. Pineapple: di-*p*-coumaroylquinic acid in $g \cdot g^{-1}$ FW. (From Dieudonné, G., Thèse Doct. Ing., Université Pierre et Marie Curie, Paris, 1977.) Pear and apple: ▨ caffeoylquinic, ☐ *p*-coumaroylquinic, ▩ *p*-coumaroylmalate and ■ catechin. (Adapted from Risch, B. and Herrmann, K., *Z. Lebens. Unters. Forsch.*, 186, 225, 1988.)

common distribution has been observed in blood oranges where anthocyanins are found only in the pulp.

Catechins and tannins are also frequently more abundant in external tissues. In the pear[475] and apple,[324,475] catechin and epicatechin contents are greater in the skin than in the rest of the fruit (Figure 31). Similar distribution has been observed in plum[552] and in fruits of nine red grape cultivars[72] where 14% of catechins are located in skins and only 1% in pulp; the largest proportion (64%) is found in seeds. Distribution of oligomeric procyanidins is practically the same (21% in skin, 1% in pulp, and 57% in seeds). In skin, procyanidin B-1 is the most abundant (34%) followed by the trimers (24%), the tetramers (23%), and

TABLE 37
The Compartmentation of Caffeoylquinic Esters in Fruits

Fruits	"External tissue"	"Internal tissue" (pulp)[a]	Ref.
Apple (mg/100 g FW)	Peel:		
Calville blanc	19.1	15.1	342
Jonathan	3.7	8.9[b]	475
Cherry (mg/g DW)	Epicarp:		
Green	36.5	11.7	381
Pink	3.1	1.5	
Pear	Peel:		
Passe-Crassanne (mg/g DW)	2	1	54
Comtesse de Paris (mg/100 g FW)	133.4	9.9	475
Tomato (mg/g DW)	Pericarp:		
Cherry			
Green	1.45	3.7	158
Ripe	0.48	0.62	
Eurocross BB (Mature green)	0.26	0.74	620
Grape (μmol/g FW)	Skin: 0.2—2.5	0.03—0.35	70

[a] Except core.
[b] 456 in the core.

finally the other dimers B-2, B-3, and B-4 (5 to 8%). This distribution is also found in white grapes where catechins and proanthocyanidins predominate in the skins, conducting tissue, and seeds.[240,528] In comparison with the fruit, seeds are generally very rich in tannins, particularly in grape[488] and the consequences of this distribution can be important during winemaking (Chapter 5). Very uneven proanthocyanidin distributions are also found in banana, where the concentration is 3 to 5 times as higher in skin as in pulp.[427] The highly vascularized internal zones are also rich in banana and Japanese persimmon,[281,282,427] as is tissue close to the stone in plum.[558] In ripe oleaster fruit (*Elaeagnus multiflora* var. *gigantea*), proanthocyanidins are more abundant in the stone than in pulp, but the levels are more or less the same in the two parts of the fruit of *Elaeagnus umbellata*.[495] In addition to tissue compartmentation, tannins are sometimes stored in large specialized cells with special morphological characteristics as in persimmon fruit[655,656] or in the latex vessels in the pulp and skin of banana.[164] Study of these "tannin cells" in Japanese persimmon fruit showed that the morphology of the walls of these cells displayed changes associated with the deastringency process.[655,656]

The highest levels of hydroxycinnamic and especially caffeic derivatives are often found in the external part of the ripe fruit (Table 37). This distribution concerns:

- Quinic esters, such as chlorogenic acid in pear[475] (Figure 31) and 3′-caffeoylquinic acid in cherry. However, in apple, distribution appears to vary with the cultivar. There is a higher content in pulp and core in Jonathan (Figure 31) whereas the level is higher in the skin of other cultivars.[342] The high *p*-coumaric acid content in the skin of fruits is probably correlated with its role in the structure of cutin[272] and appears to be directly correlated with resistance to fungal infection in apple.[409] Tomato is one of the better-known examples. Chlorogenic acid is more abundant in pulp than in whole pericarp[158,620] although caffeic acid contents (determined after hydrolysis) are higher in the skin (5.8 mg/100 g FW) than in the rest of the fruit (2.9 mg/100 g FW).[620] In cherry tomato,

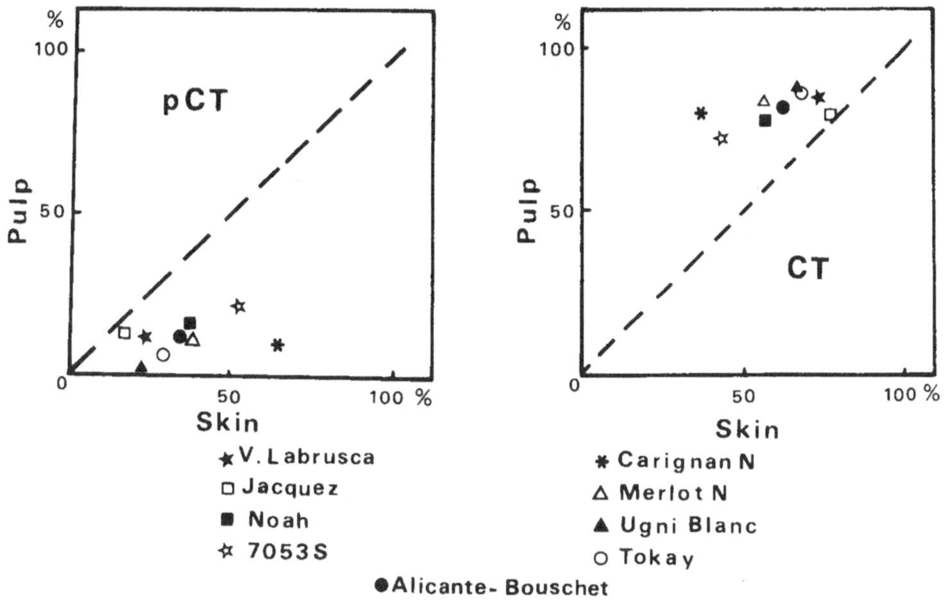

FIGURE 32. Relations between *p*-coumaroyltartaric (pCT) or caffeoyltartaric (CT) acids percentages in skin and pulp of nine grape cultivars. (Adapted from Boursiquot, J. M., Thèse Doct. Ing., ENSA, Montpellier, France, 1987.)

the distribution of hydroxycinnamic derivatives has been studied in space and also in time in relation to physiological stages (Chapter 3). Total quinic esters content in the ripe fruit was found to be higher in the pulp than in the pericarp, whereas the opposite was found for glucose derivatives. Some of these compounds are characteristic of a certain part of the fruit: neochlorogenic acid and caffeoylglucose have only been found in the pulp whereas sinapoylglucose and feruloylglucose were only present in the pericarp.

- Tartaric esters in grapes. Although the level was always higher in skin than in pulp, the percentage of caffeoyltartaric acid was higher in the pulp and the opposite was true for *p*-coumarolytartaric acid[70,71] (Figure 32).
- *p*-Coumaroylgalactaric and feruloylgalactaric acids, two compounds recently reported in Citrus and which are more abundant in the outer part of flavedo and albedo.[477]

Distribution of hydroxycinnamic derivatives is even more complex in certain cases, such as the pineapple. In addition to the gradients between inside and outside, there are very important longitudinal gradients (Figure 31).[128] This is probably related to different stages of maturity of each of the individual fruits which make up the pineapple.

Among flavanone glucosides, naringin localization has been particularly well studied in *Citrus* because of its role in fruit bitterness. An original radioimmunological assay technique used in the determination of this compound[289] showed that the highest naringin contents in grapefruit *(Citrus paradisi)* are found in flavedo and albedo on the one hand and in membranes and pith on the other.[290] Hesperidin, another characteristic flavanone in *Citrus* is also three to seven times as abundant in skin as in pulp in orange, clementine, and lemon.[184] Determination using the radioimmunoassay technique also showed that the distribution of hesperidin in the tissues of *C. sinensis* is remarkably similar to that of naringin in *C. paradisi* (Table 38) although the occurrence of the two compounds is for the most part mutually exclusive.[40] Naringenin is also present in the skin of tomatoes and the level increases considerably during maturation.[620] In this fruit, naringenin and chalconaringenin are cutin-bound in the membranes.[272] Other compounds also studied in *Citrus* are more abundant in

TABLE 38
Distribution of Hesperidin in Fruit and Vegetative Tissue of Parson Brown Orange Compared to the Distribution of Naringin in Duncan Grapefruit

Tissue	$\bar{X}$ ppm[a] Hesperidin[b]	s.d.	$\bar{X}$ ppm Naringin
Flavedo	4.177	523	4.882
Albedo	15.114	2.752	11.647
Juice vesicles	942	282	471
Membrane	8.005	1.719	7.412
Pith	5.924	1.911	9.647
Seeds	355	78	411
Young leaves	8.957	3.475	
Old leaves	5.776	3.386	

[a] ppm = μg/g fresh weight.
[b] Each value represents the average of three samples.

From Barthe, G. A., Jourdan, P. S., McIntosh, C. A., and Mansell, R. L., *Phytochemistry*, 27, 249, 1988. With permission.

the flavedo: coumarin glycosides[378] and polymethoxylated flavones which are present in the peel oil and can be used for taxonomic purposes.[187]

Little is known about the localization of certain phenolic compounds. This is the case for benzoic acids which, although common in fruits, have received little investigation except in a few cases. They are distributed in skin and pulp in tomato and melon[507] or found only in skin as in grape.[530] Compounds with very limited distribution sometimes display characteristic tissue localization. Thus, dopamine may be located in specialized cells in banana[46] and has been reported to occur at a higher concentration in peel (700 μg/g FW) than in pulp (8 μg/g FW).[427] In addition, large amounts of phloridzin were initially found in the pips of Golden Delicious apple, but this compound did not appear to be present in pulp.[109,110] Nevertheless, it has now been detected in cv. Spartan skin[127] and in flesh and skin of cv. Rhode Island Greening[119] and in apple juice.[320]

Phenolic compounds are thus found in all fruit tissues, but the greatest accumulation is in the outer part (skin, pericarp, albedo, flavedo, etc.). Is there a *raison d'être* for this distribution? There is no precise answer to this question, although several hypotheses can be put forward. The first is the attraction that pigmented fruits (due to anthocyanins and flavonols to a lesser extent) may have on the consumer — man or certain animals. In addition, these compounds may play the role of UV radiation filter[584,586] and thus protect fragile chemical structures. Finally, the relatively external distribution of certain phenolic compounds in the fruits should be related to the role that they may play in resistance to parasite infection.[47,64,273,409] The existence of this distribution and accurate knowledge of it are also important in handling of processes in which fruits are raw material, in particular in the preparation of wines and fruit juices. These aspects are discussed in Chapter 5.

At cell level, vacuolar compartmentation of soluble phenolic compounds on the one hand, and the existence of specialized cells on the other, such as tannin cells in a number of fruits, make it possible to separate oxidation enzymes (polyphenoloxidases and peroxidases) and their substrates; under normal physiological conditions they protect the fruit from biochemical transformations, such as browning. In addition, the presence of simple phenolic compounds at cutin level provides protection for fruit, as does the possible formation of lignins in certain cell walls (lignified endocarp, fruit tissues during healing). Still at cell wall level, it is probable that fruits contain complexes made up of various polysaccharidic fractions and ferulic acid, as has been shown in other parts of plants.[174] The study of such complexes in relation to the natural evolution of pectins during fruit maturation remains a very interesting subject for study.

The compartmentation of PC observed at cellular and tissue levels raises in particular the problem of the relationships that may exist between biosynthesis zones and accumulation zones within the fruit itself and also more generally between the fruit and the rest of the mother plant. These questions are discussed within the framework of the regulation of the phenolic metabolism of fruits.

Chapter 2

PHENOLIC COMPOSITION OF INDIVIDUAL FRUITS

I. INTRODUCTION

A brief summary is given for each fruit to enable rapid identification of the phenolic compounds present. This complements the information in Chapter 1. Unless otherwise specified, data are for ripe, fleshy fruits. Numerous fruits of Umbelliferae, Rutaceae, Leguminosae, Compositae, and other families may be very rich in coumarins; these can be found in Tables 12 to 14 and in Murray et al.[402] Likewise, Table 16 shows the anthocyanin composition of many fruits of secondary economic interest or fruits for which we have no data on the other phenolic groups. The classification used is by plant family with individual fruits listed by Latin names of genus and species.

II. ABBREVIATIONS

The following abbreviations are used:

Phenolic acids — pHB: *p*-hydroxybenzoic acid; mHB: *m*-hydroxybenzoic acid; Sa: salicylic acid; ProC: Protocatechuic acid; Ga: gallic acid; Ge: gentisic acid; V: vanillic acid; Sy: syringic acid; Cin: cinnamic acid; pC: *p*-coumaric acid; mC: *m*-coumaric acid; oC: *o*-coumaric acid; C: Caffeic acid; F: ferulic acid; S: sinapic acid; GaQ: Galloylquinic; pCQ: *p*-coumaroylquinic; CQ: caffeoylquinic; 5′-CQ: 5′ caffeoylquinic = chlorogenic acid; FQ: feruloylquinic; di-pCQ: di*p*-coumaroylquinic; di-CQ: dicaffeoylquinic.

Anthocyanins — Pg: pelargonidin; Cy: cyanidin; Pn: peonidin; Dp: Delphinidin; Pt: petunidin Mv: malvidin.

Flavonols — K: kaempferol; Q: quercetin; M: myricetin; iR: isorhamnetin.

Miscellaneous — ? = structure or presence to be confirmed. The names in bold type point out the most abundant compounds.

III. ACTINIDIACEAE

Actinidia chinensis KIWIFRUIT

Phenolic Acids:[178] 5′-CQ.
Tannins: see Table 31.

IV. ANACARDIACEAE

Anacardium occidentale CASHEW APPLE

Phenolic Acids:[500] pHB; ProC; Ga.
Flavonols:[500] Q; M; Q 3-galactoside.
Flavanones:[221,449] naringenin; naringenin 7- (6″-*p*-coumaroylglucoside).
Tannins:[500] condensed tannins.
Miscellaneous[588] anacardic acids[a]; cardol[a].

Comment: [a] In the shell of cashew nut.

Mangifera indica MANGO FRUIT

Phenolic Acids:[494] *m*-digallic acid.
Anthocyanins[a]:[441] Pn 3-galactoside.
Tannins:[141,494] β-glucogallin; gallotannins.
Xanthones:[494] mangiferin.

Comment: [a] Var Haden from Florida was used for analysis of skin anthocyanins.

Pistacia vera PISTACHIO nut-shell

Tannins:[67] gallotannins.
Miscellaneous:[652] anacardic acids.

V. AQUIFOLIACEAE

Ilex sp.

Phenolic acids:[279] 5′-CQ; di-CQ; CG.
Anthocyanins:[279]
Pg 3-xylosylglucoside; Cy 3-xylosylglucoside.
Cy 3-glucoside[a]
Flavonols:[279]K, Q[b]

Comments: [a] In some species only (see Table 16).
[b] After hydrolysis. The quantities were in general very low.

VI. BROMELIACEAE

Ananas comosus PINEAPPLE

Phenolic acids:[a] [128]
pC; F; **pCQ; di-pCQ;** pC glucose; C glucose;
F glucose; S glucose; CQ esters not completely identified.
Miscellaneous:[347] tyramine in pineapple juice.

Comment: [a] Many other pC derivatives are present.

VII. CAPRIFOLIACEAE

Sambucus nigra ELDERBERRY

Anthocyanins[a]:[79,217]
Cy 3-glucoside; Cy 3-sambubioside;
Cy 3,5-diglucoside; Cy 3-sambubioside-5-glucoside.
Flavonols:[205,548] Q 3-rutinoside (= rutin); Q after hydrolysis.

Comment: [a] Pomace was used for extraction of anthocyanins of *S. nigra;* Cy -sambubioside and 3-sambubioside-5-glucoside were also reported in the fruits of *S. ebulus*.[265]

Viburnum lantana

Anthocyanins:[209,278] Cy 3-glucoside.
Flavonols:[209] K 3-diglucoside; K and Q after hydrolysis.

V. tinus

Anthocyanins:[195]
Cy 3-glucoside; Cy 3-xylosylglucoside (= Cy 3-sambubioside)
Flavonols:[195]
Q 3-glucoside; Q 3-rhamnoside; Q 3-rutinoside;
Q 3,7-diglucoside.

VIII. CARICACEAE

Carica papaya PAPAYA

4-hydroxyphenyl-2-ethyl-β-D-glucoside.[513]

IX. EBENACEAE

Diospyros kaki PERSIMMON

Phenolic acids:[657] Ga.
Flavan-3-ols:[657] catechin.
Tannins: see Chapter 1.

X. EMPETRACEAE

Empetrum nigrum CROWBERRY

Anthocyanins:[217,296,297]
Cy 3-galactoside; Cy 3-arabinoside; Pn 3-galactoside;
Pn 3-arabinoside; **Dp 3-galactoside;** Dp 3-glucoside;
Dp 3-arabinoside; Pt 3-galactoside; Pt 3-glucoside;
Pt 3-arabinoside; **Mv 3-galactoside;** Mv 3-arabinoside.

Comment: Analysis concerns the northern Crowberry (*E. nigrum* ssp. *hermaphroditum*).

XI. ERICACEAE

Vaccinium corymbosum HIGHBUSH BLUEBERRY

Phenolic acids:[459,512]
pHB glucoside; ProC glucoside; Ga glucoside;
5′-pCQ; 3′-CQ; 4′-CQ; **5′-CQ;** 5′-FQ; pC glucose;
C glucose; F glucose; pC glucoside; C glucoside;
F glucoside.
Anthocyanins:[28-31,353,575]
Cy 3-glucoside; Cy 3-galactoside; Cy 3-arabinoside;
Pn 3-glucoside; Pn 3-arabinoside;
Dp 3-glucoside; Dp 3-galactoside; Dp 3-arabinoside;
Pt 3-glucoside; Pt 3-galactoside; Pt 3-arabinoside;
Mv 3-glucoside; Mv 3-galactoside; Mv 3-arabinoside.
Flavonols:[57] quercetin after hydrolysis.
Tannins: see Table 31.

Comment: Peonidin 3-galactoside was absent in ripe fruit but present in very small quantity in unripe fruit.

V. angustifolium LOWBUSH BLUEBERRY

Anthocyanins:[168,593]
Cy, Pn, Dp, Pt and Mv-glucosides, 3-galactosides, and 3-arabinosides.

Comment: Several diglycosides were also present, but in quantities too low for identification.[168]

V. macrocarpon CRANBERRY

Phenolic Acids[a]:[106,363;459]
pC; pC glucose; F glucose; S glucose;
pC glucoside;[b] glucosides[b] of pC glucose, C glucose, F glucose;
diglucoside[b] of C glucose
Anthocyanins:[180-182,213,323]
Cy 3-galactoside; Cy 3-arabinoside; Cy 3-glucoside;
Pn 3-galactoside; Pn 3-arabinoside; Pn 3-glucoside;
Dp 3-galactoside; Dp 3-arabinoside; Dp 3-glucoside.
Flavonols:[57,323,445]
Q 3-galactoside; Q 3-rhamnoside; Q 3-arabinoside;
M 3-arabinoside; M 3-digalactoside; K derivatives.
Flavan-3-ols:[619] (+)-catechin; (−)-epicatechin.
Tannins: see Table 31.

Comments: [a] No quinic derivatives.
[b] The number of glucose units has not been determined.

V. myrtillus WHORTLEBERRY (BILBERRY)

Phenolic Acids[a]:[22,76,172]
pHB; mHB; ProC; Ga; V; Sy; pC; mC; oC; C; F; 5′-CQ.
Anthocyanins[b]:[25,575,603,604]
Cy 3-glucoside; Cy 3-galactoside; Cy 3-arabinoside;
Pn 3-glucoside; Pn 3-galactoside; Pn 3-arabinoside;
Dp 3-glucoside; Dp 3-galactoside; Dp 3-arabinoside;
Pt 3-glucoside; Pt 3-galactoside; Pt 3-arabinoside;
Mv 3-glucoside; Mv 3-galactoside; Mv 3-arabinoside.
Flavonols[c]:[21,22,172]
K 3-glucoside; Q 3-glucoside (= isoquercitrin);
Q 3-rhamnoside; Q 3-galactoside; Q 3-arabinoside?.
Flavan-3-ols:[550] (+)-catechin; (−)-epicatechin.
Tannins:[76] Procyanidins B1; B2; B3; B4.

Comments: [a] In fresh juice.
[b] Timberlake and Bridle[575] reported the presence of Dp 5-glucoside and 3-rhamnosyl glucoside (From Troyan and Borukh, *Chem. Abstr.*, 73, 119394, 1968).
[c] After hydrolysis, Starke and Herrmann[548] identified: K, Q, and M whereas no glycosides of M have been found (cf. quoted works). Discrepancy may be connected to varietal differences.

V. vitis-idaea COWBERRY

Anthocyanins:[11,180,593]
Cy 3-galactoside; Cy 3-arabinoside; Cy -3-glucoside;
Dp 3-glucoside; Cy 3-xylosylglucoside?;[a]Mv 3,5-diglycoside?[a]

Comment: [a] The two diglycosides mentioned by Timberlake and Bridle[575] (From Troyan, A. V. and Borukh, I. F., *Chem. Abstr.*, 73, 119394, 1968), have not been found in recent studies using HPLC and DCCC.

Note on the Anthocyanins from *Vaccinium* sp.:

Numerous North American species have been analyzed,[28,32] Species in *Cyanococcus, Herpothamnus,* and *Pyxothamnus* sections contain monosides of the five aglycones (Cy, Pn, Dp, Pt, and Mv). Species in *Polycodium* and *Oxycoccoides* sections contain more than 90% of Cy, small percentage of Pn and Mv and no Dp.[32]

V. japonicum:[12] Cy 3-arabinoside; Cy 3-galactoside;
Pg 3-arabinoside; Pg 3-galactoside.

V. uliginosum:[13] Cy, Pn, Dp, Pt, Mv 3-glucosides, 3-galactosides,
3-arabinosides.

V. bracteatum:[280] Cy, Dp, and Pt glycosides.

V. smallii:[280] Cy, Pt and Mv glycosides.

XII. EUPHORBIACEAE

Mabea caudata

Flavanones:[39]
naringenin,
naringenin 7-O-β-(3-pC glucoside);
naringenin 7-O-β-(3,6-di-pC glucoside).

XIII. GUTTIFERAE

Garcinia mangostana MANGOSTEEN

Anthocyanins[a]:[134] **Cy 3-sophoroside;** Cy 3-glucoside.
Xanthones:[27,516,517]
garcinones A, B, and C; mangostin; γ-mangostin;
gartanins; BR-xanthone-A; BR-xanthone-B.

Comment: [a] Pigments extract from the rind which is dense, thick, and purplish-brown in color.

Kielmeyera pumila

Phenylcoumarins (Neoflavanoids):[405] Mammeigin; isomammeigin.

XIV. JUGLANDACEAE

Juglans regia WALNUT
Juglaus sp.

Quinones:[324,399,400]
juglone (= 5-hydroxy-1, 4-naphtoquinone);[a] 1,4-naphtoquinone;
4-hydroxy-1-naphtalenylglucoside; 4,8-dihydroxy-1-naphtalenylglucoside.

Comment: [a] Juglone is also present in other Juglandaceae, pecan (*Carya illinoensis*), and hickories (*Carya* sp.).[65,66,243,244]

XV. LAURACEAE

Aniba riparia

2,6-dihydroxybenzoic; benzylsalicylate; benzyl-2,6-dihydroxybenzoate; methyltyramine; benzoyltyramine and 5 phenylethylamides of benzoic acids; 3 styryl-α-pyrones; neolignans.[36]

Licaria armeniaca

11 neolignans.[35]

Persea americana AVOCADO

Phenolic Acids[a]:[196,438,452,580]
pHB; pyrocatechuic; β-resorcylic; γ-resorcylic; α-resorcylic;
ProC; Ga; isoV; V; Sy; oC; mC; **pC;** C; **F;** S; pCQ; CQ.
Flavan-3-ols:[50] catechin; epicatechin.
Miscellaneous:[347] dopamine; serotonin.

Comment: [a] Determined after hydrolysis in most cases.

XVI. LEGUMINOSAE

Tephrosia pumila

Isoflavones:[117] pumilaisoflavone A; pumilaisoflavone B.
Chalcone:[117] praecansone-A.

XVII. MORACEAE

Ficus carica FIG

Anthocyanins[a,b]:[265,443,573]
Cy 3-rutinoside; Cy 3-glucoside;
Cy 3,5-diglucoside; Pg 3-rutinoside.
Flavonols:[525]
K 3-glucoside; K 3-rutinoside;
Q 3-rutinoside; Q 3-glucoside.
C-Glycosides:[525]
apigenin-6-C-glucosyl-8-C-arabinoside;
apigenin-6-C-arabinosyl-8-C-glucoside.

Comments: [a] Puech et al.[443] analyzed the var. Mission, the peel of which is of a deep dark blue. For other varieties the main pigment would be the Cy 3-glucoside.[573]
[b] The anthocyanins were analyzed in the fruits of four other species of *Ficus*:[278,280] *F. erecta:* Cy 3-rutinoside and Cy 3-glucoside; *F. wightiana:* Cy 3-rutinoside and Cy 3-glucoside; *F. nipponica:* Cy 3-rutinoside and Cy 3-glucoside; Cy 3,5-diglucoside; *F. pumila:* Cy 3-glucoside.

Morus nigra MULBERRY
Morus alba

Anthocyanins[a]:[265,278] **Cy 3-glucoside;** Cy 3-rutinoside.

Comment: [a] The analysis corresponds to the fruits of *Morus alba* (deep-purple fruit). For *Morus nigra* (Black mulberry), the only indication is the presence of Cy 3-glucoside. According to Timberlake and Bridle[575] and Timberlake,[573] the fruits of *Morus alba* would contain other anthocyanosides with in particular Cy and Dp (Maki, Z. and Inamoto, H., *Chem. Abstr.*, 78, 82073, 1972.)

XVIII. MUSACEAE

Musa sapientum BANANA
Musa sp.

Anthocyanins:[217,526]
Pn and Mv glycosides (in red cultivars);
3-deoxyanthocyanidins (after irradiation of the fruit).
Miscellaneous:[18,347,427] dopamine; serotonin; tyramine; vanillin; Eugenol.
Tannins: see Table 31:

XIX. MYRISTICACEAE

Iryanthera laevis

Dihydrochalcones:[186]
2′, 4′-dihydroxy-4, 6′-dimethoxydihydrochalcone.
2′,4′,6′-trihydroxy-4-methoxydihydrochalcone; allylphenyl derivatives.

Virola elongata
Virola sebifera

6 lignans including eudesmin, epieudesmin, and fargesin; 19 neolignans including magnostellins A and C.[298,327]Acylresorcinols.[299]

XX. OLEACEAE

Forsythia sp.

Caffeoyl glycosides of 3,4-dihydroxyphenylethanol: forsythiaside; acteoside; suspensaside; β-hydroxyacteoside.[303]
Rutin.
Lignans and lignan-glucosides.[303]

Olea europea OLIVE

Phenolic acids[a]:[159,315,600] 5′-CQ;[b] **verbascoside;** C glucose.
Anthocyanins[c]:[345,355,573,600]
Cy 3-glucoside; **Cy 3-rutinoside;** Cy 3-caffeyl-rhamnosylglucoside;
Cy 3-glucosylrutinosides (linear and branched);
Cy 3-glucosyl-rutinoside (branched) acylated with caffeic acid;
Dp 3-rhamosylglucoside-7-xyloside?[d]
Flavonols:[600]Q 3-rutinoside.
Flavones:[8,600]
luteolin 7-glucoside; luteolin 5-glucoside; apigenin 7-glucoside.

Miscellaneous:[9,311,429,447,448,540,601]
oleuropein;[e] demethyloleuropein;[f] oleuropein aglycone; dihydroxyphenylethanol;[g] dihydroxyphenylethanol glucoside; hydroxyphenylethanol (= tyrosol); ligstroside.

Comments: [a] In oil and margines: free hydroxybenzoic acids and hydroxycinnamic acids (pHB; ProC, V, Sy, pC, oC, C, hydroC acids); dihydroxyphenylethanol and its monoglucoside; Q; apigenin; luteolin.
[b] Identification must be confirmed in most cultivars, because of possible confusion with verbascoside.
[c] Timberlake and Bridle[575] reported the work of Luh, B. and Mahecha, G., *Chem. Abstr.*, 77, 60298, 1972 and point out the presence of Pn glycosides, with acylation by *p*-coumaric acid.
[d] Mentioned by Marekov.[355] Differences in anthocyanin composition between cultivars have been noted.
[e] Principal responsible for bitterness.
[f] Seldom present in fruits, but abundant after debittering.
[g] Only in the ripe fruits of some cultivars.

XXI. PALMAE

Phoenix dactylifera DATE

pHB; ProC; V; Sy; pC; C; F; **C shikimic** isomers (= dactylifric acid); likely F shikimic; flavans; flavonol glycosides; condensed tannins.[350,352,458]

XXII. PIPERACEAE

Piper nigrum PEPPER

N transferuloyl tyramine; *N* transferuloyl piperidine;
N-5-(4-hydroxy-3-methoxyphenyl)-2*E*, 4*E*-pentadienoyl piperidine;
N-5-(4-hydroxyphenyl)-2*E*, 4*E*-pentadienoyl piperidine;
N-5-(4-hydroxy-3-methoxyphenyl)-2*E*-pentenyl piperidine:
3,4-dihydroxyphenylethanol glucoside[a].[277,408,599]

Comment: [a] Present in fresh green pepper berries.[599]

Piper aduncum, Piper hispidum

4-methoxy-3,5-bis(3′-methyl-2′-butenyl)-benzoic acid;
5-hydroxy-7-methoxyflavanone;
2,6-dihydroxy-4-methoxydihydrochalcone.
Phenyl propene: 1-allyl-2,3-(methylenedioxy)-4,5-dimethoxybenzene.[90]

Piper trichostachyon

Lignans:[308](−)-hinokinin; (−)-cubebin; (−)-dihydrocubebin and three other lignans.

XXIII. PUNICACEAE

Punica granatum POMEGRANATE

Anthocyanins:[136,265]
Pg 3-glucoside; Pg 3,5-diglucoside; **Cy 3-glucoside**[a]; Cy 3,5-diglucoside; Dp 3-glucoside; Dp 3,5-diglucoside.

Comment: [a] Cy 3-glucoside is the major anthocyanin in fruit peel. The two delphinidin glycosides are localized in seed coats.

XXIV. ROSACEAE

Aronia melanocarpa CHOKEBERRY

Anthocyanins:[423]
Cy 3-galactoside; Cy 3-glucoside; Cy 3-arabinoside; Cy 3-xyloside.

Cydonia oblonga QUINCE

β-glucosides of procyanidins.[436]

Fragaria ananasia STRAWBERRY
Fragaria sp.

Phenolic acids[a]:[251,459,512,549]
ProC[b]; V[b]; pHB glucoside; 5′-GaQ;
pC; **pC glucose;** pC glucoside; C glucose;
F glucose.
Anthocyanins[c]:[179,217,265,575,647]
Pg 3-glucoside; Pg 3-α-glucofuranose?; Cy 3-glucoside.
Flavonols:[107,246,491]
K 3-glucuronide; K 3-glucoside[d]; K 3-galactoside;
K 3-xylosylglucuronide; K 3-xylosylglucoside;
K 7-glucoside?[e]; K 7-glycoside?; **Q 3-glucuronide;**
Q 3-glucoside; Q 3-galactoside; Q 3-xylosylglucuronide.
Flavan-3-ols:[549] **(+)-catechin;** (−)-epicatechin; (+)-gallocatechin[f].
Tannins: see Table 31.

Comments: [a] Hydroxycinnammic acids exist only in the glycosylated forms.
[b] After hydrolysis.
[c] Presence in very small quantities of other pelargonidin derivatives: monoside, bioside, and trioside.[179]
[d] K 3-glucoside and Q 3-glucuronide are not separated by HPLC.[246]
[e] 7-Glycosides are reported by Ryan,[491] whereas Henning[246] concludes that "several 3,7-diglycosides could be tentatively identified, though present in minor amounts."
[f] (+-Gallocatechin is not present in all cultivars. Out of the 14 cultivars analyzed, only 6 contain (+)-catechin.

Malus pumila APPLE

Phenolic acids:[95,251,252,269,338,344,387,388,391,475,615]
3′-pCQ; 4′-pCQ; 5′-pCQ; **5′-CQ;** 4′-CQ; 3′-CQ; di-CQ;
pC glucose; C glucose; F glucose; S glucose.[a]
Anthocyanins[b]:[265,557,573-575,593]
Cy 3-galactoside; Cy 3-glucoside; Cy 3-arabinoside;
Cy 3-xyloside; Cy 7-arabinoside?[c]; acylated derivatives
of Cy 3-glucoside, Cy 3-galactoside, Cy 3-arabinoside and
Cy 3-xyloside.[d]
Flavonols[e]:[127,146,217,419,568]
Q 3-galactoside; Q 3-arabinoside; Q 3-rhamnoside; Q 3-xyloside;
Q 3-rutinoside; Q 3-glucoside, Q 3-diglucoside?

Flavan-3-ols:[395,475]
(+)-catechin; **(−)-epicatechin;** (+)-gallocatechin[f]; (−)-epigallocatechin[f].
Tannins: see Table 31.
Dihydrochalcones:[126,127,271,419] phloridzin; phloretin xyloglucoside.

Comments: [a] Presence of S glucose in cell suspension cultures.[309,434]
[b] Anthocyanins are present in the varieties with red-colored peel.
[c] Identification of the cyanidin-7-arabinoside is much debated (cf. Timberlake[573]).
[d] The nature of acids intervening in acylation is not clearly given.
[e] Q 3-diglucoside is indicated only by Fisher.[146] We must notice the absence of K derivatives though they are mentioned by Van Buren[593].
[f] Present in only two cultivars.

Prunus armeniaca APRICOT

Phenolic acids:[251,387-389,459,475]
3′-pCQ; 4′-pCQ; 5′-pCQ; **3′-CQ;** 4′-CQ; 5′-CQ; 3′-FQ;
5′-FQ; pC glucose; F glucose; coumarin; scopoletin.
Flavonols:[247,253]
K 3-rutinoside; K 3-glucoside[a]; **Q 3-rutinoside;**
Q 3-glucoside.
Flavan-3-ols:[395,475] (+)-catechin; **(−)-epicatechin.**

Comments: Anthocyanins are absent.
[a] K 3-glucoside is detected in only one of three varieties analyzed.

Prunus avium SWEET CHERRY

Phenolic acids[a]:[304,381,387-389,459,475,552]
pHB, V; pHB glucoside; ProC glucoside; V glucoside; 3′-pCQ;
4′-pCQ; 5′-pCQ; **3′-CQ;** 4′-CQ; 5′-CQ; 3′-FQ; 5′-FQ; di-CQ;
pC glucose; F glucose.
Anthocyanins[b]:[331,417,418,573,575]
Cy 3-rutinoside; Cy 3-glucoside;
Pn 3-rutinoside; Pn 3-glucoside.
Flavonols:[249,380,382]
K 3-rutinoside; K 3-glucoside; K 3-galactoside; **K 3-galactosyl-7-diglucoside**?; K 3-rutinosyl-4′-diglucoside; K 3-glucosyl-7-diglucoside?; K 7-glucoside?[c]; K 3-(2[G]-glucosylrutinoside)??[c]; **Q 3-rutinoside;** Q 3-glucoside; Q 3-rhamnoside;
Q 3-rutinosyl-4′-diglucoside; iR 3-rutinoside.[c]
Flavan-3-ols:[395,475,552] (+)-catechin; **(−)-epicatechin.**

Comments: [a] In addition to the mentioned phenolic acids, cv. Bigarreau Napoléon would contain: di-pCQ; pC fructose and sinapic derivatives.[381]
[b] Pn 3-glucoside is not detected in all cultivars. In cv. Bigarreau Napoléon, the main pigment is Pn 3-rutinoside.[417]
[c] K 7-glucoside and K 3-(2[G]-glucosylrutinoside) are found only in cv. Bigarreau Napoléon.[382] This cultivar seems to be a special one, due to the presence of isorhamnetin 3-rutinoside as main flavonol.

Prunus cerasus SOUR CHERRY

Phenolic acids:[387-389,459,475,552]
V; 3′-pCQ; 4′-pCQ; 5′-pCQ; **3′-CQ;** 4′-CQ; 5′-CQ;
3′-FQ; 5′-FQ; di-CQ; pC glucose.

Anthocyanins[a]:[119,120,519,573]

Cy 3-(2G-glucosylrutinoside); Cy 3-rutinoside; Cy 3-sophoroside; Cy 3-glucoside; Cy 3-(2^G-xylosylrutinoside); Pn 3-rutinoside.

Flavonols:[249,506,520,593]

K 3-rutinoside; K 3-glucoside; **K 3-galactosyl-7-diglucoside?;** K 3-rutinosyl-4′-diglucoside; K 3-galactoside; K 4′-glucoside?[b] K-3-rhamnosyl-4′-galactoside?[b]; **Q 3-rutinoside;** Q 3-glucoside; Q 3-glucosyl-7-diglucoside?; Q 3-galactosyl-7-diglucoside?; Q 3-rutinosyl-4′-diglucoside; Q 3-galactoside; Q 4′-glucoside?[b]

Flavan-3-ols:[395,475,552]

(+)-catechin; **(−)-epicatechin;** (+)-gallocatechin; (−)-epigallocatechin.

Coumarins:[552] aesculetin.

Comments: [a] The analysis of anthocyanins was mainly done in cv. Montmorency. Cy 3-(2^G-xylosylrutinoside) and Pn 3-rutinoside were identified in small amount in this cultivar.[519]

[b] These glycosides were reported in cv. Montmorency,[519] but their analysis needs to be confirmed.

Prunus domestica PLUM

Phenolic acids[a]:[304,387-389,444,457,459,475,552]

V[b]; pHB glucoside; ProC glucoside; V glucoside; Sy glucoside; Sa glucoside; 3′-pCQ; 4′-pCQ; 5′-pCQ; **3′-CQ;** 4′-CQ; 5′-CQ; 3′-FQ; 5′-FQ; di-CQ; pC glucose; C glucose; F glucose.

Anthocyanins[c,d]:[217,575,593]

Cy 3-glucoside; Cy 3-rutinoside; Pn 3-glucoside; Pn 3-rutinoside.

Flavonols[e]:[248,548]

K 3-rutinoside[f]; K 3-glucoside; K 3-galactoside; K 3-arabinosyl-7-rhamnoside; **Q 3-rutinoside**[f]; Q 3-glucoside[f]; Q 3-galactoside[f]; Q 3-xyloside; Q 3-rhamnoside.

Flavan-3-ols:[395,475,552] (+)-catechin; (−)-epicatechin.

Coumarins:[552] scopoletin.

Comments: [a] The phenolic acid pattern of *Prunus spinosa* is identical to *P. domestica* (Ramos and Macheix, unpublished results). However, there does not seem to be any K derivatives.

[b] After hydrolysis.

[c] Anthocyanins in cultivars with deep purple-colored peel. Anthocyanin composition of plums is close to sweet cherries (*P. avium*).

[d] The fruits of *P. salicina* contain only the two Cy derivatives.[278]

[e] In *P. salicina:* K 3-arabinosyl-7-rhamnoside, K 3-rutinoside, Q 3-rutinoside, Q 3-glucoside, and a special flavonol glycoside: K 3,7-dirhamnoside.[248]

[f] Present in the six analyzed cultivars.

Prunus persica PEACH

Phenolic acids:[89,251,388,389,459,475]

3′-pCQ; 5′-pCQ; 3′-CQ; 4′-CQ; **5′-CQ;** 3′-FQ; 5′-FQ; di-CQ; pC glucose.

Anthocyanins[a]:[278,575,592,593] **Cy 3-glucoside;** Cy 3-rhamnoglucoside.

Flavonols[b].[247]

K 3-glucoside; K 3-galactoside; K 3-rutinoside; Q 3-glucoside; Q 3-galactoside; Q 3-rutinoside; Q 3-xyloside[c].
Flavan-3-ols:[395,475] **(+)-catechin;** (−)-epicatechin; (+)-gallocatechin; (−)-epigallocatechin.

Comments: [a] Anthocyanins are present in the peel of some red-colored peach cultivars. According to Van Blaricom and Senn,[592] beside the main pigment, there were very faint bands of other red pigments, but these were not considered important in the investigation; the presence of Cy 3-rhamnoglucoside is mentioned by Ishikura.[278]
[b] Present only in a very small amount. No glycosides are clearly dominant.
[c] Present in one cultivar only.

Prunus sargentii — ORNAMENTAL CHERRY

Anthocyanins:[135,573]
Pg 3-glucoside; Pg 3-rutinoside; **Cy 3-glucoside; Cy 3-rutinoside;** Cy 3-glycoside?; Cy 3-diglucoside?; Aliphatic acylated anthocyanins?[a]

Comment: [a] The nature of the aliphatic acids is not indicated. Fruits of *P. sargentii* are different from cherries (*P. avium* and *P. cerasus*) by the presence of pelargonidin.

Note on the Prunus species:
Anthocyanins:
P. mume: Cy 3-rhamnoglucoside.[278]
P. yedoensis and *P. jamasakura:* Cy-3-glucoside.[265]
P. tomentosa: Pg 3-rhamnoglucoside; Cy 3-rhamnoglucoside.[278]
Coumarins:[402,658]
P. spinosa: aesculetin.
P. mahaleb: umbelliferone; aesculetin; herniarin; scopoletin; mahaleboside.

Pyrus communis — PEAR

Phenolic acids:[56,176,251,388,389,475,589,612,617]
4′-pCQ?; 5′-pCQ; di-pCQ; pC malate; 3′-CQ; 4′-CQ; **5′-CQ;** di-CQ; C malate; 5′-FQ.
Anthocyanins[a]:[165,574,575] **Cy 3-galactoside;** Cy 3-arabinoside.
Flavonols:[138,176,253,412,593,612]
Q 3-glucoside; Q 3-diglucoside?; Q 7-xyloside[b]; iR 3-rutinoside; iR 3-glucoside; iR 3-rhamnosyl-galactoside; iR 3-galactoside[b]; malonyl-glucosides of K, Q and iR.
Flavan-3-ols:[395,475] (+)-catechin; **(−)-epicatechin.**

Comments: [a] Anthocyanins only in some varieties, the peel of which is red. Minor pigments were negligible in the pears examined.
[b] Q 7-xyloside and iR 3-galactoside are mentioned only in 4 cultivars.[138]

Rubus fructicosus — BLACKBERRY

Phenolic acids:[459,512]
pHB, ProC; Ga; pHB glucoside; ProC glucoside; Ga glucoside; 3′-pCQ; 5′-pCQ; **3′-CQ;** 4′-CQ; 5′-CQ; 3′-FQ; 5′-FQ; pC glucose; C glucose; F glucose.
Anthocyanins[a]:[37,225,287,498,579]
Cy 3-glucoside; Cy 3-rutinoside; Cy 3-xyloside?; two dicarboxylic acid-acylated derivatives of Cy 3-glucoside.

Flavonols[b]:[57,613]
K 3-glucuronide; K 3-glucoside; K 3-galactoside;
K 3-xylosylglucuronide; Q 3-glucuronide; Q 3-glucoside;
Q 3-galactoside; Q 3-xylosylglucuronide; Q 3-rutinoside;
Q 3-[6″-(3-hydroxy-3-methylglutaryl)-β-galactoside].
Flavan-3-ols:[396] (+)-catechin; **(−)-epicatechin.**
Tannins: see Table 31

Comments: [a] The blackberry group corresponds to many species of *Rubus*. The anthocyanins of 16 species were studied by Torre and Barrit,[579] specially: *Rubus ursinus* (western trailing blackberry), *R. laciniatus* (evergreen blackberry), *R. procerus* (Himalayan blackberry), *R. plicatus*, etc. Their characteristic is the absence of pelargonidin (except *R. caucasus*), of branched triglycosides and of sophoroside (difference with raspberries).
[b] Bilyk and Sapers[57] identified Q and K after hydrolysis in different cultivars and clones of Thornless Blackberry fruit.

Rubus idaeus RED RASPBERRY

Phenolic acids:[251,459,512]
5′-GaQ; Ga glucose; **pHB glucoside;** ProC glucoside;
Ga glucoside; 5′-pCQ; 3′-CQ; 5′-CQ; 5′-FQ; pC glucose;
C glucose; F glucose; pC glucoside; F glucoside.
Anthocyanins[a]:[38,169,225,287,385,544,573]
Pg 3-sophoroside[b]; Pg 3-(2^G-glucosylrutinoside)[b];
Pg 3-glucoside[b]; Pg 3-rutinoside[b]; **Cy 3-sophoroside;**
Cy 3-glucoside; Cy 3-rutinoside; Cy 3-(2^G-glucosylrutinoside);
Cy 3,5-diglucoside.
Flavonols:[246,492]
K 3-glucuronide; K 3-glucoside; K 3-galactoside;
K 3-xylosylglucuronide; K 3-xylosylglucoside[c];
Q 3-glucuronide; Q 3-glucoside; Q 3-galactoside;
Q 3-xylosylglucuronide.
Flavan-3-ols:[396] (+)-catechin; **(−)-epicatechin;**
Tannins: see Table 31.

Comments: [a] There is an important variability in the composition in anthocyanins of 37 cultivars and selections of *R. idaeus* (Barrit and Torre[38]); Cy 3-sophoroside is the major pigment in most cultivars. Absence of xylose derivatives except for a selection (SHRI 6626/41): difference with *R. occidentalis*. Boysenberry and loganberry, raspberry × blackberry hybrids have the same pigment composition as *R. idaeus*.
[b] Pg derivatives are present in very small amount.
[c] K 3-xylosylglucoside is present in two cultivars only out of the five analyzed and always in small amount (<3%).

Rubus occidentalis BLACK RASPBERRY

Anthocyanins[a]:[225,287,573,579]
Cy 3-glucoside; Cy 3-rutinoside;
Cy 3-sambubioside; Cy 3-(2^G-xylosylrutinoside).

Comments: [a] Cyanidin 3-(2^G-xylosylrutinoside) seems to be specific of black raspberries. It is absent in red raspberries. Black raspberry group corresponds to two species: *R. occidentalis* and *R. leucodermis*.[579]

Note on the anthocyanins of Rubus sp.:[278,579]
R. parviflorus: Pg 3-glucoside; Pg 3-rutinoside; Cy 3-glucoside.

R. crataegifolius, R. illecebrosus, R. morifolius: Pg 3-glucoside; Cy 3-glucoside.
R. neo-mexicanus: Cy 3-rutinoside; Cy 3-glucoside; Cy 3,5-diglucoside and three not identified glycosides.
R. hirsutus: Pg 3-glucoside.

XXV. RUBIACEAE

Coffea arabica (pulp) COFFEE

Phenolic acids:[108,328,451]
ProC; pC; F; 3′-CQ;4′-CQ; **5′-CQ;**
di-CQ and other isochlorogenic fractions.
Anthocyanins:[659,660] Cy 3-glucoside; Cy 3-rutinoside.
Flavonols:[451] Q 3-rutinoside (rutin).
Flavan-3-ols:[451] catechin; epicatechin.

XXVI. RUTACEAE

Casimiroa edulis

7-Deoxyflavones:[221,642]
5,6,2′-trimethoxyflavone; 5-hydroxy-6,2′,6′-trimethoxyflavone; 5,6,2′,6′-tetramethoxyflavone (= zapotin).

Citrus aurantium SOUR ORANGE

Flavanones:[202,217,602]
5,6,7,8,3′,4′-hexamethoxyflavone (nobiletin);
5-hydroxy-6,7,8,3′,4′-pentamethoxyflavone;
3,6,7,8,4′-pentamethoxyflavone (auranetin); 5-hydroxy-auranetin?
5,6,7,3′,4′-pentamethoxyflavone (sinensetin);
5,6,7,4′-tetramethoxyflavone; apigenin 7-rutinoside.
Flavanones:[59,489]
hesperetin 7-neohesperidoside (= neohesperidin);
naringenin 7-neohesperidoside (= naringin);
hesperidin[a]; narirutin[a]; neoeriocitrin[a].
Coumarins: see Table 12.

Comment: [a] Present in very small quantity in juice.[689]

Citrus limon LEMON

Phenolic acids:[459,551,627]
Sa; Ge; **pC**[a]; C[a]; F[a]; S[a];
pC glucose; F glucose; S glucose.
Flavones:[217,221,598]
luteolin 7-rutinoside;
diosmetin 7-rutinoside (= diosmin).
Flavonols:[202,217,598]
rutin
3,5,7,4′-tetrahydroxy-6,8,3′-trimethoxyflavone = limocitrol;
3,5,7,4′-tetrahydroxy-8,3′-dimethoxyflavone = limocitrin;
3,5,7,3′-tetrahydroxy-6,8,4′-trimethoxyflavone = isolimocitrol.

Flavanones:[59,430]
hesperetin 7-rutinoside (hesperidin);
eriodictyol 7-rutinoside (eriocitrin).
C-Glycosylflavones:[430]
6,8-diC-glucosyl-apigenin; 6,8-diC-glucosyl-luteolin; 6,8-diC-glucosyl-chrysoeriol; 6,8-diC-glucosyl-diosmetin.
Coumarins: see Table 12.
Miscellaneous[b]:[347] tyramine; synephrine; octopamine.

Comment: [a] Free acids identified after hydrolysis.
[b] In juices.

Citrus paradisi GRAPEFRUIT

Phenolic acids:[459,551,627]
Sa; Ge; pC[a]; C[a]; **F**[a]; S[a];
F glucose; S glucose; F putrescine.
Flavonols:[351] K, Q, iR after hydrolysis.
Flavones:[202,217,351,602]
tangeretin; heptamethoxyflavone; 5,7,8,4′-tetramethoxyflavone; 5,7,8,3′,4′-pentamethoxyflavone; apigenin after hydrolysis.
Flavanonols:[351] dihydrokaempferol after hydrolysis.
Flavanones:[59,148,289,290,489]
naringenin 7-neohesperidoside (= naringin); naringenin 7-rutinoside (= narirutin); hesperidin; neohesperidin; naringenin 4′-glucoside and isosakuranetin in the form of 7-neohesperidosides and of 7-rutinosides[b].
Coumarins: see Table 12.
Miscellaneous:[565] coniferin.

Comment: [a] Free acids identified after hydrolysis.
[b] Compounds nondetected in juice.[489]

Citrus reticulata MANDARIN
CLEMENTINE
TANGERINE

Flavones:[51,187,188,562,578,641]
sinensetin; nobiletin; tangeretin; 3,5,6,7,3′,4′-hexamethoxyflavone; 3,5,6,7,8,3′,4′-heptamethoxyflavone; 5,6,7,4′-tetramethoxyflavone (tetramethylscutellarein); 7,8-dimethoxyflavone; 5,7,4′-trimethoxyflavone; 7,8,4′-trimethoxyflavone; 6,8,3′-trimethoxyflavone.
Flavanones:[59]
naringenin 7-rutinoside; isosakuranetin 7-rutinoside;
hesperetin 7-rutinoside.
Coumarins: see Table 12.
Miscellaneous[a]:[347]
tyramine; methyltyramine; synephrine; octopamine; hordenine.

Comment: [a] In juices.

Citrus sinensis SWEET ORANGE

Phenolic acids:[459,476,477,522,551,627]
Sa; Ge; pC[a]; C[a]; F[a]; S[a]; 5′-CQ; pC glucose;
F glucose; S glucose; F putrescine; pC galactaric;
F galactaric; pC glucaric; F glucaric; diF glucaric.

Anthocyanins[b];[334,335]
Pg 3 glucoside[c]; Pg 3,5-diglucoside[c];
Cy 3-glucoside; Cy 3,5-diglucoside; Cy 3-(acetyl)-glucoside;
Cy 3-(ferulyl)-glucoside; Cy 3-(coumaryl, ferulyl)-glucoside
Cy 3-(sinapyl)-glucoside; Pn 3,5-diglucoside;
Pn 3-(coumaryl)-glucoside; Pt 3-glucoside;
Pt 3,5-diglucoside; Dp 3-glucoside; Dp 3,5 diglucoside.

Flavones:[51,187,188,202,211,245,430,562,578]
5,6,7,3′,4′-pentamethoxyflavone (sinensetin); 5,6,7,8,3′4′-hexamethoxyflavone (nobiletin); 5,6,7; 8; 4′-pentamethoxyflavone (tangeretin); 3,5,6,7,3′,4′-hexamethoxyflavone; 3,5,6,7,8,3′,4′-heptamethoxyflavone; 5,6,7,4′-tetramethoxyflavone (tetramethyl scutellarein); 5,7,8,3′,4′-pentamethoxyflavone (isosinensetin); 5,7,8,4′-tetramethoxyflavone; 6,7,8,4′-tetramethoxy-5-hydroxyflavone[d].

Flavanones:[40,59,217,430,489]
hesperetin 7-rutinoside (hesperidin); naringenin 7-rutinoside (narirutin); eriodictyol 7-rutinoside (eriocitrin)[e]; isosakuranetin 7-rutinoside[e]; narirutin-4′-glucoside[e].

Flavanonols:[100] dihydrokaempferol 4′-methylether-7-0-rhamnoside.

Coumarins: see Table 12.

Miscellaneous:[347,502,565]
methyltyramine[f]; synephrine[f]; coniferin[g]; syringin[g];
dehydroconiferylalcohol-4-β-glucoside[g]; citrusins A, B, and C[g].

Comments:
[a] Free acids identified after hydrolysis.
[b] In some pigmented orange varieties (Moro, Tarocco, Sanguinello).
[c] In Moro orange,[335] but the aglycones obtained by hydrolysis of total anthocyanidin glycosides are Cy, Pn, Pt, Dp but not Pg.[334]
[d] Present in small quantity in peel only.[211]
[e] Not detected in juice.[489]
[f] In juices.
[g] Also present in peel of Citrus Hassaku.[502]

Citrus sudachi

Flavones:[256,257]
5,7,4′-trihydroxy-6,8,3′-trimethoxyflavone (sudachitin);
5,7,4′-trihydroxy-6,8-dimethoxyflavone (demethoxysudachitin);
sudachitin-4′-β-D-glucoside (sudachiin A); 7-methyl-sudachitin;
xanthomicrol; dinatin; three sudachitin glycosides = sudachiins B, C, and D.

Coumarins: see Table 12.

Citrus unshiu

Flavonols:[368]
limocitrin 3-β-D-glucose; limocitrin 3-α-L-rhamnose;
Q 3-rutinoside; iR 3-rutinoside.

Flavanones:[368] naringenin 7-β-rutinoside (narirutin).

C-Glycosylflavones:[368] 3, 6-di-C-glucosylapigenin.

Coumarins: see Table 12.

Note on *Citrus* sp.:[5,59,202,217,221,602]

C. bergamia (bergamot oil): 5-hydroxy-7,8,3′,4′-tetramethoxyflavone.
C. grandis: Naringenin-7-neohesperidoside.
C. jambhiri: Hesperetin-7-rutinoside; hesperetin-7-neohesperidoside.
C. medica: Hesperetin-7-rutinoside.
C. mitis: 5,6,7,8,3′,4′-hexamethoxyflavone (citromitin); 5-demethylcitromitin; hesperetin-7-rutinoside.
C. natsudaidai: 3-hydroxy-5,6,7,8,3′,4′-hexamethoxyflavone (= natsudaidain).
Presence in citrus fruits of C-glycosylflavones: vitexin, vitexin-2″-xyloside; 6-glucosyldiosmetin; 8-glucosyldiosmetin; 6,8-diglucosyldiosmetin.[221]

Merrillia caloxylon

Flavones:[2,170,171]
5,7,3′,4′,5′-pentamethoxyflavone; 5,7,3′,4′-tetramethoxyflavone; 5-hydroxy-6,7,3′,4′-tetramethoxyflavone; 5-hydroxy-6,7,3′,4′,5′-pentamethoxyflavone; 5,3′-dihydroxy-6,7,4′-trimethoxyflavone (eupatorin).
Chalcones:[170]
6′-hydroxy-2′,4′,3,4-tetramethoxychalcone;
6′-hydroxy-2′,4′,3,4,5-pentamethoxychalcone;
6′,3-dihydroxy-2′,4′,4-trimethoxychalcone.

Murraya omphalocarpa

Flavonols:[649]
5,4′-dihydroxy-3,6,7,3′,5′-pentamethoxyflavone (murrayanol);
3,5,6,7,3′,4′,5′-heptamethoxyflavone.

Poncirus trifoliata PONCIRUS LEMON

Flavanones:[59,217]
hesperetin 7-neohesperidoside;
isosakuranetin 7-neohesperidoside (poncirin);
naringenin 7-neohesperidoside.

Sargentia greggii

7-Deoxyflavones:[221,642]
5,6-dimethoxyflavone; 5,6,3′,5′-tetramethoxyflavone (= cerrosillin).

XXVII. SAPINDACEAE

Litchi chinensis LITCHI

Anthocyanins[a]:[439]
Pg 3-rhamnoside; Pg 3,5-diglucoside;
Cy 3-galactoside; Cy 3-glucoside.
Flavonols:[439] K glycoside; Q 3-rhamnoside.

Comment: [a] Analysis on the fruit pericarp. A misprint exists in Prasad and Jha's article:[439] Pg 3-rhamnoside has been reported as Pg 3-glucoside. This error is also found in Timberlake[573] and Hrazdina.[265]

XXVIII. SAPOTACEAE

Achras sapota SAPOTA FRUIT

Ga; CQ; Catechins, proanthocyanidins.[365]

Synsepalum dulcificum MIRACLE FRUIT

Anthocyanins:[88]
Cy 3-galactoside; Cy 3-glucoside; Cy 3-arabinoside;
Dp 3-galactoside; Dp 3-arabinoside.
Flavonols[a]:[88] K 3-glucoside; Q 3-galactoside; M 3-galactoside.

Comment: [a] Free aglycones (K, Q, M) were also isolated and identified in the analysis of the peel fruit.

XXIX. SAXIFRAGACEAE

Ribes nigrum BLACKCURRANT

Phenolic acids:[305,460,512]
pHB; Sa: ProC; Ga; 3,4-diHB; Sy;5′-GaQ; pHB glucoside;
ProC glucoside; Ga glucose; Ga glucoside; 3′-pCQ; 4′-pCQ;
3′-CQ; 4′-CQ; 5′-CQ; 3′-FQ; 4′-FQ; 5′-FQ; pC glucose;
C glucose; F glucose; pC-glucoside; C glucoside; F glucoside.
Anthocyanins[a]:[169,305,575]
Cy 3-glucoside; **Cy 3-rutinoside;**
Dp 3-glucoside; **Dp 3-rutinoside.**
Flavonols:[305,523,548]
K 3-glucoside; K 3-rutinoside; Q 3-glucoside;
Q 3-rutinoside; M 3-glucoside; M 3-rutinoside.
Flavan-3-ols[b]:[550]
(+)-catechin; (−)-epicatechin; (+)-gallocatechin;
(−)-epigallocatechin.
Tannins: see Table 31.

Comments: [a] Anthocyanins: According to Hrazdina[265] and references therein, the fruits of *Ribes nigrum* would contain Pg 3-rutinoside added to the above mentioned pigments. This was not confirmed by later works.
[b] Flavan-3-ols are present in very small amount.

Ribes rubrum REDCURRANT

Phenolic acids:[460,512]
pHB; ProC; Ga; pHB glucoside; ProC glucoside; 5′GaQ; 3′-pCQ;
3′-CQ; 5′-CQ; 5′-FQ; pC glucose; C glucose; F glucose;
pC glucoside; C glucoside; F glucoside.
Anthocyanins[a]:[225,425]
Cy 3-glucoside; Cy 3-rutinoside; Cy 3-sambubioside;
Cy 3-sophoroside; Cy 3-(2^G-xylosylrutinoside);
Cy 3-(2^G-glucosylrutinoside).
Flavonols:[524,525,548]
K 3-rutinoside; Q 3-rutinoside; Q 3-glucoside;
Q-3-(2^G-rhamnosylrutinoside)[b]; M 3-glucoside.

Flavan-3-ols:[550]
(+)-catechin; (−)-epicatechin; (+)-gallocatechin;
(−)-epigallocatechin.
Tannins: see Table 31.

Comments: [a] A total of 29 species and hybrid species and 4 cultivars of the *Ribes* genus were examined by Harborne and Hall;[225] many of these species are designated by name of Redcurrant (cf. Chapter 1).
[b] The complete structure of the branched triglycoside is as follow: quercetin-3-*O*-(2″-*O*-α-L-rhamnosyl-6″-*O*-α-L-rhamnosyl)-β-D-glucoside.[524,525]

Ribes grossularia GOOSEBERRY

Phenolic acids:[251,460,512]
ProC; Ga glucose; pHB glucoside; ProC glucoside; Ga glucoside;
3′-pCQ; 3′-CQ; 5′-CQ; 3′-FQ; pC glucose; **C glucose;** F glucose;
pC glucoside; C glucoside; F glucoside.
Anthocyanins:[217,225] Cy 3-rutinoside; Cy 3-glucoside.
Flavan-3-ols[a]:[550] **(+)-catechin;** (−)-epicatechin.

Comment: [a] Flavan-3-ols were analyzed in green, yellow, and red varieties; (−)-epicatechin was absent in yellow var.

XXX. SCHIZANDRACEAE

Schizandra chinensis

35 lignans including schizandrin and isoschizandrin[33,276,407]

XXXI. SOLANACEAE

Capsicum annuum PEPPER (Sweet and Hot)
Capsicum fructescens

C glucose; F glucose; S glucose.[639]
Capsaicin, dihydrocapsaicin and other capsaicinoids in Hot Pepper.[177,214,493]

Cyphomandra betacea TAMARILLO FRUIT

Anthocyanins[a]:[646]
Pg 3-rutinoside; Pg 3-glucoside; **Cy 3 rutinoside;**
Cy 3-glucoside; Dp 3-rutinoside; Dp 3-glucoside.

Comment: [a] Cy 3-rutinoside is the major pigment in the skin, and Dp-3-rutinoside is the major pigment in seed jelly. In the peel, possible presence of acylated anthocyanins with *p*-coumaric, caffeic, and ferulic acids, but in the form of traces.

Lycopersicum esculentum TOMATO

Phenolic acids:[150,158,272,507,620,639]
Sa; V; 5′-pCQ; 3′-CQ; 4′-CQ; **5′-CQ;** 5′-FQ; pC glucose;
C glucose; F glucose; S glucose; pC glucoside; C glucoside;
F glucoside; pC bound to cutin.
Flavonols:[150,254,620] K 3-rutinoside; Q 3-rutinoside; Q 3-glucoside.

Flavonones:[254,272,620]

naringenin; naringenin 7-glucoside; naringenin 7-rutinoside.

Chalcones:[272] 2′,4,4′,6′-tetrahydroxychalcone (= chalconaringenin).

Miscellaneous:[347] Tyramine; serotonin.

Comments: Anthocyanins and flavan-3-ols are absent. The ripe fruit is poor in flavonoids. Rutin is localized in the pericarp but disappears during the maturation giving way to a new nonidentified compound.[150]. According to Wardale,[620] naringenin occurs in the free state only, but later works show clearly the presence of naringenin glycosides.[272]

Solanum guineese GARDEN HUCKLEBERRY

Anthocyanins:[167,217,573]

Pt-(*p*-coumarylrutinoside)-5-glucoside[a]; Pt 3-rutinoside-5-glucoside;
Pt 3-(di-*p*-coumarylrutinoside)-5-glucoside[b];
Mv 3-(*p*-coumarylrutinoside)-5-glucoside[c].

Comments: [a] Major pigment = petanin.
[b] = Guineesin[573]. The exact position of the two acyl groups in the molecule still remain to be determined.
[c] = Negretin.

Solanum melongena EGGPLANT

Phenolic acids[a]:[639]

5′-pCQ; 3′-CQ; 4′-CQ; 5′CQ;
5′-FQ; C glucose; F glucose; S glucose.

Anthocyanins:[571,573,575,621]

Dp 3-(p-coumarylglucoside)-5-glucoside; **Dp 3-(p-coumarylrutinoside)-5-glucoside**[b]; Dp 3-(caffeyldiglucoside)-5-glucoside; Dp 3-rutinoside; Dp 3-rutinoside-5-glucoside; Dp 3-glucoside; Dp 3,5-diglucoside; Dp 3-diglucoside-5-glucoside (?).

Coumarins:[17] scopoletin, scoparone.

Miscellaneous:[347] tyramine.

Comment: [a] pC, C, F, and S acids were mentioned after hydrolysis.[314]
[b] = Nasunin. The nasunin is the major pigment in the purple-colored peel varieties. Whereas in the varieties with a wine-red peel, delphinidin 3-rutinoside is major one. For nasunin, *p*-coumaric acid is fixed on the C4 of the rhamnose and the structure given by Watanabe et al.[621] is as follows; delphinidin 3-[4-(*p*-coumaroyl)-L-rhamnosyl(1→6)D-glucosido]-5-glucoside.

Solanum tuberosa POTATO BERRY

pC glucose; C glucoside; 7-hydroxycoumarin (umbelliferone).[111,224]

XXXII. UMBELLIFERAE (APIACEAE)

Phenolic acids:[130]

Pimpinella anisum (Anise): 3′pCQ; 5′pCQ; 3′-CQ; 4′-CQ; **5′-CQ;** 3′-FQ; 5′FQ.
Foeniculum vulgare, var. *dulce* (Fennel): 3′-pCQ; 5′-pCQ; 3′-CQ; 4′-CQ; **5′-CQ;** 3′-FQ; 4′-FQ; 5′-FQ.
Coriandrum sativum (Coriander): 3′-CQ; 4′-CQ; **5′-CQ;** 4′-FQ; 5′FQ.
Carum cawi (Caraway): 5′-pCQ; 3′-CQ; 4′-CQ; **5′-CQ.**
Illicum verum (Staranise): 3′pCQ; **4′-pCQ;** 5′-CQ.

4-(β-D-glucopyranosyloxy)benzoic acid is also present in these fruits and other Umbelliferae ones.[129]
Coumarins: see Chapter 1.

XXXIII. VERBENACEAE

Vitex agnus-castus

Flavonols:[643]
3,6,7,4′-tetramethylether of 6-hydroxyquercetin (casticin);
3,6,7,-trimethylether of 6-hydroxyquercetin (chrysosplenol-D);
3,6,7-trimethylether of 6-hydroxykaempferol (penduletin);
3,6,7,4′-tetramethylether of 6-hydroxykaempferol.

XXXIV. VITACEAE

Vitis vinifera GRAPE

Phenolic acids:[34,404,420,421,459,463,464,481,486,508,530,533]
pHB; oHB; Sa; Ga; Cin; pC tartaric acid; **C tartaric acid**[a];
F tartaric acid; pC glucose[b,c]; F glucose[c]; glucose ester of pC tartaric acid[d].
Anthocyanins[e]:[14,26,198,242,404,465,471,483,484,650]
Cy 3-glucoside; Cy 3-acetylglucoside; Cy 3-*p*-coumarylglucoside;
Pn 3-glucoside; Pn 3-acetylglucoside; Pn 3-*p*-coumarylglucoside;
Dp 3-glucoside; Dp 3-acetylglucoside; Dp 3-*p*-coumarylglucoside;
Pt 3-glucoside; Pt 3-acetylglucoside; Pt 3-*p*-coumarylglucoside;
Mv 3-glucoside; Mv 3-acetylglucoside; Mv 3-*p*-coumarylglucoside;
Mv 3-caffeylglucoside; Pn 3-caffeylglucoside?[f].
Flavonols[g]:[102,103,253,465,467,593]
K 3-glucoside; K 3-glucuronide; K 3-glucosylarabinoside?;
K 3-galactoside; Q 3-glucoside; Q 3-glucuronide; Q 3-rutinoside;
Q 3-glucosylgalactoside?; Q 3-glucosylxyloside?;
M 3-glucoside; M 3-glucuronide; iR 3-glucoside.
Flavan-3-ols:[251,322,466,530,554,593]
(+)-catechin; (−)-epicatechin; (+)-gallocatechin;
(−)-epigallocatechin; epicatechin-3-*O*-gallate
catechin-gallate; catechin-catechin-gallate.
Flavanonols:[587]
Dihydroquercetin 3-rhamnoside (= Astilbin).
Dihydrokaempferol-3-rhamnoside (= Engeletin).
Tannins: see Table 31.

Comments: [a] pC tartaric acid (= coutaric acid) may be more abundant than C tartaric acid (= caftaric acid) in some wine cultivars (see chapter 1); pC tartaric and C tartaric contents and their ratio seem to be a stable character in a chemotaxonomic point of view.[71,533]
[b] In some red cultivars.[459]
[c] In some white cultivars.[459]
[d] Pointed out recently in some cultivars.[541,542]
[e] Acylated anthocyanins may be missing in some cultivars of *Vitis vinifera* (e.g., Pinot noir[527]) as well as in *V. rotundifolia*. Presence of 3,5-diglucosides in the American *Vitis* species and their hybrids with *V. vinifera* (cf. Chapter 1).
[f] Pn 3-caffeylglucoside is indicated in Tempranillo variety, where the two anthocyanins acylated with caffeic acid represent no more than 0.2% of the total.[242]
[g] The white cultivars analyzed contain only derivatives of K and Q.[102,103]

XXXV. ZYGOPHYLLACEAE

Tribulus terrestris

Flavonols:[49] K; K 3-glucoside; K 3-rutinoside; tribuloside[a].

Comment: [a] The structure initially proposed was K 3-β-D-(6″-*p*-coumaroyl) glucoside, but it would appear to be a mixture of tiliroside and isorhamnetin derivatives.[229]

REFERENCES

1. **Abou Aziz, A. B., Abdel-Wahab, F. K., and El-Ghandour, M. A.,** Effect of different storage temperatures on phenolic compounds in banana and mango fruits, *Sci. Hortic.*, 4, 309, 1976.
2. **Adams, J. H. and Lewis, J. R.,** Eupatorin, a constituent of *Merrillia caloxylon, Planta Med.*, 32, 86, 1977.
3. **Afek, U., Szteinberg, A., and Carmely, S.,** 6,7-Dimethoxycoumarin, a citrus phytoalexin conferring resistance against *Phytophtora gummosis, Phytochemistry,* 25, 1855, 1986.
4. **Agrawal, G. D., Rizvi, S. A. I., Gupta, P. C., and Tewari, J. D.,** Structure of fistulic acid a new colouring matter from the pods of *Cassia fistula, Planta Med.*, 21, 150, 1972.
5. **Albach, R. F. and Redman, G. H.,** Composition and inheritance of flavanones in citrus fruit, *Phytochemistry,* 8, 127, 1969.
6. **Albach, R. F. and Redman, G. H., Cruse, R. R., and Petersen, H. D.,** Seasonal variation of bitterness components, pulp, and vitamin C in Texas commercial citrus juices, *J. Agric. Food Chem.*, 29, 805, 1981.
7. **Amiot, M. J.,** Degradation de l'oleuropeine de l'Olive *(Olea europaea* L.): Aspect s physiologiques, chimiques et technologiques, Thèse Doctorat, Université de Montpellier, France, 1986.
8. **Amiot, M. J., Fleuriet, A., and Macheix, J. J.,** Importance and evolution of phenolic compounds in olive during growth and maturation, *J. Agric. Food Chem.*, 34, 823, 1986.
9. **Amiot, M. J., Fleuriet, A., and Macheix, J. J.,** Accumulation of oleuropein derivatives during olive maturation, *Phytochemistry,* 28, 67, 1989.
10. **Andary, C., Roussel, J. L., Rascol, J. P., and Privat, G.,** Microméthode d'analyse des esters hétérosidiques de l'acide caféique, *J. Chromatogr.*, 303, 312, 1984.
11. **Andersen, O. M.,** Chromatographic separation of anthocyanins in Cowberry (Lingonberry) *Vaccinium vitis-idaea* L., *J. Food Sci.*, 50, 1230, 1985.
12. **Andersen, O. M.,** Anthocyanins in fruits of *Vaccinium japonicum, Phytochemistry,* 26, 1220, 1987.
13. **Andersen, O. M.,** Anthocyanins in fruits of *Vaccinium uliginosum* L. (Bog Whortleberry), *J. Food Sci.*, 52, 665, 1987.
14. **Anderson, D. W., Gueffroy, D. E., Webb, A. D., and Kepner, R. E.,** Identification of acetic acid as an acylating agent of anthocyanin pigments in grapes, *Phytochemistry,* 9, 1579, 1970.
15. **Aoki, S., Yahagi, Y., and Tamura, T.,** The use of polyvinylpyrrolidone column for the quantitative determination of chlorogenic acid in apple fruit, *J. Jpn. Soc. Food Sci. Technol.*, 31, 110, 1984.
16. **Arakawa, O., Hori, Y., and Ogata, R.,** Relative effectiveness and interaction of ultraviolet-B red and blue light in anthocyanin synthesis of apple fruit, *Physiol. Plant,* 64, 323, 1985.
17. **Ashy, M. A. and El-Tawil, B. A. H.,** Constituents of local plants. IX. Distribution of some coumarins in different plant families grown in Saudi-Arabia, *Herba Hung.*, 20, 197, 1981.
18. **Askar, A., Rubach, K., and Schormueller, J.,** Thin layer chromatographic separation of the amine fraction of banana, *Chem. Mikrobiol. Technol. Lebensmn.*, 1, 187, 1972.
19. **Aubert, S.,** L'Aubergine (*Solanum melongena* L.). I. Composition et facteurs de qualité, *Ann. Technol. Agric.*, 20, 241, 1971.
20. **Aubert, S. and Pochard, E.,** Problème de conservation en frais de l'aubergine (*Solanum melongena* L.). I. Etude bibliographique, *P.H.M. Rev. Hortic.*, 125, 33, 1981.
21. **Azar, M.,** Contribution à l'étude des compos és polyphénoliques du jus fermenté de Myrtille *(Vaccinium myrtillus),* Thèse Doctorat, Université de Montpellier, France, 1986.
22. **Azar, M., Verette, E., and Brun, S.,** Identification of some phenolic compounds in bilberry juice *Vaccinium myrtillus, J. Food Sci.*, 52, 1255, 1987.
23. **Baijal, M., Singh, S., Shukla, R. N., and Sanwal, G. G.,** Enzymes of the banana plant: optimum conditions for extraction, *Phytochemistry,* 11, 929, 1972.

24. **Bailey, J. A. and Mansfield, J. W.**, *Phytoalexins*, Blackie, Glascow, 1982.
25. **Baj, A., Bombardelli, E., Gabetta, B., and Martinelli, E. M.**, Qualitative and quantitative evaluation of *Vaccinium myrtillus* anthocyanins by high-resolution gas chromatography and high-performance liquid chromatography, *J. Chromatogr.*, 279, 365, 1983.
26. **Bakker, J. and Timberlake, C. F.**, The distribution of anthocyanins in grape skin extracts of Port wine cultivars as determined by high performance liquid chromatography, *J. Sci. Food Agric.*, 36, 1315, 1985.
27. **Balasubramanian, K. and Rajagopalan, K.**, Novel xanthones from *Garcinia mangostana*, structures of BR-xanthone-A and BR-xanthone-B, *Phytochemistry*, 27, 1552, 1988.
28. **Ballinger, W. E., Galletta, G. J., and Maness, E. P.**, Anthocyanins of fruits of *Vaccinium, sub-genera Cyanococcus* and *Polycodium*, *J. Am. Soc. Hortic. Sci.*, 104, 554, 1979.
29. **Ballinger, W. E. Maness, E. P., and Kushman, L. J.**, Anthocyanins in ripe fruit of the highbush blueberry *Vaccinium corymbosum* L., *J. Am. Soc. Hortic. Sci.*, 95, 283, 1970.
30. **Ballinger, W. E., Maness, E. P., Galletta, G. J., and Kushman, L. J.**, Anthocyanins of ripe fruit of a "pink-fruited" hybrid of highbush blueberries, *Vaccinium corymbosum* L., *J. Am. Soc. Hortic. Sci.*, 97, 381, 1972.
31. **Ballington, J. R., Ballinger, W. E., and Maness, E. P.**, Interspecific differences in the percentage of anthocyanins aglycones and aglycone-sugars in the fruit of seven species of blueberries, *J. Am. Soc. Hortic. Sci.*, 112, 859, 1987.
32. **Ballington, J. R., Kirkman, W. B., Ballinger, W. E., and Maness, E. P.**, Anthocyanin, aglycone, and aglycone-sugar content in the fruits of temperate North American species of four sections in *Vaccinium*, *J. Am. Soc. Hortic. Sci.*, 113, 746, 1988.
33. **Bao, T. T., Liu, G. T., Song, Z. Y., Xu, G. I., and Sun, R. H.**, A comparison of the pharmacological effects of seven constituents isolated from the fruit of Schizandrae, *Chin. Med. J.*, 93, 41, 1980.
34. **Baranowski, J. D. and Nagel, C. W.**, Isolation and identification of the hydroxycinnamic acid derivatives in white Riesling wine, *Am. J. Enol. Vitic.*, 32, 5, 1981.
35. **Barbosa-Filho, J. M., Yoshida, M., and Gottlieb, O. R.**, Neolignans from the fruits of *Licaria armeniaca*, *Phytochemistry*, 26, 319, 1987.
36. **Barbosa-Filho, J. M., Yoshida, M., Gottlieb, O. R., Barbosa, R. C. S. B. C., Giesbrecht, A. M., and Young, C. M.**, Benzoyl esters and amides, styrylpyrones and neolignans from the fruits of *Aniba riparia*, *Phytochemistry*, 26, 2615, 1987.
37. **Barrit, B. H. and Torre, L. C.**, Cellulose thin-layer chromatographic separation of *Rubus* fruit anthocyanins, *J. Chromatogr.*, 75, 151, 1973.
38. **Barrit, B. H. and Torre, L. C.**, Fruit anthocyanin pigments of red raspberry cultivars, *J. Am. Soc. Hortic. Sci.*, 100, 98, 1975.
39. **Barros, D. A. D., De Alvarenga, M. A., Gottlieb, O. R., and Gottlieb, H. E.**, Naringenin coumaroylglucosides from *Mabea caudata*, *Phytochemistry*, 21, 2107, 1982.
40. **Barthe, G. A., Jourdan, P. S., McIntosh, C. A., and Mansell, R. L.**, Radioimmunoassay for the quantitative determination of hesperidin and analysis of its distribution in *Citrus sinensis*, *Phytochemistry*, 27, 249, 1988.
41. **Bate-Smith, E. C.**, Haemanalysis of tannins: the concept of relative astringency, *Phytochemistry*, 12, 907, 1973.
42. **Bate-Smith, E. C.**, Phytochemistry of proanthocyanidins, *Phytochemistry*, 14, 1107, 1975.
43. **Bate-Smith, E. C.**, Astringent tannins of *Acer* species, *Phytochemistry*, 16, 1421, 1977.
44. **Bate-Smith, E. C.**, Astringent tannins of the leaves of *Geranium* species, *Phytochemistry*, 20, 211, 1981.
45. **Beart, J. E., Lilley, T. H., and Haslam, E.**, Plant polyphenols. Secondary metabolism and chemical defence: some observations, *Phytochemistry*, 24, 33, 1985.
46. **Beckman, C. H. and Mueller, W. C.**, Distribution of phenols in specialized cells of banana roots, *Phytopathology*, 60, 79, 1970.
47. **Bell, E. A.**, The physiological role(s) of Secondary (Natural) products, in *The Biochemistry of Plants*, Vol. 7, *Secondary Plant Products*, Conn, E. E., Ed., Academic Press, New York, 1981, 1.
48. **Berenbaum, M. R., Zangerl, A. R., and Nitao, J. K.**, Furanocoumarins in seeds of wild and cultivated parsnips, *Phytochemistry*, 23, 1809, 1984.
49. **Bhutani, S. P., Chibber, S. S., and Seshadri, T. R.**, Flavonoids of the fruits and leaves of *Tribulus terrestris:* constitution of tribuloside, *Phytochemistry*, 8, 299, 1969.
50. **Biale, J. B. and Young, R. E.**, The avocado pear, in *The Biochemistry of Fruits and Their Products*, Vol. 2, Hulme A. C., Ed., Academic Press, London, 1971, chap. 1.
51. **Bianchini, J. P. and Gaydou, E. M.**, Separation of polymethoxylated flavones by straight-phase high-performance liquid chromatography, *J. Chromatogr.*, 190, 233, 1980.
52. **Billot, J.**, Analyse des pigments anthocyaniques des feuilles, *Ann. Univ. Madagascar*, 6, 123, 1968.
53. **Billot, J.**, Méthode spectrophotométrique pour l'estimation des composés phénoliques. Application aux phénols de la Poire Passe-crassane, *Physiol. Vég.*, 18, 655, 1980.

54. **Billot, J.**, Dosage des composés phénoliques dans les extraits végétaux par oxydation cuivrique en milieu non aqueux, *Anal. Chim. Acta*, 119, 367, 1980.
55. **Billot, J.**, Evolution des composés phénoliques au cours de la maturation de la poire Passe-Crassane, *Physiol. Vég.*, 21, 527, 1983.
56. **Billot, J., Hartmann, C., Macheix, J. J., and Rateau, J.**, Les composés phénoliques au cours de la croissance de la poire Passe-Crassane, *Physiol. Vég.*, 16, 693, 1978.
56a. **Billot, J.**, unpublished results.
57. **Bilyk, A. and Sapers, G. M.**, Varietal differences in the quercetin, kaempferol and myricetin contents of Highbush blueberry, cranberry and thornless blackberry fruits, *J. Agric. Food Chem.*, 34, 585, 1986.
58. **Bisson, J.**, Application de l'étude des matières colorantes du raisin noir à la sélection variétale, Thèse 3ème cycle, Université Bordeaux II, France, 1980.
59. **Bohm, B. A.**, Flavanones and dihydroflavonols, in *The Flavonoids*, Harborne, J. B., Mabry, T. J., and Mabry, H., Eds., Chapman and Hall, London, 1975, 560.
60. **Bohm, B. A.**, Chalcones, aurones and dihydrochalcones, in *The Flavonoids*, Harborne, J. B., Mabry, T. J., and Mabry, H., Eds., Chapman and Hall, London, 1975, 442.
61. **Bohm, B. A.**, The minor flavonoids, in *The Flavonoids: Advances in Research*, Harborne, J. B. and Mabry, T. J., Eds., Chapman and Hall, London, 1982, 313.
62. **Bombardelli, E., Bonati, A., Gabetta, B., Martinelli, E. M., and Mustich, G.**, Gas-liquid chromatographic and mass spectrometric identification of anthocyanidins, *J. Chromatogr.*, 120, 115, 1976.
63. **Bombardelli, E., Bonatti, A., Gabetta, B., Martinelli, E. M., and Mustich, G.**, Identification of anthocyanins by gas-liquid chromatography and mass spectrometry, *J. Chromatogr.*, 139, 111, 1977.
64. **Bompeix, G. and Bousquet, J. F.**, Les composés phénoliques dans la résistance des pommes à l'infection par les *Pezicula malicortis* et *P. alba*, *Fruits*, 29, 693, 1974.
65. **Borazjani, A., Graves, C. H., Jr., and Hedin, P. A.**, A survey of Juglone levels among walnuts and hickories, *Pecan Q.*, 17, 9, 1983.
66. **Borazjani, A., Graves, C. H., Jr., and Hedin, P. A.**, Occurrence of juglone in various tissues of Pecan and related species, *Phytopathology*, 75, 1419, 1985.
67. **Bostock, R. M., Thomas, C. S., Ogawa, J. M., Rice, R. E., and Uyemoto, J. K.**, Relationship of wound-induced peroxidase activity to epicarp lesion development in maturing Pistachio fruit, *Phytopathology*, 77, 275, 1987.
68. **Boudet, A. and Gadal, P.**, Sur l'inhibition des enzymes par les tanins des feuilles de *Quercus sessilis* Ehrh. Isolement des tanins, *C.R. Acad. Sci. Paris*, 260, 4057, 1965.
69. **Boudet, A. M., Alibert, G., and Marigo, G.**, Vacuoles and tonoplast in *the regulation of cellular metabolism, Ann. Proc. Phytochem. Soc. Europe*, Vol. 24, Boudet, A.M., Alibert, G., Marigo, G., and Lea, P. J., Eds., 1984, 29.
70. **Boursiquot, J. M.**, Contribution à l'étude des esters hydroxycinnamoyltartriques chez le genre *Vitis*. Recherche d'application taxonomique, Thèse Doct. Ing., ENSA, Montpellier, France, 1987.
71. **Boursiquot, J. M., Sapis, J. C., and Macheix, J. J.**, Les esters hydroxycinnamiques chez le genre *Vitis*. Essai d'application taxonomique. Premiers résultats, *C.R. Acad. Sci., Paris Ser. III*, 302, 177, 1986.
72. **Bourzeix, M., Weyland, D., and Heredia, N.**, Etude des catechines et des procyanidols de la grappe de raisin, du vin et d'autres dérivés de la vigne, *Bull. O.I.V.*, 669—670, 1171, 1986.
73. **Bradfield, A. E., Flood, A. E., Hulme, A. C., and Williams, A. H.**, Chlorogenic acid in fruit trees, *Nature (London)*, 170, 168, 1952.
74. **Brandl, W. and Herrmann, K.**, Analytical and preparative high-performance liquid chromatography of hydroxycinnamic acid esters, *J. Chromatogr.*, 260, 447, 1983.
75. **Brasseur, Th. and Angenot, L.**, Un réactif de choix pour la révélation des flavonoides: le mélange diphénylborate d'aminoethanol — PGE 400, *Bull. Liaison Groupe Polyphénols*, 13, 139, 1986.
76. **Brenneisen, V. R. and Steinegger, E.**, Zur analytik des polyphenole der fruchte von *Vaccinium myrtillus* L. unterschiedlichen Reifegrades, *Pharm Acta Helv.*, 56, 341, 1981.
77. **Broadhurst, R. B. and Jones, W. T.**, Analysis of condensed tannins using acidified vanillin, *J. Sci. Food Agric.*, 29, 788, 1978.
78. **Bronnum-Hansen, K. and Flinck, J. M.**, Anthocyanin colorants from Elderberry (*Sambucus nigra* L.). IV. Further studies on production of liquid extracts, concentrates and freeze dried powder, *J. Food Technol.*, 21, 605, 1986.
79. **Bronnum-Hansen, K. and Hansen, S. H.**, High-performance liquid chromatographic separation of anthocyanins of *Sambucus nigra* L., *J. Chromatogr.*, 262, 385, 1983.
80. **Brouillard, R.**, Chemical structure of Anthocyanins, in *Anthocyanins as Food Colors*, Markakis, P., Ed., Academic Press, New York, 1982, 1.
81. **Brouillard, R.**, The *in vivo* expression of anthocyanin colour in plants, *Phytochemistry*, 22, 1311, 1983.
82. **Brouillard, R.**, Organisation des anthocyanes dans leurs sites naturels, *Bull. Liaison Groupe Polyphénols*, 13, 76, 1986.

83. **Brouillard, R. and Delaporte, B.**, Chemistry of anthocyanin pigments. II. Kinetic and thermodynamic study of proton transfer, hydration and tautomeric reactions of malvidin 3-glucoside, *J. Am. Chem. Soc.*, 99, 8461, 1977.
84. **Brouillard, R. and Delaporte, B.**, Dégradation thermique des anthocyanes, *Bull. Liaison Groupe Polyphénols*, 8, 305, 1978.
85. **Brown, S. A.**, Coumarins, in *The Biochemistry of Plants*, Vol. 7, Conn, E.E., Ed., Academic Press, New York, 1981, 269.
86. **Brown, S. A.**, Recent advances in the biosynthesis of coumarins, in *Ann. Proc. Phytochem. Soc. Europe*, Vol. 25, Van Sumere, C.F. and Lea, P.J., Eds., Clarendon Press, Oxford, 1985, 257.
87. **Brunet, G. and Ibrahim, R. K.**, Tissue culture of citrus peel and its potential for flavonoid synthesis, *Z. Planzenphysiol.*, 69, 152, 1973.
88. **Buckmire, R. E. and Francis, F. J.**, Anthocyanins and flavonols of miracle fruit, *Synsepalum dulcificum*, Schum., *J. Food Sci.*, 41, 1363, 1976.
89. **Bureau, D., Macheix, J. J., and Rouet-Mayer, M. A.**, Relations entre la teneur en o-diphénols, l'activité polyphénoloxydasique et l'aptitude au brunissement de quelques variétés de Pêches, *Lebensm. Wiss. Technol.*, 10, 211, 1977.
90. **Burke, B. and Nair, M.**, Phenylpropene benzoic acid and flavonoid derivatives from fruits of Jamaican piper species, *Phytochemistry*, 25, 1427, 1986.
91. **Cabanne, F., Martin-Tanguy, J., and Martin, C.**, Phénolamines associées à l'induction florale et à l'état reproducteur de *Nicotiana tabacum* var. Xanthi n.c., *Physiol. Vég.*, 15, 429, 1977.
92. **Cabanne, F., Dalebroux, M. T., Martin-Tanguy, J., and Martin, C.**, Hydroxycinnamic acid amides and ripening to flower of *Nicotiana tabacum* var. xanthi nc., *Physiol. Plant.*, 53, 399, 1981.
93. **Carbonnier-Jarreau, M. C., Carbonnier, J., Farille, M., and Molho, D.**, Etude de la composition en coumarines de quelques *Heracleum* (Pastinaceae: Umbelliferae) pendant leur développement, *Bull. Mus. Natl. Hist. Nat. Sci. Phys. Chim.*, 19, 3, 1978.
94. **Carpenter, I., Locksley, H. D., and Scheinmann, F.**, Xanthones in higher plants: biogenetic proposals and a chemotaxonomic survey, *Phytochemistry*, 8, 2013, 1969.
95. **Cartwright, R. A., Roberts, E. A. H., Flood, A. E., and Williams, A. H.**, The suspected presence of *p*-coumaryl-quinic acids in tea, apple and pear, *Chem. Ind.*, 1062, 1955.
96. **Ceska, O., Chaudhary, S. K., Warrington, P. J., and Ashwood-Smith, M. J.**, Photoactive furocoumarins in fruits of some Umbellifers, *Phytochemisty*, 26, 165, 1987.
97. **Challice, J. S.**, Phenolic compounds of the subfamily Pomoideae: a chemotaxonomic survey, *Phytochemistry*, 12, 1095, 1973.
98. **Challice, J. S. and Westwood, M. N.**, Phenolic compounds of the genus *Pyrus*, *Phytochemistry*, 11, 37, 1972.
99. **Charrière-Ladreix, Y.**, La sécrétion flavonique chez *Populus nigra* L.: signification fonctionnelle des remaniements ultrastructuraux glandulaires, *Physiol. Vég.*, 15, 619, 1977.
100. **Chatterjee, P. and Chatterjee, A.**, Dihydroflavonol-O-glycoside of *Citrus sinensis*, *Phytochemistry*, 27, 946, 1988.
101. **Cheminat, A. and Brouillard, R.**, RMN Spectrometry of anthocyanins, *Bull. Liaison Groupe Polyphénols*, 13, 142, 1986.
102. **Cheynier, V. and Rigaud, J.**, HPLC Separation and characterization of flavonols in the skins of *Vitis vinifera* var. Cinsault, *Am. J. Enol. Vitic.*, 37, 248, 1986.
103. **Cheynier, V. and Rigaud, J.**, Identification et dosage de flavonols du Raisin, *Bull. Liaison Groupe Polyphénols*, 13, 442, 1986.
104. **Chopin, J. and Bouillant, M. L.**, C-Glycosylflavonoids, In *The Flavonoids*, Harborne, J. B., Mabry, T. J., and Mabry, H., Eds., Chapman and Hall, London, 1975, 632.
105. **Chopin, J., Bouillant, M. L., and Besson, E.**, C-glycosylflavonoids, in *The Flavonoids: Advances in Research*, Harborne, J. B. and Mabry, T. J., Eds., Chapman and Hall, London, 1982, 449.
106. **Chu, N. T., Clydesdale, F. M., and Francis, F. J.**, Isolation and identification of some fluorescent phenolic compounds in cranberries, *J. Food Sci.*, 38, 1038, 1973.
107. **Co, H. and Markakis, P.**, Flavonoid compounds in the strawberry fruit, *J. Flood Sci.*, 33, 281, 1968.
108. **Colonna, J. P.**, L'acide chlorogénique et les depsides de divers cafeiers africians et malgaches: leur participation au métabolisme et leur signification biologique, Travaux Documents ORSTOM, Paris, 102, 1, 1979.
109. **Côme, D.**, Evolution de la teneur en composés phénoliques des graines de Pommier Pirus malus L. au cours de leur développment, *Ann. Sci. Nat. Bot.*, 12e S, IX, 511, 1968.
110. **Côme, D., Fleuriet, A., and Macheix, J. J.**, Etude comparée de la croissance ponderale des jeunes graines de Pommier et de l'évolution de leur teneur en phloridzine, *C.R. Acad. Sci., Paris Ser. D*, 272, 1638, 1977.
111. **Corner, J. J. and Harborne, J. B.**, Cinnamic acid glucose derivatives in fruits, *Chem. Ind.*, 76, 1960.
112. **Corse, J., Lundin, R. E., and Waiss, A. C.**, Identification of several components of isochlorogenic acid, *Phytochemistry*, 4, 527, 1965.

113. **Cotter, D. J.,** A review of studies on Chili, *Bull. N.M. State Univ. Agric. Stn.*, 673, 1980.
114. **Crippen, D. D., Jr. and Morrisson, J. C.,** The effects of sun exposure on the phenolic content of Cabernet Sauvignon berries during development, *Am. J. Enol. Vitic.*, 37, 243, 1986.
115. **Czochanska, Z., Foo, L. Y., and Porter, L. J.,** Compositional changes in lower molecular weight flavans during grape maturation, *Phytochemistry*, 18, 1819, 1979.
116. **Czochanska, Z., Foo, L. Y., Newman, R. H., and Porter, L. J.,** Polymeric proanthocyanidins. Stereochemistry, structural units, and molecular weight, *J. Chem. Soc., Perkin I*, 2278, 1980.
117. **Dagne, E., Dinku, B., Gray, A. I., and Waterman, P. G.,** Pumilaisoflavones A and B from the seed pods of *Tephrosia pumila, Phytochemistry*, 27, 1503, 1988.
118. **Dean, B. B. and Kolattukudy, P. E.,** Synthesis of suberin during wound-healing in jade leaves, tomato fruit, and bean pods, *Plant Physiol.*, 58, 411, 1976.
119. **Dekazos, E. D.,** Anthocyanin pigments in red tart cherries, *J. Food Sci.*, 35, 237, 1970.
120. **Dekazos, E. D. and Worley, J. F.,** Effect of succinic acid 2,2-dimethyl hydrazide on anthocyanin metabolism and cell wall carbohydrates in red tart cherries, *Prunus cerasus* L., cv Montmorency, *J. Am. Soc. Hortic. Sci.*, 95, 703, 1970.
121. **Delaporte, N. and Macheix, J. J.,** Sur une méthode spectrophotométrique de dosa ge de l'acide ascorbique, de l'acide chlorogénique et des catéchines dans les pommes, *Chim. Anal.*, 50, 187, 1968.
122. **Delle Monache, F. D., Ferrari, F., Poce-Tucci, A., and Marini-Bettolo, G. B.,** Catechins with (+)-epi-configuration in nature, *Phytochemistry*, 11, 2333, 1972.
123. **Deshpande, S. S., Cheyran, M., and Salunkhe, D. K.,** Tannins analysis of food product s, *C.R.C. Crit. Rev. Food Sci. Nutr.*, 24, 401, 1986.
124. **Dewick, P. M.,** Isoflavonoids, in *The Flavonoids: Advances in Research*, Harborne, J. B. and Mabry, T. J., Eds., Chapman and Hall, London, New York 1982, 535.
125. **Diaz, A. M. and Martin, T.,** Premières recherches sur les pigments présents dans les fruits de *Phillyrea latifolia, 6ème Coll. Int. Plantes Medicinales et Substances d'Origine Naturelle*, Angers, 1988.
126. **Dick, A. J.,** Apple fruit polyphenols and fruit softening, *Bull. Liaison Groupe Polyphénols*, 13, 358, 1986.
127. **Dick, A. J., Redden, P. R., de Marco, A. C., Lidster, P. D., and Grindley, T. B.,** Flavonoid glycosides of Spartan apple peel, *J. Agric. Food Chem.*, 35, 529, 1987.
128. **Dieudonné, G.,** Les dérivés hydroxycinnamiques de l'ananas Comosus variété Cayenne Lisse (Côte d'Ivoire) et leurs relations avec le brunissement interne, Thése Doct. Ing., Université Pierre et Marie Curie, Paris, 1977.
129. **Dirks, U. and Herrmann, K.,** 4-(β-D-Glucopyranosyloxy) benzoic acid, a characteristic phenolic constituent of the Apiaceae, *Phytochemistry*, 23, 1811, 1984.
130. **Dirks, U. and Herrmann, K.,** Hochleistungsflüssigkeitschromatographic der hydroxycinnamoylchinasäuren und der 4(β-D-Glucopyranosyloxy) benzoesaüre in gewurzen. 10. Uber Gewurzphenole, *Z. Lebensm. Unters. Forsch.*, 179, 12, 1984.
131. **Dixon, R. A., Dey, P. M., and Lamb, C. J.,** Phytoalexins: enzymology and molecular biology, *Adv. Enzymol.*, 55, 1, 1983.
132. **Domon, B. and Hostettmann, K.,** Mass spectrometric studies of underrivatized polyphenolic glycosides, *Phytochemistry*, 24, 575, 1985.
133. **Donnelly, D. M. X.,** Neoflavanoids, in *The Flavonoids*, Harborne, J. B., Mabry, T. J., and Mabry, H., Eds., Chapman and Hall, London, 1975, 801.
134. **Du, C. T. and Francis, F. J.,** Anthocyanins of mangosteen, *Garcinia mangostana, J. Food Sci.*, 42, 1667, 1977.
135. **Du, C. T., Wang, P. L., and Francis, F. J.,** Anthocyanins of ornamental cherries, *J. Food Sci.*, 40, 1142, 1975.
136. **Du, C. T., Wang, P. L., and Francis, F. J.,** Anthocyanins of pomegranate, *Punica granatum, J. Food Sci.*, 40, 417, 1975.
137. **Dubois, G. E., Crosby, G. A., and Saffron, P.,** Nonnutritive sweeteners: taste-structure relationships for some new simple dihydrochalcones, *Science*, 195, 397, 1977.
138. **Duggan, M. B.,** Identity and occurrence of certain flavonol glycosides in four varieties of pears, *J. Agric. Food Chem.*, 17, 1098, 1969.
139. **Dull, G. G.,** The Pineapple: general, in *Biochemistry of Fruits and Their Products*, Vol. 2, Hulme, A. C., Ed., Academic Press, London, 1971, 303.
140. **Ebel, J. and Hahlbrock, K.,** Enzymes of flavone and flavonol glycoside biosynthesis. Coordinated and selective induction in cell suspension culture of *Petrosilum hortense, Eur. J. Biochem.*, 75, 201, 1977.
141. **El. Ansari, M. A., Reddy, K. K., Sastry, K. N. S., and Nayudamma, Y.,** Polyphenols of *Mangifera indica, Phytochemistry*, 10, 2239, 1971.
142. **El Sissi, H. I., Saleh, N. A. M., El Negoumy, S. I., Wagner, H., Apremeya Iyengar, and Seligmann, O.,** Prunin-O-6″-Gallat, aus *Acacia farnesiana, Phytochemistry*, 13, 2843, 1974.
143. **Faragher, J. D.,** Temperature regulation of anthocyanin accumulation in apple skin, *J. Exp. Bot.*, 34, 1291, 1983.

144. **Faragher, J. D. and Brohier, R. L.**, Anthocyanin accumulation in apples skin during ripening: regulation by ethylene and phenylalanine ammonialyase, *Sci. Hortic.*, 22, 89, 1984.
145. **Fernandez Diez, M. J.**, The olive, in *The Biochemistry of Fruits and Their Products*, Vol. 2, Hulme, A.C., Ed., Academic Press, London, 1971, 255.
146. **Fisher, D. J.**, Phenolic compounds of the apple fruit cuticle, *A.R. Long Ashton Agric. Hortic. Res. Stat.*, 255, 1965—1966.
147. **Fisher, J. F.**, A high-performance liquid chromatographic method for the quantitation of hesperidin in orange juice, *J. Agric. Food Chem.*, 26, 1459, 1978.
148. **Fisher, J. F. and Wheaton, T. A.**, A high-pressure liquid chromatographic method for the resolution and quantitation of naringin rutinoside in grapefruit juice, *J. Agric. Food Chem.*, 24, 898, 1976.
149. **Fletcher, A. C., Porter, L. J., Haslam, E., and Gupta, R. K.**, Plant proanthocyanidins. Part 3. Conformational and configurational studies of natural procyanidins, *J. Chem. Soc. Perkin I*, 1628, 1977.
150. **Fleuriet, A.**, Evolution des composés phénoliques au cours de la croissance et de la maturation des fruits de tomates "cerise" (*Lycopersicum esculentum* var. cerasiforme), *Fruits*, 31, 117, 1976.
151. **Fleuriet, A.**, Expression et régulation du métabolisme des dérivés hydroxycinnamiques au cours de la croissance, de la maturation et de la réaction aux blessures du fruit de la tomate "cerise" (*Lycopersicum esculentum*, var. cerasiforme), Thèse Doct. Etát, Montpellier, France, 1982.
152. **Fleuriet, A. and Deloire, A.**, Aspects histochimiques et biochimiques de la cicatrisation des fruits de tomate blessés, *Z. Pflanzenphysiol.*, 107, 259, 1982.
153. **Fleuriet, A. and Macheix, J. J.**, Séparation et dosage par chromatographie en phase gazeuse de l'acide chlorogénique et des catéchines des fruits, *J. Chromatogr.*, 74, 339, 1972.
154. **Fleuriet, A. and Macheix, J. J.**, Effet de conditions anaérobies sur les composés phénoliques des fruits de tomates "cerise", *Physiol. Vég.*, 14, 407, 1976.
155. **Fleuriet, A. and Macheix, J. J.**, Libération des acides hydroxycinnamiques à partir de leurs esters dans des fruits soumis à un traitement anaérobie, *Bull. Liaison Groupe Polyphénols*, 8, 389, 1978.
156. **Fleuriet, A. and Macheix, J. J.**, Dérivés glucosés des acides hydroxycinnamiques de la tomate: évolution au cours de la vie du fruit et formation *in vitro*, *Phytochemistry*, 19, 1955, 1980.
157. **Fleuriet, A. and Macheix, J. J.**, Quinyl esters and glucose derivatives of hydroxycinnamic acids during growth and ripening of tomato fruit, *Phytochemistry*, 20, 667, 1981.
158. **Fleuriet, A. and Macheix, J. J.**, Tissue compartimentation of phenylpropanoid metabolism in tomatoes during growth and maturation, *Phytochemistry*, 24, 929, 1985.
159. **Fleuriet, A., Macheix, J. J., Andary, C., and Villemur, P.**, Mise en évidence et dosage par chromatographie liquide à haute performance du verbascoside dans le fruit de six cultivars d'*Olea europaea* L., *C.R. Acad. Sci. Paris Ser. III*, 299, 253, 1984.
160. **Foo, L. Y.**, Proanthocyanidins: gross chemical structures by infrared spectra, *Phytochemistry*, 20, 1397, 1981.
161. **Foo, L. Y. and Porter, L. J.**, The phytochemistry of proanthocyanidin polymers, *Phytochemistry*, 19, 1747, 1980.
162. **Foo, L. Y. and Porter, L. J.**, The structure of tannins of some edible fruits, *J. Sci. Food Agric.*, 32, 711, 1981.
163. **Foo, L. Y., Jones, W. T., Porter, L. J., and Williams, V. M.**, Proanthocyanidin polymers of fodder legumes, *Phytochemistry*, 21, 933, 1982.
164. **Forsyth, W. G. C.**, Tannin in solid foods, in *the Quality of Foods and Beverages*, Charalambous, G. and Inglett, G., Eds., Vol. 1, Academic Press, New York, 1981, 377.
165. **Francis, F. J.**, Anthocyanins in pears, *HortScience*, 5, 42, 1970.
166. **Francis, F. J.**, Analysis of anthocyanins, in *Anthocyanins as Food Colors*, Markakis, P., Ed., Academic Press, New York, 1982, 182.
167. **Francis, F. J. and Harborne, J. B.**, Anthocyanins of the garden Huckleberry, *Solanum guineese*, *J. Food Sci.*, 31, 524, 1966.
168. **Francis, F. J., Harborne, J. B., and Barker, W. G.**, Anthocyanins in the Lowbush Blueberry, *Vaccinium angustifolium*, *J. Food Sci.*, 31, 583, 1966.
169. **Francis, G. W. and Andersen, O. M.**, Droplet counter-current chromatography of anthocyanins, *J. Chromatogr.*, 283, 445, 1984.
170. **Fraser, A. W. and Lewis, J. R.**, Some phenolic constituents of the Rutaceae, *Phytochemistry*, 11, 868, 1972.
171. **Fraser, A. W. and Lewis, J. R.**, Flavonoids from *Merrillia caloxylon*, *Phytochemistry*, 13, 1561, 1974.
172. **Friedrich, H. and Schonert, J.**, Untersuchungen uber Einige Inhaltsstoffe der Blatter und Fruchte von *Vaccinium myrtillus*, *Planta Med.*, 24, 90, 1973.
173. **Fry, S. C.**, Phenolic components of the primary cell wall and their possible role in the hormonal regulation of growth, *Planta*, 146, 343, 1979.
174. **Fry, S. C.**, Feruloylated pectins from the primary cell wall: their structure and possible functions, *Planta*, 157, 111, 1983.

175. **Frylinck, L., Dubery, I. A., and Schabort, J. C.,** Biochemical changes involved in stress response and ripening behaviour of γ-irradiated Mango fruits, *Phytochemistry,* 26, 681, 1987.
176. **Fuertes-Lasala, M. E., Fernandez, M., Garcia-Mina, M. C., Vega, F. A., and Martinez Valls, L.,** Componentes fenolicos en variedades cultivadas de *Pyrus communis* L. I. Estudio qualitativo, *Anal. Bromatol.,* 26, 59, 1974.
177. **Fujiwake, H., Suzuki, T., and Iwai, K.,** Intracellular localization of capsaicin and its analogues in Capsicum fruit. II. The vacuole as the intracellular accumulation site of capsaicinoid in the protoplast in Capsicum fruit, *Plant Cell Physiol.,* 21, 1023, 1980.
178. **Fuke, Y. and Matsuoka, H.,** Changes in content of pectic substances, ascorbic acid and polyphenols and activity of pectinesterase in Kiwi fruit during growth and ripening after harvest, *J. Jpn. Soc. Food Sci. Technol.,* 31, 31, 1984.
179. **Fuleki, T.,** The anthocyanins of strawberry, rhubarb, radish and onion, *J. Food Sci.,* 34, 365, 1969.
180. **Fuleki, T.,** Individual anthocyanin composition of American cranberry cultivars, *Bull. Liaison Groupe Polyphénols,* 13, 374, 1986.
181. **Fuleki, T. and Francis, F. J.,** The co-occurence of monoglucosides and monogalactosides of cyanidin and peonidin in the American cranberry, *Vaccinium macrocarpon, Phytochemistry,* 6, 1705, 1967.
182. **Fuleki, T. and Francis, F. J.,** Quantitative methods for anthocyanins. IV. Determination of individual anthocyanins in cranberry and cranberry products, *J. Food Sci.,* 33, 471, 1968.
183. **Galensa, R. and Herrmann, K.,** Analysis of flavonoids by high-performance liquid chromatography, *J. Chromatogr.,* 189, 217, 1980.
184. **Galensa, R., and Herrmann, K.,** Hochdruckflüssigkeitschromatographische Bestimmungvon Hesperidin in Orangensäften, *Dtsch. Lebensm. Rundsch.,* 76, 270, 1980.
185. **Garcia-Mina, M. C., Fuertes-Lasala, M. E., Fernandez, M., Vega, F. A. and Martinez-Valls, L.,** Componentes fenolicos en variedades cultivadas de *Pyrus communis* L. II. Estimaciones quantitativas, *Anal. Bromatol.,* 26, 177, 1974.
186. **Garzon, L. V., Cuca, L. E. S., Martinez, J. C. V., Yoshida, M., and Gottlieb, O. R.,** Flavonolignoid from the fruit of *Iryanthera laevis, Phytochemistry,* 26, 2835, 1987.
187. **Gaydou, E. M., Bianchini, J. P., and Ramananarivo, S. B.,** Les flavones polymehtoxylées des Citrus. Application de l'analyse de données à la classification des huiles d'oranges et de mandarines, *Bull. Liaison Groupe Polyphénols,* 12, 129, 1984.
188. **Gaydou, E. M., Bianchini, J. P., and Randriamiharisoa, R. P.,** Orange and mandarin peel oils differentiation using polymethoxylated flavone composition, *J. Agric. Food Chem.,* 35, 525, 1987.
189. **Gazit, S. and Adato, I.,** Effect of carbon dioxide atmosphere on the course of astringency disappearance of persimmon (*Diospyros kaki* L.) fruits, *J. Food Sci.,* 37, 815, 1972.
190. **Geiger, H. and Quin, C.,** Biflavonoids, in *The Flavonoids,* Harborne, J B., Mabry, T. J., and Mabry, H., Eds., Chapman and Hall, London, 1975, 692.
191. **Geiger, H. and Quin, C.,** Biflavonoids, in *The Flavonoids: Advances in Research,* Harborne, J. B. and Mabry, T. J., Eds., Chapman and Hall, London, 1982, 505.
192. **Ghosal, S. and Chakrabarti, D. K.,** Differences in phenolic and steroidal constituents between healthy and infected florets of *Mangifera indica, Phytochemistry,* 27, 1339, 1988.
193. **Ghosal, S., Biswas, K., and Chattopadhyay, B. K.,** Differences in the chemical constituents of *Mangifera indica,* infected with *Aspergillus niger* and *Fusarium moniliformae, Phytochemistry,* 17, 689, 1978.
194. **Glazener, J. A.,** Accumulation of phenolic compounds in cells and formation of lignin like polymers in cell walls of young tomato fruits after inoculation with *Botrytis cinerea, Physiol. Plant Pathol.,* 20, 11, 1982.
194a. **Glories,** personal communication.
195. **Godeau, R. P., Pelissier, Y., and Fouraste, I.,** Constituants de *Viburnum tinus* L. III-Composés anthocyaniques et composés flavoniques des fruits, *Plant. Med. Phytother.,* 13, 37, 1979.
196. **Golan, A., Khan, V., and Sadovski, A. Y.,** Relationship between polyphenols and browning in avocado mesocarp. Comparison between the Fuerte and Lerman cultivars, *J. Agric. Food Chem.,* 25, 1253, 1977.
197. **Goldstein, J. L. and Swain, T.,** Changes in tannins in ripening fruit, *Phytochemistry,* 2, 371, 1963.
198. **Goldy, R. G., Ballinger, W. E., and Maness, E. P.,** Fruit anthocyanin content of some *Euvitis* × *Vitis rotundifolia* hybrids, *J. Am. Soc. Hortic. Sci.,* 111, 955, 1986.
199. **Gonzalez, A. G., Fraga, B. M., and Pino, O.,** Pulverin a new chromone from the fruits of *Neochamaelea pulverulenta, Phytochemistry,* 14, 1656, 1975.
200. **Goto, T., Kondo, T., Tamura, H., and Takase, S.,** Structure of malonylawobanin, the real anthocyanin present in blue-colored flower petals of *Commelina communis, Tetrahedron Lett.,* 24, 4863, 1983.
201. **Gottlieb, O. R.,** Chemosystematics of the Lauraceae, *Phytochemistry,* 11, 1537, 1972.
202. **Gottlieb, O. R.,** Flavonols, in *The Flavonoids,* Harborne, J. B., Mabry, T. J., and Mabry, H., Eds., Chapman and Hall, London, 1975, 296.
203. **Gray, A. I.,** Structural diversity and distribution of coumarins and chromones in the Rutales, in *Chemistry and Chemical Taxonomy of the Rutales,* Waterman, P. G. and Grundon, M. F., Eds., Academic Press, London, 1983, 97.

204. **Gray, A. I. and Waterman, P. G.,** Coumarins in the Rutaceae, *Phytochemistry,* 17, 845, 1978.
205. **Green, A.,** Soft fruits, in *The Biochemistry of Fruits and Their Products,* Vol. 2, Hulme, A. C., Ed., Academic Press, London, 1971, 375.
206. **Grisebach, H.,** Lignins, in *The Biochemistry of Plants,* Vol. 7, Conn, E. E., Ed., Academic Press, New-York, 1981, 457.
207. **Grisebach, H.,** Biosynthesis of anthocyanins, in *Anthocyanins as Food Colors,* Markakis, P., Ed., Academic Press, New York, 1982, 69.
208. **Guern, J., Renaudin, J. P., and Brown, S. C.,** The compartmentation of secondary metabolites in plant cell cultures, in *Cell Culture in Phytochemistry,* Vol. 4, Constabel, F. and Vasil, I. K., Eds., Academic Press, London, 1987, 43.
209. **Guichard, J. P., Regerat, F., and Pourrat, H.,** Flavones et anthocyanes des fruits de *Viburnum lantana* L., *Plant. Med. Phytother.,* 10, 105, 1976.
210. **Haddock, E. A., Gupta, R. K., Al-Shafi, S. M. K., Layden, K., Haslam, E., and Magnolato, D.,** The metabolism of gallic acid and hexahydroxydiphenic acid in plants: biogenetic and molecular taxonomic considerations, *Phytochemistry,* 21, 1049, 1982.
211. **Hadj-Mahammed, M. and Meklati, B. Y.,** Qualitative determination of polymethoxylated flavones in Valencia orange peel oil and juice by LC-UV/VIS and LC-MS techniques, *Lebensm. Wiss. Technol.,* 20, 111, 1987.
212. **Hahlbrock, K. and Grisebach, H.,** Biosynthesis of flavonoids, in *The Flavonoids,* Harborne, J. B., Mabry, T. J., and Mabry, H., Eds., Chapman and Hall, London, 1975, 866.
213. **Hale, M. L., Francis, F. J., and Fagerson, I. S.,** Detection of enocyanin in cranberry juice cocktail by HPLC anthocyanin profile, *J. Food Sci.,* 51, 1511, 1986.
214. **Hall, R. D., Holden, M. A., and Yeoman, M. M.,** Studies on the regulation of capsaicin biosynthesis in immobilized cell cultures and the developping fruit of the chili pepper, *Capsicum frutescens,* in *Secondary Metabolism in Plant Cell Cultures,* Morris, P., Scragg, A. H., Stafford, A., and Fowler, M. W., Eds., Cambridge University Press, Cambridge, 1986, 121.
215. **Harborne, J. B.,** Phenolic glycosides and their natural distribution, in *Biochemistry of Phenolic Compounds,* Harborne, J. B., Ed., Academic Press, London, 1964, 129.
216. **Harborne, J. B.,** *Biochemistry of Phenolic Compounds,* Academic Press, London, 1964.
217. **Harborne, J. B.,** *Comparative Biochemistry of the Flavonoids,* Academic Press, London, 1967.
218. **Harborne, J. B.,** Phenolics compounds, in *Phytochemical Methods, a Guide to Modern Techniques of Plant Analysis,* Harborne, J. B., Ed., Chapman and Hall, London, 1973, chap. 2.
219. **Harborne, J. B.,** Functions of flavonoids in plants, in *Chemistry and Biochemistry of Plant Pigments,* Goodwin, T. W., Ed., Academic Press, London, 1976, 736.
220. **Harborne, J. B.,** Plant Phenolics, in *Encyclopedia of Plant Physiology, New Series,* Vol. 8, Bell, E. A., and Charlwood, B. V., Eds., Springer-Verlag, Berlin, 1980, 329.
221. **Harborne, J. B.,** The flavonoids of the Rutales, in *Chemistry and Chemical Taxonomy of the Rutales,* Waterman, P. G. and Grundon, M. F., Eds., Academic Press, London, 1983, 147.
222. **Harborne, J. B.,** The natural distribution in Angiosperms of anthocyanins acylated with aliphatic dicarboxylic acids, *Phytochemistry,* 25, 1887, 1986.
223. **Harborne, J. B. and Boardley, M.,** The widespread occurrence in nature of anthocyanins as zwitterions, *Z. Naturforsch.,* 40c, 305, 1985.
224. **Harborne, J. B. and Corner, J. J.,** Plant polyphenols, 4-hydroxycinnamic acid-sugar derivatives, *Biochem. J.,* 81, 242, 1961.
225. **Harborne, J. B. and Hall, E.,** Plant polyphenols. XIII. The systematic distribution and origin of anthocyanins containing branched trisaccharides, *Phytochemistry,* 3, 453, 1964.
226. **Harborne, J. B. and Mabry, T. J.,** *The Flavonoids: Advances in Research,* Chapman and Hall, London, New York, 1982.
227. **Harborne, J. B. and Williams, C. A.,** Flavonoid patterns in the fruits of the Umbelliferae, *Phytochemistry,* 11, 1741, 1972.
228. **Harborne, J. B. and Williams, C. A.,** Flavone and flavonol glycosides, in *the Flavonoids,* Harborne, J. B., Mabry, T. J., and Mabry, H., Eds., Chapman and Hall, London, 1975, 376.
229. **Harborne, J. B. and Williams, C. A.,** Flavone and flavonol glycosides, in *The Flavonoids: Advances in Research,* Harborne, J. B. and Mabry, T. J., Eds., Chapman and Hall, London, 1982, 261.
230. **Harborne, J. B., Mabry, T. J., and Mabry, H.,** *The Flavonoids,* Chapman and Hall, London, 1975.
231. **Harborne, J. B., Williams, C. A., Greenham, J., and Moyna, P.,** Distribution of charged flavones and caffeylshikimic acid in Palmae, *Phytochemistry,* 13, 1557, 1974.
232. **Haslam, E.,** Natural proanthocyanidins, in *The Flavonoids,* Harborne, J. B., Mabry, T. J., and Mabry, H., Eds., Chapman and Hall, London, 1975, 505.
233. **Haslam, E.,** Symmetry and promiscuity in procyanidin biochemistry, *Phytochemistry,* 16, 1625, 1977.
234. **Haslam, E.,** Vegetable tannins, in *Recent Advances in Phytochemistry,* Vol. 12, Swain, T., Harborne, J. B., and Van Sumere, C. F., Eds., Plenum Press, New York, 1979, 475.

235. **Haslam, E.,** In vino veritas: oligomeric procyandins and the ageing of red wines, *Phytochemistry,* 19, 2577, 1980.
236. **Haslam, E.,** Vegetable tannins, in *The Biochemistry of Plants,* Vol. 7, Conn, E. E., Ed., Academic Press, New York, 1981, 527.
237. **Haslam, E.,** Proanthocyanidins, in *The Flavonoids: Advances in Research,* Harborne, J. B. and Mabry, T. J., Eds., Chapman and Hall, London, 1982, 417.
238. **Haslam, E. and Lilley, T. H.,** New polyphenols for old tannins, in *Ann. Proc. of Phytochem. Soc. Europe,* Vol. 25, Van Sumere, C. F. and Lea, P. J., Eds., Clarendon Press, Oxford, 1985, 237.
239. **Haslam, H. and Lilley, T. H.,** Natural astringency in foodstuffs. A molecular interpretation, *C.R.C. Crit. Rev. Food Sci. Nutr.,* 27, 1, 1988.
240. **Hawker, J. S., Buttrose, M. S., Soeffky, A., and Possigham, J. V.,** A simple method for demonstrating macroscopically the location of polyphenolic compounds in grape berries, *Vitis,* 11, 189, 1972.
241. **Hayashi, K.,** The anthocyanins, in *The Chemistry of Flavonoid Compounds,* Geissman, T. A., Ed., Pergamon Press, Oxford, 1962, 248.
242. **Hebrero, E., Santos-Buelga, C., and Rivas-Gonzalo, J. C.,** High performance liquid chromatography-diode array spectroscopy identification of anthocyanins of *Vitis vinifera* variety Tempranillo, *Am. J. Enol. Vitic.,* 39, 227, 1988.
243. **Hedin, P. A., Langhans, V. E., and Graves, C. H., Jr.,** Identification of juglone in pecan as a possible factor of resistance to *Fusicladium effusum, J. Agric. Food Chem.,* 27, 92, 1979.
244. **Hedin, P. A., Collum, D. H., Langhans, V. E., and Graves, C. H., Jr.,** Distribution of juglone and related compounds in pecan and their effect on *Fusicladium effusum, J. Agric. Food Chem.,* 28, 240, 1980.
245. **Heimhuber, B., Galensa, R., and Herrmann, K.,** High-performance liquid chromatographic determination of polymethoxylated flavones in orange juice after solid-phase extraction, *J. Chromatogr.,* 439, 481, 1988.
246. **Henning, W.,** Flavonolglykoside der Erdbeeren (*Fragaria x ananassa* Duch.), Himbeeren (*Rubus idaeus* L.) und Brombeeren (*Rubus fructicosus* L.), *Z. Lebensm. Unters. Forsch.,* 173, 180, 1981.
247. **Henning, W. and Herrmann, K.,** Flavonolglykoside der Aprikosen (*Prunus armenia ca* L.) und der Pfirsiche (*Prunus persica* Batsch), *Z. Lebensm. Unters. Forsch.,* 171, 183, 1980.
248. **Henning, W. and Herrmann, K.,** Flavonolglykoside der Pflaumen der species *Prunus domestica* L. und *Prunus salicina* Lindley, *Z. Lebensm. Unters. Forsch.,* 171, 111, 1980.
249. **Henning, W. and Herrmann, K.,** Flavonolglykoside der Susskirschen (*Prunus avium* L.) und der Sauerkirschen (*Prunus cerasus* L.), *Z. Lebensm. Unters. Forsch.,* 170, 433, 1980.
250. **Herrman, K.,** Über Oxydation ferment und Phenolische substrate in Gemüse und Obst: Catechin, Oxyzimtsaüre und O-phenol Oxydase, *Z. Lebensm. Unters. Forsch.,* 108, 152, 1958.
251. **Herrmann, K.,** Die phenolischen Inhaltsstoffe des Obstes. I. Bisherige Kenntnis se uber Vorkommen, Gehalte sowie Veranderungen wahrend des Fruchtwachstums, *Z. Lebensm. Unters. Forsch.,* 151, 41, 1973.
252. **Herrmann, K.,** Hydroxyzimtsaüren und Hydroxybenzoesaüren enthaltende Naturstoffe in Pflanzen, *Progr. Chem. Org. Nat. Prod.,* 35, 73, 1978.
253. **Herrmann, K.,** Flavonoide Inhalsstoffe in Obst, Gemüse und Gewürzen, *Lebensmittelchem. Gerichtl. Chemie,* 33, 4, 1979.
254. **Herrmann, K.,** Übersicht über die inhaltsstoffe der Tomaten, *Z. Lebensm. Unters. Forsch.,* 169, 179, 1979.
255. **Hobson, G. E. and Davies, J. N.,** The tomato, in *The Biochemistry of Fruits and Their Products,* Vol. 2, Hulme A. C., Ed., Academic Press, London and New York, 1971, chap. 13.
256. **Horie, T., and Nakayama, M.,** Flavones from *Citrus sudachi, Phytochemistry,* 20, 337, 1981.
257. **Horie, T., Tsukayama, M., Yamada, T., Miura, I., and Nakayama, M.,** Three flavone glycosides from *Citrus sudachi, Phytochemistry,* 25, 2621, 1986.
258. **Horowitz, R. M.,** Citrus flavonoids, in *The Orange. Its Biochemistry and Physiology,* Sinclair, W. R., Ed., University of California Press, Berkeley, 1961, 334.
259. **Horowitz, R. M.,** Relations between the taste and structure of some phenolic glycosides, in *Biochemistry of Phenolic Compounds,* Harborne, J. B., Ed., Academic Press, London, 1964, 545.
260. **Horowitz, R. M. and Gentili, B.,** Flavonoid constituents of Citrus, in *Citrus Science and Technology,* Nagy, S., Shaw, P. E., and Veldhuis, M. K., Eds., Vol. 1, AVI Publishing, Westport, CT, 1977, 397.
261. **Horvat, R. J. and Senter, S. D.,** A gas-liquid chromatographic method for analysis of phenolic acids in plants, *J. Agric. Food Chem.,* 28, 1292, 1980.
262. **Hostettmann, K.,** Recent advances in methodology for the isolation and purification of plant phenolics, in *Ann. Proc. Phytochem. Soc. Eur.,* Vol. 25, Van Sumere, C. F. and Lea, P. J., Eds., Clarendon Press, Oxford, 1985, 1.
263. **Hostettmann, K. and Hostettmann, M.,** Isolation techniques for flavonoids, in *The Flavonoids: Advances in Research,* Harborne, J. B. and Mabry, T. J., Eds., Chapman and Hall, London, New York, 1982, chap. 1.
264. **Hrazdina, G.,** Anthocyanins and their role in food products, *Lebensm. Wiss. Technol.,* 14, 283, 1981.

265. **Hrazdina, G.**, Anthocyanins, in *The Flavonoids: Advances in Research*, Harborne, J. B. and Mabry, T. J., Eds., Chapman and Hall, London, 1982, 135.
266. **Hrazdina, G. and Wagner, G. J.**, Metabolic pathways as enzyme complexes: evidence for the synthesis of phenylpropanoids and flavonoids on membrane associated enzyme complexes, *Arch. Biochem. Biophys.*, 237, 88, 1985.
267. **Hrazdina, G., Wagner, G. J., and Siegelman, H. W.**, Subcellular localisation of enzymes of anthocyanin biosynthesis in protoplasts, *Phytochemistry*, 17, 53, 1978.
268. **Huet, R.**, Constituants des agrumes à effet pharmacodynamique: les citroflavonoides, *Fruits*, 37, 267, 1982.
269. **Hulme, A. C.**, The isolation of chlorogenic acid from the apple fruit, *Biochem. J.*, 53, 337, 1953.
270. **Hulme, A. C.**, The Mango, in *The Biochemistry of Fruits and Their Products*, Vol. 2, Hulme, A. C., Ed., Academic Press, London, 1971, 233.
271. **Hulme, A. C. and Rhodes, M. J. C.**, Pome fruits, in *The Biochemistry of Fruits and Their Products*, Vol. 2., Hulme, A. C., Academic Press, London, 1971, 363.
272. **Hunt, G. M. and Baker, E. A.**, Phenolic constituents of tomato fruit cuticles, *Phytochemistry*, 19, 1415, 1980.
273. **Hwang, B. K.**, Contents of sugars, fruit acids, amino acid and phenolic compounds of apple fruit in relation to their susceptibility to *Botryosphaeria ribis*, *Phytopath. Z.*, 108, 1, 1983.
274. **Iacobucci, G. A. and Sweeny, J. G.**, The chemistry of anthocyanins, anthocyanidins and related flavylium salts, *Tetrahedron*, 39, 3005, 1983.
275. **Ibrahim, R. K., De Luca, V., Khouri, H., Latchinian, L., Brisson, L., and Charest, P. M.**, Enzymology and compartmentation of polymethylated flavonol glucosides in *Chrysosplenium americanum*, *Phytochemistry*, 26, 1237, 1987.
276. **Ikeya, Y., Taguchi, H., Mitsuhashi, H., Takeda, S., Kase, Y., and Aburada, M.**, A lignan from *Schizandra chinensis*, *Phytochemistry*, 27, 569, 1988.
277. **Inatani, R., Nakatani, N., and Fuwa, H.**, Structure and synthesis of new phenolic amides from *Piper nigrum* L., *Agric. Biol. Chem.*, 45, 667, 1981.
278. **Ishikura, N.**, A survey of anthocyanins in fruits of some Angiosperms. I., *Bot. Mag. Tokyo*, 88, 41, 1975.
279. **Ishikura, N.**, A further survey of anthocyanins and other phenolics in *Ilex* and *Euonymus*, *Phytochemistry*, 14, 743, 1975.
280. **Ishikura, N. and Sugahara, K.**, A survey of anthocyanins in fruits of some Angiosperms, II., *Bot. Mag. Tokyo*, 92, 157, 1979.
281. **Ito, S.**, The persimmon, in *Biochemistry of Fruits and Their Products*, Vol. 2, Hulme, A. C., Ed., Academic Press, London, 1971, 281.
282. **Itoo, S.**, Persimmon, in *Handbook of Fruit Set and Development*, Monselise, S. P., Ed., CRC Press, Boca Raton, FL, 1986, 355.
283. **Itoo, S., Matsuo, T., Ibushi, Y., and Tamari, N.**, Seasonal changes in the levels of polyphenols in Guava fruit and leaves and some of their properties, *J. Jpn Soc. Hortic. Sci.*, 56, 107, 1987.
284. IUPAC, Nomenclature of cyclitols, *Biochem. J.*, 153, 23, 1976.
285. **Iwai, K., Lee, K. R., Kobashi, M., and Suzuki, T.**, Formation of pungent principles in fruits of sweet pepper, *Capsicum annum L.* var. grossum, during post-harvest ripening under continous light, *Agric. Biol. Chem.*, 41, 1873, 1977.
286. **Jacques, D., Opie, C. T., Porter, L. J., and Haslam, E.**, Plant proanthocyanidins. Part 4. Biosynthesis and observations on the metabolism of cyanidin in plants, *J. Chem. Soc., Perkin I*, 1637, 1977.
287. **Jennings, D. L. and Carmichael, E.**, Anthocyanin variation in the genus *Rubus*, *New Phytol.*, 84, 505, 1980.
288. **Jerumanis, J.**, Methodes modernes de fractionnement et d'analyse des composés phénoliques, *Bull. Liaison Groupe Polyphénols*, 11, 309, 1982.
289. **Jourdan, P. S., Weiler, E. W., and Mansell, R. L.**, Naringin levels in Citrus tissues. I. Comparison of different antibodies and tracers for the radioimmunoassay of naringin, *Plant Physiol.*, 77, 896, 1985.
290. **Jourdan, P. S., McIntosh, C. A., and Mansell, R. L.**, Naringin levels in Citrus tissues. II. Quantitative distribution of naringin in *Citrus paradisi* Mac Fad, *Plant Physiol.*, 77, 903, 1985.
291. **Jurd, L.**, The detection of aromatic acids in plant extracts by the ultra-violet absorption spectra of their ions, *Arch. Biochem. Biophys.*, 66, 284, 1957.
292. **Kahnt, C.**, *Trans-cis* equilibrium of hydroxycinnamic acids during irradiation of aqueous solutions at different pH, *Phytochemistry*, 6, 755, 1967.
293. **Kalemba, D., Kurowska, A., and Gora, J.**, Polyphenols in *Aronia melanocarpa* fruits, *Bull. Liaison Groupe Polyphénols*, 12, 132, 1984.
294. **Kamya, S., Esaki, S., and Knoishi, F.**, Flavonoids in citrus hybrids, *Agric. Biol. Chem.*, 43, 1529, 1979.
295. **Kaneta, M., Hikichi, H., Endo, S., and Sugiyama, N.**, Identification of flavones in nineteen Rosaceae species, *Agric. Biol. Chem.*, 43, 657, 1979.
296. **Kärppä, J.**, Analysis of anthocyanins during ripening of crowberry, *Empetrum nigrum* Coll., *Lebensm. Wiss. Technol.*, 17, 175, 1984.

297. **Kärppä, J., Kallio, H., Peltonen, I., and Linko, R.,** Anthocyanins of crowberry, *Empetrum nigrum* coll., *J. Food Sci.*, 49, 634, 1984.
298. **Kato, M. J., Fo, H. F. P., Yoshida, M., and Gottlieb, O. R.,** Neolignans from fruits of *Virola elongata*, *Phytochemistry*, 25, 279, 1986.
299. **Kato, M. J., Lopes, L. M. X., Fo, H. F. P., Yoshida, M., and Gottlieb, O. R.,** Acylresorcinols from *Virola sebifera* and *Virola elongata*, *Phytochemistry*, 24, 533, 1985.
300. **Kaul, B. and Staba, J.,** *Ammi visnaga* (L) Lam. tissue cultures, multilitre suspension growth and examination for furanochromones, *Planta Med.*, 15, 145, 1967.
301. **Kefford, J. B. and Chandler, B. V.,** *The Chemical Constituents of Citrus Fruits*, Academic Press, New York, 1970, 113.
302. **Kikuchi, M., and Yamauchi, Y.,** Studies on the constituents of Ligustrum species. VIII. On the components of the fruits of *Ligustrum obtusifolium* Sieb et Zucc., *Yakugaku Zhassi*, 103, 360, 1983.
303. **Kitagawa, S., Hisada, S., and Nishibe, S.,** Phenolic compounds from Forsythia leaves, *Phytochemistry*, 23, 1635, 1984.
304. **Klick, S. and Herrmann, K.,** Bestimmung von Hydroxybenzoesaüre-Verbindungen in Gewürzen und weiteren planzlichen Lebensmitteln, *Z. Lebensm. Unters. Forsch.*, 187, 444, 1988.
305. **Koeppen, B. H. and Herrmann, K.,** Flavonoid glycosides and hydroxycinnamic acid esters of blackcurrants, *Z. Lebensm. Unters. Forsch.*, 164, 263, 1977.
306. **Kolattukudy, P. E.,** Structure, biosynthesis and biodegradation of cutin and suberin, *Annu. Rev. Plant Physiol.*, 32, 539, 1981.
307. **Komiyama, Y., Harakawa, M., and Tsuji, M.,** Polyphenol contents and enzymatic browning of plums harvested in Japan, *Nippon Shokuhin Kogyo Gakkaishi*, 26, 325, 1979.
308. **Koul, S. K., Taneja, S. C., Pushpangadan, P., and Dhar, K. L.,** Lignans of *Piper trichostachyon*, *Phytochemistry*, 27, 1479, 1988.
309. **Koumba-Koumba, D. and Macheix, J. J.,** Biosynthesis of hydroxycinnamic derivatives in apple fruit cell suspension culture, *Physiol. Vég.*, 20, 137, 1982.
310. **Kramling, T. E. and Singleton, V. L.,** An estimate of the nonflavonoid phenols in wines, *Am. J. Enol. Vitic.*, 20, 86, 1969.
311. **Kubo, I. and Matsumoto, A.,** Molluscicides from olive *Olea europaea* and their efficient isolation by countercurrent chromatographies, *J. Agric. Food Chem.*, 32, 687, 1984.
312. **Lakshminarayana, S., Mathew, A. G., and Parpia, H. A. B.,** Changes in polyphenols of sapota fruit (*Achras zapota* L.) during maturation, *J. Sci. Food Agric.*, 20, 651, 1, 1969.
313. **Lamikanra, O.,** Development of anthocyanin pigments in muscadine grapes, *HortScience*, 23, 597, 1988.
314. **Lattanzio, V.,** High-performance liquid chromatography of free and bound phenolic acids in the eggplant (*Solanum melongena* L.), *J. Chromatogr.*, 250, 143, 1982.
315. **Lavee, S. and Avidan, N.,** Growth responses of tree callus to chlorogenic acid and related phenolic substances, *Proc. 5th Int. Cong. Plant Tissue Cell Culture*, Japanese Assoc. Plant Tissue Culture, Tokyo, 1982, 165.
316. **Lea, A. G. H.,** The phenolics of ciders: oligomeric and polymeric procyanidins, *J. Sci. Food Agric.*, 29, 471, 1978.
317. **Lea, A. G. H.,** High performance liquid chromatography of cider procyanidins, *J. Sci. Food Agric.*, 30, 833, 1979.
318. **Lea, A. G. H.,** Reversed-phase gradient high-performance liquid chromatography of procyanidins and their oxidation products in ciders and wines optimised by Snyder's procedures, *J. Chromatogr.*, 194, 62, 1980.
319. **Lea, A. G. H.,** Reversed-phase high-performance liquid chromatography of procyanidins and other phenolics in fresh and oxidising apple juices using a pH shift technique, *J. Chromatogr.*, 238, 253, 1982.
320. **Lea, A. G. H. and Timberlake, C. F.,** The phenolics of ciders. I. Procyanidins, *J. Sci. Food Agric.*, 25, 1537, 1974.
321. **Lea, A. G. H., Bridle, P., Timberlake, C. F., and Singleton, V. L.,** The procyanidins of white grapes and wines, *Am. J. Enol. Vitic.*, 30, 289, 1979.
322. **Lee, C. Y. and Jaworski, A.,** Phenolic compounds in white grapes grown in New York, *Am. J. Enol. Vitic.*, 38, 277, 1987.
323. **Lees, D. H., and Francis, F. J.,** Effect of gamma radiation on anthocyanin and flavonol pigments in cranberries (*Vaccinium macrocarpon* Ait.), *J. Am. Soc. Hortic. Sci.*, 97, 128, 1972.
324. **Leistner, E.,** Biosynthesis of plant quinones, in *The Biochemistry of plants*, Vol. 7, Conn, E. E., Ed., Academic Press, New York, 1981, 403.
325. **Levand, O. and Heinicke, R. M.,** Ferulic acid as a component of a complex carbohydrate polymer of bromelain, *Phytochemistry*, 7, 1659, 1968.
326. **Linko, R., Kärppä, J., Kallio, H., and Ahtonen, S.,** Anthocyanin contents of Crowberry and Crowberry juice, *Lebensm. Wiss. Technol.*, 16, 343, 1983.
327. **Lopes, L. M. X., Yoshida, M., and Gottlieb, O. R.,** Aryltetralone and arylindanone neolignans from *Virola sebifera*, *Phytochemistry*, 23, 2021, 1984.

328. **Lopes, C. R., Musche, R., and Hec, M.,** Identificao de pigmentos flavonoides e acidos fenolicos nos cultivares Bourbon e Cattura de *Coffea arabica* L., *Rev. Brasil. Genet.*, 7, 657, 1984.
329. **Lowry, J. B.,** Anthocyanins of the Melastomataceae, Myrtaceae and some allied families, *Phytochemistry*, 15, 513, 1976.
330. **Luckner, M.,** Expression and control of secondary metabolism, in *Encyclopedia of Plant Physiol., New Series*, Vol. 8, *Secondary Plant Products*, Bell, E. A. and Charlwood, B. V., Eds., Springer-Verlag, Berlin, 1980, 23.
331. **Lynn, D. Y. C. and Luh, B. S.,** Anthocyanin pigments in bing cherries, *J. Food Sci.*, 29, 735, 1964.
332. **Mabry, T. J. and Markham, K. R.,** Mass spectrometry of flavonoids, in *The Flavonoids*, Harborne, J. B., Mabry, T. B., and Mabry, H., Eds., Chapman and Hall, London, 1975, 78.
333. **Mabry, T. J., Markham, K. R., and Thomas, M. B.,** *The Systematic Identification of Flavonoids*, Springer-Verlag, New York, 1970.
334. **Maccarone, E., Maccarone, A., and Rapisarda, P.,** Acylated anthocyanins from oranges, *Ann. Chim. Rome*, 75, 79, 1985.
335. **Maccarone, E., Maccarone, A., Perrini, G., and Rapisarda, P.,** Anthocyanins of the Moro orange juice, *Ann. Chim. Rome* 73, 533, 1983.
336. **Macheix, J. J.,** Quelques observations sur les composés phénoliques des pommes: recherches préliminaires à une étude particulière de l'acide chlorogénique, *Fruits*, 23, 13, 1968.
337. **Macheix, J. J.,** Role de divers facteurs intervenant dans le brunissement enzymatique des pommes pendant la croissance, *Physiol. Veg.*, 8, 585, 1970.
338. **Macheix, J. J.,** Présence dans les pommes de plusieurs dérivés de l'acide para-coumarique, *C.R. Acad. Sci. Paris*, Ser. D, 272, 1097, 1971.
339. **Macheix, J. J.,** Evolution de la teneur du fruit de Pirus Malus L. en esters des acides *p*-coumarique at caféique au cours de la croissance, *Physiol. Vég.*, 12, 25, 1974.
340. **Macheix, J. J.,** Les esters hydroxycinnamiques de la pomme: identification, variations au cours de la croissance du fruit et métabolisme, Thèse Doct. Etat, Université Pierre et Marie Curie, Paris, 1974.
341. **Macheix, J. J.,** Etude de quelques caractéristiques de l'accumulation du *p*-coumaryl glucose dans le parenchyme de la pomme (*Pirus Malus* L.), *Physiol. Vég.*, 13, 591, 1975.
342. **Macheix, J. J. and Delaporte, N.,** Etude de la répartition dans le fruit de Pirus Malus L., des principaux o-diphénols par dosage oxydimétrique en milieu non aqueux, *Lebensm. Wiss. Technol.*, 6, 19, 1973.
343. **Macheix, J. J. and Fleuriet, A.,** Les dérivés hydroxycinnamiques des fruits, *Bull. Liaison Groupe Polyphénols*, 13, 337, 1986.
344. **Macheix, J. J., Rateau, J., Fleuriet, A., and Bureau, D.,** Les esters hydroxycinnamiques des fruits, *Fruits*, 32, 397, 1977.
344a. **Macheix, J. J.,** unpublished data.
345. **Maestro Duran, R. and Vasquez Roncero, A.,** Colorantes antocianicos de las aceitunas manzanillas maduras, *Grasas Aceites*, 27, 237, 1976.
346. **Maga, J. A.,** Simple phenols and phenolic compounds in food flavor, *CRC Crit. Rev. Food Nutr.*, 10, 323, 1978.
347. **Maga, J. A.,** Amines in foods, *CRC Crit. Rev. Food Sci. Nutr.*, 10, 373, 1978.
348. **Maier, V. P.,** Chemistry and significance of selected citrus limonoids and flavonoids, in *Chemistry and Chemical Taxonomy of the Rutales*, Waterman, P. G., Grundon, M. F., Eds., Academic Press, London, 1983, 319.
349. **Maier, V. P. and Metzler, D. M.,** Quantitative changes in date polyphenols and their relation to browning, *J. Food Sci.*, 30, 80, 1965.
350. **Maier, V. P. and Metzler, D. M.,** Changes in individual date polyphenols and their relation to browning, *J. Food Sci.*, 30, 747, 1965.
351. **Maier, V. P. and Metzler, D. M.,** Grapefruit phenolics. II. Principal aglycones of endocarp and peel and their possible biosynthetic relationship, *Phytochemistry*, 6, 1127, 1967.
352. **Maier, C. P., Metzler, D. M., and Huber, A. F.,** 3-O-Caffeoylshikimic acid (dactylifric acid) and its isomers, a new class of enzymic browning substrates, *Biochem. Biophys. Res. Commun.*, 14, 124, 1964.
353. **Makus, D. J. and Ballinger, W. E.,** Characterization of anthocyanins during ripening of fruit of *Vaccinium corymbosum* L. Cv. Wolcott, *J. Am. Soc. Hortic. Sci.*, 98, 99, 1973.
354. **Mansell, R. L., McIntosh, C. A., and Vest, S. E.,** An analysis of the limonin and naringenin content of grapefruit juices samples collected from Florida state test houses, *J. Agric. Food Chem.*, 31, 156, 1983.
355. **Marekov, N. L.,** Polyphenols Research in Bulgaria, *Bull. Liaison Groupe Polyphénols*, 12, 31, 1984.
356. **Marigo, G.,** Sur une méthode de fractionnement et d'estimation des composés phénoliques chez les végétaux, *Analusis*, 2, 106, 1973.
357. **Markakis, P.,** Stability of anthocyanins in foods, in *Anthocyanins as Food Colors*, Markakis, P., Ed., Academic Press, New York, 1982, 163.
358. **Markham, K. R.,** Isolation techniques for flavonoids, in *The Flavonoids*, Harborne, J. B., Mabry, T. J., and Mabry, H., Eds., Chapman and Hall, London, 1975, 1.

359. **Markham, K. R. and Mabry, T. J.,** Ultraviolet-visible and proton magnetic resonance spectroscopy of flavonoids, in *The Flavonoids,* Harborne, J. B., Mabry, T. J., and Mabry, H., Eds., Chapman and Hall, London, 1975, 45.
360. **Markham, K. R., Chari, V. M., and Mabry, T. J.,** Carbon-13 NMR spectroscopy of flavonoids, in *The Flavonoids: Advances in Research,* Harborne, J. B., and Mabry, T. J., Eds., Chapman and Hall, London, 1982, 19.
361. **Martin-Tanguy, J.,** The occurrence and possible function of hydroxycinnamoyl acid amides in plants, *Plant Growth Regul.,* 3, 381, 1985.
362. **Martin-Tanguy, J., Cabanne, F., Perdrizet, E., and Martin, C.,** The distribution of hydroxycinnamic acid amides in flowering plants, *Phytochemistry,* 17, 1927, 1978.
363. **Marwan, A. G. and Nagel, C. W.,** Separation and purification of hydroxycinnamic acid derivatives in Cranberries, *J. Food Sci.,* 47, 585, 1982.
364. **Marwan, A. G. and Nagel, C. W.,** Identification of the hydroxycinnamic acid derivatives in Cranberries, *J. Food Sci.,* 47, 774, 1982.
365. **Mathew, A. G. and Lakshminarayana, S.,** Polyphenols of immature sapota fruit, *Phytochemistry,* 8, 507, 1969.
366. **Mathew, A. G. and Parpia, H. A. B.,** Food browning as a polyphenol reaction, *Adv. Food Res.,* 19, 75, 1971.
367. **Matile, P.,** Biochemistry and function of vacuoles, *Annu. Rev. Plant Physiol.,* 29, 193, 1978.
368. **Matsubara, Y., Kumamoto, H., Iizuka, Y., Murakami, T., Okamoto, K., Miyake, H., and Yokoi, K.,** Structure and hypotensive effect of flavonoid glycosides in *citrus unshiu* peelings, *Agric. Biol. Chem.,* 49, 909, 1985.
369. **Matsuo, T. and Itoo, S.,** The chemical structure of Kaki-tannin from immature fruit of the persimmon, *Agric. Biol. Chem.,* 42, 1937, 1978.
370. **Matsuo, T. and Itoo, S.,** A simple and rapid purification method of condensed tannins from several young fruits, *Agric. Biol. Chem.,* 45, 1885, 1981.
371. **Matsuo, T. and Itoo, S.,** Comparative studies of condensed tannins from several young fruits, *J. Jpn. Soc. Hortic. Sci.,* 50, 262, 1981.
372. **Matsuo, T., Shinohara, J., and Ito, S.,** An improvement on removing astringency in persimmon fruits by carbon dioxide gas, *Agric. Biol. Chem.,* 40, 215, 1976.
373. **Mayer, A. M. and Harel, E.,** Polyphenol oxidases in plants, *Phytochemistry,* 18, 193, 1979.
374. **Mayer, A. M. and Harel, E.,** Polyphenol oxidases in fruits. Changes during ripening, in *Recent Advances in the Biochemistry of Fruits and Vegetables,* Friend, J. and Rhodes, M. J. C., Eds., Academic Press, London, 1981, 161.
375. **Mayer, W., Gorner, A., and Andra, K.,** Punicalagin und punicalin, zwei Gerbstoffe aus den Schalen der Granatapfel, *Liebigs Ann. Chem.,* 1976, 1977.
376. **McClure, J.,** The physiology of phenolic compounds in plants, *Recent Advances in Phytochemistry,* Vol. 12, Swain, T., Harborne, J. B., and Van Sumere, C. F., Eds., Plenum Press, New York, 1979, 525.
377. **McDonald, R. E. and Kushad, M. M.,** Accumulation of putrescine during chilling injury of fruits, *Plant Physiol.,* 82, 324, 1986.
378. **McHale, D., Khopkar, P. P., and Sheridan, J. B.,** Coumarin glycosides from citrus flavedo, *Phytochemistry,* 26, 2547, 1987.
379. **McManus, J. P., Davis, K. G., Beart, J. E., Gaffney, S. H., Lilley, T. H., and Haslam, E.,** Polyphenol interactions. I. Introduction. Some observations on the reversible complexation of polyphenols with proteins and polysaccharides, *J. Chem. Soc., Perkin Trans II,* 1429, 1985.
380. **Melin, C.,** Les composés phénoliques au cours de la croissance et de la maturation de la cerise, *Prunus avium* (L.) variété "Bigarreau Napoléon", Thèse Doctorat 3ème cycle, Université d'Orléans, France, 1976.
381. **Melin, C., Billot, J., and Dupin, J. F.,** Les composés phénoliques de la Cerise Bigarreau Napoléon I. Les dérivés hydroxycinnamiques, *Physiol., Vég.,* 17, 557, 1979.
382. **Melin, C., Billot, J., and Dupin, J. F.,** Les composés phénoliques de la cerise Bigarreau Napoléon II. Les flavonoides, *Physiol. Vég.,* 17, 573, 1979.
383. **Merlin, J. C., and Brouillard, R.,** Resonance Raman spectrometry of anthocyanins, *Bull. Liaison Groupe Polyphénols,* 13, 145, 1986.
384. **Merlin, J. C., Statoua, A., and Brouillard, R.,** Investigation of the *in vivo* organization of anthocyanins using resonance Raman microspectrometry, *Phytochemistry,* 24, 1575, 1985.
385. **Misic, P. D.,** The anthocyanin pigments of some red raspberry (*Rubus idaeus* L.) fruits, *Hortic. Res.,* 13, 45, 1973.
386. **Molgaard, P. and Ravn. H.,** Evolutionary aspects of caffeoyl ester distribution in dicotyledons, *Phytochemistry,* 27, 2411, 1988.
387. **Möller, B. and Herrmann, K.,** Analysis of quinic acid esters of hydroxycinnamic acids in plant material by capillary gas chromatography and high-performance liquid chromatography, *J. Chromatogr.,* 241, 371, 1982.

388. **Möller, B., and Herrmann, K.,** Ester der hydroxyzimtsaüren mit Chinasaüre als Inhaltsstoffe des Stein und Kernobstes, *Lebensmittelchem. Gerichtl. Chem.*, 36, 135, 1982.
389. **Möller, B. and Herrmann, K.,** Quinic acid esters of hydroxycinnamic acids in stone and pome fruit, *Phytochemistry*, 22, 477, 1983.
390. **Montedoro, G.,** I costituenti fenolici presenti negli oli vergini di oliva. Nota I. Identificazione di alcuni acidi fenolici e loro potere antiossidante, *Riv. "S et TA"*, 3, 177, 1972.
391. **Monties, B.,** Nature et propriétés chimiques des principaux polyphénols de la pomme. Propriétés chimiques des polyphénols, application à la technologie du jus de pommes et de ses dérivés, *Ann. Physiol. Vég.*, 8, 101, 1966.
392. **Monties, B.,** Les lignines, in *Les Polymères Végétaux*, Monties, B., Ed, Gauthier-Villars, Paris, 1980, 122.
393. **Monties, B.,** Recent advances on lignin inhomogeneity, in *Ann. Proc. Phytochem. Soc. Eur.*, Vol. 25, Van Sumere, C. F. and Lea, P. J., Eds., Clarendon Press, Oxford, 1985, 161.
394. **Morita, H. and Montgomery, V. G.,** Gas chromatography of chloromethylsilylated phenolic acids, *J. Chromatogr.*, 123, 454, 1976.
395. **Mosel, H. D. and Herrmann, K.,** The phenolics of fruits. III. The contents of catechins and hydroxycinnamic acids in pome and stone fruits, *Z. Lebensm. Unters. Forsch.*, 154, 6, 1974.
396. **Mosel, H. D. and Herrmann, K.,** Die phenolischen Inhaltsstoffe der Obstes. IV. Die phenolischen Inhaltsstoffe der Brombeeren und Himbeeren und deren Veränderun gen während Wachstum und Reife der Früchte, *Z. Lebensm. Unters. Forsch.*, 154, 324, 1974.
397. **Moskowitz, A. H. and Hrazdina, G.,** Vacuolar contents of fruit subepidermal cells from *Vitis* species, *Plant Physiol.*, 68, 686, 1981.
398. **Muller, W. U. and Leistner, E.,** 1,4-naphtoquinone, an intermediate in juglone (5-hydroxy-1,4-naphtoquinone) biosynthesis, *Phytochemistry*, 15, 407, 1976.
399. **Muller, W. U. and Leistner, E.,** Metabolic relation between naphtalene derivatives in Juglans, *Phytochemistry*, 17, 1735, 1978.
400. **Muller, W. U. and Leistner, E.,** Aglycones and glycosides of oxygenated naphtalenes and a glycosyltransferase from Juglans, *Phytochemistry*, 17, 1739, 1978.
401. **Mulvena, D., Webb, E. C., and Zerner, B.,** 3-4-dihydroxybenzaldehyde, a fungistatic substance from green Cavendish bananas, *Phytochemistry*, 8, 393, 1969.
402. **Murray, R. D. H., Mendez, J., and Brown, S. A.,** *The Natural Coumarins. Occurrence, Chemistry and Biochemistry*, J. Wiley-Interscience, Chichester, 1982.
403. **Nadudvari-Markus, V. and Vamos-Vigyazo, L.,** Enzymic browning substrates in apple cultivars, *Acta Aliment.*, 13, 97, 1984.
404. **Nagel, C. W. and Wulf, L. W.,** Changes in the anthocyanins, flavonoids and hydroxycinnamic acid esters during fermentation and aging of Merlot and Cabernet Sauvignon, *Am. J. Enol. Vitic.*, 30, 111, 1979.
405. **Nagem, T. J. and De A. e Silva, M.,** Xanthones and phenylcoumarins from Kielmeyera pumila, *Phytochemistry*, 27, 2961, 1988.
406. **Nakabayashi, T.,** Studies on tannin of fruits and vegetables. VII. Difference of the compound of tannin between the astringent and non astrigent persimmon fruits (Japanese), *J. Jpn. Soc. Technol.*, 18, 33, 1971.
407. **Nakajima, K., Taguchi, H., Ikeya, Y., Endo, T., and Yosioka, I.,** The constituents of *Schizandra chinensis* Baill. XIII. Quantitative analysis of lignans in the fruits of *Schizandra chinensis* Baill. by high performance liquid chromatography, *Yakugaku Zhassi*, 103, 743, 1983.
408. **Nakatoni, N., Inatani, R., and Fuwa, H.,** Structures and synthesis of two phenolic amides from *Piper nigrum* L., *Agric. Biol. Chem.*, 44, 2831, 1980.
409. **Ndubizu, T. O. C.,** Relations of phenolic inhibitors to resistance of immature apple fruits to rot, *J. Hortic. Sci.*, 51, 311, 1976.
410. **Neu, R.,** Ein neues Reagens zum Nachweis und zur Unterscheidung von Flavonen im Papierchromatogram, *Naturwiss.* 43, 82, 1956.
411. **Niemann, G. J., Heerma, W., and De Koster, C. G.,** Fast atom bombardment mass spectrometry of flavonols and aromatic hydroxyacids, *Bull. Liaison Groupe Polyphénols*, 13, 95, 1986.
412. **Nortje, B. K. and Koeppen, B. H.,** The flavonol glycosides in the fruit of *Pyrus communis* L. cultivar Bon Chrétien, *Biochem. J.*, 97, 209, 1965.
413. **Nothmann, J., Rylski, I., and Spigelman, M.,** Effects of air and soil temperatures on colour of eggplant fruits (*Solanum melongena* L.), *Exp. Agric.*, 14, 189, 1978.
414. **Oh, H. I. and Hoff, J. E.,** Fractionation of grape tannins by affinity chromatography and partial characterization of the fractions, *J. Food Sci.*, 44, 87, 1979.
415. **Oh, H. I., Hoff, J. E., and Haff, L. A.,** Immobilized condensed tannins and their interaction with proteins, *J. Food Sci.*, 50, 1652, 1985.
416. **Okamura, S. and Watanabe, M.,** High-performance liquid chromatography analyses of phenolic compounds in grape must and wine, *J. Agric. Chem. Soc. Japan*, 53, 165, 1979.

417. **Okombi, G.,** Les pigments de la Cerise, *Prunus avium* (L.), variété "Bigarreau Napoléon": variations au cours de la croissance, de la maturation et de la conservation, Thése Doctorat 3ème Cycle, Université d'Orléans, France, 1979.
418. **Olden, E. J. and Nybom, N.,** On the origin of *Prunus cerasus* L., *Hereditas,* 59, 327, 1968.
419. **Oleszek, W., Lee, C. Y., Jaworski, A. W., and Price, K. R.,** Identification of some phenolic compounds in apples, *J. Agric. Food Chem.,* 36, 430, 1988.
420. **Ong, B. Y., and Nagel, C. W.,** Hydroxycinnamic acid-tartaric acid ester content in mature grapes and during the maturation of White Riesling grapes, *Am. J. Enol. Vitic.,* 29, 277, 1978.
421. **Ong, B. Y. and Nagel, C. W.,** High-pressure liquid chromatographic analysis of hydroxycinnamic acid-tartaric acid esters and their glucose esters in *Vitis vinifera, J. Chromatogr.,* 157, 345, 1978.
422. **Osawa, Y.,** Copigmentation of anthocyanins, in *Anthocyanins as Food Colors,* Markakis, P., Ed., Academic Press, New York, 1982, 41.
423. **Oszmianski, J. and Sapis, J. C.,** Anthocyanins in fruits of *Aronia melanocarpa* (Chokeberry), *J. Food Sci.,* 53, 1241, 1988.
424. **Oszmianski, J., Ramos, T., and Bourzeix, M.,** Fractionation of phenolic compounds in red wine, *Am. J. Enol. Vitic.,* 39, 259, 1988.
425. **Oydvin, J.,** Inheritance of four cyanidin-3-glycosides in red currant, *Hortic. Res.,* 14, 1, 1974.
426. **Ozawa, T., Lilley, T. H., and Haslam, E.,** Polyphenol interactions: astringency and the loss of astringency in ripening fruit, *Phytochemistry,* 26, 2937, 1987.
427. **Palmer, J. K.,** The banana, in *The Biochemistry of Fruits and Their Products,* Vol. 2, Hulme, A. C., Ed., Academic Press, London, 1971, 65.
428. **Panizzi, L., Scarpati, M. L., and Oriente, E. G.,** Costituzione della oleuropeina, glucoside amaro e ad azione ipotensiva dell'olivo, Nota II, *Gazz. Chim. Ital.,* 90, 1449, 1960.
429. **Panizzi, L., Scarpati, M. L., and Trogolo, C.,** Sulla oleuropeina, Nota III, *Gazz. Chim. Ital.,* 95, 1279, 1965.
430. **Park, G. L., Avery, S. M., Byers, J. L., and Nelson, D. B.,** Identification of bioflavonoids from citrus, *Food Technol.,* 37, 98, 1983.
431. **Pecket, R. C. and Small, C. J.,** Occurrence, location and development of anthocyanoplasts, *Phytochemistry,* 19, 2571, 1980.
432. **Peynaud, E. and Ribereau-Gayon, P.,** The grape, in *The Biochemistry of Fruits and Their Products,* Vol. 2, Hulme, A. C., Ed., Academic Press, London, 1971, 172.
433. **Pezet, R. and Pont, V.,** Mise en évidence de ptérostilbène dans les grappes de *Vitis vinifera, Plant Physiol. Biochem.,* 26, 603, 1988.
434. **Phan, C. T. and Macheix, J. J.,** Relation entre la structure anatomique et le métabolisme des tissus végétaux. I. Différences entre la composition qualitative et quantitative des substances phénoliques d'explantats de pomme et des cals et des cellules qui en sont issus par culture, *Rev. Can. Biol.,* 38, 17, 1979.
435. **Poessel, J. L.,** Composés phénoliques et peroxydases de l'Abricotier (*Prunus armeniaca* L.): Etude comparative de deux variétés (Luizet et Canino) en relation a vec l'incompatibilité au greffage, Thése 3ème Cycle, Montpellier, France, 1983.
436. **Porter, L. J., Foo, L. Y., and Furneaux, R. H.,** Isolation of three naturally occurring *O*-beta-glucopyranosides of procyanidin polymers, *Phytochemistry,* 24, 567, 1985.
437. **Porter, L. J., Hrstich, L. N., and Chan, B. G.,** The conversion of procyanidins and prodelphinidins to cyanidin and delphinidin, *Phytochemistry,* 25, 223, 1986.
438. **Prabhaa, T. N. and Patwardhan, M. V.,** Polyphenols of avocado (*Persea americana*) and their endogenous oxidation, *J. Food Sci. Technol.,* 17, 215, 1980.
439. **Prasad, U. S. and Jha, O. P.,** Changes in pigmentation patterns during Litchi ripening: flavonoid production, *Plant Biochem. J.,* 5, 44, 1978.
440. **Proctor, J. T. A.,** Color stimulation in attached apples with supplementary light, *Can. J. Plant Sci.,* 54, 499, 1974.
441. **Proctor, J. T. and Creasy, L. L.,** The anthocyanin of the mango fruit, *Phytochemistry,* 8, 2108, 1969.
442. **Prusky, D., Kobiler, I., and Jacoby, B.,** Involvement of epicatechin in cultivar susceptibility of avocado fruits to *Colletotrichum gloeosporoides* after harvest, *J. Phytopathol.,* 123, 140, 1988.
443. **Puech, A. A., Rebeiz, C. A., Caltin, P. B., and Crane, J. C.,** Characterization of anthocyanins in fig (*Ficus carica* L.) fruits, *J. Food Sci.,* 40, 775, 1975.
444. **Puech, J. L. and Jouret, C.,** Les acides à noyau aromatique de la prune d'Ente et du pruneau d'Agen, *C.R. Séances Acad. Agric. Fr.,* 60, 92, 1974.
445. **Puski, G. and Francis, F. J.,** Flavonol glycosides in cranberries, *J. Food Sci.,* 32, 527, 1967.
446. **Pyysalo, H. and Kuusi, T.,** Phenolic compounds from the berries of Mountain Ash, *Sorbus aucuparia, J. Food Sci.,* 39, 636, 1974.
447. **Ragazzi, E. and Veronese, G.,** Indagini sulla presenza di demetiloeuropeina in olive di cultivar diverse, *Ann. Chim.,* 63, 21, 1973.
448. **Ragazzi, E., Veronese, G., and Guiotto, A.,** Demetiloeuropeina, Nuevo glucoside isolato da olive mature, *Ann. Chem.,* 63, 13, 1973.

449. **Rahman, W., Ishratullah, K., Wagner, H., Seligmann, O., Charii, V. M., and Osterdahl, B. G.,** Prunin-6″-*O*-*p*-coumarate, a new acylated flavonone glycoside from *Anacardium occidentale, Phytochemistry,* 17, 1064, 1978.
450. **Ramina, A. and Masia, A.,** Levels of extractable para-coumaric acid in the exomesocarp and seed of persisting and abscising peach fruit, *Sci. Hortic.,* 16, 375, 1982.
451. **Ramirez-Martinez, J. R.,** Phenolic compounds in coffee pulp: quantitative determination by HPLC, *J. Sci. Food Agric.,* 43, 135, 1988.
452. **Ramirez-Martinez, J. R. and Luh, B. S.,** Phenolic compounds in frozen avocados, *J. Sci. Food Agric.,* 24, 219, 1973.
453. **Ranadive, A. S. and Haard, N. F.,** Changes in polyphenolics on ripening of selected pear varieties, *J. Sci. Food Agric.,* 22, 86, 1971.
454. **Rao, G. S. R. L., Willison, J. H. M., and Ratnayake, W. M. N.,** Suberin production by isolated tomato fruit protoplasts, *Plant Physiol.,* 75, 716, 1984.
455. **Rao, G. S. R. L., Willison, J. H. M., Ratnayake, W. M. N., and Ackman, R. G.,** Phenolics of suberized envelopes generated by isolated tomato locule protoplasts, *Phytochemistry,* 24, 2127, 1985.
456. **Raven, J. A.,** The role of vacuoles, *New Phytol.,* 106, 357, 1987.
457. **Raynal, J.,** Modifications structurales et biochimiques de la prune d'entre au début du séchage. Effets de diverses températures sur la structure cellulaire et ses incidences sur les composés pectiques et phénoliques, Thèse Doctorat, I.N.P., Toulouse, 1987.
458. **Regnaut-Roger, C., Hadidane, R., Biard, J. F., and Boukef, K.,** Determination of phenolic compounds in the palm (*Phoenix dactilifera*) products, date and sape extracts ("Legmi"), *Bull. Liaison Groupe Polyphénols,* 13, 494, 1986.
459. **Reschke, A. and Herrmann, K.,** Vorkommen von 1-*O*-Hydroxicinnamyl beta-D-glucosen im Obst. 15. Phenolische Inhaltsstoffe des Obstes, *Z. Lebensm. Unters. Forsch.,* 173, 458, 1981.
460. **Reschke, A. and Herrmann, K.,** Vorkommen von 1-O-Hydroxicinnamyl-beta-D-glucosen im Gemüse 1. Phenolcarbonsaüre-Verbindungen der Gemüses, *Z. Lebensm. Unters. Forsch.,* 174, 5, 1982.
461. **Reznik, H. and Egger, K.,** Benedikts Reagens als Indicator fur phenolische ortho-Dihydroxygruppen, *Z. Anal. Chem.,* 183, 196, 1961.
462. **Rhodes, M. J. C.,** The physiological significance of plant phenolic compounds, in *Ann. Proc. Phytochem. Soc. Europe,* Vol. 25, Van Sumere, C. F. and Lea, P. J., Eds., Clarendon Press, Oxford, 1985, 99.
463. **Ribereau-Gayon, P.,** Les acides phénols de *Vitis vinifera, C.R. Acad. Sci., Paris,* 256, 4108, 1963.
464. **Ribereau-Gayon, P.,** Les composés phénoliques du raisin et du vin. I. Les acides phénols, *Ann. Physiol. Vég.,* 6, 119, 1964.
465. **Ribereau-Gayon, P.,** Les composés phénoliques du raisin et du vin. II. Les flavonosides et les anthocyanosides, *Ann. Physiol. Vég.,* 6, 210, 1964.
466. **Ribereau-Gayon, P.,** Les composés phénoliques du raisin et du vin. III. Les tannins, *Ann. Physiol. Vég.,* 6, 259, 1964.
467. **Ribereau-Gayon, P.,** Les flavonosides de la baie dans le genre *Vitis, C.R. Acad. Sci., Paris,* 258, 1335, 1964.
468. **Ribereau-Gayon, P.,** Identification d'esters des acides cinnamiques et de l'acide tartrique dans les limbes et les baies de *V. vinifera, C.R. Acad. Sci, Paris,* ser D, 260, 341, 1965.
469. **Ribereau-Gayon, P.,** *Plant Phenolics,* Oliver and Boyd, Edinburgh, 1972.
470. **Ribereau-Gayon, P.,** Evolution des composés phénoliques au cours de la maturation du raisin. II. Discussion des résultats obtenus en 1969, 1970 et 1971, *Conn. Vigne Vin,* 6, 161, 1972.
471. **Ribereau-Gayon, P.,** The anthocyanins of grapes and wines, in *Anthocyanins as Food Colors,* Markakis, P., Ed., Academic Press, New York, 1982, 209.
472. **Riley, R. G. and Kolattukudy, P. E.,** Evidence for covalently attached *p*-coumaric and ferulic acid in cutins and suberins, *Plant Physiol.,* 56, 650, 1975.
473. **Riov, J.,** 6,7-Dimethoxycoumarin in the peel of gamma-irradiated grapefruit, *Phytochemistry,* 10, 1923, 1971.
474. **Riov, J., Goren, R., Monselise, S. P., and Kahan, R. S.,** Effect of gamma radiation on the synthesis of scopoletin and scopolin in grapefruit peel in relation to radiation damage, *Rad. Res.,* 45, 326, 1971.
475. **Risch, B. and Herrmann, K.,** Die gehalte an Hydroxyzimtsäure-Verbindungen und Catechinen in Kern- und Steinobst, *Z. Lebens. Unters. Forsch.,* 186, 225, 1988.
476. **Risch, B., Herrmann, K., and Wray, V.,** (E)-*O*-*p*-coumaroyl-, (E)-*O*-feruloyl-derivatives of glucaric acid in Citrus, *Phytochemistry,* 27, 3327, 1988.
477. **Risch, B., Herrmann, K., Wray, V., and Grotjahn, L.,** 2′-(E)-*O*-*p*-coumaroylgalactaric acid and 2′-(E)-*O*-feruloylgalactaric acid in citrus, *Phytochemistry,* 26, 509, 1987.
478. **Rivas, N. and Luh, B. S.,** Polyphenolic compounds in canned tomato pastes, *J. Food Sci.,* 33, 358, 1968.
479. **Robberecht, R. and Caldwell, M. M.,** Leaf epidermal transmittance of ultraviolet radiation and its implications for plant sensitivity to ultraviolet-radiation induced injury, *Oecologia,* 32, 277, 1978.
480. **Robbin, R. C.,** Regulatory action of phenylbenzo-γ-pyrone (PBP) derivatives on blood consitutents affecting rheology in patients with coronary hearth disease (CHD), *Int. J. Vit. Nutr. Res.,* 46, 338, 1976.

481. **Robertson, G. L.**, Salicylic acid in grapes, *Am. J. Enol. Vitic.*, 34, 42, 1983.
482. **Robertson, G. L. and Kermode, W.**, Salicylic acid in fresh and canned fruit and vegetables, *J. Sci. Food Agric.*, 32, 833, 1981.
483. **Roggero, J. P., Coen, S., and Larice, J. L.**, Etude comparative de la composition anthocyanique des cépages. Essai de classificatión, *Bull. Liaison Groupe Polyphénols*, 13, 380, 1986.
484. **Roggero, J. P., Coen, S., and Ragonnet, B.**, High performance liquid chromatography survey on changes in pigment content in ripening grapes of Syrah. An approach to anthocyanin metabolism, *Am. J. Enol. Vitic.*, 37, 77, 1986.
485. **Roggero, J. P., Ragonnet, B., and Coen, S.**, Utilisation de la chromatographie liquide à haute performance dans l'analyse des anthocyanes des vins et des baies de raisin; étude de quelques cépages méridionaux, *Bull. Liaison Groupe Polyphénols*, 12, 594, 1984.
486. **Romeyer, F. M., Macheix, J. J., Goiffon, J. P., Reminiac, C. C., and Sapis, J. C.**, The browning capacity of grapes. 3. Changes and importance of hydroxycinnamic-acid-tartaric esters during development and maturation of the fruit, *J. Agric. Food Chem.*, 31, 346, 1983.
487. **Romeyer, F. M., Sapis, J. C., and Macheix, J. J.**, Hydroxycinnamic esters and browning potential in mature berries of some grape varieties, *J. Sci. Food Agric.*, 36, 728, 1985.
488. **Romeyer, F., Macheix, J. J., and Sapis, J. C.**, Changes and importance of oligomeric procyanidins during maturation of grape seeds, *Phytochemistry*, 25, 219, 1986.
489. **Rouseff, R. L., Martin, S. F., and Youtsey, C. O.**, Quantitative survey of narirutin, naringin, hesperidin and neohesperidin in *Citrus*, *J. Agric. Food Chem.*, 35, 1027, 1987.
490. **Rudnicki, R., Hammond, R. K., and Bukovac, M. J.**, Endogenous plant growth substances in developing fruit of *Prunus cerasus* L. II. Levels of extractable para-coumaric acid in the pericarp, *J. Am. Soc. Hortic. Sci.*, 98, 225, 1973.
491. **Ryan, J. J.**, Flavonol glycosides of the cultivated strawberry, *J. Food Sci.*, 36, 867, 1971.
492. **Ryan, J. J. and Coffin, D. E.**, Flavonol glucuronides from red raspberry, *Rubus idaeus* (Rosaceae), *Phytochemistry*, 10, 1675, 1971.
493. **Rylski, I.**, Pepper (Capsicum), in *Handbook of Fruit Set and Development*, Monselise, S. P., Ed., CRC Press, Boca Raton, FL, 1986, 341.
494. **Saeed, A. R., Karamalla, K. A., and Khattab, A. H.**, Polyphenolic compounds in the pulp of *Mangifera indica* L., *J. Food Sci.*, 41, 959, 1976.
495. **Sakamura, F. and Suga, T.**, Changes in chemical components of ripening oleaster fruits, *Phytochemistry*, 26, 2481, 1987.
496. **Sapers, G. M., Phillips, J. G., Rudolf, H. M., and Di Vito, A. M.**, Cranberry quality: selection procedures for breeding programs, *J. Am. Soc. Hortic. Sci.*, 108, 241, 1983.
497. **Sapers, G. M., Jones, S. B., Kelley, M. J., Phillips, J. G., and Stone, E. G.**, Breeding strategies for increasing the anthocyanin content of cranberries, *J. Am. Soc. Hortic. Sci.*, 111, 618, 1986.
498. **Sapers, G. M., Hicks, K. B., Burgher, A. M., Hargrave, D. L., Sondey, S. M., and Bilyk, A.**, Anthocyanin patterns in ripening thornless blackberries, *J. Am. Soc. Hortic. Sci.*, 111, 945, 1986.
499. **Sarkanen, K. V.**, Precursors and their polymerization, in *Lignins, Occurrence, Formation, Structure and Reaction*, Sarkanen, K. V. and Ludwig, C. H., Eds., Wiley Interscience, New York, 1971, 95.
500. **Satyanarayana, D., Mythirayee, C., Krishnamurty, V., and Madhavakrishna, W.**, Studies on the polyphenols of cashew apple *(Anacardium occidentale)*, *Leather Sci.*, 25, 51, 1978.
501. **Saunders, J. A. and McClure, J. W.**, The distribution of flavonoids in chloroplasts of twenty-five species of vascular plants, *Phytochemistry*, 15, 809, 1976.
502. **Sawabe, A., Matsubara, Y., Kumamoto, H., Iizuka, Y., and Okamoto, K.**, Structure and physiological activity of phenyl propanoid glycosides of Hassaku (*Citrus hassaku* Hort.) and Orange (*Citrus sinensis* Osbeck.) Peelings, *Nippon Nogeikagaku Kaishi*, 8, 593, 1986.
503. **Scarpati, M. L. and Esposito, P.**, Neo-chlorogenic acid and "Band 510" structure, *Tetrahedon Lett.*, 18, 1147, 1963.
504. **Scarpati, M. L. and Guiso, M.**, Structure of the three dicaffeoyl-quinic acids of coffee (isochlorogenic acid), *Tetrahedron Lett.*, 2851, 1964.
505. **Scarparti, M. L. and Esposito, P.**, Isomerizzazione per transesterificazione degli acide 3-,4-e 5-mono-caffeil-chinici e del loro metileri, *Ann. Chim. Roma*, 54, 51, 1964.
506. **Schaller, D. R. and Von Elbe, J. H.**, Polyphenols in Montmorency cherries, *J. Food Sci.*, 35, 762, 1970.
507. **Schmidtlein, H. and Herrmann, K.**, Quantitative analysis for phenolic acids by thin-layer chromatography, *J. Chromatogr.*, 115, 123, 1975.
508. **Schreier, P.**, Flavor composition of wines: a review, *CRC Crit. Rev. Food Sci. Nutr.*, 12, 59, 1979.
509. **Schulten, H. R.**, Recent advances in soft ionization mass spectrometry and its applications to plant phenolics, in *The Biochemistry of Plant Phenolics*, Van Sumere, C. F. and Lea, P. J., Eds., Clarendon Press, Oxford, 1985, 75.
510. **Schulz, J. M. and Herrmann, K.**, Analysis of hydroxybenzoic and hydroxycinnamic acids in plant material. I. Sample preparation and thin-layer chromatography, *J. Chromatogr.*, 195, 85, 1980.

511. **Schulz, J. M. and Herrmann, K.**, Analysis of hydrobenzoic and hydroxycinnamic acids in plant material. II. Determination by gas-liquid chromatography, *J. Chromatogr.*, 195, 95, 1980.
512. **Schuster, B. and Herrman, K.**, Hydroxybenzoic and hydroxycinnamic acid derivatives in soft fruits, *Phytochemistry*, 24, 2761, 1985.
513. **Schwab, W. and Schreier, P.**, Aryl beta-D-glucosides from *Carica papaya* fruit, *Phytochemistry*, 27, 1813, 1988.
514. **Scienza, A., Visai, C., Conca, E., and Romano, F.**, Il profilo antocianico delle uve quale mezzotassonomico per il riconoscimento dei vitigni rossi, *4ème Symp. Int. Genet. Vigne*, Vérone (Italie), 17, 1985.
515. **Scora, R. W. and Kumamoto, J.**, Chemotaxonomy of the genus citrus, in *Chemistry and Chemical Taxonomy of the Rutales*, Watermann, P. G. and Grundon, M. F., Eds., Academic Press, London, 1983, 343.
516. **Sen, A. K., Sarkar, K. K., Mazumber, P. C., and Banerji, N.**, Minor xanthones of *Garcinia mangostana*, *Phytochemistry*, 20, 183, 1981.
517. **Sen, A. K., Sarkar, K. K., Mazumder, P. C., Banerji, N., Uusvuori, R., and Hase, T. A.**, The structure of garcinones A, B and C: three new xanthones from *Garcinia mangostana*, *Phytochemistry*, 21, 1747, 1982.
518. **Sharma, V. and Strack, D.**, Vacuolar localization of l-sinapoylglucose: L-malate sinapoyltransferase in protoplasts from cotyledons of *Raphanus Sativus*, *Planta*, 163, 563, 1985.
519. **Shrikhande, A. J. and Francis, F. J.**, Anthocyanin pigments of sour cherries, *J. Food Sci.*, 38, 649, 1973.
520. **Shrikhande, A. J. and Francis, F. J.**, Flavonol glycosides of sour cherries, *J. Food Sci.*, 38, 1035, 1973.
521. **Shrivastava, S. P.**, A flavonone glycoside from *Prunus cerasoides*, *Phytochemistry*, 21, 1464, 1982.
522. **Siddiqi, A. I. and Freedman, S. O.**, Identification of chlorogenic acid in castor bean and oranges, *Can. J. Biochem.*, 11, 947, 1963.
523. **Siewek, F., Galensa, R., and Herrmann, K.**, Nachweis eines Zusatzes von Roten zu Schwarzen Johannisbeer-Erzeugnissen úber die hochdruckflússigchromatographische Bestimmung der Flavonolglykoside, *Z. Lebensm. Unters. Forsch.*, 79, 315, 1984.
524. **Siewek, F., Galensa, R., and Herrmann, K.**, Isolation and identification of a branched quercetintriglycoside from *Ribes rubrum* (Saxifragaceae), *J. Agric. Food Chem.*, 32, 1291, 1984.
525. **Siewek, F., Herrmann, K., Grotjahn, L., and Wray, V.**, Isomeric di-C-glucosylflavones in fig (*Ficus carica* L.), *Z. Naturforsch.*, 40c, 8, 1985.
526. **Singh, N. S.**, Influence of UV light on the biosynthesis of anthocyanin-like pigments in ripening bananas, *Phytochemistry*, 11, 163, 1972.
527. **Singleton, V. L.**, Grape and wine phenolics, background and prospects, *C.R. Coll. Centenaire Paris*, 215, 1980.
528. **Singleton, V. L. and Esau, P.**, *Phenolic Substances in Grapes and Wine, and Their Significance*, Academic Press, New York, 1969.
529. **Singleton, V. L. and Rossi, J. A., Jr**, Colorimetry of total phenolics with phosphomolybdic-phosphotungstic acid reagents, *Am. J. Enol. Vitic.*, 16, 144, 1965.
530. **Singleton, V. L. and Trousdale, E.**, White wine phenolics: varietal and processing differences as shown by HPLC, *Am. J. Enol. Vitic.*, 34, 27, 1983.
531. **Singleton, V. L., Timberlake, C. F., and Lea, A. G. H.**, The phenolic cinnamates of white grapes and wine, *J. Sci. Food Agric.*, 29, 403, 1978.
532. **Singleton, V. L., Timberlake, C. F., and Whiting, G. C.**, Chromatography of natural phenolic cinnamate derivatives on Sephadex LH-20 and G-25, *J. Chromatogr.*, 140, 120, 1977.
533. **Singleton, V. L., Zaya, J., and Trousdale, E.**, Caftaric and coutaric acids in fruit of *Vitis*, *Phytochemistry*, 25, 2127, 1986.
534. **Singleton, V. L., Salgues, M., Zaya, J., and Trousdale, E.**, Caftaric acid disappearance and conversion to products of enzymic oxidation in grape must and wine, *Am. J. Enol. Vitic.*, 36, 50, 1985.
535. **Singleton, V. L., Zaya, J., Trousdale, E., and Salgues, M.**, Caftaric acid in grapes and conversion to a reaction product during processing, *Vitis*, 23, 113, 1984.
536. **Slinkard, K. W. and Singleton, V. L.**, Total phenol analysis: automation and comparison with manual methods, *Am. J. Enol. Vitic.*, 28, 49, 1977.
537. **Slinkard, K. W. and Singleton, V. L.**, Phenol content of grape skins and the loss of ability to make anthocyanins by mutation, *Vitis*, 23, 175, 1984.
538. **Small, C. J. and Pecket, R. G.**, The ultrastructure of anthocyanoplasts in red cabbage, *Plant*, 154, 97, 1982.
539. **Smith, T. A., Negrel, J., and Bird, C. R.**, The cinnamic acid amides of the di and polyamines, *Adv. Polyamine Res.*, 4, 347, 1983.
540. **Solinas, M., di Giovacchino, L., and Cucurachi, A.**, I polifenoli delle olive e del l'olio di oliva. Nota 1: Variazoni che subiscono alcuni polifenoli delle olive col procedere della maturazione, *Publicazione 18*, *Ann. Instit. Sperim. Elaiotecn.*, 5, 3, 1975.

541. **Somers, T. C., Verette, E., and Pocock, F.,** Hydroxycinnamate esters of *Vitis vinifera:* changes during white vinification, and effect of exogenous enzymic hydrolysis, *J. Sci. Food Agric.*, 40, 67, 1987.
542. **Somers, T. C., Verette, E., Pocock, K. F., and Strauss, C. R.,** Spectral characteristics of hydroxycinnamate esters of *Vitis vinifera, Bull. Liaison Groupe Polyphénols,* 13, 502, 1986.
543. **Sondheimer, E.,** On the distribution of caffeic acid and the chlorogenic acid isomers in plants, *Arch. Biochem. Biophys.*, 74, 131, 1958.
544. **Spanos, G. A. and Wrolstad, R. E.,** Anthocyanin pigment, nonvolatile acid, and sugar composition of red raspberry juice, *J. Assoc. Off. Anal. Chem.*, 70, 1036, 1987.
545. **Spencer, C. M., Cai, Y., Martin, R., Gaffney, S. H., Goulding, P. N., Magnolato, D., Lilley, T. H., and Haslam, E.,** Polyphenol complexation. Some thoughts and observations, *Phytochemistry,* 27, 2397, 1988.
546. **Stafford, H. A. and Lester, H. H.,** Flavan-3-ol biosynthesis. The conversion of (+)-dihydroxymyricetin to its flavan-3,4-diol (leucodelphinidin) and to (+)-gallocatechin by reductases extracted from tissue cultures of *Ginkgo biloba* and *Pseudotsuga menziessii, Plant Physiol.*, 78, 791, 1985.
547. **Stafford, H. A., Kreitlow, K. S., and Lester, H. H.,** Comparison of proanthocyanidins and related compounds in leaves and leaf-derived cell cultures of *Ginkgo biloba* L., *Pseudotsuga menziesii* Franco, and *Ribes sanguineum* Pursh, *Plant Physiol.*, 82, 1132, 1986.
548. **Starke, H. and Herrmann, K.,** Die phenolischen Inhaltsstoffe des Obstes VIII. Veranderungen des Flavonolgehaltes wahrend der Fruchtenwicklung, *Z. Lebensm. Unters. Forsch.*, 161, 131, 1976.
549. **Stohr, H. and Herrmann, K.,** Die phenolischen Inhaltsstoffe des Obstes V. Die phenolischen Inhaltsstoffe der Erdbeeren und deren Veranderungen wahrend Wachstrum und Reife der Fruchte, *Z. Lebensm. Unters. Forsch.*, 159, 341, 1975.
550. **Stohr, H. and Herrman, K.,** Die phenolischen Inhaltsstoffe des Obstes VI. Die phenolischen Inhaltssoffe der Johannisbeeren, Stachelbeeren und Kulturheidelbeeren. Veranderungen der Phenolsauren und Catechine wahrend Wachstum und Reife von Swarzen Johannisbeeren, *Z. Lebens. Unters, Forsch.*, 159, 31, 1975.
551. **Stohr, H. and Hermann, K.,** On the occurrence of derivatives of hydroxycinnamic acids, hydroxybenzoic acids and hydroxycoumarins in citrus fruits, *Z. Lebensm. Unter. Forsch.*, 159, 305, 1975.
552. **Stohr, H., Mosel, H. D., and Herrmann, K.,** The phenolics of fruits. VII. The phenolics of cherries and plums and the changes in catechins and hydroxycinnamic acid derivatives during the development of fruits, *Z. Lebensm. Unters. Forsch.*, 159, 85, 1975.
553. **Strack, D. and Sharma, V.,** Vacuolar localization of the enzymatic synthesis of hydroxycinnamic acid esters of malic acid in protoplasts from *Raphanus sativus* leaves, *Physiol. Plant,* 65, 45, 1985.
554. **Su, C. T. and Singleton, V. L.,** Identification of three flavan-3-ols from grapes, *Phytochemistry,* 8, 1553, 1969.
555. **Subramanyam, H., Krishnamusthy, S., and Parpia, H. A. B.,** Mango fruit in *Advances in Food Research,* Chichester, C. O., Mrak, E. M., and Stewart, G. F., Eds., Vol. 21, Academic Press, New York, 1975, 224.
556. **Sugiura, A. and Tomana, T.,** Relationships of ethanol production by seeds of different types of Japanese persimmons and their tannin content, *HortScience,* 18, 319, 1983.
557. **Sun, B. H. and Francis, F. J.,** Apple anthocyanins: identification of cyanidin-7-arabinoside, *J. Food Sci.*, 32, 647, 1967.
558. **Swain, T. and Hillis, W. E.,** The phenolic constituents of Prunus domestica. The quantitative analysis of phenolic constituents, *J. Sci. Food Agric.*, 10, 63, 1959.
559. **Szilagyi, I., Tetenyi, P., Antus, S., Seligmann, O., Chari, V. M., Seitz, M., and Wagner, H.,** Isolation and structure of silymonin ans silandrin, two new flavanolignans from *Silybum marianum* (L) Gaertn. flore albo, in *Flavonoids and Bioflavonoids, 1981,* Farkas, L., Gabor, M., Kallay, F., and Wagner, H., Eds., Elsevier, Amsterdam, 1982, 345.
560. **Takeda, K., Harborne, J. B., and Self, R.,** Identification and distribution of malonated anthocyanins in plants of the Compositae, *Phytochemistry,* 25, 1337, 1986.
561. **Takeda, K., Harborne, J. B., and Self, R.,** Identification of malonated anthocyanins in the Liliaceae and Labiatae, *Phytochemistry,* 25, 2191, 1986.
562. **Tatum, J. H. and Berry, R. E.,** Six new flavonoids from citrus, *Phytochemistry,* 11, 2283, 1972.
563. **Tatum, J. H. and Berry, R. E.,** 6,7-Dimethoxycoumarin in the peels of citrus, *Phytochemistry,* 16, 1091, 1977.
564. **Tatum, J. H. and Berry, R. E.,** Coumarins and psoralens in grapefruit peel oil, *Phytochemistry,* 18, 500, 1979.
565. **Tatum, J. H. and Berry, R. E.,** Coniferin. An astringent glycoside in citrus, *Phytochemistry,* 26, 322, 1987.
566. **Tatum, J. H., Berry, R. E., and Hearn, C. J.,** Characterization of citrus cultivars and separation of nucellar and zygotic seedlings by thin layer chromatography, *Proc. Fla. State Hortic. Soc.*, 87, 75, 1974.
567. **Tatum, J. H., Hearn, C. J., and Berry, R. E.,** Characterization of citrus cultivars by chemical differentiation, *J. Am. Soc. Hortic. Sci.*, 103, 492, 1978.

568. **Teuber, H. and Herrmann, K.,** Flavonolglykoside der Apfel (*Malus silvestris* Mill.), *Z. Lebensm. Unters. Forsch.*, 166, 80, 1978.
569. **Teusch, M. and Forkmann, G.,** Malonyl-coenzyme A: anthocyanidin 3-glucoside malonyltransferase from flowers of *Callistephus chinensis*, *Phytochemistry*, 26, 2181, 1987.
570. **Thompson, R. S., Jacques, D., Haslam, E., and Tanner, R. J. N.,** Plant proanthocyanidins. Part 1. Introduction: the isolation, structure and distribution in nature of plant proanthocyanidins, *J. Chem. Soc., Perkin I*, 1837, 1972.
571. **Tigchelaar, E. C., Janick, J., and Erickson, H. T.,** The genetics of anthocyanin coloration in eggplant (*Solanum melongena* L.), *Genetics*, 60, 475, 1968.
572. **Timberlake, C. F.,** Anthocyanins — occurrence, extraction and chemistry, *Food Chem.*, 5, 69, 1980.
573. **Timberlake, C. F.,** Anthocyanins in fruits and vegetables, in *Recent Advances in the Biochemistry of Fruits and Vegetables*, Friend, J., and Rhodes, M. J. C., Eds., Academic Press, London, 1981, 221.
574. **Timberlake, C. F. and Bridle, P.,** The anthocyanins of apples and pears: the occurrence of acyl derivatives, *J. Sci. Food Agric.*, 22, 509, 1971.
575. **Timberlake, C. F., and Bridle, P.,** Anthocyanins, in *The Flavonoids*, Harborne, J. B., Mabry, T. J., and Mabry, H., Eds., Chapman and Hall, London, 1975, 214.
576. **Timberlake, C. F., Bakker, J., Bridle, P., Loeffler, R. T. S., Saito, N., and Self, R.,** Fast atom bombardment mass spectrometry (FABMS) and nuclear magnetic resonance (NMR) of anthocyanins, *Bull. Liaison Groupe Polyphénols*, 13, 96, 1986.
577. **Ting, S. V. and Attaway, J. A.,** Citrus fruits, in *The Biochemistry of Fruits and Their Products*, Vol. 2, Hulme, A. C., Ed., Academic Press, London, 1971, 107.
578. **Ting, S. V., Rouseff, R. L., Dougherty, M. H., and Attaway, J. A.,** Determination of some methoxylated flavones in citrus juices by high performance liquid chromatography, *J. Food Sci.*, 44, 69, 1979.
579. **Torre, L. C. and Barrit, B. H.,** Quantitative evaluation of Rubus fruit anthocyanin pigments, *J. Food Sci.*, 42, 488, 1977.
580. **Torres, A. M., Mau-Lastovicka, T., and Rezaaiyan, R.,** Total phenolics and high performance liquid chromatography of phenolic acids of avocado, *J. Agric. Food Chem.*, 35, 921, 1987.
581. **Towers, G. H. N. and Yamamoto, E.,** Interactions of cinnamic acid and its derivatives with light, *Ann. Proc. Phytochem. Soc. Europe*, Vol. 25, Van Sumere, C. F., and Lea, P. J., Eds., Clarendon Press, Oxford, 1985, 271.
582. **Treutter, D.,** Separation of naturally occuring mixtures of phenolic compounds from various Prunus tissues by reversed-phase high-performance liquid chromatography, *J. Chromatogr.*, 436, 490, 1988.
583. **Tronchet, J.,** Evolution des flavonols dans la peau des tomates (*Lycopersicum esculentum* var. Marmande) cultivees en plein air, de l'ovaire au fruit mur, *Ann. Sci. Univ. Besancon, Bot.*, 7, 27, 1970.
584. **Tronchet, J.,** Les dérivés flavoniques des fruits, localisation, répartition et évolution, *Bull. Soc. Bot. Fr.*, 119, 25, 1972.
585. **Tronchet, J.,** Les flavonols des fruits de Lycopersicum esculentum Mill, *95e Cong. Nat. Soc. Savantes, Biol. Vég.*, 339, 1975.
586. **Tronchet, J.,** Effets de talures sur l'équipement flavonique superficiel de fruits: dégradation de l'écran filtre épidermique, *97e Cong. Natl. Soc. Sav.*, 4, 347, 1976.
587. **Trousdale, E. K. and Singleton, V. L.,** Astilbin and engeletin in grapes and wine, *Phytochemistry*, 22, 619, 1983.
588. **Tyman, J. H. P.,** Non-isoprenoid long chain phenols, *Chem. Soc. Rev.*, 8, 499, 1979.
589. **Vamos-Vigyazo, L. and Nadudvari-Markus, V.,** Enzymatic browning, polyphenol content, polyphenol oxidase and peroxidase activities in pear cultivars, *Acta Aliment.*, 11, 157, 1982.
590. **Vamos-Vigyazo, L., Gajzago, I., Nadudvari-Markus, V., and Magdolna-Kerek, M.,** Changes in some enzyme activities and browning rate of apricots during storage, *Qual. Plant. Plant Foods Hum. Nutr.*, 35, 139, 1985.
591. **Vamos-Vigyazo, L., Schuster-Gajzago, I., Kiss-Kutz, N., Hoschke, A., Toth, E., Erdos, Z., and Bartucz-Kovacs, O.,** Variations of polyphenols content and polyphenol oxidase activity during ripening of plums, *Bull. Liaison Groupe Polyphénols*, 12, 196, 1984.
592. **Van Blaricom, L. O., and Senn, T. L.,** Anthocyanin pigments in freestone peaches grown in the Southeast, *Proc. Am. Soc. Hortic. Sci.*, 90, 541, 1967.
593. **Van Buren, J.,** Fruit phenolics, in *The Biochemistry of Fruits and Their Products*, Vol. 1, Hulme, A. C., Ed., Academic Press, London, 1970, 269.
594. **Van Lelyveld, L. J., Gerrish, C., and Dixon, R. A.,** Enzyme activities and polyphenols related to mesocarp discolouration of avocado fruit, *Phytochemistry*, 23, 1531, 1984.
595. **Van Sumere, C. F., Van Brussel, W., Vande Casteele, K., and Van Rompaey, L.,** Recent advances in the separation of plant phenolics, in *Biochemistry of Plant Phenolics*, Swain, T., Harborne, J. B., and Van Sumere, C. F., Eds., Plenum Press, New York, 1979, 1.

596. **Van Sumere, C. F., Vande Casteele, K., De Loose, R., and Heursel, J.,** Reversed phase-HPLC analysis of flavonoids and the biochemical identification of cultivars of evergreen Azalea, in *Ann. Proc. Phyto-Chemical Soc. Europe,* Vol. 25, Van Sumere, C. F. and Lea, P. J., Eds., Clarendon Press, Oxford, 1985, 27.
597. **Van Teeling, C. G., Cansfield, P. E., and Gallop, R. A.,** Chromatography on anthocyanins on columns of insoluble polyvinylpyrrolidone, *J. Chromatogr. Sci.,* 9, 505, 1971.
598. **Vandercook, C. E. and Tisserat, B.,** Flavonoid changes in developing lemons grown *in vivo* and *in vitro, Phytochemistry,* 28, 799, 1989.
599. **Variyar, P. S., Pendharkar, M. B., Banerjee, A., and Bandyopadhyay, C.,** Blackening in green pepper berries, *Phytochemistry,* 27, 715, 1988.
600. **Vasquez Roncero, A., Graciani Constante, E., and Maestro Duran, R.,** Componentes fenolicos de la aceituna. I. Polifenoles de la pulpa, *Grasas Aceites,* 25, 269, 1974.
601. **Vasquez Roncero, A., Maestro Duran, R., and Graciani Constante, E.,** Componentes fenolicos de la aceituna. II. Polifenoles del alpechin, *Grasas Aceites,* 25, 3, 41, 1974.
602. **Venkataraman, K.,** Flavones, in *The Flavonoids,* Harborne, J. B., Mabry, T. J., and Mabry, H., Eds., Chapman and Hall, London, 1975, 267.
603. **Verette, E., Brun, S., and Cabanis, J. C.,** Identification des anthocyanes monomères de la myrtille, *"Vaccinium myrtillus", Bull. Liaison Groupe Polyphénols,* 12, 143, 1984.
604. **Verette, E., Brun, S., Benet, F., and Cabanis, J. C.,** Contribution a l'étude des constituants anthocyanosidiques de la Myrtille, *Bull. Liaison Groupe Polyphénols,* 11, 442, 1982.
605. **Villeneuve, F., Abravanel, G., Moutounet, M., and Alibert, G.,** General scheme of analysis of phenolic compounds in plant extracts by reversed-phase high-performance liquid chromatography, *J. Chromatogr.,* 234, 131, 1982.
606. **Wagner, G. J.,** Compartmentation in plant cells: the role of the vacuole, *Recent Advances in Phytochemistry,* Vol. 16, Creasy, L. L. and Hrazdina, G., Eds., Plenum Press, New York, 1982, 1.
607. **Wagner, G. J. and Hrazdina, G.,** Endoplasmic reticulum as a site of phenylpropanoid and flavonoid metabolism in *Hippeastrum, Plant Physiol.,* 74, 901, 1984.
608. **Wagner, H.,** The antihepatotixic principle of *Silybum marianum* Gaertn., in *Recent Flavonoid Research, Recent Development in the Chemistry of Natural Carbon Compounds,* Vol. 5, Bognar, R., Brucker, V., and Szantay, C. S., Eds., Akademiai Kiado, Budapest, 1973, 51.
609. **Wagner, H.,** Flavonolignans from *Silybum marianum, Bull. Liaison Groupe Polyphénols,* 5, 1974.
610. **Wagner, H.,** New structures in the flavonoid series, in *Flavonoids and Bioflavonoids: Current Research Trends,* Farkas, L., Gabor, M., and Kallay, F., Eds., Elsvier, Amsterdam, 1977, 29.
611. **Waiss, A. C., Lundin, R. E., and Corse, J. W.,** The N.M.R. spectra of chlorogenic acids, *Chem. Ind.,* 1984, 1964.
612. **Wald, B., Wray, W., Galensa, R., and Hermann, K.,** Malonated flavonol glycosides and 3,5-dicaffeoylquinic acid acid from pears, *Phytochemistry,* 28, 663, 1989.
613. **Wald, B., Galensa, R., Herrmann, K., Grotjahn, L., and Wray, V.,** Quercetin 3-0-(6″-(3-hydroxy-3-methylglutaroyl)-beta-galactoside) from blackberries, *Phytochemistry,* 25, 2904, 1986.
614. **Walker, J. R. L.,** Studies on the enzymic browning of apple fruit, *N.Z. J. Sci.,* 5, 316, 1962.
615. **Walker, J. R. L.,** A note on the polyphenol content of ripening apples, *N.Z. J. Sci.,* 6, 492, 1963.
616. **Walker, J. R. L.,** Inhibition of the apple phenolase system through infection by *Penicillium expansum, Phytochemistry,* 8, 561, 1969.
617. **Wang, C. Y. and Mellenthin, W. M.,** Relationship of friction discoloration to phenolic compounds in d'Anjou pears, *HortScience,* 8, 321, 1973.
618. **Wang, J., Kasai, R., Sakimori, M., Miyakoshi, M., Tanaka, O., Jia, M. R., and Ling, Y. K.,** Flavonol glycosides from the fruits of *Rhammus leptophylla, Phytochemistry,* 27, 3995, 1988.
619. **Wang, P. L., Du, C. T., and Francis, F. J.,** Isolation and characterization of polyphenolic compounds in cranberries, *J. Food Sci.,* 43, 1402, 1978.
620. **Wardale, D. A.,** Effect of phenolic compounds in *Lycopersicum esculentum* on the synthesis of ethylene, *Phytochemistry,* 12, 1523, 1973.
621. **Watanabe, S., Sakamura, S., and Obata, Y.,** The structures of acylated anthocyanins in eggplant and *Perilla* and the position of acylation, *Agric. Biol. Chem.,* 30, 420, 1966.
622. **Weinges, K.,** The occurrence of catechins in fruits, *Phytochemistry,* 3, 263, 1984.
623. **Weinges, K. and Nader, F. W.,** Proanthocyanidins, in *Anthocyanins as Food Colors,* Markakis, P., Ed., Academic Press, New York, 1982, 93.
624. **Weinges, K. and Piretti, M. V.,** Isolation of procyanidin B1 from grapes, *Justus Liebig Ann. Chem.,* 748, 218, 1971.
625. **Weinges, K., Bahr, W., Ebert, W., Goritz, K., and Marx, H. D.,** Konstitution, Entstehung und Bedentung der Flavonoid-Gerhstoffe, *Fortschr. Chem. Org. Naturstoffe,* 27, 158, 1969.
626. **Weissenbock, Q., Plesser, A., and Trinks, K.,** Flavonoidgehalt und Enzymaticaktivitaten isolierter. Haferschloroplasten (*Avena sativa* L.), *Ber. Deustch. Bot. Ges.,* 89, 457, 1976.

627. **Wheaton, T. A. and Stewart, I.,** Feruloylputrescine: isolation and identification from citrus leaves and fruit, *Nature (London)*, 206, 620, 1965.

628. **Wheaton, T. A. and Stewart, I.,** Biosynthesis of synephrine in Citrus, *Phytochemistry*, 8, 85, 1969.

629. **Whiting, G. C. and Coggins, R. A.,** Estimation of the monomeric phenolics of ciders, *J. Sci. Food Agric.*, 26, 1833, 1975.

630. **Whiting, G. C. and Coggins, R. A.,** 4-*p*-Coumarylquinic acid in apple fruits, *Phytochemistry*, 14, 593, 1975.

631. **Wiermann, R.,** Secondary plant products and cell and tissue differentiation, in *The Biochemistry of Plants*, Vol. 7, Conn, E.E., Ed., Academic Press, New York, 1981, 86.

632. **Wilkinson, M., Sweeny, J. G., and Iacobucci, G. A.,** High-pressure liquid chromatography of anthocyanidins, *J. Chromatogr.*, 132, 349, 1977.

633. **Williams, A. H.,** *p*-Coumaryl-quinic acid from apple fruit, *Chem. Ind.*, 1200, 1958.

634. **Williams, A. H.,** The distribution of phenolic compounds in apple and pear trees, in *Phenolic in Plants in Health and Disease*, Pridham, J.B., Ed., Pergamon Press, London, 1960, 3.

635. **Williams, A. H.,** Chemical evidence from the flavonoids relevant to the classification of Malus species, *Bot. J. Linnean Soc.*, 84, 31, 1982.

636. **Williams, M., Hrazdina, G., Wilkinson, N. M., Sweeny, J. G., and Iacobucci, G. A.,** High-pressure liquid chromatographic separation of 3-glycosides, 3,5-diglucosides, 3-(6-*0*-*p*-coumaryl)glucosides and 3-(6-*0*-*p*-coumarylglucoside)-5-glucosides of anthocyanidins, *J. Chromatogr.*, 155, 389, 1978.

637. **Wilson, E. L.,** High-pressure liquid chromatography of apple juice phenolic compounds, *J. Sci. Food Agric.*, 32, 257, 1981.

638. **Winter, M. and Herrmann, K.,** Analysis of hydroxycinnamic acid esters and their glucosides by reversed-phase high-performance liquid chromatography after polyamide separation, *J. Chromatogr.*, 315, 243, 1984.

639. **Winter, M. and Herrmann, K.,** Esters and glucosides of hydroxycinnamic acids in vegetables, *J. Agric. Food Chem.*, 34, 616, 1986.

640. **Winter, M., Brandl, W., and Herrmann, K.,** Bestimmung von Hydroxyzimtsaüre-Derivaten in Gemüse, *Z. Lebensm. Unters. Forsch.*, 184, 11, 1987.

641. **Wollenweber, E.,** Flavones and flavonols, in *The Flavonoids: Advances in Research*, Harborne, J. B. and Mabry, T. J., Eds., Chapman and Hall, London, 1982, chap. 4.

642. **Wollenweber, E. and Dietz, V. H.,** Occurrence and distribution of free flavonoid aglycones in plants, *Phytochemistry*, 20, 869, 1981.

643. **Wollenweber, E. and Mann, K.,** Flavonole aus Fruchten von *Vitex agnus castus*, *Planta Med.*, 48, 126, 1983.

644. **Wong, E.,** Isoflavonoids, in *The Flavonoids*, Harborne, J. B., Mabry, T. J., and Mabry, H., Eds., Chapman and Hall, London, 1975, 743.

645. **Woodward, J. R.,** Physical and chemical changes in developing strawberry fruits, *J. Sci. Food Agric.*, 23, 465, 1972.

646. **Wrolstad, R. E. and Heatherbell, D. A.,** Identification of anthocyanins and distribution of flavonoids in Tamarillo fruit (*Cyphomandra betaceae* (Cav.) Sendt.), *J. Sci. Food Agric.*, 25, 1221, 1974.

647. **Wrolstad, R. E., Hildrum, K. I., and Amos, J. F.,** Characterization of an additional anthocyanin pigment in extracts of strawberries, *Fragaria*, *J. Chromatogr.*, 50, 311, 1970.

648. **Wrolstad, R. E., Putnam, T. P., and Varseveld, G. W.,** Color quality of frozen strawberries: effects of anthocyanin, pH total acidity and ascorbic acid variability, *J. Food Sci.*, 35, 448, 1970.

649. **Wu, T. S., Tien, H. J., Arisawa, M., Shimizu, M., and Morita, N.,** Flavonols and coumarins from the fruit of *Murraya omphalocarpa*, *Phytochemistry*, 19, 2227, 1980.

650. **Wulf, L. W. and Nagel, C. W.,** High-pressure liquid chromatographic separation of anthocyanins of *Vitis vinifera*, *Am. J. Enol. Vitic*, 29, 42, 1978.

651. **Wulf, L. W. and Nagel, C. W.,** Identification and changes of flavonoids in Merlot and Cabernet Sauvignon wines, *J. Food Sci.*, 45, 479, 1980.

652. **Yalpani, M. and Tyman, J. H. P.,** The phenolic acids of *Pistacia vera*, *Phytochemistry*, 22, 2263, 1983.

653. **Yamaki, S.,** Isolation of vacuoles from immature apple fruit flesh and compartmentation of sugars, organic acids, phenolic compounds and amino acids, *Plant Cell Physiol.*, 25, 151, 1984.

654. **Yonemori, K., and Matsushima, J.,** Chemical characteristics of tannins from nonastringent and astringent type fruits of Japanese persimmon (*Diospyros kaki*) with particular reference to ultra centrifugal behavior, *J. Japan Soc. Hortic. Sci.*, 53, 121, 1984.

655. **Yonemori, K. and Matsushima, J.,** Morphological characteristics of tannin cells in Japanese persimmon fruit, *J. Am. Soc. Hortic. Sci.*, 112, 812, 1987.

656. **Yonemori, K. and Matsushima, J.,** Changes in tannin cell morphology with growth and development of Japanese persimmon fruit, *J. Am. Soc. Hortic. Sci.*, 112, 818, 1987.

657. **Yonemori, K., Matsushima, J., and Sugiura, A.,** Differences in tannins of non-astringent and astringent type fruits of Japanese persimmon (*Diospyros kaki* Thunb.), *J. Jpn. Soc. Hortic. Sci.*, 52, 135, 1983.

658. **Yu, K. and Carlson, R. F.,** Paper chromatographic determination of phenolic compounds occurring in the leaf, bark and root of *Prunus avium* and *P. mahaleb*, *J. Am. Soc. Hortic. Sci.*, 100, 536, 1975.
659. **Zualaga Vasco, J. and Tabacchi, R.,** Contribution à l'étude des composés phénoliques de la pulpe de Café, *Séminaire C.P.C.I.A., Narbonne*, France, 17 to 19 Juin, 1981.
660. **Zuluaga Vasco, J. and Tabacchi, R.,** Contribution à l'étude de la composition chimique de la pulpe de Café, in *9e Coll. Sci. Int. sur le Café*, Londres, ASIC, Paris, 1980, 33.

Chapter 3

CHANGES AND METABOLISM OF PHENOLIC COMPOUNDS IN FRUITS

I. INTRODUCTION

Accumulation. of phenolic compounds varies strongly in relation to the physiological state of the fruit, and is a result of an equilibrium between biosynthesis and further metabolism including turnover and catabolism. The most important control mechanisms in the phenolic metabolism involve control of the amount of enzymes, regulation of enzyme activity, compartmentation of enzymes, availability of precursors and intermediates, and integration in the differentiation and development programs of the producer organism.[238] All these aspects are still far from being well known as regards the phenolic metabolism of fruits, but considerable progress has been made over the past 20 years and this is, to the best of our knowledge, the first time that a synthesis of data on the subject has been compiled. It has enabled us to show a number of points which are common to all fruits, although existing data are still too fragmentary. Three aspects are generally approached: (1) accumulation of the various classes of compounds during growth and maturation; the compounds have generally been chosen in relation to their abundance in fruits and their importance in sensory characteristics; (2) comparison of the rate of synthesis of these compounds *in vivo* with activity of specific enzymes measured *in vitro*; and (3) regulation of enzyme activity by a certain number of external and internal factors, such as light, temperature, hormones, nutrients, etc.

Although, as it is seen below, it is clear that the phenolic metabolism is perfectly integrated in the development phases of the fruit life to the point of being considered as a possible marker of maturation, many points remain to be elucidated. These include in particular the regulation of amounts of enzymes and gene expression, the study of enzyme synthesis by radioactive labeling, or by mRNA translation *in vitro*. An example of this type of study (as far as we know the only one of its kind to have been carried out on a fruit) is a very recent investigation of phenylalanine ammonialyase (PAL), a key enzyme in the phenolic metabolism, in strawberry.[128,129]

II. CHANGES IN PHENOLIC COMPOUNDS DURING GROWTH AND MATURATION

Considerable variations are generally observed in the levels of phenolic compounds in fruits during growth and maturation. Several parameters can *a priori* account for these variations, i.e.,

- The availability of precursors of phenolic molecules and in particular phenylalanine, whose regulatory role has frequently been reported[65,257]
- The activity of enzymes involved in the supply of these precursors (in particular that of the shikimic acid pathway), in interconversions between phenolic molecules (hydroxylation, methylation, glycosylation, condensation, etc.), or in their degradation
- The energizing coupling required in a large number of these reactions, resulting, for example, in the contribution of activated forms (CoA esters of hydroxycinnamic acids, UDPG, etc.)
- The reducing potential available, in particular, for the accumulation of certain reduced forms (hydroxycinnammic alcohols, flavans, etc.)

- Modifications occurring in the cell structures in fruits during maturation, insofar as certain enzymes involved in the phenolic metabolism are located on membranes
- The possible contribution of phenolic molecules or of their more or less immediate precursors brought by sap movements
- The vacuolar accumulation capacities of phenolic compounds whose regulation may be different from that of biosynthesis[147]
- The effect of external factors (physical, chemical, and biological) which may modify each of the parameters above

Although most of these aspects have been studied, it is fairly rare or even exceptional for several of them to have been investigated simultaneously in the same fruit. It is thus difficult today to obtain an overall idea of the phenomena which regulate accumulation of phenolic compounds in fruits. In addition, although variations in phenolic compound levels during growth and maturation of fruits have been the subject of numerous studies, most of the older works only approached the problem of total phenols or of a single category of phenolic compounds (anthocyanins, tannins, etc.). In rarer cases, variation concerned one of the major compounds (e.g., chlorogenic acid in apple,[185]) frequently determined using methods that were not specific. Evolution of techniques, in particular, the progressive introduction of high performance liquid chromatography (HPLC), has now made it possible to monitor the individual evolution of certain phenolic compounds. Nevertheless, much work remains to be done here because recent studies have only covered a limited group of fruits. However, we outline the main points concerning variations of phenolic compounds and certain enzymes in their metabolism during the life of the fruit using the best-studied examples.

A. PRELIMINARY REMARKS

Growth of fleshy fruits can be quantified fairly easily using classic criteria such as fresh weight (FW), dry weight (DW), or diameter.[26] The fruit accumulates considerable dry matter and water in a relatively short time (approximately 40 g DW and 160 g of water in a few months in an apple with an FW of 200 g), and during this growth period the fruit forms a very active sink for assimilates from neighboring leaves or directly from root absorption. In all cases, although direct photosynthetic assimilation in fruit exists, it only makes a small contribution to the accumulation of organic substances. The regulation of transport of carbon, nitrogen, and minerals between the vegetative part of the plant and the fruit is thus a major concern for physiologists. Numerous studies have been devoted to the question, for example, in the tomato.[179] However, most of the works are on the distribution and accumulation of primary metabolites and no attention has been paid to secondary metabolites, some of which (e.g., tannins) may be very plentiful in fruits. What is the source of the considerable quantities of phenolic molecule which accumulate in fruit? Are they the result of *in situ* biosynthesis and from what precursors, or are they brought at least in part by sap flow? The experimental data available on the question are still very fragmentary. They cover on the one hand, the study of the respective evolution of phenolic compounds during growth, and on the other, the demonstration of biosynthetic activity (possibly degradation) of phenolic molecules in the fruit itself. These two aspects are examined below.

A fruit is an heterogeneous entity for a number of reasons. First, it contains seeds with which are set up a number of trophic and hormonal correlations. Second, under hormonal control by the seeds, its structure can be derived from a variety of tissues ranging from ovary (most fruits, e.g., tomato, cherry, etc.) to neighboring tissue such as the receptacle (apple, strawberry), the conceptacle (fig), the inflorescence (pineapple), etc. In addition, tissue with very different characteristics can appear during fruit development (meso and endocarp for example); this results in different growth curves for different fruits:[26] classic sigmoid curves in the case of tomato and apple and double sigmoid for peach and olive. In the two latter fruits, the presence of a lignified endocarp (characteristic of drupes) corresponds to tissue in which the phenolic metabolism is *a priori* very different to that of fleshy pericarp which never lignifies, except under exceptional conditions discussed below.

One of the remarkable aspects of the growth of fruits is the dissociation of the phenomena of meresis and auxesis.[26] In the apple, for example, the number of cells (excepting slight subepidermal divisions) is fixed during the 3 to 4 weeks which follow fruit set. This period is limited to a week in tomato. One of the results is the extraordinary increase in volume of each cell, some of them being expanded about 350,000 times from their meristematic volume.[26] Expansion results from intake of dry matter and water in proportions which can vary during the development of the fruit. At maturity, the moisture content of most fleshy fruits is about 80 to 85%. The vacuolar compartment is thus particularly important in the ripe fruit and it contains almost all the soluble phenolic compounds, whereas any lignins present accumulate in the cell wall.

This naturally leads to discussing the different modes of expression of phenolic compound contents:

- Expression on the basis of FW is often used. Given the quasi-exclusive preferential location of soluble phenolic compounds in the vacuole and the importance of water in this compartment, it is a fairly good representation of the vacuolar concentration of these phenolic compounds. It is important to know this parameter since it is directly involved in the regulation of certain physiological phenomena and in the speed of the chemical reactions in which phenolic compounds play a role. However, when the fruit is expanding rapidly, utilization of the expression in terms of FW requires the taking of a number of precautions, since the values obtained may stem from the simultaneous evolution of two parameters (phenol content and FW) or even from the evolution of FW alone. For correct interpretation of the evolutions observed, it should always be accompanied by the two other types of expression.
- Expression based on DW makes it possible to eliminate variations in water between two samplings and thus shows the quantitative importance of the phenolic compounds in relation to the whole of the organic and mineral matter accumulated by the fruit. This expression is nevertheless somewhat artificial since a large proportion of this matter accumulates in compartments (cytoplasm, nucleus, membranes, cell walls) which do not contain any soluble phenolic compounds (or only very small quantities). This type of expression is also of interest when fruits destined for enzyme analysis are freeze-dried beforehand; the water is removed and the possible inhibition of enzymes by phenols[186] is a direct function of the value given by this expression.
- Expression in terms of amount per fruit should always accompany the two preceding ones. It shows the content of the whole fruit and the value can fairly easily be converted into amount per cell, at least in the case of numerous fruits whose number of cells is fixed very early on, as mentioned above. Balances for the various physiological stages are easy to obtain using this type of expression by showing the increase in phenolic compounds in the fruit (biosynthesis, import or export). In addition, it provides complementary information about the intervarietal or interspecific differences identified using the preceding expressions. Thus, in the comparison of different olive cultivars,[4] expression per fruit reduced variations very considerably because of a fairly close inverse correlation between the size of fruits and their oleuropein concentration (Figure 33). In contrast, in *Vitis*,[384] expression per fruit clearly separated wild species (with small berries which, therefore, accumulated fairly small amounts of hydroxycinnamoyltartaric esters) from those of *Vitis vinifera* cultivars. However, the expression per fruit cannot provide information about the concentration of phenolic compounds in water, upon which their physiological and technological importance depends.

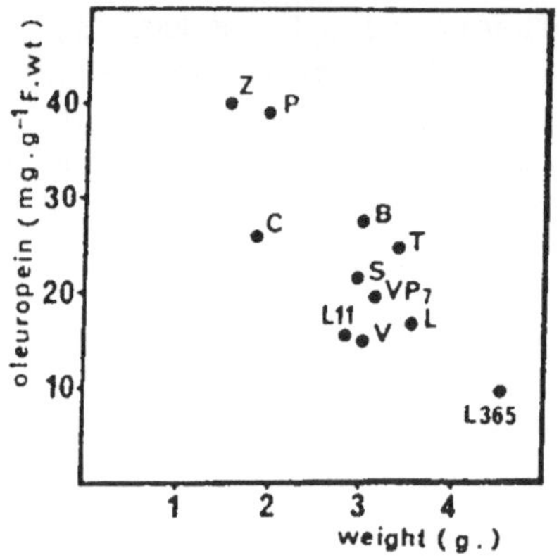

FIGURE 33. Relationship between oleuropein content and fresh weight in the fruit of 11 olive varieties: P, Picholine; L, Lucques; S, Salonenque; B, Bouteillan; V, Verdale; Z, Zrappola; C, Cailletier; T, Tanche; L11; L365; VP7. (From Amiot, M. J., Fleuriet, A., and Macheix, J. J., *J. Agric. Food Chem.*, 34, 823, 1986. With permission.)

As has been reviewed above, one frequently finds that not all tissues in a fruit contain uniform quantities (or any at all) of the various phenolic compounds studied. In the case of anthocyanins, siting may be at the surface only (apple peel, skin of most black grapes) and levels can be expressed in terms of unit surface area rather than unit DW. In this case, expression in amount per fruit indicates only "the part of the fruit" which contains the anthocyanins. This expression is found to be a better characteristic for judging differences between grape cultivars.[385]

Finally, several problems of expression may occur in the case in which phenolic compounds are studied in *in vitro* cultures from fruits. Although the expression of results from the calli does not display any particular features, the case of cell suspensions is very different.[223,248] It is difficult to determine fresh matter in this case and it is preferable to express the results either in relation to a cell or to the total cells in the culture. It is also necessary to analyze the liquid medium in which certain phenols may accumulate after either excretion by living cells or after the death of certain cells and the release of vacuole contents.

B. TOTAL PHENOLICS

The inaccuracy of this type of result and the difficulties in interpretation that may occur have already been mentioned. Nevertheless, examination of different fruits leads to preliminary comparisons. Since the initial work of Swain and Hillis[408] on plum, numerous investigations have confirmed that phenol concentrations are generally higher in young fruits.

Expression per unit of FW or DW can often be used to reveal two successive phases separated by a maximum during the early weeks in the life of a fruit. These correspond, respectively, to the accumulation of phenolic compounds followed by their dilution in the vacuole. Such variations have been reported in apple,[185] grape (Figure 34),[210] pear,[21] cherry,[273] plum,[403,408] persimmon,[263,454] sapota fruit,[225] and bottle gourd fruit.[294] Only the second phase can be seen when fruits are not sampled at sufficiently young stages.[124,187,259,364] The total phenol content (mg/g FW) then falls steadily during growth, and it is only towards maturation that two distinct cases can be observed. Either the level continues to fall steadily in white species and varieties, e.g., white grape cultivars (Figure 35), mango,[123] and banana,[226] or it rises again in red fruits in which anthocyanins[210,310,341] or other flavonoids[442] accumulate.

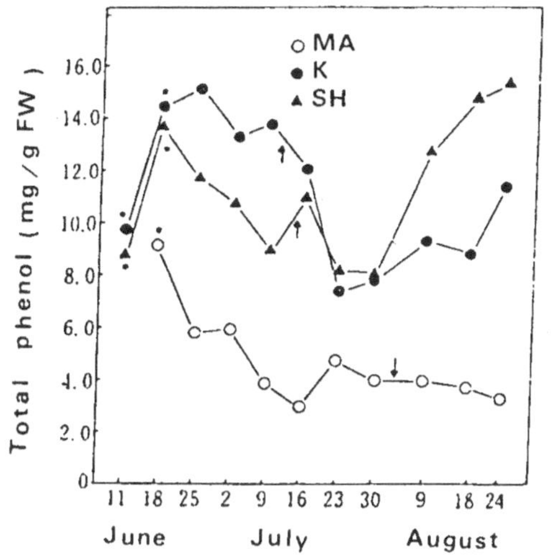

FIGURE 34. Changes in total phenols in the skin of three grape cultivars. MA: Muscat of Alexandria; K: Kyoho; SH: Super Hamburg. Whole berry was used for extraction. Arrows denote the beginning of veraison. (Adapted from Kataoka, I., Kubo, Y., Sugiura, A., and Tomana, T., *J. Jpn. Soc. Hortic. Sci.*, 52, 273, 1983.)

In avocado, ripening caused a slight decrease in total phenolics in the peel and a slight increase in pulp and seed coat.[318]

The conclusions which can be drawn from these data nevertheless remain very general and may cover elementary phenomena — on the one hand, individual variations of each of the phenolic compounds, and, on the other, the appearance or disappearance of some of these during the life of the fruit. In tomato, for example,[442] the increase in total phenol content during maturation is caused mainly by a strong increase in naringenin and unidentified compounds in the skin of the fruit which mask the fall in hydroxycinnamic acid content in the flesh.

C. PHENOLIC ACIDS

Whereas these variations have been the subject of very numerous studies in certain fruits (apple, tomato, grape, peach, etc.), data are very incomplete for many others. In addition, as has been mentioned in relation to the study of ripe fruits (Chapter 1), the accuracy of the results depends to a great extent on the quality of the analytical approaches used.

A preliminary comparison between fruits can be carried out taking into account the hydroxycinnamic acid contents determined after hydrolysis of combined forms.[279-281,370,400-402] Caffeic acid concentrations are generally high in young fruits (Figure 35), fall rapidly, and then more slowly during maturation and postharvest storage. The same situation is found for *p*-coumaric and ferulic acids, except in the tomato where these substances accumulate steadily, as does sinapic acid. When these results are expressed per fruit, they generally reveal accumulation which either extends throughout the life of the fruit or, in the case of caffeic acid, which is usually the most abundant, peaks during growth or in early maturation (Figure 35). The situation may also be different according to the location of the tissue studied in the fruit. A fall in caffeic acid content in the flesh (peeled fruit) has been observed in tomato while the substance continued to accumulate in the skin during maturation.[442]

Data on benzoic acids are very fragmentary for these fruits and it is difficult to pinpoint the general features of their variations. In contrast, in Japanese persimmon, evolution of the

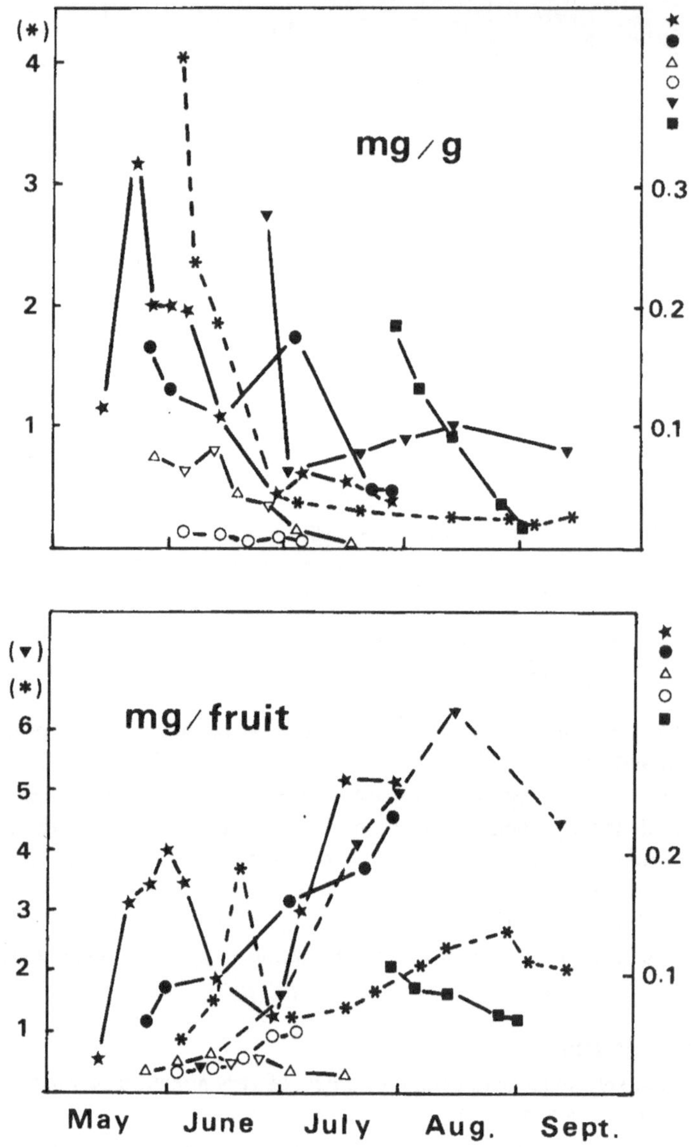

FIGURE 35. Changes in caffeic acid levels during growth and maturation of sweet cherry (★), sour cherry (●), strawberry (○), blackcurrant (△), blackberry (■), tomato (▼), and plum (***). (Adapted from References 281, 370, 401—403.)

gallic acid content is very characteristic;[454] in pollination constant nonastringent type cultivars, gallic acid was detected in the flesh of the fruit only at the earlier stage whereas in other cultivars it increased greatly and reached a peak (1.5 mg/g DW in the richest cultivar) in late June and then decreased to a very low level in late July (Figure 36). Somewhat similar evolution has been reported in different grape cultivars.[81]

Variations in chlorogenic acid and other hydroxycinnamic derivatives have been studied in many fruits (Reference 429 and references therein), often in relation with the browning capacity of the fruits or with certain phytopathological aspects. Fruits include apple,[185,241,243,287] pear,[21,327,440] peach,[43] cherry,[273] and kiwi.[124] It is very common to find in each of these fruits the variations previously found in total phenols or caffeic acid. Thus, in apple cv. Calville blanc,[243] accumulation of chlorogenic acid continued for several weeks in the whole fruit whereas maximum concentrations were found in very young fruits weighing

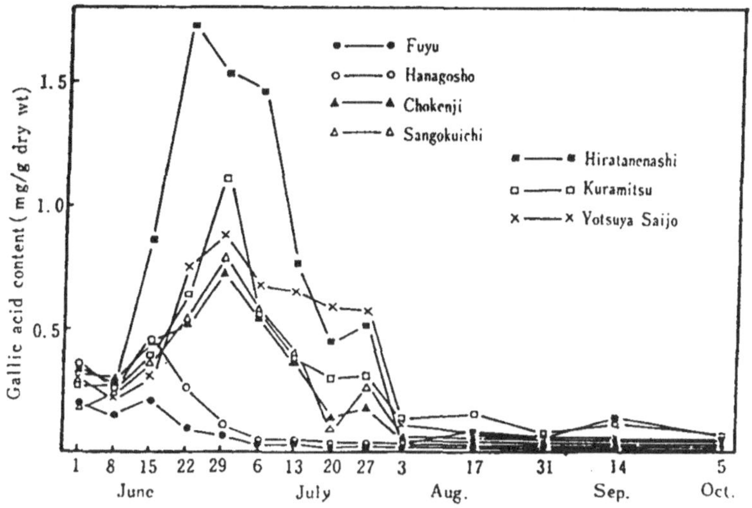

FIGURE 36. Seasonal changes of gallic acid content in the fruit flesh of seven persimmon cultivars. (From Yonemori, K., Matsushima, J., and Sugiura, A., *J. Jpn. Soc. Hortic. Sci.*, 53, 121, 1983. With permission.)

approximately 6.5 g (Figure 37). This variations are very similar from one year to the next with the only differences being the position of the maximum in time and the values observed. Similar evolution is observed in the levels of *p*-coumaroylquinic acid and *p*-coumaroylglucose, the two other main hydroxycinnamic derivatives in apple, but the peaks may be slightly staggered in time.[243] These evolutions make it possible to divide the life of a fruit into two main periods. During the first (lasting for approximately two months in apple), hydroxycinnamic derivatives accumulate in the fruit with a positive balance between *in situ* biosynthesis, migration, and possible reutilization. However, in the second period, this balance becomes negative and the overall content falls in the fruit (Figure 37). This second phase is not always as marked, depending on the cultivars studied.[178] The initial period of accumulation of hydroxycinnamic derivatives in apple covers two phases consisting successively of an increase and a reduction in concentrations. They indicate the relative rates of accumulation of phenolic compounds, water, and other constituents of the dry matter in the fruit. The relative proportions of the various constituents may be modified during these changes; for example, the ratio of *p*-coumaric esters to chlorogenic acid falls steadily during growth of Calville blanc apple (Figure 38).

A very complete analysis of hydroxycinnamic derivatives was carried out on tomato cv. cerasiforme.[108,113,114,116] Growth and maturation of the fruit is accompanied by qualitative and quantitative variations which reveal considerable diversification in hydroxycinnamic derivatives. Thus, whereas very young green fruits contain only chlorogenic acid, other derivatives appear during growth and the ripe fruit contains 11 different ones (Table 39). Some compounds are characteristic of a physiological stage; this is the case of neochlorogenic acid, caffeoylglucose, sinapolylglucose, and feruloylglucose, which only appear at the end of growth just before veraison. Keep in mind that these compounds are also characteristic of parts of the fruit; the first two are only found in pulp and the others are present in the pericarp. Analysis of the whole fruit[114] shows that chlorogenic acid forms 76% of total hydroxycinnamic derivatives in green fruits halfway through the growth stage; the figure then falls to 15% in ripe fruits. In contrast, glucosides as a whole form 23 and 84% at the same periods, which may suggest certain metabolic relations between these compounds. Variations in overall levels of these derivatives in cherry tomato are similar to those described above for other fruits, but they mask very important differences on the one hand in the

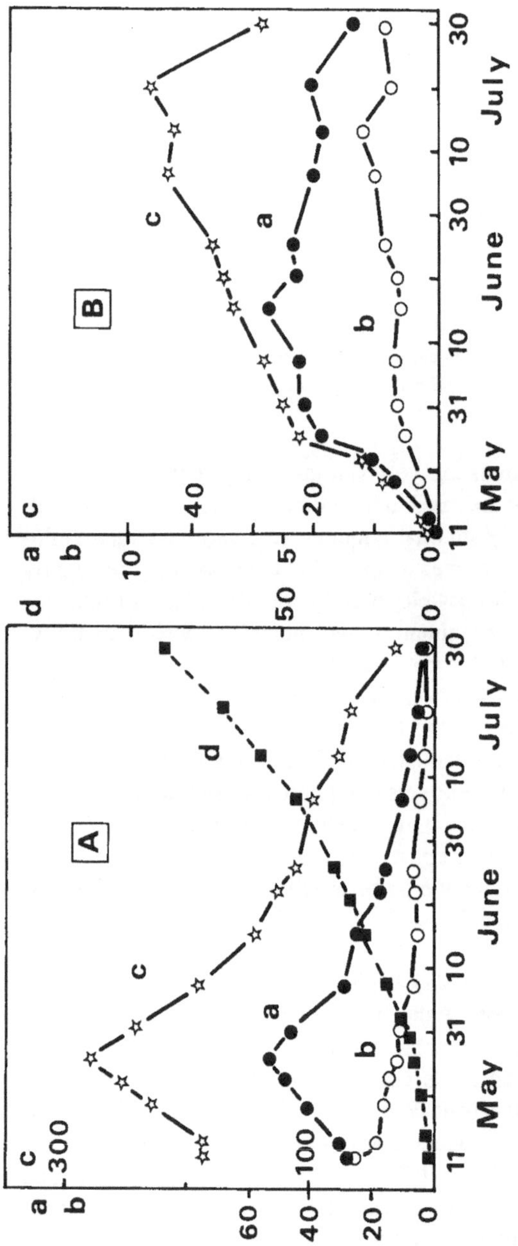

FIGURE 37. Changes in the levels of: (a) *p*-coumaroylquinic acid; (b) *p*-coumaroylglucose; (c) chlorogenic acid; and (d) fresh weight (in g) during apple growth. A: mg/100 g FW; B; mg/fruit. (Adapted from Macheix, J. J., *Physiol. Vég.*, 12, 25, 1974.)

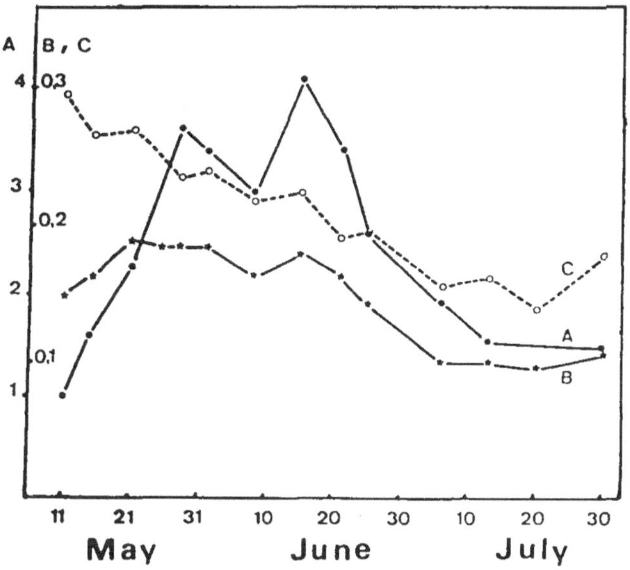

FIGURE 38. Changes in three ratios during apple growth: (A) *p*-coumaroylquinic acid/*p*-coumaroylglucose; (B) *p*-coumaroylquinic acid/chlorogenic acid; (C) total *p*-coumaric esters/chlorogenic acid. (Adapted from Macheix, J. J., *Physiol. Vég.*, 12, 25, 1974.)

individual pattern of evolution of each of the constituents and on the other in their location in the fruit. Chlorogenic acid is the only quinic ester found in substantial quantities in very young fruits. 3-Caffeoyl, *p*-coumaroyl, and feruloylquinic acids accumulate mainly during maturation (Figure 39A and B) when the chlorogenic acid content falls strongly although the latter compound is predominant at all times. Among glucose derivatives of hydroxycinnamic acids, glucosides are always much more plentiful than esters (Figure 39C). However, whereas the latter increase continuously during maturation, the glucosides of *p*-coumaric and caffeic acid (Table 39) reach a maximum at the breaker stage. Thus the growth and maturation of the fruit of *Lycopersicon esculentum* var. cerasiforme are characterized by different expressions of the metabolism of hydroxycinnamic derivatives. In growing pulp it is mainly oriented towards the accumulation of quinic derivatives, whereas glucose derivatives (particularly glucosides) accumulate in the pericarp during maturation (Figure 39). Thus, although quinic derivatives remain the major compounds in pulp (58%), their content becomes less than that of glucose derivatives in the pericarp (46%) (Table 39).

Individual variations of all these compounds are reflected in their relative contents expressed as ratios.[114] For example, the ratio of quinic derivatives to glucose derivatives decreased during the life of the fruit (Figure 39D). Interpretation of these variations and their possible physiological significance is discussed later.

The example of the grape can be used to compare the relative evolution of hydroxycinnamoyltartaric esters in numerous black or white varieties.[32,296,355,384] In all cases, although contents may differ greatly from one variety to another, variations identical to those of the quinic derivatives already described can be observed (Figure 40): the young berries are very rich, with a maximum observed very early on when expression is per gram FW or DW. The contents then fall steadily during growth and maturation. Expression per fruit leads on the other hand either to a certain stabilization from veraison onwards or even in certain cases to steady increase until complete maturity is reached.[384] The differences observed results mainly from the speed of growth of the berries during maturation. During this evolution, the percentage of *p*-coumaroyltartaric esters increases significantly in the skin of grapes (Table 40).

TABLE 39
Amounts (nmol/g dry wt) of Hydroxycinnamic Acid Derivatives in Pulp and Pericarp during Growth (G_{12} and G_{16} stages) and Maturation (b and R stages) of Cherry Tomato

	Pericarp				Pulp			
	Growth		Maturation		Growth		Maturation	
Hydroxycinnamic derivatives	G_{12}	G_{16}	b	R	G_{12}	G_{16}	b	R
Hydroxycinnamic derivatives	5363	4772	5091	3343	11690	10097	5081	3070
Quinic esters[a]	4116	3360	2850	1524	10535	8245	3333	1782
Glucose derivatives[a]	1247	1412	2241	1819	1155	1852	1746	1288
Quinic esters								
Chlorogenic acid	4086	3279	2718	1352	10511	8064	2992	1365
Neochlorogenic acid	—	—	—	—	—	81	193	387
p-Coumaroylquinic acid	23	50	66	100	8	100	148	30
Feruloylquinic acid	7	31	66	72	16	—	—	—
Glucose esters[a]	268	413	514	877	68	50	96	326
Caffeoylglucose	—	—	—	—	—	t	54	296
p-Coumaroylglucose	75	73	72	70	68	50	42	30
Feruloylglucose	193	300	395	721	—	—	—	—
Sinapoylglucose	—	40	47	86	—	—	—	—
Glucosides[a]	979	999	1727	942	1087	1802	1650	962
Caffeic acid	389	761	1226	726	408	1403	1102	512
p-Coumaric acid	490	238	501	216	679	399	548	450
Ferulic acid	100	—	—	—	—	—	—	—

Note: t = trace component.

[a] Total amount of hydroxycinnamic acid derivatives.

From Fleuriet, A. and Macheix, J. J., *Phytochemistry*, 24, 929, 1985. With permission.

Pineapple would seem to be an original case in which, in contrast with the features described above in other fruits, the hydroxycinnamic derivative content increases during maturation.[75] These variations, nevertheless, require confirmation for cultivars other than the one studied; they could be the expression at the phenolic level of the well-known unique structure and special physiological and biochemical features of this fruit.[14]

The olive also displays a very late increase in the amount of verbascoside, the main caffeic acid derivative in this fruit;[4] this occurs during what is known as the green maturation period when the oleuropein content has decreased strongly (Figure 41) and when demethyloleuropein, one of its derivatives, appears in some cultivars.[5]

These few examples thus underline the fact that there is — apart from a few exceptions — very great homogeneity in the quantitative variations in phenolic acids during the lifetime of a fruit. Nevertheless, the very existence of the exceptions and the diversity of the responses found when analyses are applied to individual variations in each of the compounds (as in the case of the cherry tomato), and not to the major compound only, shows the interest in broadening these observations to a larger number of fruits using the most probing analytical approaches. In addition, studies on cell suspensions isolated from fruits should make it possible to approach easily both the problems of biosynthesis and the vacuolar accumulation mechanisms of phenolic derivatives. Preliminary results[223] already show that accumulation of each hydroxycinnamic derivative is not identical during the development cycle of the culture; this is probably related to different biosynthetic capacities.[246]

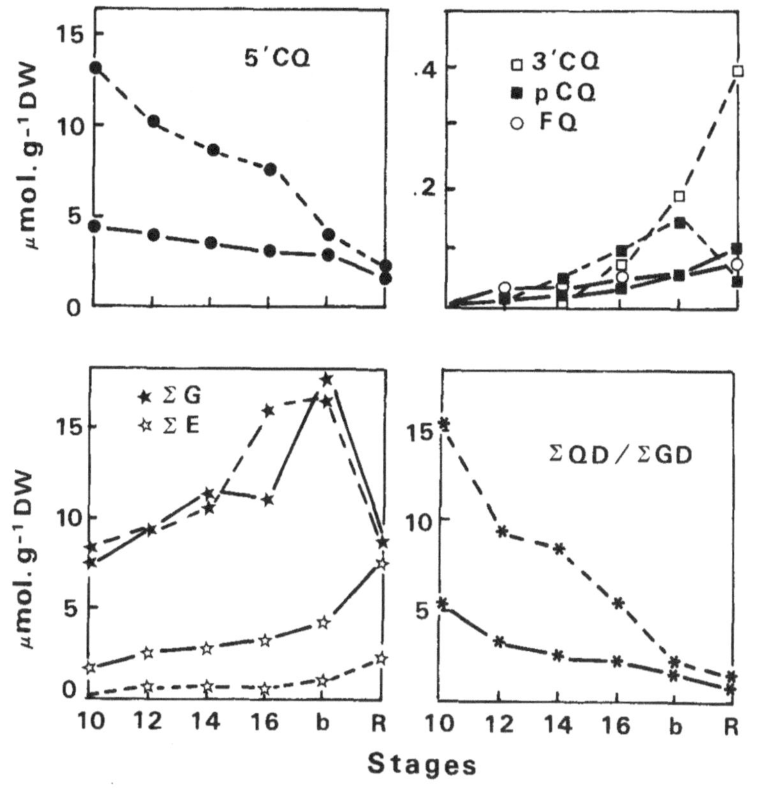

FIGURE 39. Changes in hydroxycinnamoyl derivatives in pericarp (—) and pulp (-----) during growth (10 to 16 mm diameter green fruits) and maturation (breaker and red fruits: b, R) of cherry tomato. 5'CQ: chlorogenic acid; 3'CQ: neochlorogenic acid; *p* CQ: *p*-coumaroylquinic acid; FQ: feruloylquinic acid; Σ G = total glucosides; Σ E: total glucose esters; Σ QD: total quinic esters; Σ GD: total glucose derivatives. (Adapted from Fleuriet, A., Thèse Doct. Etat., Université de Montpellier, France, 1982.)

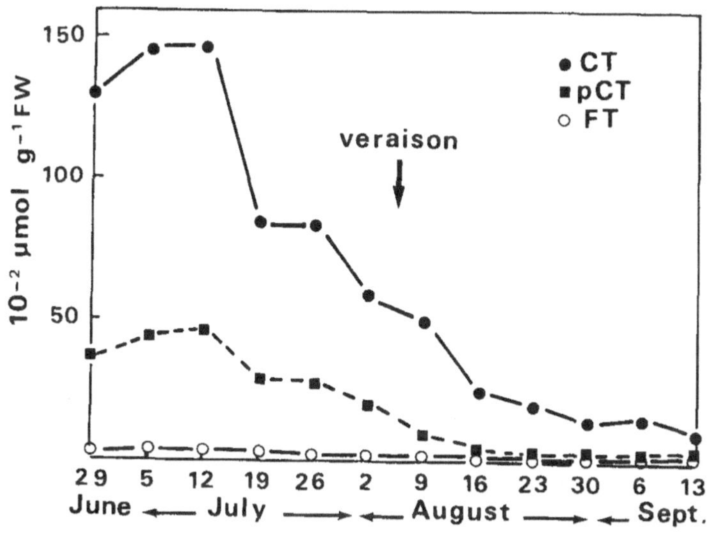

FIGURE 40. Changes in hydroxycinnamoyltartaric acids during growth and maturation of cv. Merlot grape. CT: caffeoyltartaric, *p*CT: *p*-coumaroyltartaric, FT: feruloyltartaric acids. (Adapted from Boursiquot, J. M., Thèse Doct. Ing., ENSA, Montpellier, France, 1987.)

TABLE 40
Changes in Hydroxycinnamoyltartaric Esters Percentages in Skin of Grape Berries during Maturation

		CT[a] %	pCT %	FT %
Grenache	07/2	60.0	35.5	4.5
	08/14	56.1	38.6	5.3
	09/19	53.9	38.4	7.7
Merlot	08/12	54.9	44.1	1.0
	08/26	54.7	43.4	1.9
	09/9	50.6	48.0	1.4

[a] CT: caffeoyltartaric ester, *p*CT: *p*-coumaroyltartaric ester; FT: feruloyltartaric ester.

Data from Reference 32.

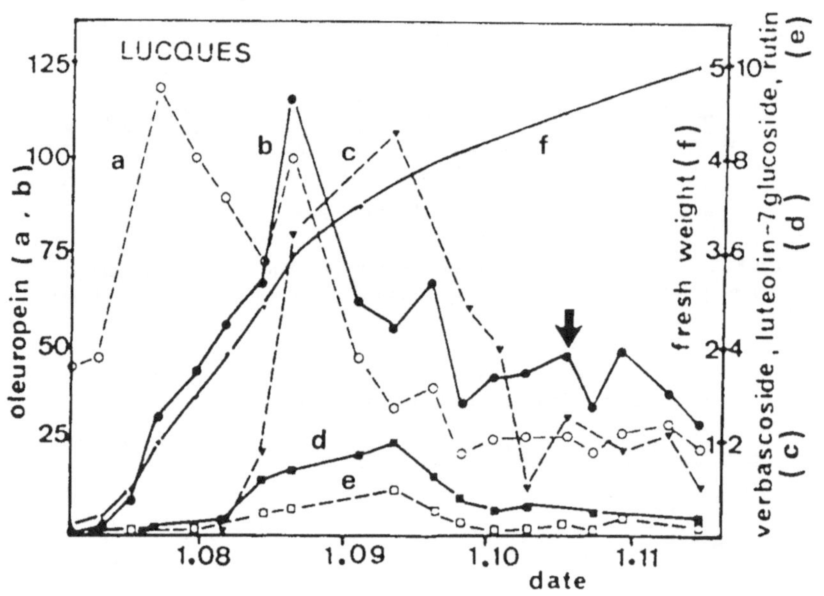

FIGURE 41. Changes in the main phenolic compounds in olive fruit (cv. Lucques) during growth and maturation. a: oleuropein (mg/fruit); b: oleuropein (mg·g^{-1} DW); c: verbascoside, d: rutin; e: luteolin 7-glucoside (mg/fruit); f: fresh weight (g). Arrow denotes the beginning of black maturation. (From Amiot, M. J., Fleuriet, A., and Macheix, J. J., *J. Agric. Food Chem.*, 34, 823, 1986. With permission.)

D. ANTHOCYANINS

One of the characteristic aspects of the maturation of red fruits is the replacement of the initial green color by red, violet, or blackish color frequently caused by accumulation of anthocyanins. This is often limited to the epiderm and subepidermal cell layers.[146] However, anthocyanins also accumulate in the pulp of numerous small red fruits and the *Vitis vinifera* varieties known as teinturier grapes during maturation.

A great many studies have characterized the increase in anthocyanin content during maturation, but they very frequently cover only variations of the major pigment or of all

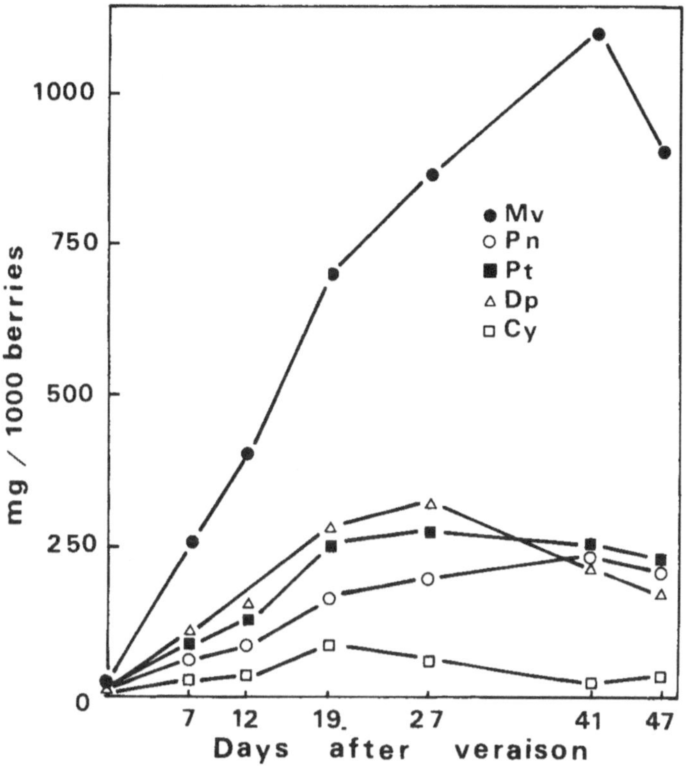

FIGURE 42. Changes in anthocyanin content during ripening of cv. Syrah grape. Mv: malvidin; Pn: peonidin; Pt: petunidin; Dp: delphinidin; Cy: cyanidin. (Adapted from Roggero, J. P., Coen, S., and Ragonnet, B., *Am. J. Enol. Vitic.*, 37, 77, 1986.)

the pigments (evaluated by measurement of absorption at about 550 nm). Steady increase in the anthocyanin content is observed from veraison onwards in grape,[183,210,270,309,310,341,342,382,383,390] cherry,[273,295] strawberry,[449] fig,[324] eggplant,[9,290] apple,[54,60,100,102] plum,[42] cranberry,[95] olive,[251] and many other red fruits. In grape berry, accumulation of anthocyanins follows a sigmoidal curve which coincides to a considerable extent with the accumulation of soluble sugars.[183,309,310,352]

The anthocyanin content generally stabilizes at maturity, but in some cases it may diminish slightly as in grape[341,342,352,390] or crowberry (*Empetrum nigrum*)[209] when maturation is too advanced. In eggplant,[9,290,292] the anthocyanin content peaks and then decreases strongly. The fruit then turns dark yellow and brown at full maturity, but it is no longer edible. The degradation of anthocyanins is probably controlled enzymatically.[363]

In fact, the most interesting data are from individual analysis of the constituents of the anthocyanin fraction during maturation. They make it possible to relate the qualitative aspects to quantitative data by revealing the different rates of accumulation of the different anthocyanins. Thus, in the thornless blackberry the proportion of cyanidin 3-glucoside increases from 43 to 80% during the initial phases of maturation and then stabilizes.[368] An interesting study on three clones of the *V. vinifera* cultivar Syrah[352] relates variation of the different anthocyanidins with their position in the biosynthesis pathway. Cyanidin 3-glucoside displays an early peak (19th d after veraison) (Figure 42); it is followed by delphinidin and petunidin glucosides and later by peonidin and malvidin derivatives (47th d). There is, thus, successive accumulation of more and more substituted molecules in the B cycle. The derivatives of malvidin, representing the most stable form, accumulate last and the percentage increases

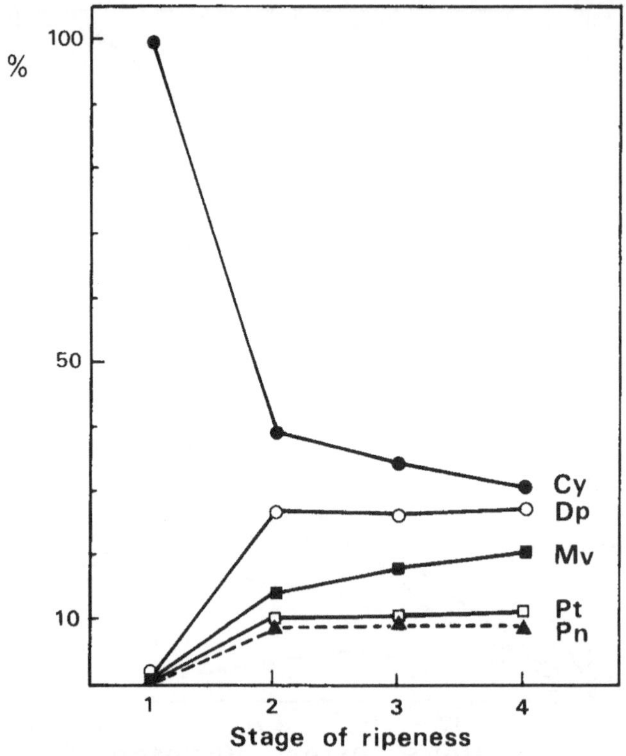

FIGURE 43.. Crowberry anthocyanins (*Empetrum nigrum*): relative amounts of various anthocyanins at different stages of ripeness (in percent of total anthocyanins). Total anthocyanin content in mg/100g FW: stage 1: 3; stage 2: 180; stage 3: 460; stage 4: 450; 1 = unripe; 2 = halfripe; 3 = ripe; 4 = overripe; Cy, Dp, Mv, Pt, Pn = cyanidin, delphinidin, malvidin, petunidin, and peonidin glycosides). (Adapted from Karppa, J., *Lebensm. Wiss. Technol.*, 17, 175, 1984.)

steadily during maturation from 33 to 58%.[352] Slightly different evolution was obtained with the cv. Cabernet-Sauvignon in which delphinidin 3-glucoside became more abundant than malvidin 3-glucoside at the end of ripening.[67] In the same cultivar, the percentage of acylated anthocyanins increased steadily during ripening.[67] This is reminiscent of observations during the growth of a grape cell suspension cultured *in vitro*.[237]

The late accumulation of the most substituted derivatives, particularly the most methylated, appears to be fairly widespread. In highbush blueberry (*Vaccinium corymbosum*), cyanidin glycosides comprised about 40% of the total anthocyanins in unripe fruit and only 2% in the ripe fruit. In contrast, the proportions of malvidin glycosides increased from 15 to 60%.[254] The phenomenon is also very marked in crowberry[209] where cyanidin glycosides (Cy-galactoside and Cy-arabinoside) are only present in unripe fruit whereas the proportion of malvidin derivatives, and to a lesser extent petunidin derivatives, increase during maturation (Figure 43). These modifications thus reveal an evolution in the biosynthetic capacity of the fruit in relation to its physiological stage, leading to maximum methoxylation of the aglycone as maturation proceeds. This results in a change in the color of the fruit and the products derived from it.

Nevertheless, such observations cannot be generalized in the present state of knowledge. Indeed, in the Montmorency sour cherry[69] and the raspberry,[13,362] there are only very small variations in the relative proportions of pigments during maturation. In addition, in fruits which contain only derivatives of cyanidin, the evolution may be in the diversity of the sugar. Thus, in redcurrant (*Ribes rubrum*), there is a tendency for synthesis of Cy-glycosides containing three sugar units (Cy-xylorutinoside and Cy-glucorutinoside) during maturation;[297] they amounted to 32% of total glycosides in the pink fruit and 53% in the ripe fruit. However, very different results were obtained for Manzanillo olives.[251] The changes in the levels of cyanidin glycosides (one monoside, two diosides and three triosides) were monitored during black maturation from October to January. The trioside content was highest (50%) at the beginning and then stabilized at about 25% in the final stages after an intermediate decrease. The monoside content remained approximately constant (15%) where as the diosides reached a maximum in midseason and then decreased while still remaining the most plentiful (60%). The example of the litchi (*Litchi chinensis*) provides a contrary example to those above:[319] cyanidin glycosides are the first to appear during ripening, followed by pelargonidin glycosides which are dominant at later stages.

In addition to these very conventional patterns of evolution, several special cases display a number of interesting variations, such as the very early presence of anthocyanins in the eggplant[289] where the ovary is already colored and in which the development of color sometimes starts a few days after the anther. On the other hand, anthocyanins may disappear during growth and maturation. Thus, certain apple cultivars (e.g., Calville blanc) and the shell pineapple may be slightly colored during the young stage;[14] this coloring then disappears at a greater or lesser rate. Likewise, in oil palm, the exocarp of immature fruits contains large quantities of anthocyanins and is almost black in color; most of the anthocyanins then disappear very rapidly before abscission and the mature fruit is a yellow color.

E. FLAVONOLS, FLAVANONES, AND FLAVONES

Few works have been devoted to variations in flavonols and their glycosides in fruits. The reason probably lies in the small concentrations encountered and in the diversity of the molecules present. Relatively old data[450] already reported a fall in the flavonol content during maturation of apple. They were confirmed recently and extended to a large number of cultivars at the approach to maturation of McIntosh apples for example.[73] Whereas individual quercetin glycosides varied considerably, total flavonol glycosides remained relatively constant until the onset of the climacteric when the levels decreased significantly while the fruit was still on the tree. In contrast, when the fruits were picked and kept in cold storage for 60 d, the levels of quercetin glycosides increased strongly. Phloridizin evolved in a very similar way in the same fruits.

A fairly complete study was made of the changes in flavonol concentrations (measured by determining the aglycones after hydrolysis of the glycosides) during the development of numerous fruits which accumulate anthocyanins during maturation.[397] In most cases (redcurrants and whitecurrants, sour cherries, plums, cultivated blueberries, elderberries), kaempferol and quercetin glycoside contents were lower in ripe fruits. Similar results were found for the litchi.[319] The highest level is found either in very young fruits, e.g., in the sweet cherry cultivar Bigarreau Napoleon[272] or, as in the sour cherry cultivar Fanal, at an intermediate stage.[397] There are a few exceptions: privet (*Ligustrum vulgare*) and certain clones of cultivated blueberries in which the quercetin and myricetin contents are higher in the ripe fruit. However, the most notable exception is the blackcurrant (*Ribes nigra* cv. Sibergieters Schwarze) in which there is a very marked increase in the quercetin and myricetin contents during maturation.[397] In a general manner, but perhaps with the exception of myricetin derivatives, there does not appear to be any relation between the accumulation of anthocyanins and that of flavonols during maturation. The grape is another case in which flavonol glycosides accumulate steadily during maturation[82] with the evolution being the same in the pink or white forms of the same cultivar. Here again, there does not appear to be a connection with the accumulation of anthocyanins.

In tomato fruit, rutin is present only in the external cell layers of the pericarp. Concentration is highest in young fruits and decreases during growth and maturation.[108] The flavanone naringenin is also present in the free state in the skin of certain cultivars of tomato fruit;[442] the concentration increases very strongly between the preclimacteric (0.11 mg/100 g FW) and the postclimacteric (6.4 mg/100 g FW). However, little naringenin is present in the peeled fruit and the level remains approximately stable.

In the grapefruit, naringin accumulates mainly at the beginning of growth[2,253] since biosynthesis is particularly active during the cell division phase. The perfecting of a specific radioimminoassay for naringin has made it possible to monitor its evolution in all the tissues of *Citrus paradisi*, including flower and fruit.[203] In the pistil, the ovary had the highest naringenin content and concentration. There was a 26-fold decrease in naringenin concentration during fruit growth related to the cell enlargment phase. However, average total naringin per fruit increased steadily and stabilized at maturity. There was a direct relationship between PAL activity and the rate of naringenin glycoside accumulation.[253] In peanut hull, total flavonoid content increased as the mesocarp color changed from yellow to black.[66] Simultaneously, relative concentrations of flavonoids also changed with maturity: eriodictyol was predominant in immature peanuts, whereas luteolin was the predominant flavonoid in mature fruits.

Hesperidin accumulates rapidly in the young lemon fruit, whereas eriocitrin increases later and diosmin gradually throughout the fruit development.[431] However, the concentrations of these compounds decreased regularly. Flavonoids of juice vesicles from *in vitro* cultured fruits show a pattern of development similar to fruit grown on the tree. These changes suggest biosynthetic relations between the various flavone glycosides present in lemon fruit.[163]

F. MONOMERIC FLAVANS

Variations in flavanols are frequently combined with those of condensed tannins, but in some cases it is possible to monitor individual variations of monomer forms of flavans. In grape,[64,82] the highest concentrations of flavans are found in the early stages of development. The level then decreases rapidly and becomes more or less stable during maturation, in a similar manner to the evolution of total phenols in the same fruit,[383] with the exception of the anthocyanins. Among the different compounds present, (−)-epicatechin-3-*O*-gallate displays the most spectacular evolution: whereas it forms over 20% of monomeric flavan-3-ols in young Gamay Beaujolais cultivar, there is only a very low concentration in the ripe fruit. The physiological evolution of the grape thus results at monomer flavan-3-ol level in both quantitative variations and modifications in the relative proportions of the different molecular forms present.

Catechin variations in the apple are similar to those already described for chlorogenic acid and are identical in several cultivars.[241,280] Concentrations are very high in young fruits, with a peak about 40 d after flowering. They then decrease rapidly and stabilize at a very low level at maturation. (+)-Catechin and (−)-epicatechin vary in a very similar manner. Expression per fruit shows a later peak followed by a decrease which is more or less marked according to the compound.

Even more pronounced variations than those above are reported in three of the main types of Japanese persimmon;[454] the very high catechin content in the young fruit at the beginning of June had disappeared completely in early July. In contrast, in the pollination constant nonastringent cultivar Fuyu, catechin was detected throughout the growing period of the fruit (Figure 44) and only decreased very slowly, whereas the decrease in condensed tannins occurred earlier.[453]

The data obtained for other fruits are generally fragmentary and mainly apply to the strawberry[401] and *Rubus* (raspberry, blackberry, other Rubus)[281] and *Prunus* fruits (cherry, plum).[403] The variations are identical in all cases with those described previously: young fruits are always richer in monomeric flavan-3-ols and the levels are low at maturity. In addition, as has been reported, the relative proportions of the different constituents vary

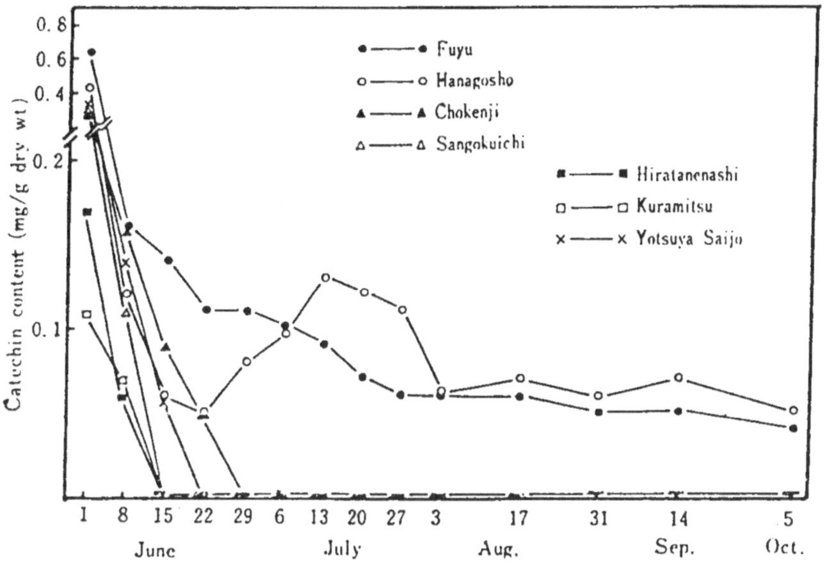

FIGURE 44. Seasonal changes of catechin content in the fruit flesh of seven persimmon cultivars. (From Yonemori, K., Matsushima, J., and Sugiura, A., *J. Jpn. Soc. Hortic. Sci.*, 53, 121, 1983. With permission.)

with the physiological stage without it being possible to ascribe a general trend to this evolution. Indeed, the (−)-epicatechin:(+)-catechin ratio decreases during the growth of strawberry, increases in cherry and remains more or less stable in grape.

G. TANNINS

Very numerous, sometimes old works have been devoted to variations in tannins during the life of fruits and on their consequences with regard to the decrease in astringency generally observed at maturation.[131,202] However, it is necessary to recall that the quantitative determination of tannins depends to a considerable degree on the extraction and assay techniques used (Chapter 1). Changes in the degree of polymerization or in the links that these compounds may form with other compounds in the fruit (proteins, polysaccharides, etc.) may strongly modify their solubility[174] and extraction. Such modifications are suggested during maturation[131,298,307,312] and can, therefore, interfere during determination with the natural quantitative variations in tannins.

Many studies have been carried out on economically important fruits, such as the apple, the grape, and the persimmon. They generally show that the soluble tannin concentrations decrease strongly during growth and reach low levels at maturity.

In the persimmon (*Diospyros kaki*) (Figure 45), there are very marked differences between astringent cultivars (PCA and PVA: respectively, pollination constant and variant astringent types) and nonastringent cultivars (PCNA and PVNA: respectively, pollination constant and variant nonastringent types).[194,195] In the latter, residual concentrations of soluble tannins are very low or nonexistent at maturity[452-454] and the contents per fruit are much lower than in the astringent cultivars. Tannins accumulate in specific cells whose size varies with the type of cultivar.[195] During the tannin accumulation period in the fruit, there is for each cultivar a very close correlation between cell size and accumulation expressed on a per fruit basis. In addition, changes in tannin cell morphology have been reported: pores in the tannin cell wall appear to be involved in the tannin accumulation.[453] They occlude naturally in PCNA and PVNA types and only after treatment with ethanol in PVA types. This occlusion could be the result of natural coagulation of tannins.

In grape, relatively differing data are found in the literature, but the discrepancies are

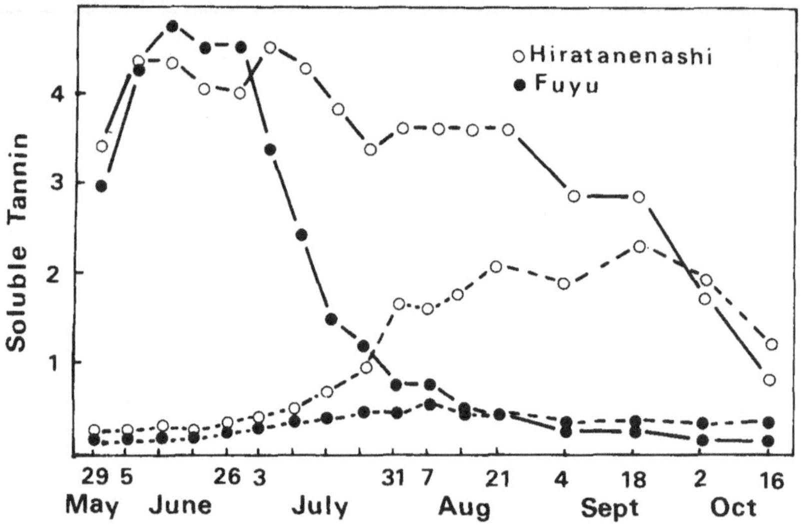

FIGURE 45. Changes in soluble tannin content during growth and maturation of two persimmon fruits. (—): percent fresh weight; (----): mg/fruit (Adapted from Yonemori, K. and Matsushima, J., *J. Am. Soc. Hortic. Sci.*, 112, 818, 1987.)

TABLE 41
Changes in Flavans during Maturation of Gamay Beaujolais Grape

Proportion of flavans in	Sampling Dates 2/2	2/17	3/1	3/8	3/15	3/22	3/29
EtOH fraction (1) (% w/w DW)	8.0	6.6	5.3	4.9	2.8	4.8	3.7
Polymer fraction (2)	12.5	6.4	7.0	9.8	3.5	5.9	4.6
Ratio (2)/(1)	1.5	1	1.5	2	1.3	1.3	1.3

From Czochanska, Z., Foo, L. Y., and Porter, L. J., *Phytochemistry*, 18, 1819, 1979. With permission.

mainly caused by the origin of the tannins studied which, during extraction, may come from the fruit itself, the seeds, or from stalks in certain types of wine-making. The flavan constituents decrease steadily during berry development when expressed on a DW basis.[64] This decrease involves both the polymeric fraction and that extracted by ethyl acetate which, in addition to gallic acid, includes monomer flavans and dimeric proanthocyanidins (Table 41). In light of the fairly considerable fluctuations between the different samples, no significant difference is observed between these two fractions during the development and maturation of the berry. In particular, the proportion of polymerized forms does not increase with maturity.

After veraison, the tannin content of grape berry skin (expressed on a per berry basis) first increases and then decreases at maturity,[341,343] but here again the degree of polymerization remains the same. In contrast, the tannin level is stationary or decreases in seeds during maturation, whereas the rate of polymerization of flavans increases.[343] Seed maturation is characterized by a very clear enrichment in procyanidins compared to monomeric forms.[354] In the quince, maturation is characterized by the appearance of glucosides of procyanidin polymers.[312]

Similar evolutions to those described above are generally found in other fruits, e.g., in the cherry,[273] the banana,[300] the apple,[429] the guava,[196] and the pistachio fruit.[429] An increase in flavanols during the climacteric rise with a decrease in total phenols has been observed in more rare cases;[123] in this case, the nature of the substances involved remains to be specified.

Variations in tannins during the life of the fruit cannot be separated from variations in astringency; the latter is an organoleptic parameter which results from interaction between tannins and salivary glycoproteins in the mouth. Two main factors are responsible for variations in astringency:[131,132,194,195,202] on the one hand, genetic factors which lead to considerable differences between species or cultivars and on the other, physiological factors which are directly linked with the stage of development of the fruit. Indeed, astringency decreases or disappears completely during maturation. In addition, certain technological treatments can also remove the astringency in certain fruits and render them edible (Chapter 4).

As is frequently reported in the literature since the old but classic works,[131,132,202] a decrease in astringency during maturation is generally accompanied by a decrease in the tannins in the fruit or in physicochemical modifications in these molecules. Thus, the degree of polymerization of tannins may play an important role in their ability to bond to proteins: the two parameters increase simultaneously as far as a certain degree of polymerization (mol wt in the order of 3000) whereas the more polymerized forms are less and less aggressive with regard to proteins. On the basis of these data, it was suggested long ago[131,202] that maturation might be connected with both a decrease in the tannin content and an increase in their molecular size, thus causing a decrease in astringency. In fact, as has been stressed by several authors,[119,120,170] few accurate experimental data support these hypotheses. In arecanut, however, monomeric, little-polymerized forms increase during maturation,[259] and in the grape, the proportion of polymerized forms of flavans does not increase with maturation[64] (Table 41). Likewise, the ^{13}C-NMR spectra of the polymers and other properties were virtually identical in ripe and unripe fruits of blueberry and strawberry.[119] According to Haslam and co-workers,[173,174] the mechanisms involved in the natural decrease in astringency might be different according to the fruit in question. We examine successively the two best-studied cases: the persimmon and *Rubus* fruits.

Case 1 — In Japanese persimmon fruits, loss of astringency appears to be caused mainly by an immobilization of tannins caused by their reaction with acetaldehyde.[260-262] The mechanism was investigated *in vitro*[262] under mild conditions (pH 6 to 8); the phloroglucinol A ring of the flavan-3-ol units of persimmon tannins reacted with electrophilic acetaldehyde to form a gel which causes insolubilization. Phosphate, malate, and citrate accelerated gel formation whereas ethanol and ascorbate prevented it. A comparative study on *in vitro* coagulation of fruit juice from Fuyu (PCNA type) and Hiratanenashi (PVA type)[451] shows the importance of the tannin concentration (Figure 46) and the role of the pH; the highest rate was observed at around pH 4.0.

Evolution of astringency during growth and maturation is thought to result from the production of acetaldehyde by the fruit under anaerobic conditions.[260,406] This production is triggered by various treatments, but in natural conditions, it is probably the result of ethanol formation by the seeds[405] and diffusion into the flesh of the fruit. Seeds of PVNA cultivars produced large amounts of ethanol and the fruits rapidly lost their astringency when the ethanol in the fruits approached 0.1 μl/g FW. PVA and PCA cultivars produce less ethanol and remain astringent. In contrast, PCNA cultivars lose their astringency although they produce only a small amount of ethanol in the seeds and accumulate only small amounts or none in the flesh. These observations for 47 different cultivars[405] suggest that two different mechanisms may be involved in the natural deastringency in the persimmon. The first is associated with PVNA, PCA, and PVA types and is dependent on the production and accumulation of ethanol and subsequently of acetaldehyde. The second involves PCNA types and is not connected with these volatile substances. It must be linked with the different physicochemical properties of the tannins in these cultivars. Indeed, their coagulation *in*

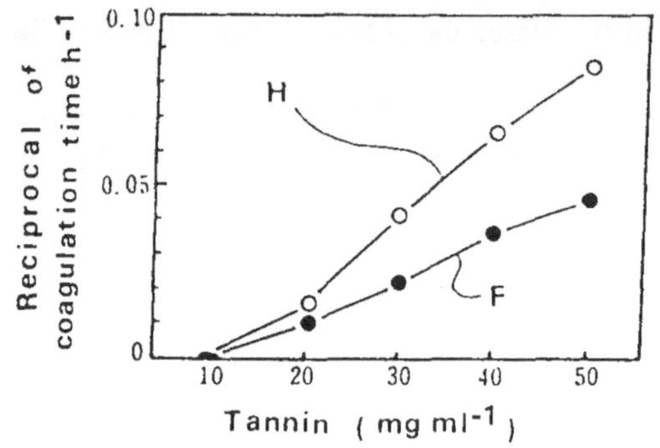

FIGURE 46. Effect of tannin concentrations on coagulation time of the fruit juice from "Fuyu" (F) and "Hiratanenashi" (H) cultivars. Each juice was incubated in acetaldehyde vapor at 30°C. (Adapted from Yonemori, K. and Matsushima, J., *J. Jpn. Soc. Hortic. Sci.*, 53, 121, 1984.)

vitro is much slower[451] and, unlike observations for all the other cultivars, the low molecular fraction of tannin in PCNA cultivars was prominent in young fruits (see Figure 27) and its conversion to a high molecular fraction was very slow.[451,454] Moreover, this low molecular fraction of tannin in PCNA cultivars was detected throughout the course of natural removal of astringency.

Case 2 — A second mechanism which may account for the loss of astringency during maturation of the fruit involves the possible chelation of tannins with macromolecules[173,174,298,374] or other phenolic compounds. An interesting aspect is that the ability of polyphenols to bind and precipitate proteins may be modified and disrupted by the presence of other molecular species, such as polysaccharides:[173,174,391] competitive chelation with polyphenols occurs which can inhibit precipitation of proteins and may be responsible for the decrease in astringency.

The principal phenols in fruits of *Rubus* sp. are based on gallic and hexahydroxydiphenic acid.[149] The reversible chelation of an enzymatic protein, β-glucosidase, with different polyphenols from tri-*O*-galloylglucose to the dimer sanguin H6 has been studied[298] in the presence of molecules well known for their possible combination with polyphenols (in particular, polysaccharides and sodium polygalacturonate which can develop hydrophobic cavities in their secondary structures). These substances compete with the polyphenolic substrates and lead to relief of the inhibition of β-glucosidase. The cell-wall structure is strongly modified during the maturation process (Figure 47) and softening occurs with the release of small water-soluble pectic fragments in which de-esterification has occurred. When the fruit is crushed in the mouth, these soluble pectin fragments can compete with mucal polysaccharides and proteins for soluble tannins (accumulated in vacuoles). A reduction in the sensation of astringency would result. This interpretation shows the interest in stressing the immediate environment of tannins during maturation rather than the modifications in their polymerization which, as has been mentioned above, are sometimes weak or nonexistent. They also involve the pectic substances group whose evolution is characteristic of maturation and it is probable that the interpretation proposed within this framework might apply in a general manner to numerous fruits.

H. MISCELLANEOUS

Briefly reported here are only a few examples which either reveal particular variations or concern infrequently observed compounds.

In banana, whereas total phenols decrease steadily throughout growth and maturation,[226]

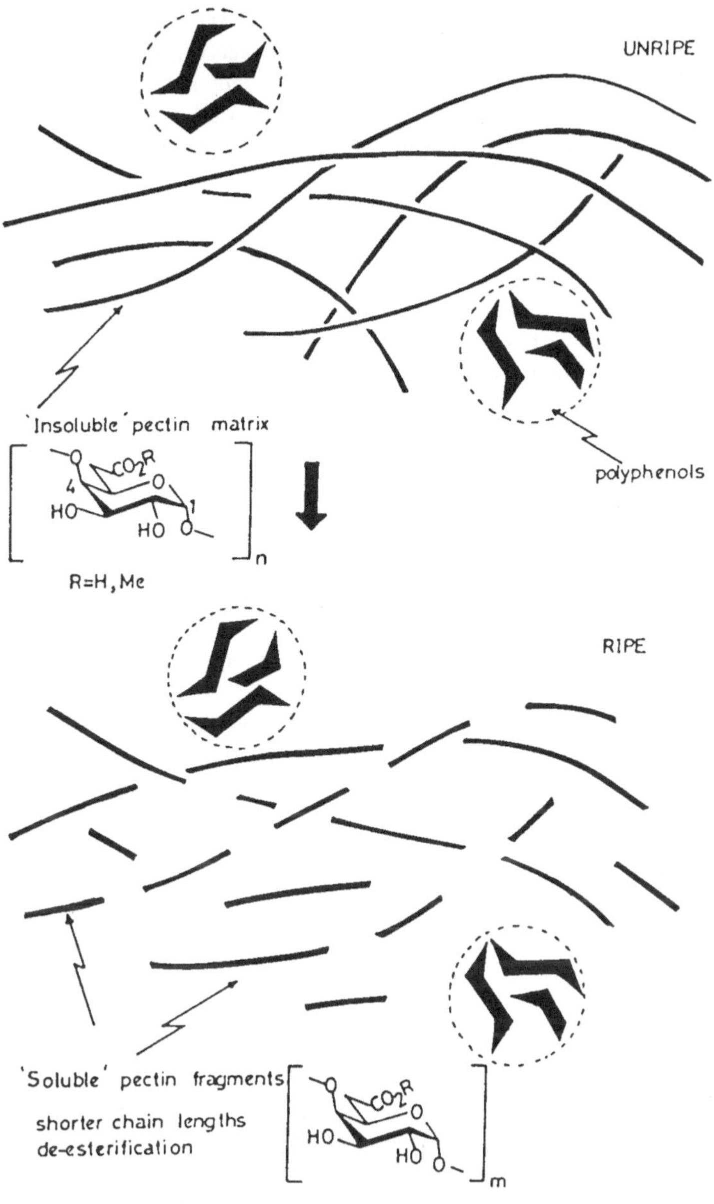

FIGURE 47. Loss of astringency in blackberry (*Rubus* sp.) during ripening (From Haslam, E. and Lilley, T. H., *CRC Crit. Rev. Food Sci. Nutr.*, 27, 1, 1988.

dopamine accumulates at an early stage in the skin of growing fruit:[300] the amount present at harvesting accumulates from the end of the first month following fruit emergence. However, a sharp increase is still observed during maturation with contents 30 to 60% greater than those found at harvest. When banana skin turns brown, dopamine is rapidly degraded under the effect of *o*-diphenoloxidase. Browning may also occur on green or yellow banana skin under the effect of chilling injury, to which the fruits are very sensitive, after transport or storage or even at the plantation if the temperature falls below 11°C.

Capsaicin provides another example of accumulation during maturation of chili pepper *Capsicum fructescens* and *C. annuum*.[57,157,158,359] A significant negative correlation was found between fruit size and capsaicin concentration.[57] Accumulation begins 20 d after fertilization and coincides with decline in fruit growth; it then continues for 10 to 15 d, reaches a maximum in ripe fruit, and the capsaicin level then decreases. The accumulation kinetics observed for whole fruit are very similar to those found in cell cultures of *C. frutescens*.[157,158,235] A major part of the growth phase is dissociated from the period of capsaicin accumulation. Capsaicin and analogs also accumulate during postharvest ripening under continuous light.[197]

Unlike the cases above, juglone accumulates at an early stage during the intensive growth period of walnut (*Juglans regia*) until lignification of the endocarp.[436] Fairly similar evolution is then observed in the embryo. In contrast, in walnut seeds, serotonin accumulated mainly during the period following abscission of the fruit.[19]

There are different situations in the evolution of coumarins during fruit development (see Reference 283). The highest furanocoumarin content was found in unripe fruits of *Amni majus* or *Angelica archangelica*,[51] but the opposite was true for *Poncirus trifoliata*.[148] Unripe fruits were found to be very rich in simple coumarins, whereas furanocoumarins were predominant in ripe fruits. Certain coumarins are only found in ripe fruits: mature berries of *Luvanga scandens* yielded xanthotoxin, xanthyletin, and luvangetin which were totally absent in the unripe fruits (see Reference 283).

III. BIOSYNTHESIS OF PHENOLIC COMPOUNDS IN FRUITS

The biosynthetic pathways of phenolic compounds are now well known and many general articles are regularly devoted to them.[16,17,89,141,142,153,154,162] The two aromatic amino acids phenylalanine and tyrosine are produced in plants via the shikimate pathway[30,118] (Figure 48). However, phenylalanine is the common precursor of most phenolic compounds in the higher plants. The hydroxycinnamic acids (HCA) formed, and more particularly their coenzyme A esters, are common structural elements of various other classes of phenolic compounds, such as cinnamate esters and amides, lignin, benzoic acids, flavonoids, and condensed tannins (Figure 49).

Proofs of the existence of an active phenolic metabolism in fruits are based on the results of radioactive tracer studies and on the *in vitro* demonstration of the enzymatic activities which lead to the biosynthesis of phenolic compounds.

The most commonly used exogenous precursors are phenylalanine and cinnamic acid, generally labeled with ^{14}C. They have been supplied to whole tomato[109] and pepper[157] fruits, to apple discs,[244] citrus calli,[418] cell suspensions of apple,[223] and immobilized cell cultures of pepper fruit.[235] The most abundant labeled compound is frequently *p*-coumaric acid or its esters, but derivatives of all the HCA and flavonoids also accumulate.

Most of the enzymatic pathways which lead to biosynthesis of the various classes of phenolic compounds have been revealed *in vitro* from protein extracts of fruits, but investigation is far from being as complete as that of other plant materials. This is partly because of their low protein content[76,159] and partly because of the importance of the vacuolar compartment which is rich in various enzyme inhibitors, including the phenolic compounds themselves. Thus, after a brief description of the reactions involved and the characteristics of the enzymes involved, we stress, in particular, at each step the data obtained from fruits; these are summarized in Table 42.

Before approaching the specific pathways which lead to biosynthesis of the different classes of phenolic compounds in fruits, we describe the successive reactions which lead from phenylalanine to CoA esters of HCA which form the "core sequence"[160] or what has become commonly referred to as the "general phenylpropanoid metabolism".[90,141,153,154]

A. GENERAL PHENYLPROPANOID METABOLISM

This concerns in particular the transition from phenylalanine to *p*-coumaric acid; the enzymes catalyzing the individual steps are phenylanine ammonialyase (PAL), cinnamate

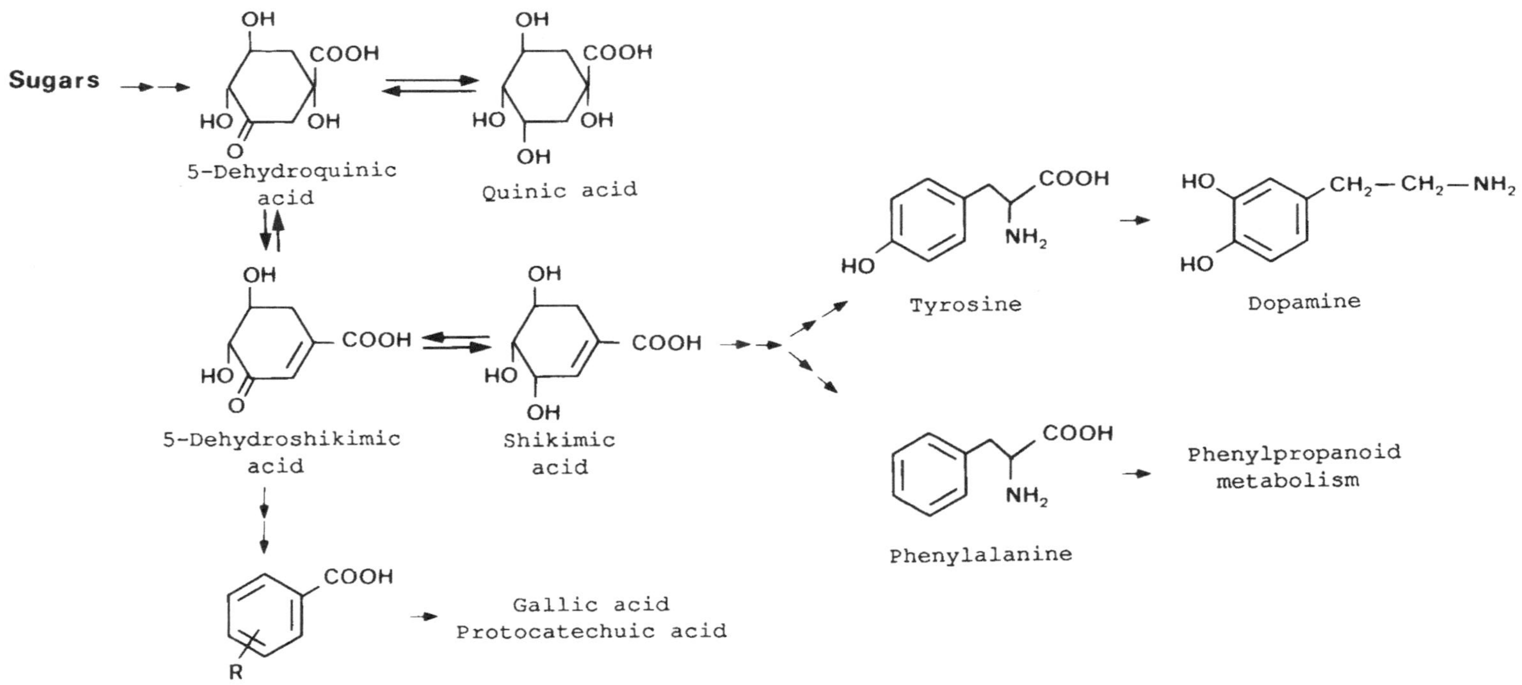

FIGURE 48. The shikimate pathway.

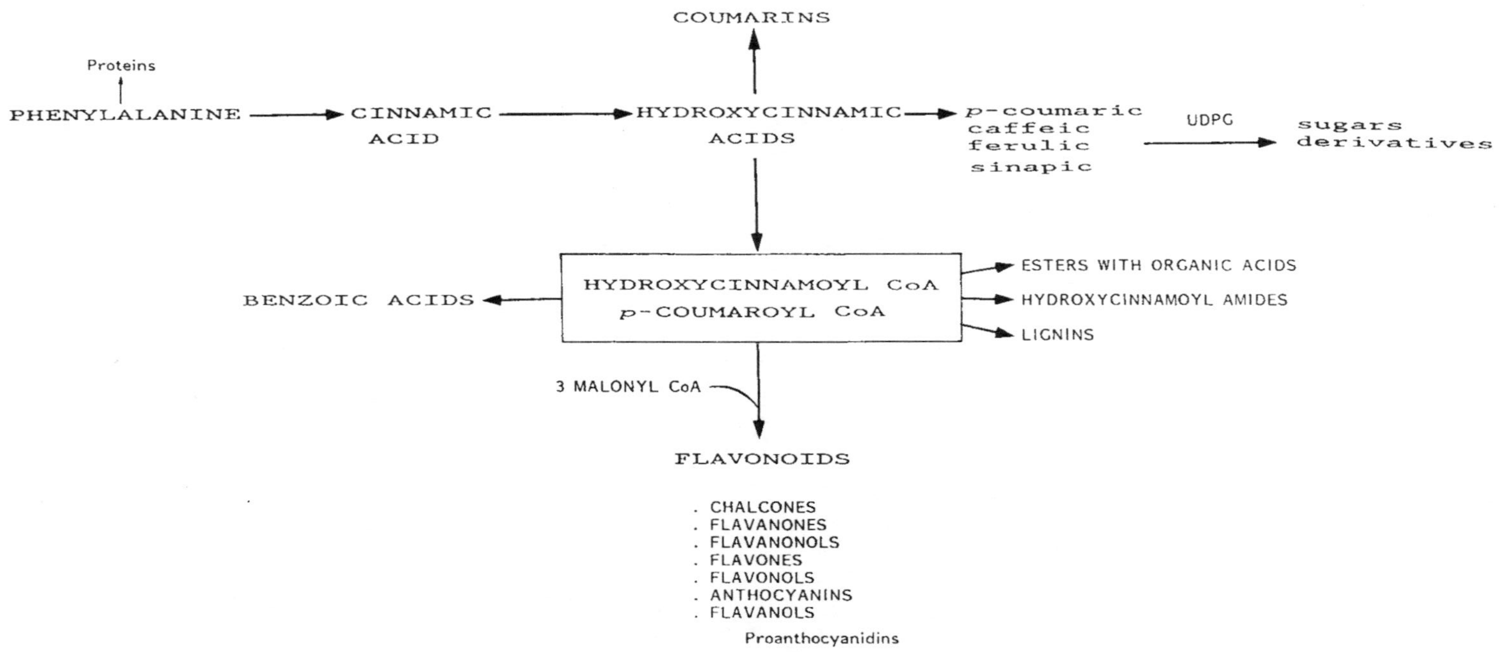

FIGURE 49. General scheme of phenolic metabolism.

TABLE 42
Occurrence of the Enzymes of Phenolic Metabolism in Fruits

Enzymes	Fruits
Phenylalanine ammonialyase (PAL) (EC 4.3.1.5)	Apple,[99-101,244,410,411] apple cell culture,[248] cherry,[273,274] citrus callus tissue,[38,418] grape,[183,210] grapefruit,[253,348] mango,[123,412] orange,[398] papaya,[412] peach,[7] pear,[21,274] pepper,[224] plum,[424] strawberry,[128,129,188] tomato[115,116,130,338]
Cinnamate 4-hydroxylase (C4H) (EC 1.14.13.11)	Apple cell culture,[248] grape,[183] grapefruit,[167] tomato[338]
Hydroxycinnamate CoA Ligase (CoAL) (EC 6.2.1.12)	Grape,[183] grape cell culture,[237] tomato[114-116,337,338]
o-Diphenol oxidase (*o*-DPO) cresolase activity (EC 1.14.18.1)	Apple,[133,242] avocado,[205] banana,[68,417] cherimoya,[258] grape,[165,427] kiwi[302]
O-Methyltransferase (OMT) (EC 2.1.1)	Apple,[106] apple cell culture[246,248] orange,[39,40] tomato[115,116]
Hydroxycinnamoyl transferase (HT) (EC 2.3.1)	Apple[a], grape[a], tomato[115,337]
Glucosyl transferase (GT) (EC 2.4.1)	Apple,[245] apple cell culture,[248] grape,[183] grapefruit,[234] strawberry[128] tomato[113,115-117]
Chalcone synthase	Grape,[183] grapefruit [253]
Chalcone isomerase (EC:5.5.1.6.)	Grape,[183] grapefruit[332]

[a] Lotfy and Fleuriet, unpublished data.

4-hydroxylase (C4H), and hydroxycinnamate: CoA ligase (CoAL). These three steps are necessary for the biosynthesis of practically all phenolic compounds, but there is in addition a close metabolic relationship between the three enzymes. This is demonstrated in particular by their coordinated induction to different parameters (such as light, elicitors, etc.).[88] In addition, it is possible that these enzymes are bonded in multienzyme systems and that all sequences producing secondary products are aggregated in such complexes.[182,393]

1. Phenylalanine Ammonialyase (PAL)

PAL is the key enzyme in phenolic biosynthesis and was first reported in *Hordeum vulgare*.[222] Many reviews have been devoted to it.[50,160,161] It catalyzes the deamination of phenylalanine (Figure 50) and produces *trans*-cinnamic acid. The enzyme is tetrameric and each subunit has a molecular weight of approximately 83000. Optimum pH for the reaction is approximately 8 to 8.7 and no cofactors are required. It is principally cytoplasmic and found in plastids, mitochondria, and microbodies.[161,271]

Its activity in plant tissues is generally much greater than that required for normal phenolic synthesis; indeed, addition of exogenous substrate generally leads to an increase in biosynthesis of phenolic compounds, which shows that the substrate may be the limiting factor.[65,257] Nevertheless, increase or induction of PAL activity generally corresponds to an increase in the accumulation of phenolic compounds (Chapter 4). Two izoenzymes of PAL have been reported in *Quercus*;[31] they were differentially inhibited by phenolic substances, one of which may be specific to the biosynthetic pathway of benzoic acids and the other to lignins. However, it does not appear to be possible to generalize these results, which have not been found in fruits.

PAL is the enzyme in the phenolic metabolism which has been most studied in a varied range of fruits: from pip fruits (citrus, tomato, grape, apple, pear, etc.), through stone fruits (cherry, plum, peach, mango, etc.) to others like strawberry (Table 42). It has also been revealed in tissue cultures of various citrus[38,418] and in cell suspension cultures of apple.[248] It is very difficult to perform quantitative comparisons of PAL activities in the various species (Table 43) partly because of differences in expression of the results, but mainly because of the large fluctuations to which this enzyme is subjected in relation to physiological stages on the one hand and to its sensitivity to numerous external factors (temperature, light,

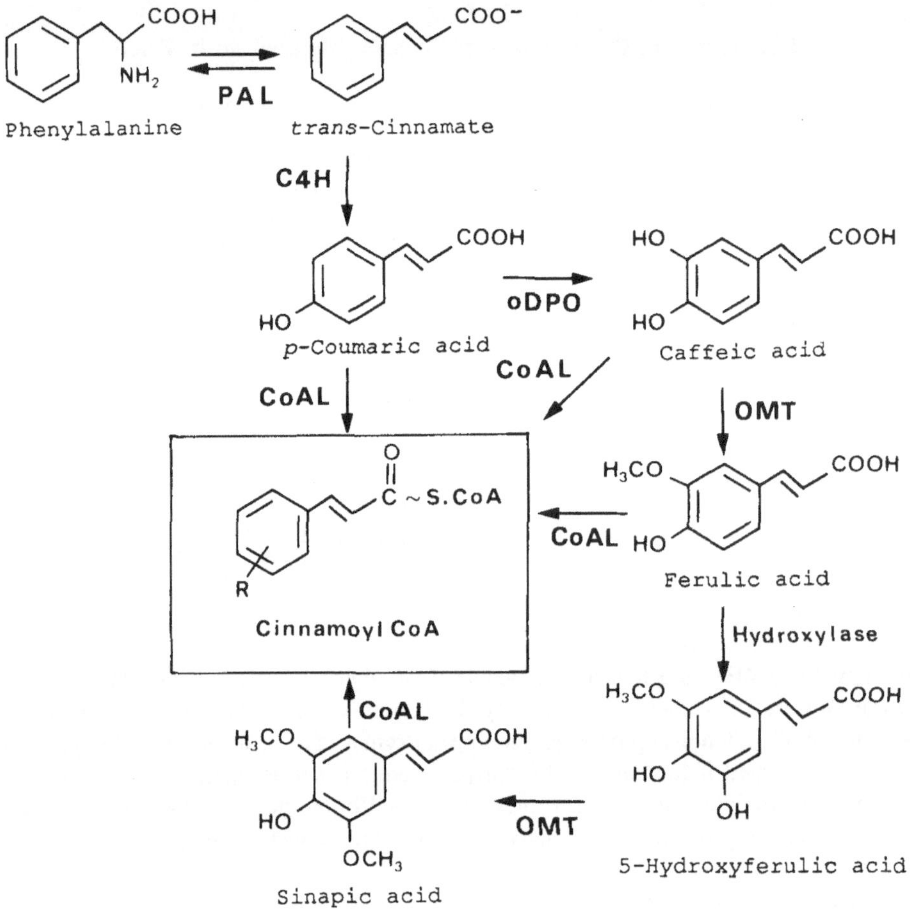

FIGURE 50. Biosynthetic pathway of cinnamoyl CoA esters. (Identification of enzymes: see text.)

various types of stress, etc.) on the other. These factors are discussed below. Thus, in the same cherry cultivar picked in two different years, PAL activity may vary from 8.3 to 25 pKat/g DW at the same physiological stage.[273] Likewise, strong variations can be observed between species[418] or between two different cultivars of ripe grape[183,210] (Table 43).

2. Cinnamate 4-hydroxylase (C4H)

The enzyme responsible for the conversion of cinnamic to 4- or *p*-hydroxycinnamic acid was first characterized from *Pisum sativum*.[358] This reaction requires molecular oxygen and NADPH (Figure 50).[45-47] Hydroxylation in ortho position leads to the formation of *o*-coumaric acid, a precursor of biosynthesis of coumarin. C4H is usually a cytochrome P-450-dependent enzyme associated with microsomal membranes. Extracts have to be treated with salts or detergent to release this membrane-bound enzyme. There are few fruit-specific examples (Table 42) and the values found are of the same order as those for PAL for the same fruit.[248,338]

3. Hydroxycinnamate CoA Ligase (CoAL)

CoAL has been isolated from various plant tissues and partially purified.[145,218,329-331,438] It enables the formation of CoA esters of hydroxycinnamic acids (Figure 50) in the presence of coenzyme A, ATP, and in most cases Mg^{2+} ions.[455] The enzyme is soluble and its subcellular location is probably cytoplasmic. CoA esters of hydroxycinnamic acids play a

TABLE 43
PAL Activity in Various Ripe Fruits and Derived *In Vitro* Cultures

Fruits	PAL Activity	Ref.
Apple (skin)	66.7 pKat/g FW	101
Grape		
De Chaunac	2.8 pKat/g FW	183
Super Hamburg (skin)	58.3 pKat/g FW	210
Grapefruit	0.27 pKat/g FW	253
Peach		
Peel	6.2 pKat/g FW	7
Pulp	3.7 pKat/g FW	
Plum	19 pKat/g FW	424
Strawberry	100 pKat/g DW	128
Sweet Cherry		
1974	8.3 pKat/g DW	273
1975	25 pKat/g DW	
Tomato		
Eurocross BB (wall)	2.5 pKat/mg prot.	338
Cherry		
Pulp	2.6 pKat/mg prot.	116
Pericarp	3.4 pKat/mg prot.	
Citrus callus (light 2 d)		
Grapefruit	0.20 pKat/g FW	
Navel orange	0.19 pKat/g FW	
Lemon	0.15 pKat/g FW	418
Citrangequat	0.28 pKat/g FW	
Mandarin	0.09 pKat/g FW	
Apple cell cultures	16.8×10^3 pKat/mg prot.	248

major role in the biosynthesis of the various groups of phenolic compounds and in particular *p*-coumaroyl CoA (pC CoA) as a precursor of flavonoids and lignins with feruloyl CoA and sinapoyl CoA. It would seem that CoAL governs the distribution of cinnamoyl CoA esters to the various specific pathways through the existence of isoenzymes with very marked specificities with regard to HCA. Two or three isoenzymes have been isolated which display a strong affinity for *p*-coumaric acid, but some show relatively high rates of conversion for typical substrates of the lignin pathway (ferulate and sinapate) whereas others have relatively high affinities for caffeate but do not activate sinapate.[48,218,329-331,438] Regulation of these different isoenzymes can be completely independent.[154]

Certain characteristics of CoAL isolated from other parts of plants are found in fruits (Table 42). In particular, there is a strong affinity for *p*-coumaric acid in tomato[337] and in cell suspensions of grape[237] (Table 44). In both cases, no activity was detected with sinapic acid, and activities with caffeic and ferulic acids are very close.

However, activities with regard to the same substrate may vary considerably from one species to another. For example with *p*-coumaric acid, 114 pKat/mg protein was recorded in tomato,[337] 2.5 in grape,[183] and 15 in grape cells.[237] There may also be substantial variation within the same species: 114 and 1.6 pKat/mg protein in two tomato cultivars (Eurocross BB and Cherry tomato) at a fully comparable physiological stage.[116,337]

In fruits, analysis of CoAL has never gone as far as the determination of possible isoenzymes. However, when tomato fruits are wounded experimentally, healing results partly from the lignification of the wounded zones,[110] and the specificity of CoAL activity with regard to its potential substrates is then different to that in nonwounded fruits. In particular, the increase in the activity of this enzyme during the hours following wounding is much more important for ferulic and *p*-coumaric acids than for caffeic acid. After wounding, there

TABLE 44
Substrate Specificity of Fruit Hydroxycinnamoyl CoA Ligase

Substrates (acids)	Relative activity in %	
	Tomato[a]	Grape cell culture[b]
Cinnamic	0	0
p-Coumaric	100	100
o-Coumaric	13	
m-Coumaric	49	
Caffeic	69[c]—77	60.4
Ferulic	87[c]—88.5	81.3
Isoferulic	86.8	
Sinapic	0	0
p-Methoxycinnamic	8.8	
m-Methoxycinnamic	6	
2-4-Dimethoxycinnamic	0	

[a] Rhodes and Wooltorton.[337]
[b] Lotfy et al.[237]
[c] Fleuriet and Macheix.[116]

is thus a change in specificity towards lignification in comparison with controls.[115] These fruits form an excellent model for investigation of the specificity of CoAL and its consequences for the supplying of the various biosynthetic pathways.

B. SPECIFIC PATHWAYS

The biosynthesis of coumarins, lignins, benzoic acids, flavonoids, or hydroxycinnamic derivatives corresponds to modifications to the skeleton formed in the general phenylpropanoid pathway and leads to native compounds.

1. Biosynthesis of Hydroxycinnamic Acid Derivatives

Before approaching the biosynthesis of the various derivatives of hydroxycinnamic acids, the formation of caffeic, ferulic, and sinapic acids from *p*-coumaric acid is described. This requires two types of reaction: hydroxylation and methylation.

a. Hydroxylation

Despite the ubiquity of caffeic acid, the enzymatic origin of the ortho-dihydroxylation pattern is still being discussed. The introduction of a second hydroxyl group into *p*-coumaric acid to give caffeic acid (Figure 50) is probably catalyzed by monophenol monooxygenases, a well-known group of plant enzymes. Different authors give them various names: tyrosinase, phenolase, catechol oxidase, polyphenoloxidase, *o*-diphenol O_2-oxidoreductase.[45-47,265-268] They are Cu enzymes, sometimes firmly bonded to membrane but usually readily solubilized. These enzymes oxidize mono (cresolase activity) or *o*-diphenols (catecholase activity) to the corresponding *o*-quinones which polymerize to dark products. However, the formation of *o*-diphenols can be observed under strongly reducing conditions. Attempts to separate hydroxylase and oxidase activities have been unsuccessful so far. Thus, the physiological reality of the intervention of this hydroxylase function in the biosynthesis of diphenols has not been definitively revealed, and some authors consider that phenolase is not involved in the synthesis of phenolic compounds in healthy, intact cells.[80,432]

In addition, hydroxylation of ferulic acid to 5-hydroxyferulic acid takes place by means of a cytochrome P-450 oxygenase,[135] like that of *p*-coumaroyl CoA to caffeoyl CoA.[208] There are thus two possible pathways in the hydroxylation of HCA or their derivatives

TABLE 45
O-Methylation of Phenylpropanoid and Flavonoid Compounds by Partially Purified Apple Cell Culture O-Methyltransferases

Substrate	Product	OMT activity 10^3·dpm/mg protein/min	pH optimum	Mol wt (kDa)
Phenylpropanoids				
5-OH Ferulic	Sinapic	28.7		
Caffeic	Ferulic	20.2	6.8	78
Esculetin	Scopoletin	14.3		
Chlorogenic	Feruloylquinic	1.2		
Flavonoids				
Quercetin (Q)	3-Me-Q	8.95		
3-Me-Quercetin	3,7-; 3,7,4′-Me-Q	6.45	7.3/8.3	47
7-Me-Q (rhamnetin)	3,7-; 3,7,4′-Me-Q	4.80		

Data from Reference 246.

catalyzed by enzymes with totally different biochemical characteristics and also different subcellular localizations.

Hydroxylation of *p*-coumaric acid into caffeic acid has been studied in apple (Table 42), and it would appear that the same enzyme is capable of hydroxylating *p*-coumaroyl quinic acid into chlorogenic acid, *p*-coumaroylglucose into caffeoylglucose,[244] and phloridzin into 3-hydroxyphloridzin.[133] Phenolase from avocado mesocarp,[205] banana pulp,[68] kiwi flesh,[302] and cherimoya epicarp[258] catalyzes the orthohydroxylation of monophenols like tyrosine, tyramine, and *p*-cresol. In grapes, the results are sometimes contradictory since some authors have revealed strong activity with monophenols,[427] and in particular *p*-hydroxycinnamic acid,[165] and others have reported total absence of activity in relation to the same substrates.[286] This contradiction, together with the absence of this cresolase activity in many preparations of phenolase from fruits and from other plant organs, is probably related to the lability of the phenolase and may be the result of changes in the structure of the protein during purification.[267]

b. Methylation

Methylation of HCA is an important stage insofar as it leads to the formation of ferulic and sinapic acids which, together with *p*-coumaric acid, form the precursors of lignins. Methylation of caffeic and 5-hydroxyferulic acids leading to ferulic and sinapic acids, respectively (Figure 50), takes place by means of an O-methyltransferase, with the methyl group from S-adenosyl methionine.[314] Two different enzymes occur in plants: the first methylates substituted cinnamic acids (CMT) whereas the other is specific for the methylation of flavonoid substrates (FMT).[314] These two enzymes display maximum activity in two very different pH ranges: 6 to 8 for cinnamic substrates and 8 to 9 for flavonoids.

They have been characterized in cell suspensions of apple cultivated *in vitro*.[246] One was specific for phenylpropanoids and the other for quercetin, and each had different characteristics (specificity in relation to substrates, optimum pH, etc.) (Table 45). A CMT was also revealed a long time ago[105,106] in apple when its action was recommended as a method for preventing oxidation.[288]

Methylation of HCA has also been studied in cherry tomato where it has been shown that the enzymes play a role in the orientation of the phenolic metabolism towards lignification during the healing of fruits after wounding.[115]

c. Formation of Esters with Hydroxyacids

The formation of HCA derivatives with organic acids is now well known.[399,455] Two

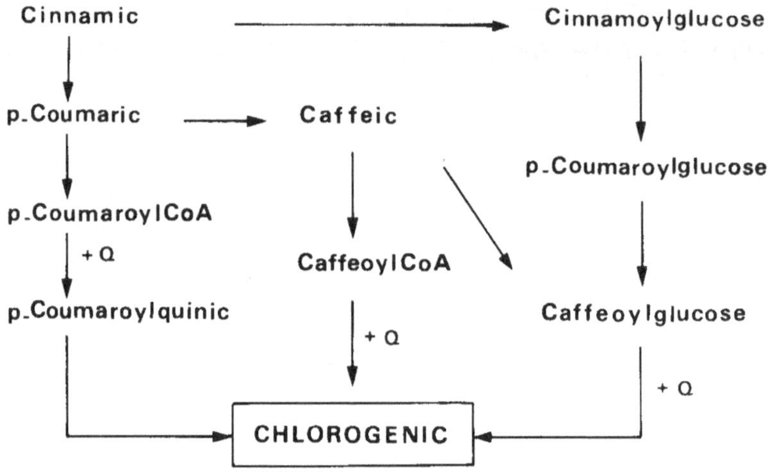

FIGURE 51. Possible biosynthetic pathways of chlorogenic acid. Q: quinic acid. (Adapted from Villegas, R. J. A. and Kojima, M., *J. Biol. Chem.*, 261, 8729, 1986.)

steps are required: the formation of CoA esters and then transesterification with hydroxyacids (quinic, malic, shikimic, etc.) catalyzed by a hydroxycinnamate transferase (HT).[337,339,404,425,426] The second step is reversible[337,338] (Figure 51A).

HT activity has been revealed in tomato fruit (cv. Eurocross BB and cherry tomato),[109,337] apple, and grape[237] (Table 46) with distinct specificity of the enzyme with quinic acid. However, although caffeoylquinic (chlorogenic) acid is the most abundant compound in these fruits, HT has much greater affinity for *p*-coumaroyl CoA than for caffeoyl CoA (Table 46). Chlorogenic acid may, therefore, be biosynthesized not from caffeoyl CoA but from *p*-coumaroyl CoA with an extra step which may consist of the hydroxylation of *p*-coumaroyl quinic acid, as in apple[242] and other plant materials.[285] Other types of compound — derivatives of cinnamic acids with sugars — may also be precursors of the biosynthesis of chlorogenic acid (Figure 51). This has not been shown in fruits but in potato where several intermediate compounds, such as cinnamoylglucose,[219,220] *p*-coumaroylglucose,[221,278] and caffeoylglucose[434] have been shown to be involved. Subsequent transesterification with quinic acid makes it possible to form caffeoylquinic acid. Another case has been studied in radish where the formation of sinapoylmalate can take place from sinapoylglucose by means of a vacuolar transferase.[372] If these pathways are those of the biosynthesis of chlorogenic acid, they are completely different to those that we have described above and they certainly do not involve CoA esters.

The formation of hydroxycinnamic derivatives with polyamines also uses the CoA esters pathway, and an HT was recently isolated in tobacco.[107] Tests on apple and grape carried out *in vitro* show that HT activity is nil with an amine (Table 46), but these fruits do not accumulate phenolamines and it would be interesting to extend this study to fruits such as citrus, in which feruloylputrescine has been revealed.

d. Formation of Sugar Derivatives

After exogenous addition of radioactive substrate, derivatives of HCA with glucose often accumulate in fruits, e.g., *p*-coumaroyl glucose in apple[244] and tomato.[109] The biosynthetic pathway is different to that of esters with hydroxyacids since it is no longer the acid that is activated but the sugar. Transfer of glucose to HCA is by means of uridine diphosphoglucose (UDPG) with a glucosyltransferase (GT). This is a nonspecific enzyme since it appears to effect glycosylation of both HCA and of flavonoids, coumarins, and lignin precursors. GT activity in relation to flavonoids is much higher than that related to HCA in

TABLE 46
Specificity of Hydroxycinnamoyl Transferase in Fruits

Substrates	Relative enzyme activity (%)		
	Tomato[a]	Apple[b]	Grape[b]
p-Coumaroyl CoA			
+ Quinic acid	100	100	0
+ Shikimic acid	2.5	11.1	0
+ Tartaric acid		0	100
+ Malic acid		0	86.3
+ Citric acid		0	85.3
+ Glucose		0	0
+ Tyramine	0	0	0
Caffeoyl CoA			
+ Quinic acid	58	22	0
+ Shikimic acid	0.7	t	0
+ Tartaric acid		0	29
+ Malic acid		0	10.5
+ Citric acid		0	0
Feruloyl CoA			
+ Quinic acid		10.4	0
+ Shikimic acid		t	0
+ Tartaric acid		0	15
+ Malic acid		0	7.2
+ Citric acid		0	0
Sinapoyl CoA			
+ all hydroxyacids		0	0

Note: t: very weak activity.

[a] Rhodes and Wooltorton;[337] Fleuriet and Macheix.[115]
[b] Lotfy and Fleuriet, unpublished data.

cell cultures of apple.[248] Apple[245] and tomato[117] are the only fruits in which GT has been studied specifically in relation to HCA (Table 42). In tomato, GT can effect the transfer of the glucose either on the carboxylic function of an HCA (ester) or on the hydroxyl group (glucoside).[117] This accounts for the accumulation of the two types of glucose derivatives in this fruit.

The biosynthesis of HCA derivatives is thus based on two main pathways: from CoA esters for the link with hydroxyacids, or directly from HCA for glucose derivatives. There does not appear to be anything in common between the two pathways insofar as the substrates of CoAL are not accepted by GT and vice versa.[117,337] The two pathways must exist in most fruits since both types of derivatives accumulate but, as has been seen, *in vitro* study has been undertaken on very few fruits. In addition, as already mentioned, biosynthesis of esters with hydroxyacids by means of glucose derivatives cannot be ruled out.

2. Biosynthesis of Benzoic Acids

Two pathways are possible for the biosynthesis of benzoic acids.[144,170,455] They can be derived directly from the shikimate pathway (Figure 48), and especially from dehydroshikimic acid; this reaction is the main route to gallic acid, a common constituent of plants. However, they can also be produced by the degradation of cinnamic acids, in a similar manner to the β-oxidation of fatty acids; the main intermediate substances in this case are cinnamoyl CoA esters.[3,455] Knowledge of the mechanisms and especially the enzymes involved in biosynthesis of benzoic acids and their derivatives is rather limited, particularly with regard to fruits, although gallic acid and its derivatives play an important role in the formation of hydrolyzable tannins.

3. Biosynthesis of Lignins

Reviews on the biosynthesis of lignin and its enzymatic controls are available.[27,138,140,154,543] Lignin is formed by oxidative polymerization of *p*-coumaryl, coniferyl, and sinapyl alcohols by peroxidases.[166] Peroxidases frequently occur in plants as isoenzymes, and these cell-wall-bound peroxidases are very probably involved in the lignification process.[125,249,250] The first stages are those of the general phenylpropanoid metabolism, and we have already reported the probable specificity of certain PAL and CoAL isoenzymes related to the lignin pathway and the existence of two distinct methyltransferases, one of which may be involved in lignin biosynthesis.

However, the reduction of cinnamoyl CoA esters to cinnamyl alcohols is specific for the lignin biosynthesis pathway.[87,256] The reaction requires two NADPH or NADP-dependent enzymes: cinnamoyl CoA: NADP oxidoreductase (E.C. 1.2.1.44) and cinnamoyl alcohol NADP oxidoreductase (E.C. 1.1.1.). These enzymes have been studied in cell suspension cultures of soybean, lignifying tissues of *Forsythia* and various other plants.[143,336,446] The highest affinity of the former enzyme is generally for feruloyl CoA and 5-hydroxyferuloyl CoA, but 4-coumaroyl CoA and sinapoyl CoA are also reduced with high efficiency. Nevertheless, other compounds may also be lignin precursors, e.g., coniferin (coniferyl alcohol β-D-glucoside) which appears to form in suspension of 'Paul Scarlet' rose by means of a specific glucosyltransferase.[191]

Biosynthesis of lignins thus involves a whole series of enzymes whose compartmentation may differ greatly, the final stage being polymerization and accumulation in the cell wall. Some experimental evidence suggests that the transport of lignin monomers from the cytoplasm to the cell wall is mediated by vesicles and that lignin precursors are polymerized when they reach the cell wall.

There are very few data for fruits on the biosynthesis of lignins since in fleshy fruits this compound is only involved in the formation of drupe stone, the teguments of some seeds, and in healing in response to infection or wounding.[115,247] No particular specificity of enzymes or isoenzymes has been clearly demonstrated for lignification in fruits. In contrast, correlations have been shown for example between stone formation and PAL activity during the development of peach,[7] the location of lignified cells in pear pulp and peroxidasic activity,[328] and the increase in activity and the change in affinity of PAL, OMT, CoAL, and peroxidase in response to wounding or infection of tomato fruit.[110,115,130] In this case, a specific lignification peroxidase bound to the cell wall of tissues during healing was revealed which was not present in the control fruits.[110]

4. Biosynthesis of Coumarins

Biosynthesis of coumarins has been reviewed at length.[36,37,283] These articles provide a full list of the relevant references.

Simple coumarins are formed by plants via the shikimate pathway, the immediate precursors being phenylalanine and cinnamic acids. *Trans*-cinnamic acid is a common precursor of all coumarins, and its ortho- or para-hydroxylation leads respectively to the formation of coumarin (and of other coumarins lacking oxygenation at C-7) and hydroxycoumarins (Figure 52). As coumarin occurs mainly in the cells as coumarinyl glucoside, the conversion of *trans*-cinnamic acid to coumarin thus involves the following sequence: ortho-hydroxylation, glucoside formation at the 2′-hydroxyl, and isomerization to the *cis*-isomer, which is a light-catalyzed reaction. A specific β-glucosidase releases coumarin by hydrolysis of glucoside at cell disruption. However, it seems clear that most coumarins exist in free form in the cell[36] where the final lactonization takes place. In the case of 7-oxygenated simple coumarins, para-hydroxylation precedes ortho-hydroxylation (Figure 52) and umbelliferone, herniarin, and scopoletin are derived from the cinnamic acids of the corresponding oxygenation pattern.

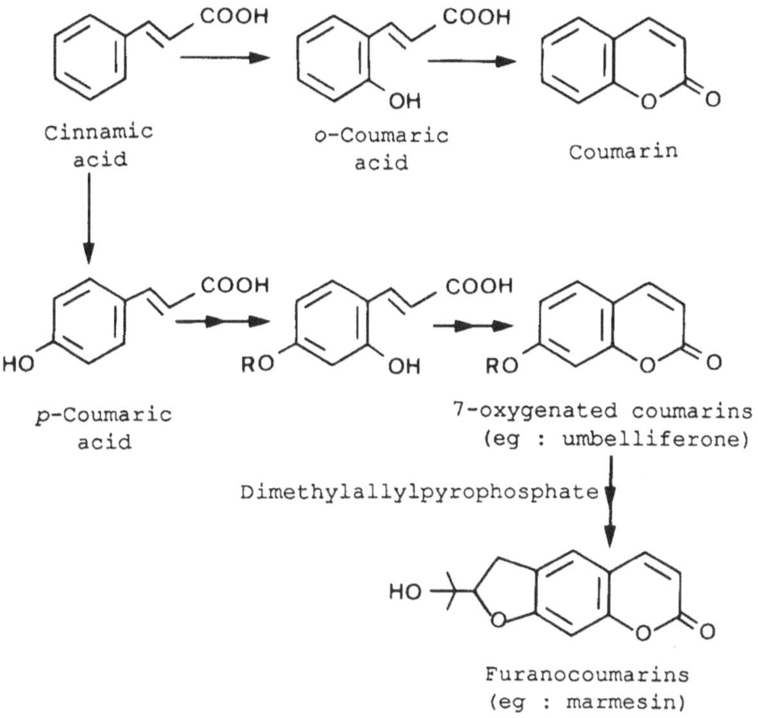

FIGURE 52. Biosynthesis of coumarins.

Nevertheless, additional oxygenation of the nucleus may sometimes occur in simple coumarins as in furanocoumarins.

A major feature of the diversification of coumarins is the presence of isoprenoid units derived from mevalonic acid and attached to a carbon atom of the nucleus or to an oxygen atom. Later, secondary modifications may occur on the prenyl side-chain via initial epoxidation. Linear and angular furanocoumarins are derived from *p*-coumaric acid and umbelliferone. The additional two-carbon side-chain is derived from mevalonic acid, with condensation of umbelliferone and dimethylallyl phosphate. Subsequent cyclization of the side-chain leads to the formation of furanocoumarin.

Subcellular sites of coumarin formation have not been clearly elucidated, but there is evidence that certain enzymes in their biosynthesis are at chloroplast sites whereas the β-glucosidase which leads to the formation of coumarin is extracellular.[36,37] Sites of coumarin biosynthesis in plants are varied[36] and include leaves, shoots, roots, and fruits. In the case of parsnip, *(Pastinaca sativa)*, accumulation of furanocoumarins is the result of synthesis in the fruit itself.[20] This has been demonstrated indirectly by reciprocal grafts between *P. sativa* and *Levisticum officinale* where fruit and root furanocoumarins are not affected by grafting. As no evidence of translocation of furanocoumarins was obtained, it is concluded that their biosynthesis takes place in the parts where accumulation can be observed.

As with many other phenolics, biosynthesis and accumulation of coumarins are stimulated in diseased plants,[283] but little work has been carried out on fruits. It has recently been shown that scoparone accumulates in the peel of various citrus fruits following inoculation with *Phytophtora citrophtora*.[1]

5. Biosynthesis of Flavonoids

Numerous reviews have dealt with the biosynthesis of this group.[89,142,152,154] There is extensive knowledge above all of the biosynthesis of flavones and flavonol glycosides thanks to work carried out on tissue culture and cell suspensions. Experiments with radioactive-

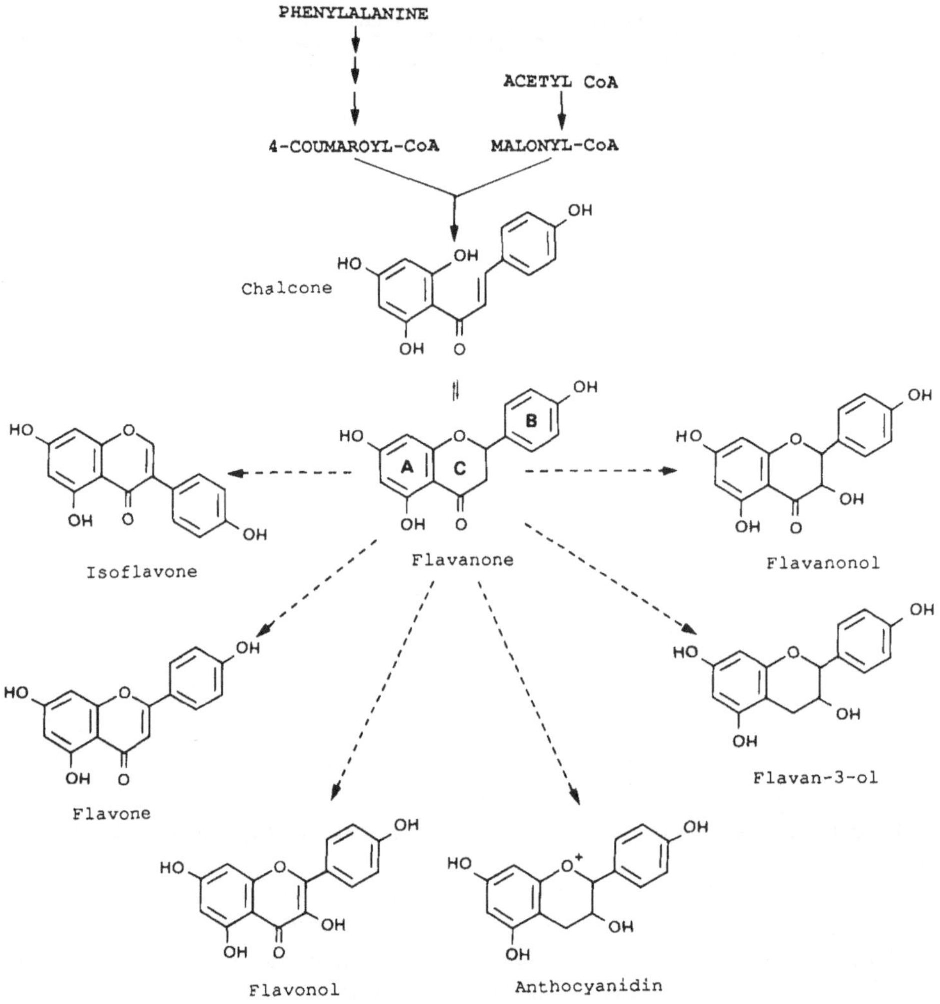

FIGURE 53. Biosynthesis of flavonoids.

labeled precursors established that the carbon skeleton of all flavonoids is derived from acetate and phenylalanine. A-ring (Figure 53) is formed from three acetate units (malonate) and phenylalanine gives rise to the B-ring and to the 3-C chain of the heterocyclic C-ring. Chalcone or the isomeric flavanone are central intermediates in the synthesis of all flavonoids (Figure 53). The enzymes involved are all those of the phenylpropanoid metabolism and 13 other enzymes which are specific to these pathways.[154]

The first step is the conversion of acetyl CoA into malonyl CoA by acetyl CoA carboxylase. In the second, the C-15 skeleton is formed by a chalcone synthase which catalyzes the condensation of a molecule of 4-coumaroyl CoA with 3 mol of malonyl CoA. This is the key reaction in flavonoid biosynthesis. For several years, the enzyme responsible for this stage was called "flavanone synthase", but the intermediate product of the reaction was identified as narigenin chalcone[176] and so it was subsequently referred to as chalcone synthase. It is specific to 4-coumaroyl CoA at pH 8 but also accepts caffeoyl CoA at pH 6.5.[366] In certain cases, feruloyl CoA is not a substrate of this enzyme,[184] whereas for others the CoA esters of *p*-coumaric, caffeic, and ferulic acids are substrates of chalcone synthase and display similar efficiency.

Chain elongation mechanisms are similar in the chalcone synthase reaction and in the synthesis of fatty acids or of other polyacetate compounds derived from malonyl CoA.

However, a feature of chalcone synthase is the use of CoA esters as intermediate substrates. Chalcone is then converted into flavanone by a chalcone flavanone isomerase. The various flavonoids then formed result from the changes which occur in the different rings — essentially hydroxylations, oxidations, methylations and glycosylations.[139] Flavonoid *O*-methyltransferase is generally different from caffeate *O*-methyltransferase with regard to the specificity of substrates and several other properties.[316,317,407] Methylation is only involved after the formation of the flavonoid ring structures. Likewise, glycosylations occur after all substitutions and modifications of the flavonoid ring structure.[153] Sequential steps should be considered which correspond to the different enzymes: 7-*O*-glucosyltransferase and 3-*O*-glucosyltransferase, whose range of specificity can vary with plant material.

In fruits, biosynthesis of flavonoids has been very little studied at enzymatic level (Table 42). Chalcone synthase and chalcone flavanone isomerase have been revealed in grape[183] and only chalcone isomerase in grapefruit.[253] In citrus, almost all hydroxyl groups can be methylated, suggesting the presence of several transferases methylating OH groups at different positions in the A-, B-, and C-rings.[39,40] One is reminded that methyltransferase, isolated from apple cell cultures (Table 45), was found to be capable of methylating 3, 7, and 4 hydroxyl groups of quercetin.[246] The same cells possess a GT which is able to glucosylate flavonols with greater activity for kaempferol than for quercetin (132 and 51 pKat/mg protein, respectively).[248] Grapefruit *(Citrus paradisi)* cells in suspension cultures were able to specifically *O*-glucosylate exogenous flavanone in position 7 even though they did not accumulate flavanone glycoside.[234] In the same fruit, it would seem that chalcones are glycosylated during their formation and that methylation occurs after cyclization of chalcones to flavanones.[332]

The biosynthesis of anthocyanins proceeds by the pathway chalcone → flavanone → 3-hydroxyflavanone (dihydroflavonol) → anthocyanidin → anthocyanin.[141] Enzymes for the conversion of dihydroflavonols into anthocyanidin are still unknown. Leucoanthocyanidins (monomeric flavan-3,4-diols) may also be intermediate substances in anthocyanin biosynthesis.[177,192] The different anthocyanins derive either from the various hydroxycinnamic acids, and in this case 4-coumarate is the precursor of pelargonidin, caffeate that of cyanidin, ferulate that of peonidin, etc., or only from 4-coumarate. Introduction of further hydroxyl groups and methylation reactions possibly occur at the chalcone/flavanone stage or later in the pathway. Both possibilities can also be envisaged for introduction of hydroxyl groups in the case of cyanidin, but the situation is less clear with methylated anthocyanidins such as peonidin and petunidin.[141] As is the case for all the classes studied, glycosylation is probably the last step in anthocyanin biosynthesis although the possibility of conversion of dihydroflavonol 3-glucoside to anthocyanidin 3-glucoside cannot be excluded.[141] In most cases, glycosyltransferases are not specific to anthocyanins but may also glycosylate flavonols. However, in some cases, such as *Silene dioica,* they are specific[206,207] and glycosylation takes place in the following order: anthocyanidin 3-glucoside → 3-rutinoside → 3-rutinosyl 5-glucoside.

Anthocyanins also occur in acylated form with acetic, *p*-coumaric, or caffeic acids. The presence of an enzyme (acyltransferase) catalyzing the transfer of the 4-coumaroyl or caffeoyl moiety of the corresponding coenzyme-A esters to the hydroxyl group of the sugar moiety of anthocyanidin has been demonstrated in petals of *Silene dioica*[207] and in *Matthiola incana.*[415]

Although there is good knowledge of anthocyanin distribution in parts of plants, there is little information concerning the subcellular location of their synthesis. In protoplasts obtained from *Hippeastrum* and *Tulipa*, 90% of the activity of chalcone synthase, chalcone isomerase, and anthocyanidin 3-*O*-glucosyltransferase was found to be located in the cytosol fraction. However, recent studies have drawn attention to the presence of anthocyanoplasts in cells and raised the possibility that these might be the site of anthocyanin biosynthesis.[306] No enzyme activity was detected in isolated vacuoles.[183]

Most of the data concerning the biosynthesis of anthocyanins comes from studies on flowers, and although these substances are very abundant in fruits, very little is known about their biosynthesis. Two enzymes studied were chalcone synthase and chalcone isomerase in grape[183] (Table 42). Study of the evolution of the various anthocyanins found in Syrah grapes was carried out during maturation of the fruit[227,352] and led the authors to propose a diagram of the relations between primitive pigments (cyanidin and delphinidin), stable pigments (peonidin and malvidin), and petunidin which lies in the middle of the chain of transformation from cyanidin to malvidin. However, this was not accompanied by a study of enzymes and can thus only serve as a basis for future investigations.

6. Biosynthesis of Tannins

There has been little experimental work to date on the biosynthesis of tannins in plants, probably because of the complexity of these molecules. However, substantial progress has been made over the past decade[168,171,172,394,395] although fruits have been little used as experimental material in spite of their richness in tannins.

Gallic acid is almost always present in plants in ester forms with different polyols including sugars; their combinations constitute hydrolyzable tannins. It has already been reminded (Figures 48 and 49) that gallic acid can be formed by different pathways,[170] with the direct pathway from dehydroshikimic acid being the most probable. Some plants principally synthesize and accumulate simple gallic acid esters with D-glucose: β-penta-*O*-galloyl-D-glucose in particular. For many others, the formation of the penta-galloyl-glucose ester is merely the point of departure from which three pathways can diverge (Figure 54).[151,172] One leads to *gallotannins* by the addition of further gallic acid molecules and the two further pathways diverge to *ellagitannins* by oxidative coupling of suitably arranged galloyl ester groups in a gallotannin. Biogenetic relationships and transformations of the hexa and nonahydroxy di and tri phenoyl esters have been discussed[151,170] in relation with polyphenol-protein complexation and phylogenetic relations of Dicotyledons.[172]

The main features of the biosynthetic pathway to proanthocyanidins and their related flavan-3-ols are known (Figure 55) although certain aspects are still the subject of serious discussion. Like other flavonoids, proanthocyanidins are synthesized both from malonyl CoA and a C_6-C_3 skeleton derived from cinnamic acid. Studies of the biosynthesis of procyanidin have been conducted with various fruit-bearing plants[170,199] and in rarer cases on the fruits themselves, e.g., the incorporation of ^{14}C-acetate in (−)epicatechin and procyanidins in seed shells of *Aesculus hippocastaneum*.[85] An interesting observation was that the two apparently identical units that form procyanidin dimer incorporate cinnamic acid in a different manner.[199,396] Several interpretations can be put forward to account for this including, for example, phenylpyruvic acid as an intermediary,[168,172] although this hypothesis seems highly unlikely,[394] or the characteristics of a condensing enzyme in a multienzyme complex controlling proanthocyanidin biosynthesis.[394]

According to Haslam,[168,170,172] a particularly important intermediary is the symmetrical flav-3-en-3-ol (Figure 55). A two-stage stereospecific reduction (in which NADPH might be the biological reducing agent) of this compound, including the formation of a carbocation, would lead to the formation of the two stereoisomeric flavan-3-ols (+)-catechin and (−)-epicatechin. Proanthocyanidins would form by condensation of the carbocation with the flavan-3-ols already formed to give dimers, trimers, and higher polymers. This set of reactions (Figure 55) makes it possible to understand the formation of the "genetically homogeneous" types of procyanidins encountered in plants.[170] Procyanidins appear as by-products of flavan-3-ols and cyanidin metabolism in plants, and the balance of procyanidins moves to the higher polymeric forms when the supply of NADPH is low and the flux to flav-3-en-3-ol high.[170,172] This situation may occur towards the end of the fruit growth and during maturation when

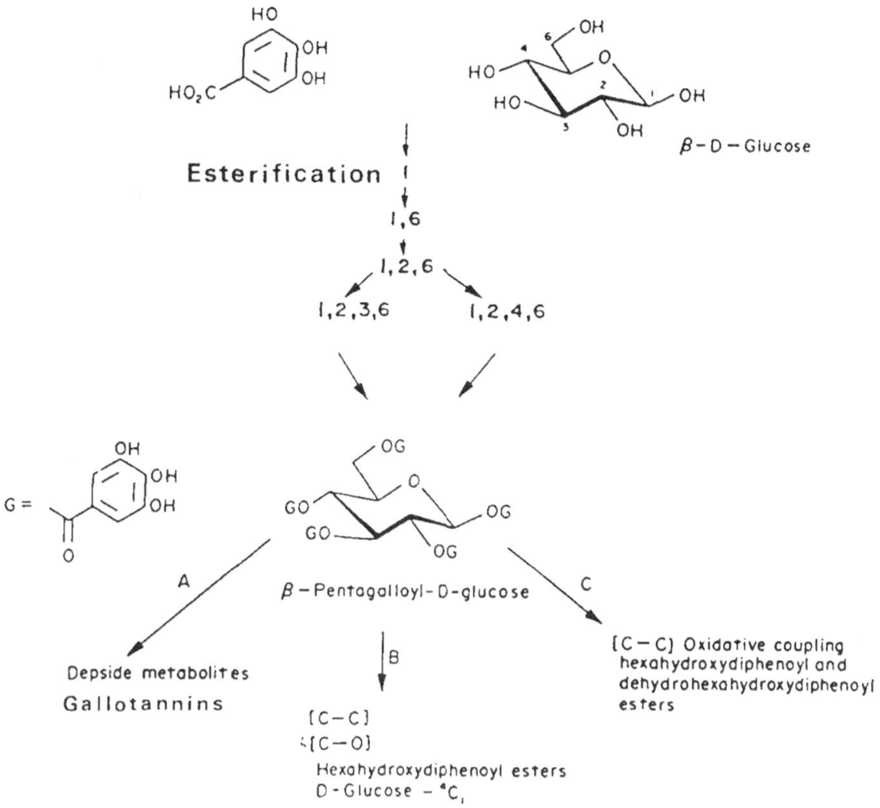

FIGURE 54. Gallic acid metabolism in higher plants: the overall pattern. (From Haddock, E. A., Gupta, R. K., Al-Shafi, S. M. K., Layden, K., Haslam, E., and Magnolato, D., *Phytochemistry*, 21, 1049, 1982. With permission.)

photosynthesis, which is the source of reducing power, has strongly diminished in the fruit. Indeed, as is discussed below, maturation can be accompanied by an increase in the degree of polymerization of condensed tannins in certain fruits, although this is not always the case.[131,169]

In contrast to Haslam's hypothesis, speculation concerning proanthocyanidins has been put forward.[394,396] Flavan-3-ols may form directly from flavan-3,4-diols by the action of an NADPH-dependent reductase which may be grouped in multienzymatic complexes which may also contain the terminal condensing enzyme. Indeed, although condensation of the stereospecific carbocation with flavan-3-ols easily occurs nonenzymatically, it is probably controlled enzymatically *in vivo*. In addition, in this system (Figure 55), the existing stereochemistry at C-3 and C-4 of (+)-dihydroquercetin would be maintained in (+)-catechin and its oligomers. (−)-Epicatechin might be formed by means of different pathways, with the most likely requiring the action of an epimerase on the (+)-dihydroquercetin.

There are thus important differences between authors on the subject of the biosynthetic systems which lead to flavan-3-ols and proanthocyanidins, one of the points of discussion being whether the condensation stages are enzymatic or not. However, there is general agreement that oligomeric proanthocyanidins are formed by the addition of a stereospecific flavanyl-4-carbocation to one of the two stereospecific flavan-3-ols or to an existing chain rather than via self-condensation of the carbocation.[394]

C. CHANGES IN ENZYME ACTIVITIES DURING FRUIT DEVELOPMENT

The ripening of many fruits is accompanied by an increase in the rate of protein synthesis

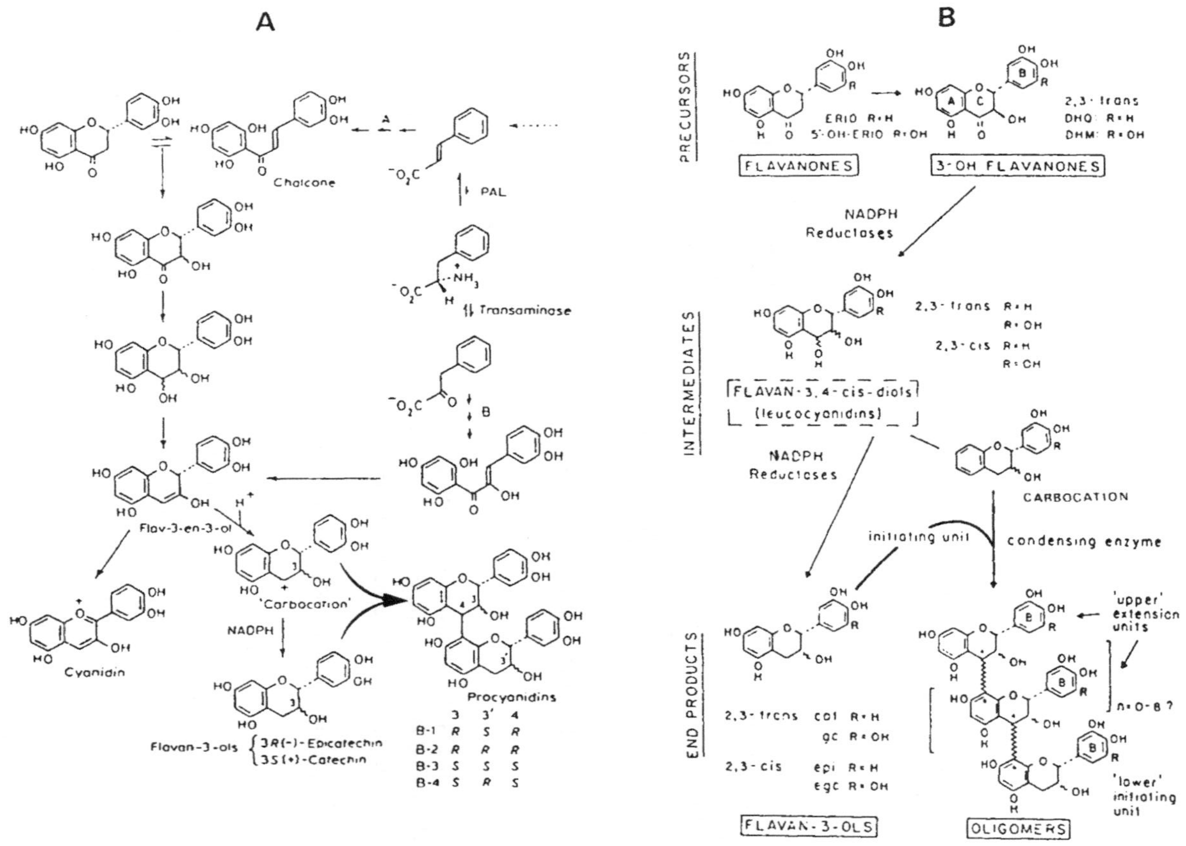

FIGURE 55. Suggested pathways of biosynthesis of (+)-catechin, (−)-epicatechin and procyanidin (A: From Haslam, E., in *The Flavonoids: Advances in Research*, Chapman and Hall, London, 1982, 417; B: from Stafford, H. A., Kreitlow, K. S., and Lester, H. H., *Plant Physiol.*, 82, 1132, 1986. With permission.)

and in the activity of numerous enzymes.[34,76] Climacteric fruits generally display an intense metabolism which is clearly defined in time. Increases in enzymatic activity may be related either to an increase in RNA and protein synthesis or to simple activation following the disappearance of inhibitory substances. The variations of phenolic compounds result from the activity of the enzymes responsible for their biosynthesis and their degradation. However, there have been very few studies of the degradation enzymes of phenolic compounds in fruits with the exception of polyphenoloxidases and peroxidases which lead to the formation of brown substances and which are discussed below (Chapter 5). Hence only the variations in the enzymes in the phenolic anabolism of fruits are described here.

Variations in PAL activity have been extensively studied during the growth and maturation of various fruits (Figure 56). Depending on the author, results are reported either per unit of fresh or dry matter or in terms of protein or again per fruit. In addition, variation criteria may be certain growth parameters of the fruit (size, weight, etc.), parameters making it possible to characterize the physiological stage (particularly color) or more simply the date of sampling. In many fruits (grape,[210] cherry,[273,274] pear,[21,274] grapefruit,[253], etc.), PAL activity is at its maximum in very young fruits and then falls very rapidly during growth (Figure 56). In contrast, in tomato (Figure 57), PAL activity only reaches maximum at the end of growth, just before the fruits change color, and then decreases. PAL increases again during maturation in fruits in which anthocyanins accumulate, such as grape,[210] plum,[424] peach,[7] cherry,[273,274] and strawberry.[128] The example of grape[210] is a perfect illustration of this behavior: whereas PAL activity is at its maximum in very young fruits and diminishes during growth, it only increases again at the beginning of maturation of black cultivars and does not vary in white varieties.

Although PAL activity is absolutely necessary for the synthesis of phenolic compounds, it is often considered to be nonlimitative at a given physiological stage. Thus, in grapefruit there is sufficient PAL activity to account for the rate of naringenin glycoside accumulation in intact fruit.[253] In contrast, its substrate phenylalanine might appear to be a limiting factor[65,257] in the production of phenolic compounds. Nevertheless, when the different physiological stages of a fruit are compared, numerous examples display good correlation between variations in PAL activity and accumulation of phenolic compounds. This is the case in grapefruit with naringenin (Figure 58),[167] in tomato[115] and in pear[21] with hydroxycinnamic derivatives, and in grape,[183,210] plum,[424] and strawberry[128,129] with anthocyanins (Figures 56 and 59). In strawberry, two lines of evidence suggest that the increase in PAL activity during ripening is directly related to the synthesis of anthocyanins: first, the simultaneous timing of the increases in the activities of PAL and glucosyltransferase (Figure 56), the terminal enzyme in anthocyanin synthesis; second, the demonstration that anthocyanin synthesis can be inhibited by the introduction of L-α aminooxy-β-phenylpropionic acid, a potent and relatively specific PAL inhibitor.[128] In the same fruit, increase in PAL activity during ripening was accompanied by *de novo* synthesis of the enzyme.[129] Therefore, as in climacteric fruit, protein synthesis plays a significant part in the ripening of a nonclimacteric fruit, such as the strawberry.

In the tomato,[116] the beginning of maturation is also marked by important changes in other enzymatic activities: coenzyme A ligase, *O*-methyltransferase, hydroxycinnamoyl transferase, and glucosyltransferase. The first three activities increase during growth (Figure 57) and then decrease during maturation, whereas GT activity increases throughout the whole of the life of the fruit and in particular during maturation. There is a very good correlation between the accumulation of quinic and glucose derivatives and the activity of two enzymes directly involved in their biosynthesis: CoAL and glucosyltransferase (Figure 58).

In grape, all the enzymatic activities studied increase and reach their maximum during the period of greatest accumulation of anthocyanins (September and October).[183] This includes enzymes which are specific to the biosynthesis of flavonoids, chalcone synthase,

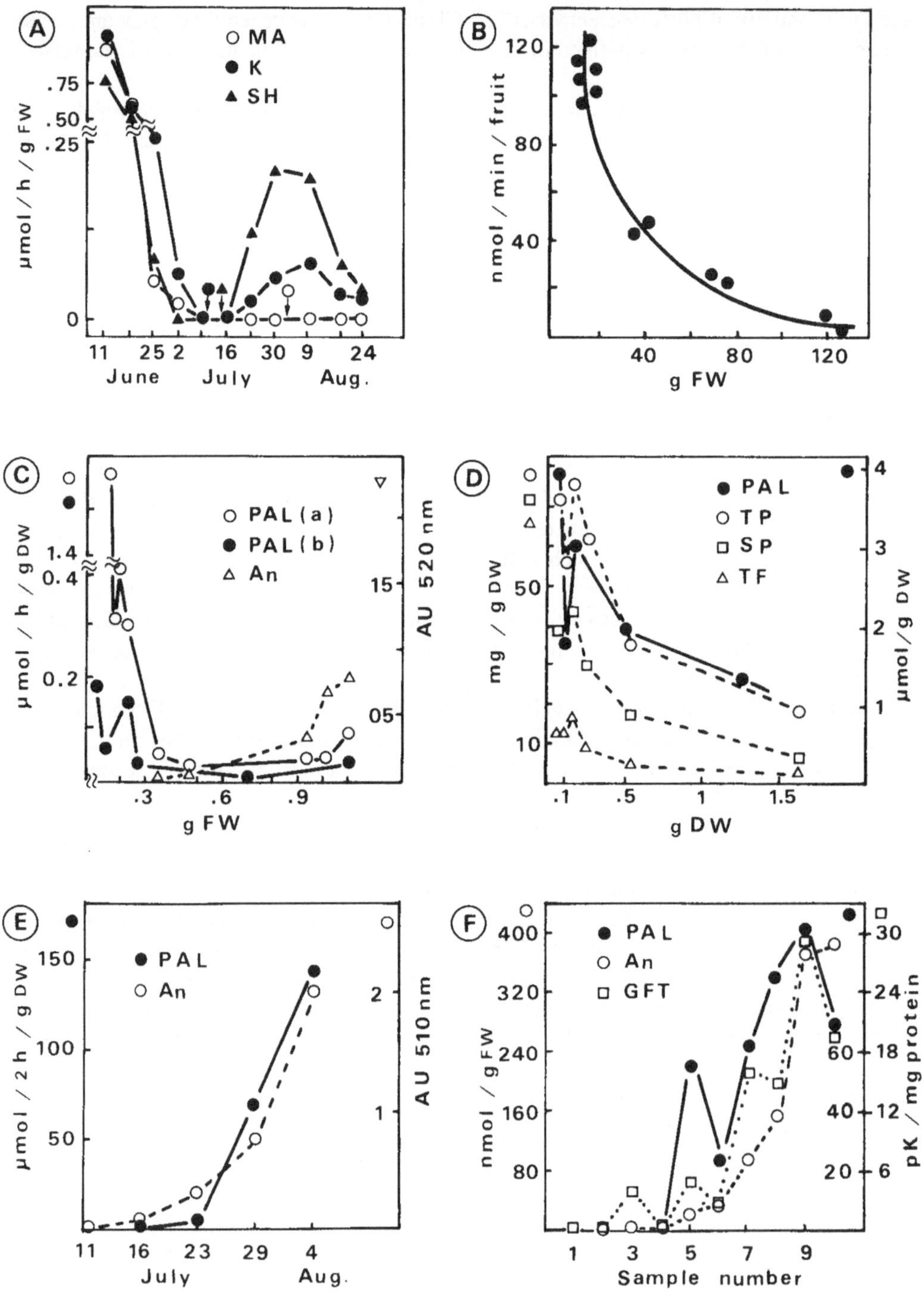

FIGURE 56. Changes in phenylalanine-ammonia lyase (PAL) activity during growth and maturation of fruits. A: grape, MA: Muscat of Alexandria, K: Kyoho, SH: Super Hamburg. Arrows denote the beginning of veraison. (Adapted from Reference 210.) B: grapefruit (Adapted from Reference 253.) C: cherry, PAL (a): 1974; PAL (b): 1975. (Adapted from Reference 273.) D: pear, TP: total phenols, SP: simple phenols, TF: total flavonoids. (Adapted from Reference 21.) E: plum (Adapted from Reference 424.) F: strawberry, GFT: flavonoid glucosyltransferase (Adapted from Reference 128.) An: Anthocyanin; AU: Absorbance unit.

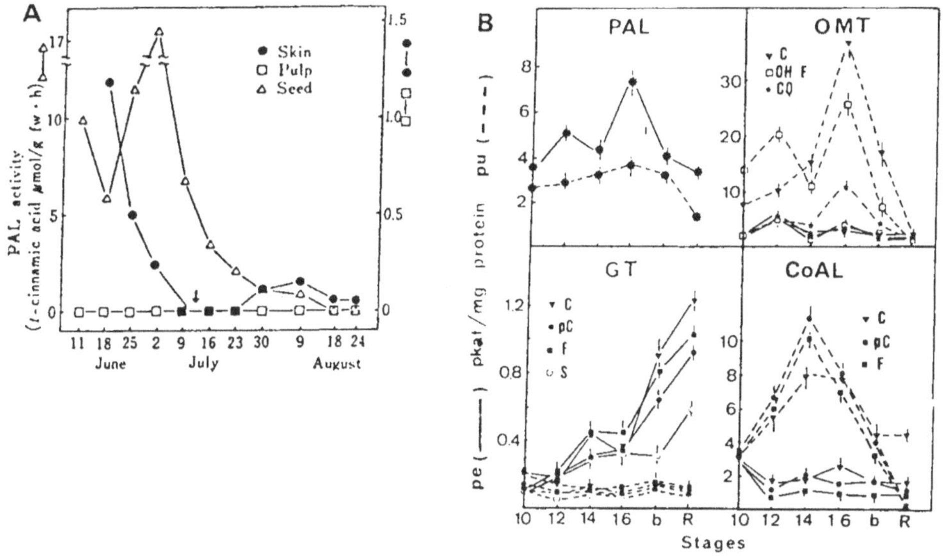

FIGURE 57. Changes in enzyme activities in different parts of grape (A) and cherry tomato (B) during growth (10 to 16 mm diameter green fruits) and maturation (breaker and red fruits: b, R). PAL: phenylalanine ammonialyase; OMT: *O*-methyltransferase; GT: glucosyltransferase; CoAL: coenzyme A ligase; pe: pericarp; pu: pulp. Substrates: caffeic (C), *p*-coumaric (pC), ferulic (F), 5-hydroxyferulic (OHF), sinapic (S), and chlorogenic (CQ) acids. (A: from Kataoka, I., Kubo, Y., Sugiura, A., and Tomana, T., *J. Jpn. Soc. Hortic. Sci.*, 52, 273, 1983; B: from Fleuriet, A. and Macheix, J. J., *Phytochemistry*, 24, 929, 1985. With permission.)

chalcone isomerase, and GT and those of the general phenylpropanoid pathway, such as PAL, CoAL, and C4H (Figure 59). This is to the best of our knowledge the only example in which there are data on the evolution of C4H, chalcone synthase, and chalcone isomerase during the life of a fruit.

Apple is one of the rare fruits in which the evolution of the hydroxylating activity of polyphenoloxidase is known during the growth period, when there is intense accumulation of hydroxycinnamic derivatives.[242] The evolution curves (Figure 60) are totally different according to the mode of expression. When expressed per milligram of proteins, activity is at maximum in July, which is fully correlated with the reduction of the *p*-coumaric/caffeic acid ratio.[243] When it is related to dry matter, most activity is observed in young fruits, corresponding to the highest chlorogenic acid content.

Changes in the activities of several enzymes in the phenolic metabolism have also been studied in *in vitro* cultures of fruits: citrus callus,[38,418] cell suspensions of apple parenchyma,[248] and of grape berry pulp.[237] In apple cell suspensions, PAL and C4H activities are at a maximum during the lag period of culture growth (Figure 61). In cell suspensions of grape (Figure 62), CoAL activity increases considerably in the hour following subculturing. This is probably related to enzyme synthesis since there is simultaneously a very large increase in proteins which is probably linked with a dilution effect already reported in other cell cultures.[155,293,299,371] The very early high CoAL activity in these grape cell cultures probably provides the *p*-coumaroyl CoA required for the synthesis of the anthocyanins which accumulate during the exponential growth phase (Figure 62). In contrast, in apple cell suspensions, OMT activity is maximal during the exponential growth phase (Figure 61) whereas GT activity exhibited a gradual but constant increase during culture growth which took place in parallel with the accumulation of phenolic compounds. The coordination between GT activity and the accumulation of hydroxycinnamoyl compounds during culture growth clearly indicates that glycosylation is a terminal step in the biosynthesis of phenolic compounds.[248] The activity pattern of GT compares well with that of soybean culture[315] but

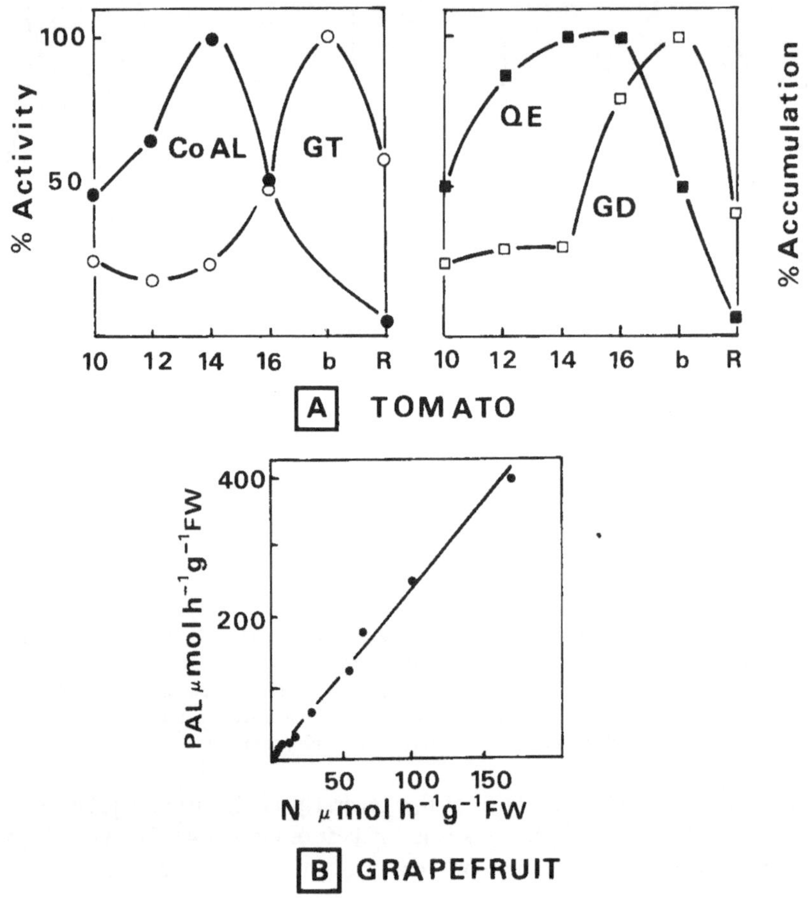

FIGURE 58. Relation between enzyme activities and phenolic accumulation in cherry tomato (A) and grapefruit (B). Enzymes and stages of growth and maturation: see Figure 57. QE: quinic esters; GD: glucose derivatives; N: naringenin. (A: Adapted from Fleuriet, A., Thèse Doct. Etat, Université de Montpellier, France, 1982; B: adapted from Hasegawa, S. and Maier, V. P., *Phytochemistry*, 11, 1365, 1972.)

is different from those of rose[191] and tobacco[190] cell cultures where maximum activity was reached during exponential growth.

In fact, there have been few complete sets of data on the evolution of enzymatic activities of the phenolic metabolism of fruits, and above all there has been no study concerning the origin of variation and the regulation of enzymes contents either at protein level or linked with the presence of inhibitors or activators.

In most of the cases mentioned, changes in enzyme activities have been studied in the whole fruit without distinction between the various parts. However, it has been seen that phenolic components are particularly abundant in the external parts of the fruit (peel, pericarp, skin, etc.), suggesting that there is active biosynthesis, probably related to the enzymatic activities which are located there. Tissue distribution of phenolic metabolism enzymes have only been studied in a few fruits: grape,[210,356] tomato,[116] and peach.[7] In grape (Figure 57A), PAL activity is ten times as high in the seeds as in the skin and zero in pulp. This is in perfect agreement with the presence of anthocyanins in skin only in nonteinturier cultivars and the abundance of catechins in seeds in particular.[210] In contrast, in the tomato (Figure 57B), although the activities of the various enzymes are approximately equivalent in pulp and pericarp of very young fruits, evolution is completely different during growth. PAL and

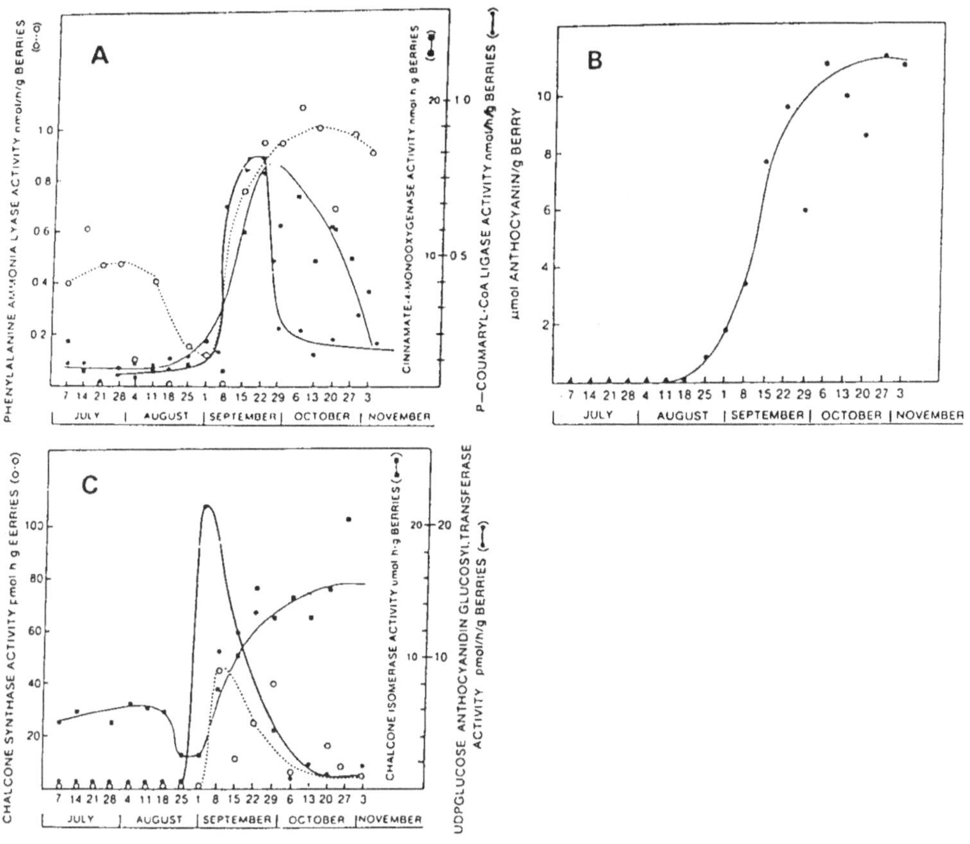

FIGURE 59. Relation between enzyme activities (A, C) and anthocyanin accumulation (B) during development and maturation of cv. De Chaunac grape berries. (From Hrazdina, G., Parsons, G. F., and Mattick, L. R., *Am. J. Enol. Vitic.*, 35, 220, 1984. With permission.)

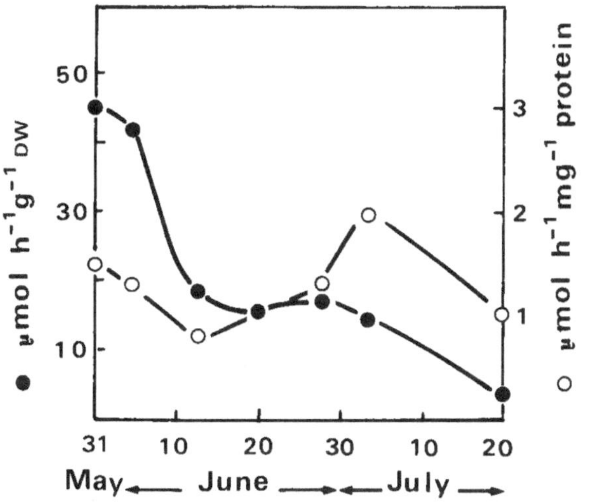

FIGURE 60. Changes in *o*-diphenoloxidase (cresolase activity) during apple growth. (Adapted from Macheix, J. J., Thèse Doct. Etat, Université Pierre et Marie Curie, Paris, 1974.)

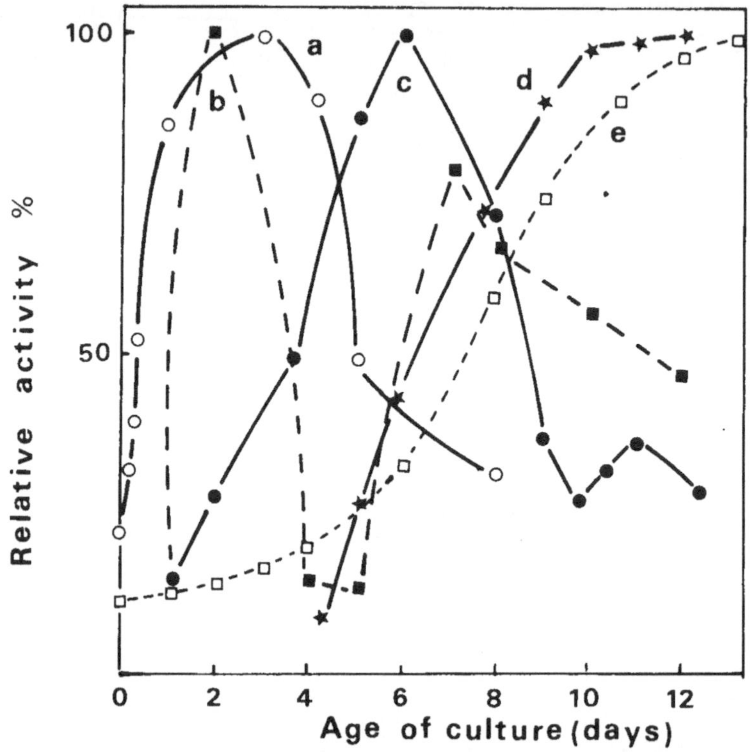

FIGURE 61. Changes in enzyme activities during growth of apple cell culture. (Relative values in percent maximum). Phenylalanine ammonia lyase (a); cinnamate 4-hydroxylase (b); *o*-methyltransferase (c); glucosyltransferase (d); fresh weight (e). (Adapted from Macheix, J. J., Suen, R., and Ibrahim, R. K., *Biochem. Physiol. Pflanzen.*, 176, 195, 1981.

GT activities become much greater in the pericarp than in the pulp, whereas the opposite applies to OMT and CoAL activities, which are greatest in the pulp.[116] This distribution in time is fully correlated with the phenolic contents in the two parts of the fruit: quinic derivatives accumulate in the pulp during growth and glucose derivatives accumulate in the pericarp during maturation. In peach,[7] PAL activity is very high in the core during growth and then in peel and pulp of reddish mature fruit. In this case, PAL activity is correlated with lignification of the stone during growth and synthesis of anthocyanins during maturation.

IV. REGULATION OF PHENOLIC METABOLISM IN FRUITS

Variations of the phenolic metabolism in relation to physiological stages depend directly on protein synthesis. However, the secondary metabolism in general, and in particular the phenolic metabolism, is also greatly dependent on a great number of factors, both external (light, temperature, etc.) and internal (hormones, nutrients, etc.). Knowledge of the role of these factors in the regulation of the phenolic metabolism is particularly interesting insofar as they can be used for practical applications to increase or decrease the synthesis of certain groups of phenolic compounds in fruits.

A. LIGHT AND PHENOLIC COMPOUNDS

Light is the environmental factor whose effect on the phenolic metabolism has been most studied. Indeed, field observations have long underlined the importance of solar radiation with regard to accumulation of anthocyanins, in particular in fruits. In addition,

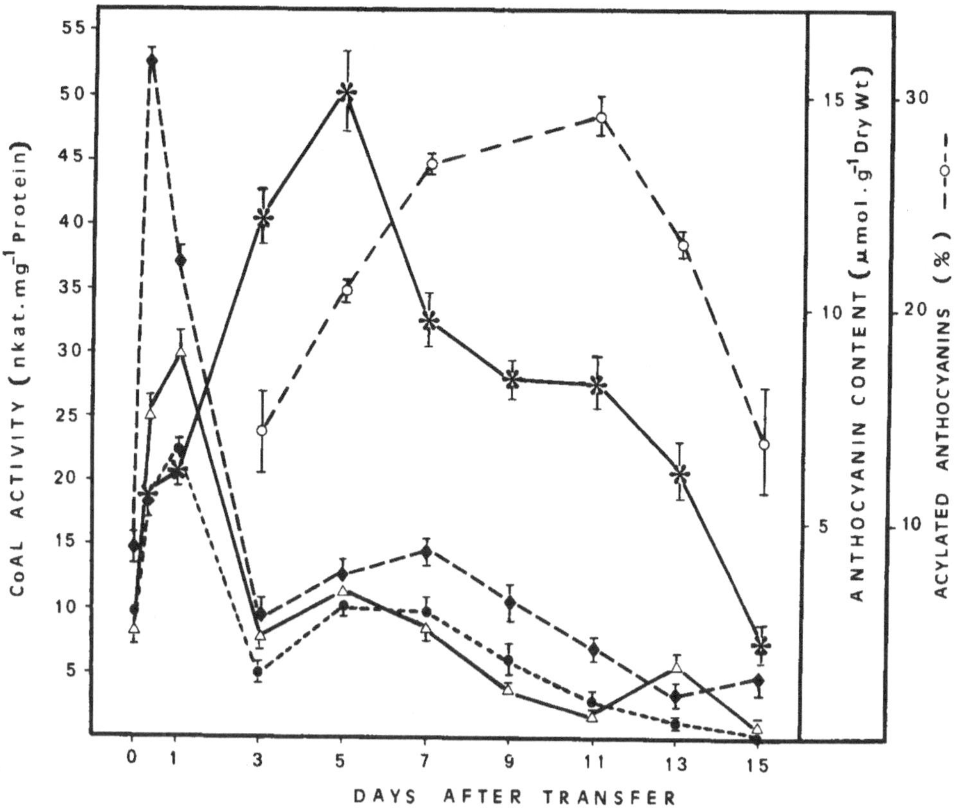

FIGURE 62. Changes in anthocyanin level (★), acylated anthocyanins (O), and coenzyme A ligase activity with *p*-coumaric (♦), ferulic (△), and caffeic (●) acids during growth of grape cell culture (From Lotfy, S., Fleuriet, A., Ramos, T., and Macheix, J. J., *Plant Cell Rep.*, 8, 93, 1989. With permission.)

numerous experiments have made it possible to specify its role in regulation of the phenolic metabolism. It can be involved, on the one hand, through the intensity of light and on the other, through the photoperiod and the wavelengths involved, numerous enzymes being controlled by these different parameters.[457] Among them, PAL has been the subject of exhaustive studies[50,386] since it is generally induced by the change from darkness to light under the control of phytochrome or possibly of other photoreceptors.[255] Detailed studies of plant cell suspensions have made it possible to describe the effect of light on numerous enzymes of the phenolic metabolism.[152,154] They have revealed two groups of enzymes with coordinated induction, one comprising the enzymes of the general phenylpropanoid metabolism (including PAL and hydroxycinnamoyl CoA ligases) and the other the enzymes of the flavonoid metabolism. In addition, light may play a direct regulatory role in cell growth by acting on the stereochemistry of ferulic acid bound to polysaccharides and causing changes in the structure and properties of the cell wall.[420]

Numerous data are available on the relations between light and phenolic compounds in fruits; they essentially concern the anthocyanin group and, to a markedly lesser extent, flavonol glycosides and hydroxycinnamic derivatives. They are derived initially from observation of natural phenomena, such as the influence of climate or more or less direct exposure of fruits to sunlight (position of the fruit in relation to foliage, sides facing south or north, etc.). There is interference with other factors in all cases — temperature in particular — and it is not always easy to identify the features which can be attributed to light alone.

In addition, all these factors can have an effect on the rate of maturation of fruit and

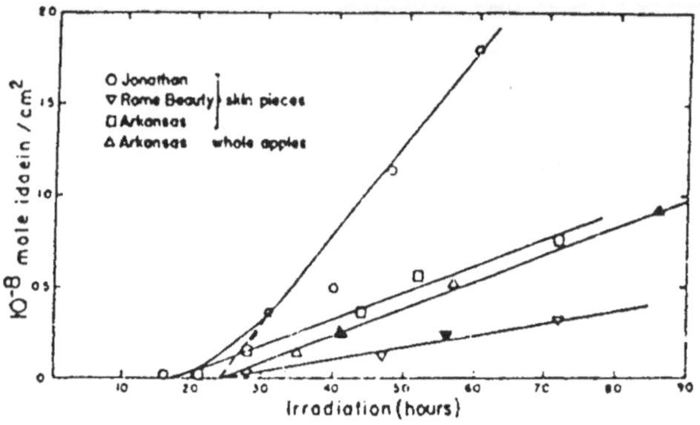

FIGURE 63. The dependence of anthocyanin formation in apple skin on the time of exposure under constant irradiance with a fluorescent light source equivalent in photochemical effectiveness to 0.6 mwatt/ cm^2 at 7000 Å. (From Siegelman, H. W. and Hendriks, S. B., *Plant Physiol.*, 33, 185, 1958. With permission.)

the accumulation of anthocyanins, which is often used as a maturity index in many red fruits, may only be the indirect consequence of the action of these factors on maturation rather than a result of direct action on the anthocyanin metabolism. Here, comparison of the response obtained in fruits and in other parts of plants (leaves, stem, seedlings, etc.) should make it possible to distinguish between the two phenomena.[308]

In fruits, the effect of light on the phenolic metabolism has been monitored using three types of experimental approaches. The first concerns whole fruits, detached from the mother plant or not, and on which the effect of natural solar light and its qualitative and quantitative modifications are studied. The second consists of experiments carried out on skin disks cut from fruits and kept under aging conditions. This type of approach is very commonly used since it is easy to study the effect of light in combination with that of other factors (temperature, chemical stimulators or inhibitors, etc.) and compare the data with results obtained for whole fruits. However, it also involves a response of the plant sample to wounding, to which the phenolic metabolism is very sensitive;[339] in apple, for example, this can lead to different regulation of the accumulation of anthocyanins by PAL.[52,410] Finally, *in vitro* cultures of fruits, which are not yet sufficiently developed, nevertheless form good models for study of the effect of light and other factors on the phenolic metabolism.[38,223,418,435]

In **apple fruit**, anthocyanin biosynthesis is highly dependent on the light conditions, as shown by many old observations.[52,411] The accumulation level is, however, principally controlled by genetic factors and can thus vary considerably according to cultivar (Figure 63).[380] Accumulation of anthocyanins in fruits kept on the tree depends both of the time of application of light treatment[380] and on light intensity. For example, for the cv. Red Delicious, the best coloration is obtained with 70 to 100% of full solar radiation, an adequate color with between 40 and 70% and inadequate color with less than 40%.[175] Supplementary artificial light in addition to natural light made it possible to determine the conditions of maximum efficiency with minimum energy requirement for anthocyanin.[320,321] A good correlation between anthocyanin synthesized and added incident energy was found (Figure 64), except when the low temperature necessary for anthocyanin synthesis did not occur. The anthocyanin content of apples at harvest was correlated with their potential for the production of anthocyanin during the growing season. It was possible to obtain maximum response after brief treatments no longer than 20 d.[322]

Synergic effects have been revealed on the accumulation of anthocyanins between the visible part of the spectrum (maximum efficiency being towards 650 nm)[380] and UV.[8,52] Anthocyanin production increased with fluence rates and the combination of UV (max 312

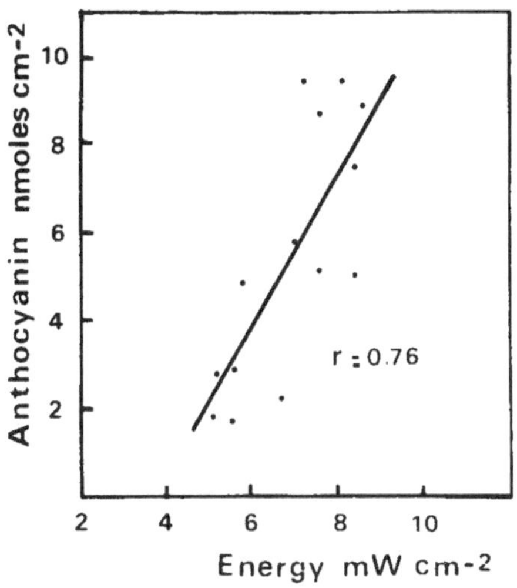

FIGURE 64. Correlation between energy levels for 48 h (August 1 to 3) and anthocyanin production in Richared Delicious apples. (Adapted from Proctor, J. T. A., *Can. J. Plant Sci.*, 54, 499, 1974.)

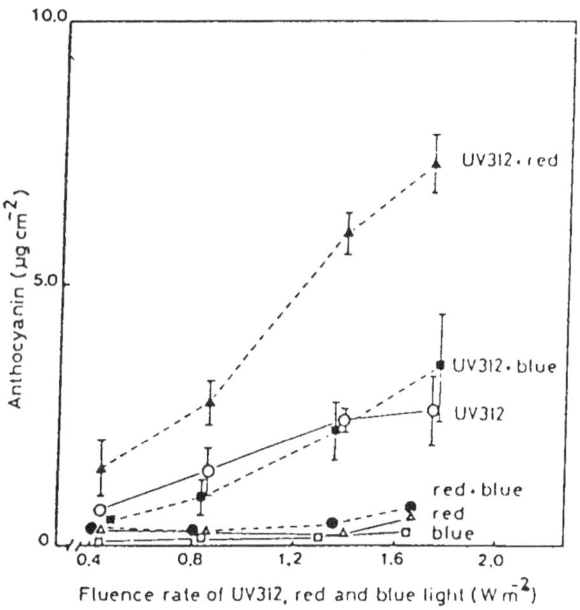

FIGURE 65. Effect of fluence rate on anthocyanin production in apple skin disks under monochromatic irradiation with red, blue, UV light, and simultaneous irradiation with two kinds of lights. Anthocyanin was assayed after 72 h light irradiation (From Arakawa, O., Hori, Y., and Ogata, R., *Physiol. Plant.*, 64, 323, 1985. With permission.)

nm) and red light led to much greater accumulation than each of the two treatments applied individually (Figure 65). In this case, blue light is not very effective alone or combined with red or UV light. As in many plants, synthesis of anthocyanins in apple skin displayed the feature of a high irradiance reaction (HIR).[77,380,457] Phytochrome is involved in the photoregulation of HIR and anthocyanin production.[255] However, in some species, it is

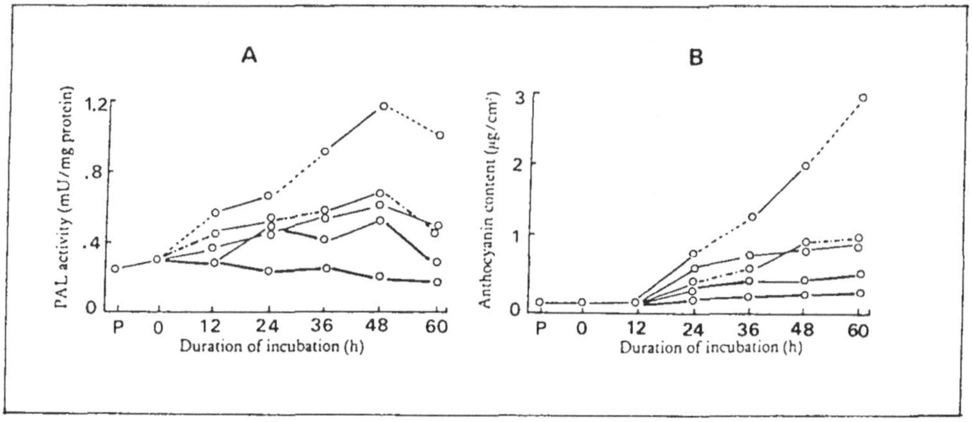

FIGURE 66. Effects of light and temperature on PAL activity (A) and anthocyanin level (B) in the skin of Red Spy whole apples. (—) light 25°C; (---) light 6°C; (—) Dark 25°C; (-·-·-·) Dark 6°C. P, preincubation. (Adapted from Tan, S. C., *Aust. J. Plant Physiol.*, 7, 159, 1980.)

possible that another photoreceptor may be involved which is specific for the UV-blue region and which is known as cryptochrome.[392] In any case, the final expression of cryptochrome-mediated response to blue or UV light depends on the state of the phytochrome system.[275] Theoretical models of phytochrome dynamics have been developed to explain the HIR in terms of interactions between photochemical and nonphotochemical reaction of phytochrome. However, experimental results and model predictions were not in complete agreement.[325]

PAL is believed to be the main limiting factor in the synthesis of flavonoids and anthocyanins.[50,52,53,101,457] Good correlation was observed in Red Spy apple between PAL activity (varying according to the two main parameters light and temperature) and accumulation of anthocyanins.[411] In fact, PAL in apple skin may be regulated by a PAL-inactivating system (PAL-IS).[410] In apple skin,[99,411] PAL was induced by exposure to light after a lag period of about 10 h, reached a maximum at between 30 and 40 h, and then declined (Figure 66). When the whole apples were kept in constant darkness, no accumulation of PAL or of anthocyanins occurred, but the level of PAL-IS increased steadily throughout incubation.[411] In contrast, in constant light, thermoperiodism (12 h at 6°C and 12 h at 25°C) gave maximum accumulation of anthocyanins correlated with the increase in PAL activity (Figure 67) and reduction in PAL-IS. Under the effect of light and temperature, regulation of anthocyanin accumulation thus appears to take place at two levels, that of PAL and PAL-IS.

Many fruits, other than apple, display relations between anthocyanin accumulation and light. In **grape**, the level of solar radiation has long been known to be an important parameter in red coloration of berries.[213,214] Reduction of this radiation results in a fall in the anthocyanin content (Table 47), whereas total soluble solids are very little affected. Nevertheless, the effect of low light (9.5% sunlight) is much less marked when the average temperature is low[214]. In addition, although the difference in anthocyanins and total phenolics between sun-exposed and shaded berries is considerable during the growing season, it diminishes strongly at harvest.[63] As in apple, it would appear that one of the major regulation features in grape is PAL activity, which occurs mainly in berry skins.[183,357] When whole isolated berries are exposed to light for 72 h, PAL activity and anthocyanins increase whereas in the dark PAL activity decreases and no accumulation of anthocyanin is observed.[357]

In **pear**, chlorogenic acid levels were some four times greater in tissues from the shaded side of the fruit in comparison with the light-exposed side.[440] In fruits affected by cork-spot, a physiological disorder which appears as the fruit approaches maturity, the differences

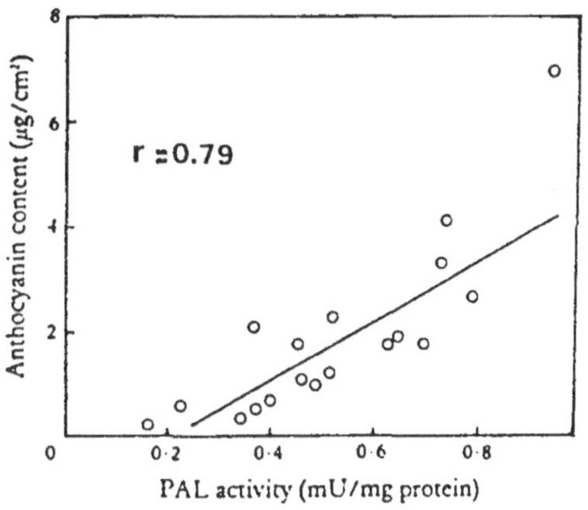

FIGURE 67. Relationship between PAL activity and anthocyanin content in the skin of Red Spy whole apples. (From Tan, S. C., *Aust. J. Plant Physiol.*, 7, 159, 1980. With permission.)

TABLE 47
Influence of Level of Solar Radiation on Coloration of Emperor Grapes[a]

Level of solar radiation (% of full sunlight)	Total soluble solids (°Brix)		Fruit coloration (mg anthocyanin/cm² berry surface)	
	Sept. 2	Sept. 14	Sept. 2	Sept. 14
100	17.42	19.85a	0.32a	0.31a
54	17.37a	19.55a	0.34a	0.33a
15	16.50ab	17.90b	0.17b	0.17b
3	16.62ab	17.67b	0.13c	0.13c
0.8	15.75b	17.15b	0.09c	0.11c
0	15.60b	16.95b	0 d	0 d

[a] Within a column, means followed by the same letter are not significantly different at the 5% level by Duncan's multiple-range test.

From Kliewer, W. M., *Am. J. Enol. Vitic.*, 28, 96, 1977. With permission.

between the two faces were even more marked (ten times as much chlorogenic acid in the cork spot tissue of the exposed face as in the shaded side). Tissue browning on the exposed side of the fruit may thus be caused by direct exposure to sunlight of sensitive cultivars of pear.

In **cranberries**, the anthocyanin content reflected differences in environmental factors and in particular light exposure combined with genetic differences and differences in ripeness.[367] However, variability in simple anthocyanin content was more dependent on differences in berry size that on differences in surface coloration. Anthocyanin contents of **crowberries** also varied in relation to weather conditions. The differences were particularly marked from one year to the next according to overall radiation and the number of hours of sunshine.[236] Southern crowberries contained more anthocyanins than northern ones (on a per milligram FW basis). However, values were almost identical in terms of DW since southern crowberries contained more dry matter.

In **fig**,[324] ethephon treatments had an effect on both fruit growth and accumulation of

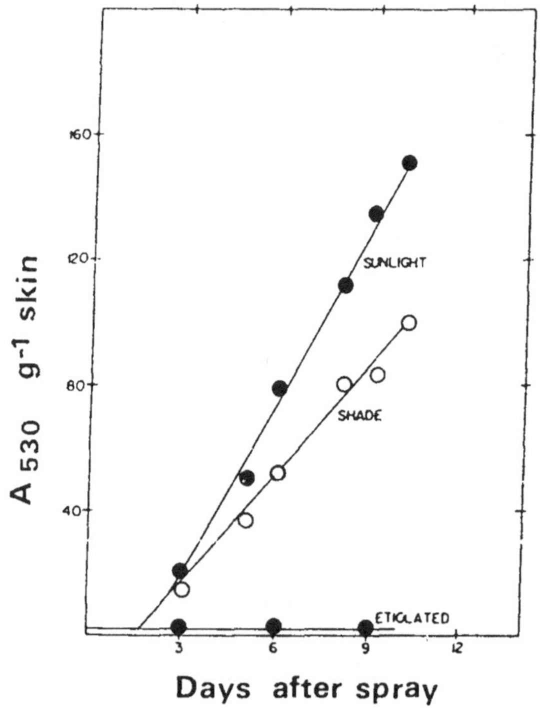

FIGURE 68. Anthocyanin synthesis in the skin of Ethephon-treated Mission fig fruits. (From Puech, A. A., Rebeiz, C. A., and Crane, J. C., *Plant Physiol.*, 57, 504, 1976. With permission.)

anthocyanins, and both of these features were strongly affected by light (Figure 68). Treatment with ethephon in light increased the rate of anthocyanin accumulation, but a smaller amount was finally accumulated by the fruit. Growth of ethephon-treated fruits was strongly stimulated in the dark, but the anthocyanin level remained near zero.

Insufficient accumulation of anthocyanins in **peach** is one of the important problems raised by high-density planting systems which increase the shading of developing fruit.[97] As the degree of shading increased, red pigment development decreased and a high correlation coefficient between light and peach skin color was found using a power equation and not a linear one.[97] Temporary exposure at the beginning of the color-change period is almost as effective as exposure just before picking as response can be rapid after only 8 d of light treatment.

Treatment of **banana** by UV irradiation can lead to reddening of the skin during storage. It would appear that the pigments concerned are 3-deoxyanthocyanidins.[381]

The phenolic metabolism of *in vitro* cultures derived from fruits is also very sensitive to light. Thus, in strawberry, growth and coloration of the fruit are regulated by light:[11] high intensities result in red coloring whereas the fruit remains pink in diffuse light. In callus cultures of orange and lemon flavedo, profuse growth, absence of cell differentiation, and a more or less constant PAL level occur in darkness.[38] In light, however, there is active cell differentiation, tissue differentiation and extensive lignification. An eightfold increase in PAL activity occurs after 24 h of illumination; this then leads to a maximum at 48 h and then rapid decrease. Simultaneously, the characteristic methoxyflavones of this plant material accumulate. In certain cases, light may inhibit the accumulation of polyphenols in *in vitro* culture,[189] but no example of this type has been reported for fruits.

In conclusion, it would thus seem that light is a determinant factor in the accumulation of most phenolic compounds in fruits, with regulation being, on the one hand, by the contribution of carbon chains derived directly from photosynthesis and, on the other, in a more specific manner by means of phenolic metabolism enzymes and PAL in particular. The level of subcellular differentiation induced by light may also be a determinant factor insofar as the phenolic metabolism is at least partially associated with membrane systems.[271,437] In all cases, responses are first under genetic control but can be considerably modulated in quality and quantity by light and all the other environmental factors.

B. TEMPERATURE

Two major types of observation reveal relations between phenolic compounds in fruits and temperature. First of all, data links the overall effects of climate and local environmental conditions with the accumulation of phenols, frequently shown by red color caused by anthocyanins. Interpretation of this type of result is always delicate since there are usually multiple variation factors (temperature, light, humidity, etc.). However, these first observations led to experiments under more strictly controlled conditions in which temperature was the only variable parameter. Another group of results derived from the frequent use of low temperatures in postharvest storage of fruits. In this case, physiological disturbances may occur even when the fruit is kept at temperatures above 0°C. Such effects are generally referred to as chilling injury and can have many causes;[18,239,333] they frequently take the form of discoloration for which phenolic compounds may be directly responsible (Chapter 5). Indeed, several examples show large variations in the level of phenolic compounds or of the enzymes of their metabolism during cold storage of fruits.

The effect of temperature on plant physiology and metabolism is seen at very varied levels, ranging from membrane scale to that of the individual as a whole.[233] One of the very general actions is the regulatory role of temperature on the rate of chemical reactions, resulting in slowing of the metabolism when the temperature falls. In these terms, the phenolic metabolism can be affected by cold, but this is only an indirect result caused in particular by a slowing of natural ripening. In contrast, some examples reveal much more specific action, indicating an original response of the phenolic metabolism to temperature variations; the latter can be seen as the direct cause of the modifications observed. All these data are reported and discussed here.

A favorable effect of low temperatures on anthocyanin accumulation is generally noted in field observations, as has long been reported in apple,[61,252] grape,[213] and many other fruits, in particular those which exceptionally accumulate anthocyanins, such as the "blood" orange in which a deep red coloration is best attained in a cool climate.[276] Increase in anthocyanin in the skin of **apple** coincides with decreasing temperature and ethylene production.[61,99] The fact that the optimum temperature for maximum rate of accumulation of anthocyanins varies with the physiological stage of the fruit is particularly interesting. It is 12°C in unripe apple but varies between 16 and 24°C in ripe fruit (Figure 69). These observations have been confirmed in several red apple cultivars remaining on the tree or picked and subjected to thermoperiodism combined with day/night (D/N) alternance.[74] Optimum alternance was 12°C/2°C for unripe fruit but shifted to 24°C/2°C or, better still, 24°C/24°C in ripe fruit. These results lead to envisaging the possibility of finding a suitable combination of temperatures for each stage of development of the fruit.

The effect of temperature on fruits during development on the plant has received particular attention in **grape**. The phenolic content of grapes grown in regions with a cool climate and a short growing season is generally higher than in other regions.[230] In warm regions, such as California, sprinkling treatments of vines from bloom to fruit maturity greatly increased the coloration of cv. 'Cardinal'.[215] However, the response depends to a considerable extent on genetic control; cv. Carignane displayed only very small increase in comparison with nonsprinkled fruits. Temperatures of about 15°C (D/N) lead to better coloration of several

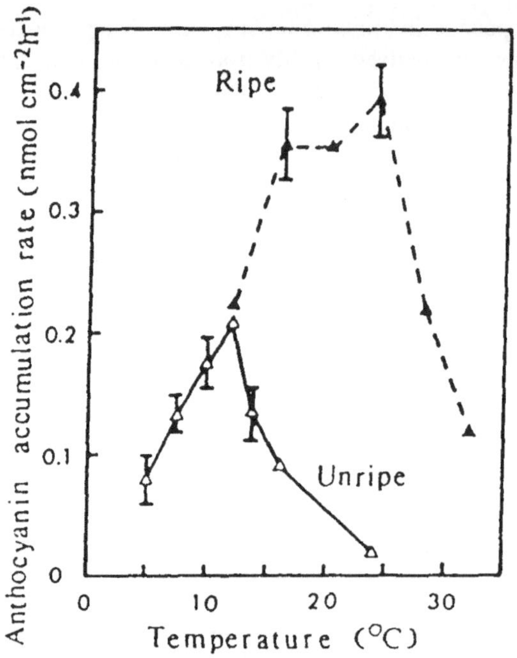

FIGURE 69. Effects of temperature and ripening on the rate of anthocyanin accumulation in detached unripe and ripe apples. (From Faragher, J. D., *J. Exp. Bot.*, 34, 1291, 1983. With permission.)

cultivars of *Vitis vinifera* in comparison with situations in which the temperature is higher (e.g., 35°C D/15°C N or 35°C D/25°C N or 15°C D/25°C N).[49] Spectacular effects can be obtained in certain sensitive varieties such as 'Tokay'[216] or 'Emperor'[214] where temperatures of over 30°C (e.g., at D/N temperatures of 35/30°C or 37/32°C) lead to complete blockage of anthocyanin synthesis. In contrast, at field temperature (average 20°C), anthocyanins are formed in high light or low light conditions.[214] The influence of thermoperiodism does not always seem to be very determinant: although greatest fruit coloration is frequently observed under day temperatures of 15 to 25°C and night temperatures of 10 to 20°C,[213,216] continuous treatment at 15°C can also be very effective in apple[74,99] or grape.[216] In the latter case, it has been shown that coloration diminishes when the difference between day and night temperatures is greater than 10°C.

The influence of interactions between temperature and other factors has been monitored in whole grape berries isolated from the mother-plant and fed through the stalk.[357] Surprisingly, in the presence of light and after treatment with ethephon, high temperature (35°C) stimulated biosynthesis of anthocyanins more effectively than low and medium temperatures (14 and 22°C). However, the effect was only studied over a short period (72 h) which might account for the results obtained.

The limiting of anthocyanin synthesis by high temperatures can be related to the accumulation of soluble sugars in the fruit. However, results are generally fairly contradictory. Whereas there may be a good correlation between the two parameters in certain grape cultivars,[214] other cases show that the total soluble solids in the juice of the fruit was highest in the absence of coloration.[216] In addition, supplying sucrose to grape berries kept at 35°C under light resulted in only a very small increase in anthocyanin level.[357] It is thus probable that the soluble sugar content is a limiting factor for anthocyanin accumulation at high temperatures. However, this is not the prime determinant, which should certainly be sought at a much more specific level in the phenolic metabolism.

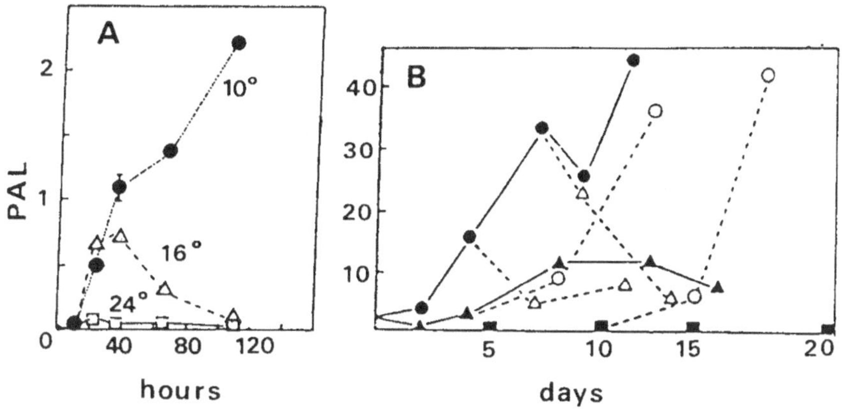

FIGURE 70. Effects of temperature on PAL activity in unripe apple skin (A) and "Sordum" plum (B), A: PAL in pKat cm^{-2}. (Adapted from Faragher, J. D., *J. Exp. Bot.*, 34, 1291, 1983.) B: PAL in nmol/2h/g FW; 3°C (■), 20°C (●), 30°C (▲); 3°C → 20°C (■----○), 30°C →20°C (▲----○), 20°C → 30°C (●----△). (Adapted from Tsuji, M., Harakawa, M., and Komiyama, Y., *J. Jpn. Soc. Food Sci. Technol.*, 30, 688, 1983.)

An important point in the effect of temperature may be the regulation of PAL activity. In fact, temperature may have a twofold effect leading to the final accumulation of anthocyanins: first, stimulation or reduction of PAL activity itself (Figure 70A) and second, modification of the level of PAL inactivators and PAL-IS in particular[99,410,411] (Figure 71). There is thus a very good correlation in apple between anthocyanin levels and PAL activity level when the latter is modified by different combinations of heat and light treatment (Figure 67). Likewise, in plums, numerous anomalies can occur in fruit coloration if the temperature is too high during growth and maturation.[150] The inhibition of increase in pulp color at 30°C in *Prunus salicina* may be caused by the decrease in PAL activity (Figure 70B).[424] In addition, keeping the temperature at 3°C blocks both PAL activity and accumulation of anthocyanins.

The negative effect of high temperatures may lead to stimulation of the synthesis of PAL inhibitors rather than acting directly on PAL itself.[99] Indeed, PAL inactivation did not occur or occurred to a lesser extent at low temperature (alternating 6/20°C) than at high temperature (constant 25°C).[411]

The action of various RNA and protein synthesis inhibitors on PAL activity or phenolic biosynthesis has been reported by numerous authors in different plant tissues (see Reference 410). It has been stated that both PAL synthesis and inactivation were sensitive to cycloheximide, with inactivation being more sensitive than synthesis.[62,410] Effectively, in whole apple, cycloheximide increased the accumulation of PAL and anthocyanin, reduced the level of PAL inactivators, and inhibited the loss of activity after transfer to darkness.[99,101,410] Regulation of the anthocyanin metabolism in apple may thus operate through the synthesis of PAL inactivators. Nevertheless, apple skin disks responded differently, with cycloheximide inhibiting PAL biosynthesis and the accumulation of anthocyanins.[410] The same occurred in grape.[357] The differences observed may stem from characteristics related to the different experimental systems (whole or wounded fruits) which probably caused very great differences in the penetration of the substances supplied. In addition, it is now well known that the action of these antibiotics is not always very specific. It has also been proposed that in apples the rise in anthocyanin content during ripening is more connected with the rise in ethylene content than with the fall in temperature.[100] This parameter may thus only have an indirect effect through its role in the biosynthesis of ethylene, although previous data[99] showed that PAL and ethylene levels varied in opposite directions in response to a change

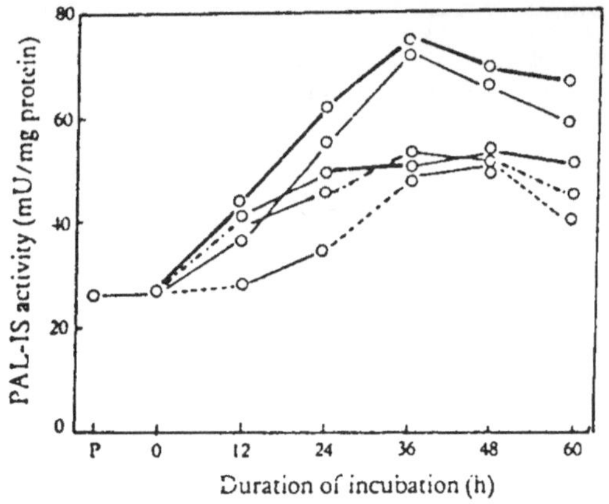

FIGURE 71. Effects of light and temperature on the activity of PAL-IS in the skin of Red Spy whole apples. Symbols are as in Figure 66. (From Tan, S. C., *Aust. J. Plant Physiol.*, 7, 159, 1980. With permission.)

in temperature. The relations of cause and effect between the two factors remain to be clarified.

Cold storage of fruits has given very interesting data on relations with the phenolic metabolism. One of the most important sets of data concerns the response of **tomato**, a fruit which is particularly sensitive to chilling injury, during cold storage.[338,340] Physiological disorders occur above freezing point but below a certain critical temperature threshold (about 12°C). Temperature-dependent phase changes in the cell membranes lead to changes in membrane permeability and activity of membrane-bound enzymes[239] causing accumulation of toxic intermediates. In tomato, very marked, fairly specific action was observed on the phenolic metabolism and in particular on the chlorogenic acid metabolism. Among the enzymes tested, levels of PAL and HQT (hydroxycinnamoyl-quinate transferase, an enzyme which enables synthesis of chlorogenic acid) increased considerably in low-temperature storage (Figure 72); maximum levels were obtained at 2 and 5°C. The effect was still very marked at 10°C but nil at 15 and 20°C. The other enzymes of the phenylpropanoid metabolism (*p*-coumarate CoAL, cinnamate-4-hydroxylase, coniferylalcohol NADP oxidoreductase) are little affected by cold treatment. Nevertheless, their activity remains higher at 2°C than at 20°C. This is probably related to the slowing of ripening in cold storage. Transfers from 2 to 20°C during treatment result, after a lag period of 8 h, to a substantial fall in HQT activity, leading after 48 h in a situation identical to that of controls kept at 20°C. In tomato, there is thus a very specific response of chlorogenic acid to cold; this is independent of ripening as the response is obtained at any stage of maturation with, however, a maximum at the "breaker" and orange stages (Figure 72). The action of cycloheximide and certain RNA synthesis inhibitors on tomato plugs suggests that the increase in activity may result from both *de novo* synthesis of HQT and the activation of preexisting forms under the effect of cold.[340] However, responses are not always very distinct since there is a wounding effect in addition to that of cold.[339]

Whereas the effects of cold on the phenolic metabolism are particularly marked in tomato during cold storage, distinctness varies in other species. In **apples**, there is either a steady fall in chlorogenic acid content during storage at 0.5°C in air or at 1.5°C in a controlled atmosphere in cv. Golden Delicious[28] or the maintaining of a fairly constant level in Calville blanc or Starking Delicious.[6,240] However, the ability to synthesize anthocyanins during conservation under controlled atmosphere and at low temperature falls strongly during storage,[22,74] whereas it does not change under the action of cold alone. Quercetin glycosides

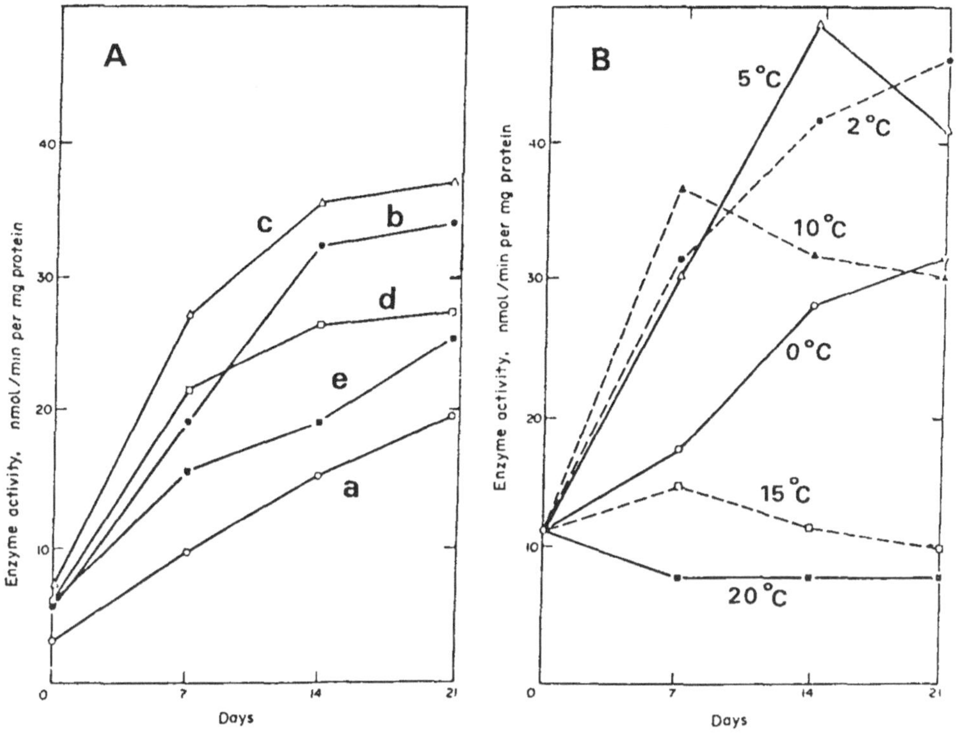

FIGURE 72. Time courses of changes in hydroxycinnamoyl-transferase activity during storage of tomato. A: storage at 2°C of tomatoes harvested at the mature-green (a), breaker (b), orange (c), pink (d), full red (e) stages of ripeness. B: breaker tomatoes stored for up to 21 d at different temperatures. (From Rhodes, M. J. C. and Wooltorton, L. S. C., *Phytochemistry*, 15, 947, 1976. With permission.)

increase significantly during the cold storage of cv. McIntosh and Golden Delicious. This can be related to the role that these compounds may play in the regulation of the firmness of ripening fruit.[73]

In the case of Anjou **pear**, as in many other varieties of pear, cold storage is necessary for satisfactory development of the maturation process. Nevertheless, during storage at 1°C, chlorogenic acid and total phenol levels increase in both fruit skin[441] (Figure 73) and flesh.[23] In both cases, these increases can be related to alteration of the fruit. Thus, catechin and chlorogenic acid levels increased by 50 and 30% when stored fruit displayed bruising,[23] a physiological disorder which occurs during conservation. These compounds may serve as substrates for oxidation enzymes and accelerate the oxidation of flavonols. Certain phenolic compounds (*p*-coumaric derivatives, (−)-epicatechin) only appear during conservation and may indicate evolution of this fruit towards senescence[23]. Increase in the chlorogenic acid content of pear skin leads to much greater sensitivity to friction discoloration although other factors are involved.[441] In contrast, in certain other pear cultivars (e.g., Passe Crassane), there is no significant increase in the level of *o*-diphenols or chlorogenic acid during conservation.[121] Indeed, the levels fall after several months of cold storage while characteristic symptoms of internal browning occur. However, evolution is more marked in controlled atmospheres when no browning is observed. In *Vitis rotundifolia* berries (cv. Muscadine or cv. Fry) conserved at 4.5 or 0°C for 24 d,[409] cold slows the metabolism as a whole and in particular that of pectins, although there is a distinct increase in phenolic compounds in berry pulp and skin in comparison with controls kept at 20°C (Figure 74). This evolution was correlated with development of decay and skin browning.

In **apricots** conserved at 5°C for 10 d,[428] enzymatic browning decreased during storage

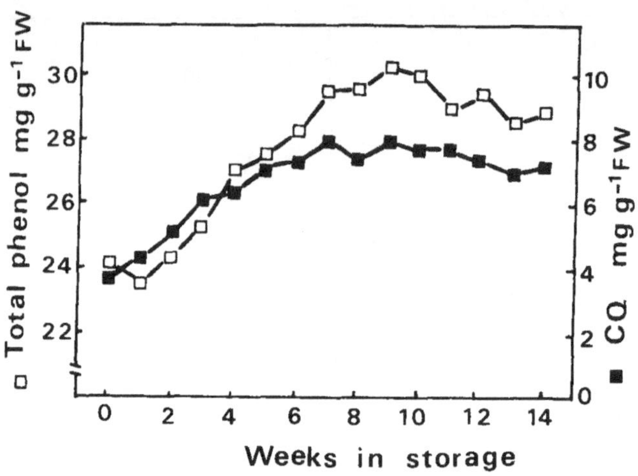

FIGURE 73. Changes in the level of total phenol and chlorogenic acid (CQ) in "d'Anjou" pear skin during storage. (Adapted from Wang, C. Y. and Mellenthin, W. M., *HortScience*, 8, 321, 1973.)

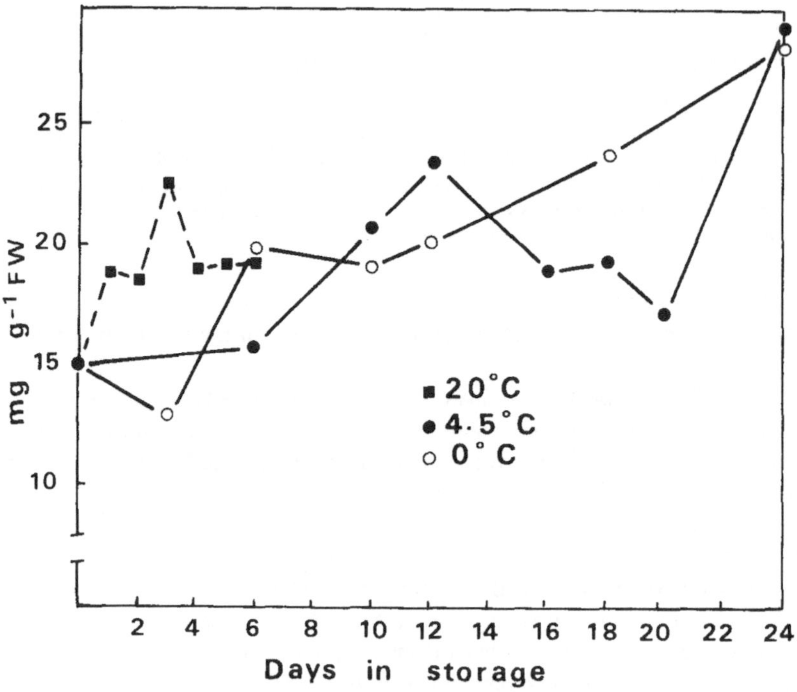

FIGURE 74. Changes in the phenol content in the skin of Fry muscadine grapes during storage at three different temperatures. (Adapted from Takeda, F., Saunders, M. S., and Saunders, J. A., *Am. J. Enol. Vitic.*, 34, 180, 1983.)

while peroxidase increased, although there was irregular variation of polyphenoloxidase and *o*-diphenols with considerable difference according to cultivar. It did not appear to be possible to identify a specific action of cold in this case.

Pineapple is a very sensitive fruit at storage temperatures of around 8°C. When it is returned to 20°C it frequently develops internal browning which renders it unfit for consumption.[414] Numerous agroclimatic parameters can modify the sensitivity of the fruit, but

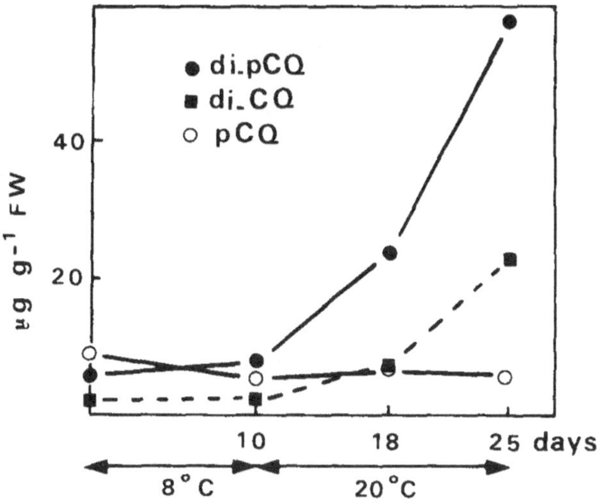

FIGURE 75. Changes in quinic derivatives during storage of pineapple. di-pCQ: di*p*-coumaroylquinic acid; di-CQ: dicaffeoylquinic acid; *p*CQ: *p*-coumaroyl quinic acid. (Adapted from Dieudonné, G., Thèse Doct. Ing., Université Pierre et Marie Curie, Paris, 1977.)

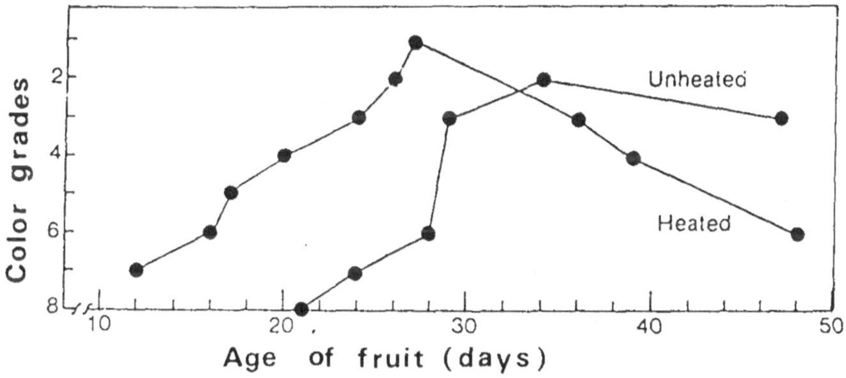

FIGURE 76. Effects of air temperature on color of eggplant during the cool season. (Adapted from Nothmann, J., Rylski, I., and Spigelman, M., *Expl. Agric.*, 16, 189, 1978.)

browning intensity is generally correlated with activity of peroxidases and phenolases[413] (Chapter 5). Nevertheless, remarkable evolution of certain endogenous phenolic compounds are observed in cv. Cayenne Lisse during storage at 8°C, and above all when temperature is returned to 20°C.[75] While most of the compounds do not vary, two *p*-coumaric and caffeic derivatives which have not been fully identified accumulate strongly; their levels are multiplied tenfold 15 d after the return of fruits to 20°C (Figure 75). Correlatively, internal browning may appear during this period with intensity varying between fruits.

Eggplant is a highly thermophilic plant commonly grown as a summer crop. However, in regions where local climatic conditions enable production all the year round, the fruits which form in winter display slow growth and anthocyanic coloration which is always inadequate.[292] In this case, winter temperatures are probably not specific with regard to the phenolic metabolism; air or soil heating during the cool season leads to improved growth rate and fruit color (Figure 76).

In conclusion, temperature appears to be an important factor in the expression of the phenolic metabolism of a fruit either through direct, more or less specific action on certain enzymes of this metabolism, or indirectly through the overall metabolism and fruit growth and maturation. Nevertheless, with a few noteworthy exceptions, few studies have been devoted to seeking the mechanisms by which this factor operates. In particular, in the very frequent case in which the consequences are observed in anthocyanin accumulation, the only regulation feature which has been approached concerns PAL whose activity also governs orientation towards all the other classes of phenolic compounds which accumulate in fruits. What role might cold play with regard to each of the phenolic metabolism enzymes which might change the orientation of this metabolism according to the method of application and the reactivity of the species studied? Some answers have been provided in the case of tomato for the chlorogenic acid metabolism[338] and in grape for the more or less substituted or acylated anthocyanins,[214] suggesting that anthocyanidins lacking an OH or OCH_3 group in the 3′ position of the B ring may be those most sensitive to unfavorable light and temperature conditions. However, these data remain to be broadly developed for different fruits and under varied experimental conditions. Finally, temperature is an interesting factor insofar as it would seem possible to start at a medium temperature (of the order of 15 to 20°C for temperate fruits) and modulate the phenolic metabolism either by raising or lowering this temperature. The responses depend on both the species in question and the metabolites which have accumulated.

Not considered in this chapter are the consequences of freezing temperatures on the phenolic metabolism; insofar as cell structures are destroyed, a return to temperatures above 0°C result in the appearance of characteristic browning in which phenolic compounds are involved. Several aspects concerning the consequences of the freezing conditions used for the conservation of fruits and vegetables with regard to browning are discussed in Chapter 5.

C. IRRADIATION

Ionizing radiation — mainly cobalt 60 γ-radiation — is used in the treatment of plant organs, including fruits, to improve conservation. The use of irradiation in the conservation of foodstuffs makes use of the bactericidal effect of radiation, but the doses required sometimes have unpleasant effects on flavor or color. The effect of γ-irradiation on storage ability has been tested on different fruits. It slows the processes of maturation and senescence, resulting in an increase in the duration of conservation.[276a] However, data on phenolic compounds are fairly scarce,[276a] with the fullest work carried out on *Citrus* and mainly on **grapefruit.**[204,345-351]

Treatment causes considerable increase in PAL activity in the flavedo of grapefruit and Shamouti orange.[348] There is also an increase in total phenol content. In mature Shamouti orange, for example, PAL activity, which is very weak in nonirradiated fruits, peaks after 32 h and remains at a high level for about 6 d (Figure 77A). The curve of phenolic accumulation ran more or less parallel to that of PAL activity in irradiated tissues (Figure 77A). Ionizing radiation also stimulates biosynthesis of two coumarins, scopoletin and scopolin (cf., Table 12) in grapefruit peel.[351] These two compounds, which are found in only small quantities in untreated fruits, accumulate strongly in 7 d (Figure 77B). Nine days after 300 krad irradiation, the scopoletin content increased from 8 to 177 μg/g FW. In addition, another compound, 6,7-dimethoxycoumarin, which is not present in nonirradiated fruits, appears after fairly heavy doses of radiation (300 and 400 krad).[345] Both scopoletin and scopolin contents and radiation damage (peel pitting) were positively correlated with radiation doses up to 300 krad.[351] This accumulation followed stimulation of PAL activity by radiation and the corresponding release of ethylene.[346,349,350] A 200 krad γ-radiation dose results in release of ethylene from grapefruit which peaks after 12 h; there is a parallel, but slightly staggered PAL activity peak.[350] It would seem that a *de novo* synthesis of PAL is

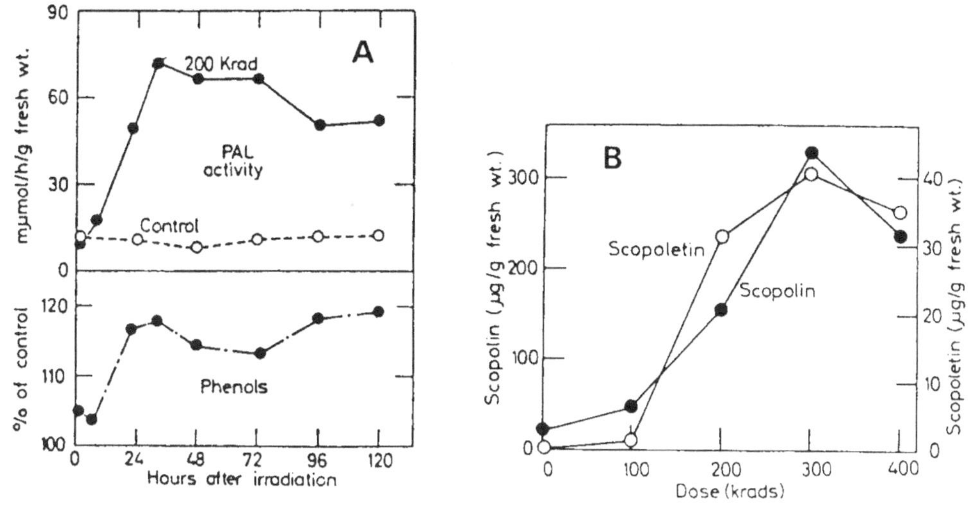

FIGURE 77. Effects of irradiation on citrus fruits. A: PAL activity and total phenols of flavedo of irradiated and non-irradiated Shamouti oranges. B: Effects of irradiation doses on scopoletin and scopolin content of grapefruit peel. (Adapted from References 348 and 351.)

induced as a consequence of irradiation,[348] although in the peel of mature grapefruit there is a linear relation between radiation doses and inhibition of synthesis of total proteins.[346]

In function of the duration of irradiation, the dose applied and physiological conditions at time of treatment, certain fruits do not tolerate ionizing treatment. Stress conditions are shown by browning of the skin and flesh tissue in banana[417] and mango[123,416] for example.

Doses of over 50 krad applied to **banana** produce appreciable radiation injury and fruits turn brown to black during storage.[417] A partially purified polyphenoloxidase (*o*-DPO) preparation from banana showed that γ-irradiation has different effects on cresolase and catecholase activities. Whereas catecholase activity increases with the radiation dose, cresolase activity increases to 100 krad and falls at higher doses. Results indicate that the effect of γ-irradiation on *o*-DPO activity is due to an activation of the enzyme rather than a *de novo* synthesis. It is possible that irradiation affects the permeability of cell membranes, resulting in increased availability of substrates of *o*-DPO, such as dopamine.[417]

In ripening **mango** fruit *(Mangifera indica)* treated in the preclimacteric phase with radiation doses of 0.75, 1.25, and 1.75 kGy, significant differences were observed in L-phenylalanine ammonia-lyase, polyphenoloxidase, peroxidase, and catalase activities, and also in total phenolic and flavanol contents (Figure 78).[123] PAL activity increased immediately after irradiation with a maximum on the 3rd or 4th d. *o*-DPO activity was higher in irradiated fruits, but above all this activity appeared much earlier, on the 6th d with a dose of 1.75 kGy, and before the postclimacteric phase. Conversion of particulate forms of *o*-DPO into soluble forms took place in the fruits after exposure to stress conditions.[266,267] Gamma radiation causes membrane damage which results in the solubilization of *o*-DPO and an earlier increase in activity than in the nonirradiated control. Irradiation resulted in almost immediate stimulation of peroxidase activity, which increased steadily during the ripening phase up to the climacteric peak; peroxidase activity then appeared to decline. Irradiation has the effect of maintaining a high concentration of total phenols whereas these decrease rapidly during maturation of the fruit. Pronounced accumulation of flavanols is also observed in irradiated fruit tissues. Similar results have been obtained for **papaya** fruits *(Carica papaya)*. PAL activity and phenolics contents were much higher in fruits which had been irradiated with 100 and 150 krad than in untreated fruits.[412]

Irradiation of **persimmon** (0.15 to 0.25 Mrad γ-rays) has been carried out to remove

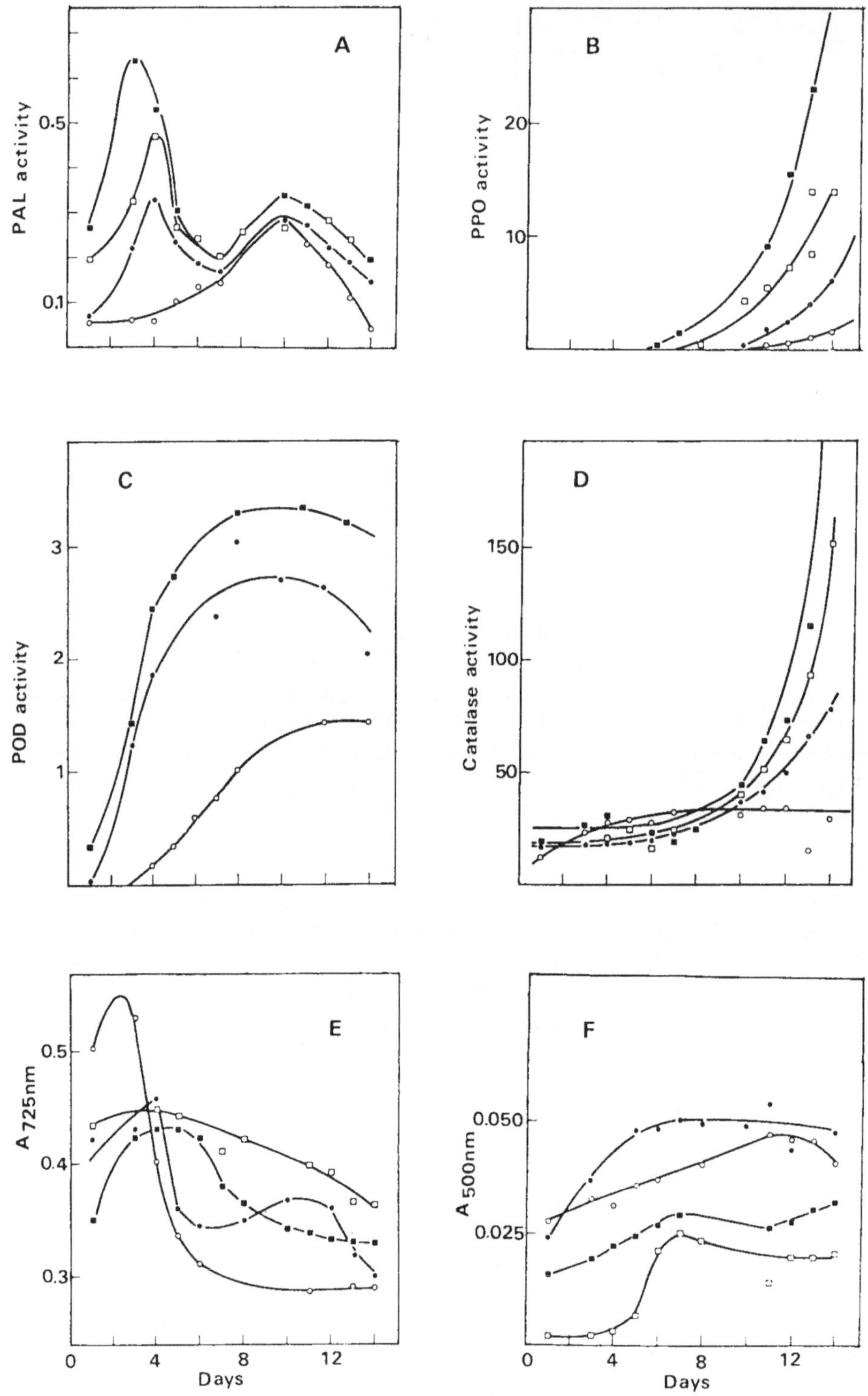

FIGURE 78. Biochemical changes during ripening of γ-irradiated mango fruit. A: PAL activity (nkat/100 g tissue); B: polyphenoloxydase activity (ΔA_{480}/min/100 g tissue); C: peroxidase activity (μkat/100 g tissue); D: catalase activity (μkat/100 g tissue); E: total phenolic concentration; F: flavanol content; 1.75 kGy (■—■), 1.25 kGy (□—□), 0.75 kGy (●—●) γ-irradiated and control (○—○). Kent mangoes at 20°C. (From Frylinck, L., Dubery, I. A., and Schabort, J. C., *Phytochemistry*, 26, 681, 1987. With permission.)

astringency. Tannin content and astringency did in fact decrease, but unfortunately the fruit softened too much.[195]

There has been little study of the effects of γ-radiation on the synthesis of anthocyanins in fruits. Irradiation had a slightly beneficial effect on pigmentation of cranberries, with anthocyanins accumulating more rapidly during maturation.[232] However, anthocyanin content after 56 d of storage was the same in fruits which had been exposed to 150 or 300 krad and nonirradiated fruits; in addition, there were no qualitative changes.[232] Higher doses of radiation can cause the decolorization of certain fruits through loss of anthocyanins. This is the case, for example, in strawberries treated with 500 and 1000 krad before storage.[10]

D. ETHYLENE

Ethylene is a hormone which plays a fundamental role in the maturation of fruits and the senescence of plant organs in general. Treatment with exogenous ethylene, which has the same effect as the triggering of endogenous ethylene synthesis, has long been used to cause and accelerate the ripening of fruits picked unripe.[276a] Ethephon ([2-choroethyl] phosphonic acid or Ethrel) is frequently used in practical applications; the molecule releases gaseous ethylene in an alkaline solution. As accumulation of anthocyanins is normally associated with the maturation of numerous fruits (Chapter 3, Section II above), it is observed that ethylene or ethephon treatments cause an increase in anthocyanin contents by stimulating enzymes involved in phenolic biosynthesis, and in particular PAL.[55,112,334,335,349,369]

Ethephon was found to be effective on **cranberries**, particularly during the first few weeks of the maturation period; it increased the production of anthocyanins in the cultivars Early Black, McFarlin, and Searles.[35,58,71,92,94,344,373] Application of ethephon (100 to 1000 mg/l) two weeks before the harvest accelerated the development of anthocyanin pigments, whereas treatments just before bloom and during peak bloom had little effect on the coloration of fruits; in addition, it reduced their size and yield (Figure 79).[71] Anthocyanin production by berries at three different stages of maturity was greatest when the fruit was treated with ethylene in the light.[58]

Trials were also carried out to concentrate the harvest of **blueberry** by using ethephon to accelerate fruit ripening.[93,181,193,443] Depending on the cultivar, effective concentrations of ethephon were 240 to 1920 ppm and optimum time of application for acceleration of the formation of anthocyanins appeared to be near the end of stage II and the beginning of stage III in the berry growth curve; this is when the color of the berries changes from red to blue.[443]

A number of works have been devoted to the effect of ethephon and ethylene on PAL and the accumulation of anthocyanins in skin of **red apples**.[24,53,96,100,137,269,282] When ethylene is applied to detached unripe Jonathan apples, PAL activity and accumulation of anthocyanins increased to levels almost similar to those observed in ripe apples (Figure 80).[100] However, ethylene applied to ripe apple has practically no effect on PAL activity and has little effect on anthocyanin synthesis. These observations suggest that ethylene initiates rapid anthocyanin accumulation during apple ripening by increasing the level of the rate-limiting enzyme, PAL, in the skin. Another ethylene-releasing compound, CGA 15281, was applied 2 or 3 weeks before the onset of autogenous ethylene production in McIntosh apple.[282] In contrast with ethephon, anthocyanin increase observed within 6 d was not accompanied by autocatalytic ethylene production or accelerated flesh softening. The most satisfactory increase in anthocyanin with minimal loss of firmness was obtained with 1 m*M* CGA 15281 applied 3 weeks prior to harvest. Results suggest that anthocyanin biosynthesis may be enhanced in apple using ethylene-releasing compounds without necessarily inducing other ripening changes.[282] Combined treatment by ethephon and auxins (NAA or 2,4,5-TP) enhances red color, particularly in fruits previously treated with Alar.[96].

In ripe **grapefruit** *(Citrus paradisi)*, PAL activity, which is weak in the flavedo, is

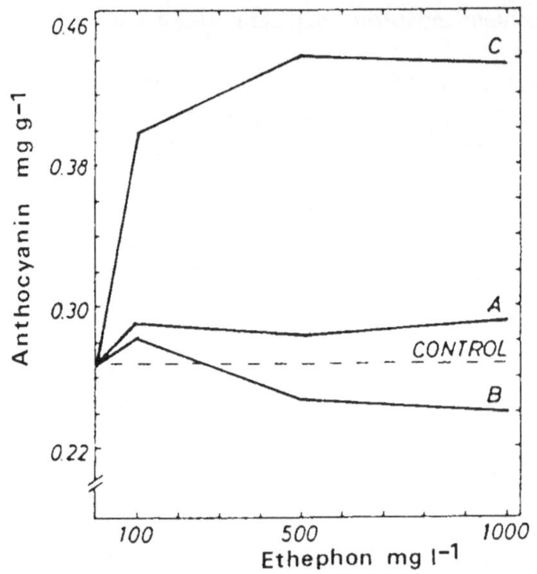

FIGURE 79. Influence of ethephon applications on anthocyanin development in cranberry. A: application just before bloom; B. application at peak bloom C: application two weeks before harvest. The broken line represents the level of anthocyanin in the control berries. Anthocyanins in mg/g fresh fruit. (From Devlin, R. M. and Demoranville, I. E., *Physiol. Plant.*, 23, 1139, 1970. With permission.)

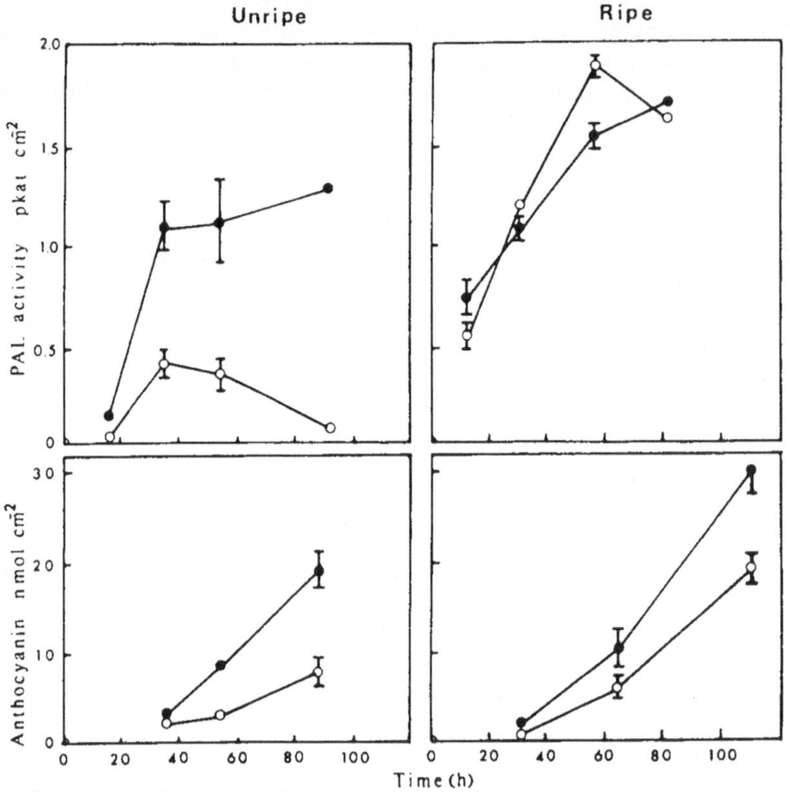

FIGURE 80. The effect of ripening and applied ethylene on PAL activity and anthocyanin accumulation in detached 'Jonathan' apples. Apples were held in glass boxes in continuous light and treated with zero (0.06 $\mu l \cdot l^{-1}$) ethylene (○—○) or with 15 $\mu l \cdot l^{-1}$ ethylene (●—●) in a continuously flowing air stream. (From Faragher, J. D. and Brohier, R. L., *Sci. Hortic.*, 22, 89, 1984. With permission.)

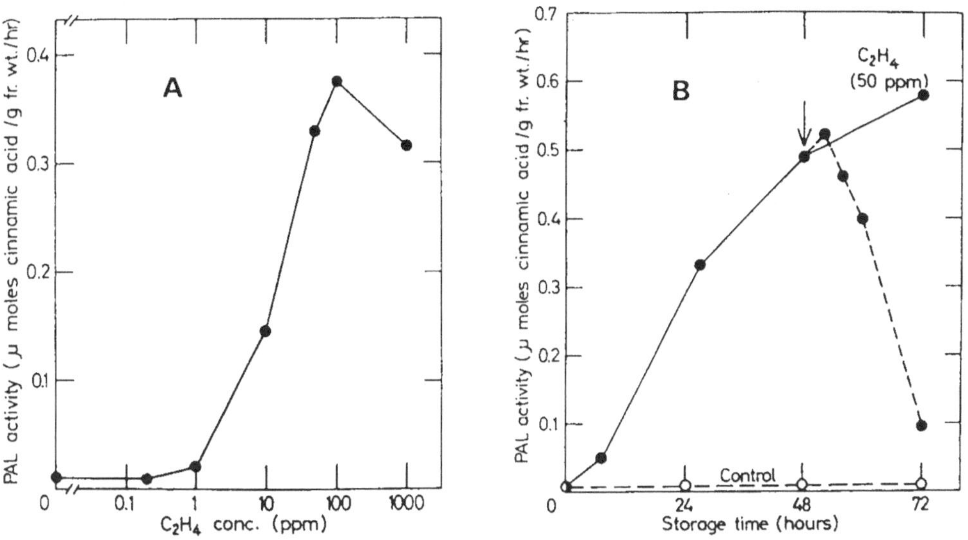

FIGURE 81. Effect of ethylene on PAL activity in flavedo of intact grapefruit; A: fruit was stored for 24 h in ethylene at 25°C, before enzyme was extracted and assayed; B: fruit was stored in 50 ppm ethylene at 25°C. Broken line: activity after fruit removal from ethylene atmosphere, at time shown by arrow. (From Riov, J., Monselise, S. P., and Kahan, R. S., *Plant Physiol.*, 44, 631, 1969. With permission.)

stimulated by ethylene (Figure 81).[349] Maximum effect is obtained with 50 to 100 ppm concentrations of ethylene applied for 48 to 72 h at 25°C. PAL activity decreases rapidly when fruit is removed from the atmosphere containing ethylene (Figure 81).

In **grape** *(Vitis vinifera)*, color development was enhanced by ethephon in most colored cultivars.[156,200,357,444,445,448] Response of berries to ethephon treatment depends on the time of application and on the cultivar.[156,445] For example, ethephon applied to cv. Ruby Cabernet at 1000 ppm on August 12, 19, or 26 stimulated anthocyanin production.[444] Optimum time for treatment was about 2 weeks after initiation of coloring.[445] On Tokay and Emperor table grapes, the most effective concentrations of ethephon were 200 ppm and 100 to 200 ppm, respectively.[200] In two cultivars, Ribier and Emperor, treated with ethephon at veraison, both light and ethephon significantly affected anthocyanins in grape skins. Total phenol contents were also increased significantly by ethephon in Ribier and by light in Emperor.[448] These problems of interaction between several factors, such as ethephon, light, temperature, and sucrose have been studied in cv. Cardinal berries removed from the mother plant and fed by the stalk: sucrose and ethephon had a synergistic effect on anthocyanin synthesis at 22°C but had little effect on total phenol accumulation in berry skin.[357]

Application of ethephon to **figs** (cv. Mission) during late period II of their development stimulated growth and ripening within 8 d[59] and caused linear anthocyanin accumulation after a latency period of 2 to 4 d.[324] No synthesis of anthocyanins occurs in the dark with or without ethephon. This indicates that ethylene can only stimulate PAL activity which is induced by light (see Figure 68).

Ethephon treatment generally gives good results in other fruits, such as blackcurrants, cherries, raspberries,[305] and *Prunus salicina* (Santa Rosa plum).[25,231] Synthesis and accumulation of anthocyanins is generally increased. In contrast, ethylene or ethephon treatment of green or half-black **olive** fruits (cv. Manzanillo) results in delayed maturation and inhibition of anthocyanin accumulation.[379] The very special reaction of the olive fruit to ethylene is not an isolated phenomenon. Other growth regulators have also been found to have a unique effect on this system (see cytokinins below).

Peaches treated with ethephon and gibberellic acid at the end of stage I of their growth displayed lessened browning phenomena at maturation, resulting from low polyphenoloxidase activity.[41,217,303,304] Lower levels of polyphenoloxidases were also found in ethephon-sprayed apples.[365]

E. GROWTH REGULATORS

Growth regulators act in different ways on the growth and maturation of fruits. After much research, exogenous application of these compounds has become common.[276a] There are fairly few precise data on the effect of these applications on the phenolic contents and metabolism of fruits. The effect is generally in the color of fruits, either through anthocyanins or *o*-diphenoloxidase and browning.

1. Auxins

In treatment of fruits, auxins are most often used in combination with other regulators, such as ethephon, and it is difficult to evaluate their precise effect on phenolic compounds, especially since published results are very fragmentary.

Indole-3-acetic acid (30 or 50 ppm) was sprayed on Early Black cranberries 2 weeks prior to harvesting but did not affect fruit color.[72,91] Treated and untreated berries were also analyzed for anthocyanin development after 7 and 14 d in common storage; the preharvest treatment did not have a significant effect on color development in storage.[72] In apple, synthetic auxins NAA (α-naphtaleneacetic acid) and 2,4,5 -TP (2-[2,4,5-trichlorophenoxy] propionic acid or fenoprop) used in combination with ethephon and Alar increased the red color of the skin of certain cultivars.[24,96,137] In olive cv. Manzanillo, application of auxins, NAA, 2,4-D, or 2,4,5-TP, at stage II (pit hardening) caused a delay in fruit ripening, but late application can prevent preharvest drop and slightly enhance coloration.[228,378] Immersion of green olive fruit for 1 h in NAA or 2,4,5-TP solutions caused after 10 d a moderate increase in anthocyanin content, particularly at high concentrations of 200 and 300 ppm, respectively.[377] As the auxins also caused fruit softening and increased CO_2 and ethylene evolution, it seems that auxin-induced coloration occurs through promotion of olive ripening, in contrast with cytokinins.

2. Cytokinins

The only usable data is on olive in which the effects of kinetin and other cytokinins on anthocyanin accumulation have been studied.[376,377] Application of exogenous cytokinins (kinetin, benzyladenine, and zeatin) to picked green mature olives caused a marked enhancement in anthocyanin accumulation without affecting the other maturation parameters, CO_2 or fruit softening. Immersion of green olives for 1 h in a cytokinin solution was enough to cause considerable increase in the anthocyanin content of fruits analyzed 10 d later. At 3.5 ppm, zeatin produced the maximum effect, while the optimum concentration of kinetin and benzyladenine was 50 ppm. Adenine and uracil did not induce pigmentation when applied separately and had only a slight effect when applied as a mixture. Kinetin had only a weak effect on accumulation of anthocyanins in half-black olives.

Kinetin did not induce coloration in apple peel and plum.[377] It is suggested that cytokinins hasten anthocyanin synthesis in the olive fruit mainly by increasing the level of free reducing sugars. The lack of effect on apple and stone fruits might be explained by high enough concentrations of simple sugars to enable anthocyanin synthesis.

3. Gibberellins

Applications of gibberellic acid (GA_3) and ethephon to peach trees prevents fruits used in canning from browning.[41,217] Foliar sprays of gibberellic acid and ethephon applied to Red Haven peach trees at 21 and 46 d after bloom caused a reduction of enzymatic browning of mature fruit tissues.[303,304] Polyphenoloxidases (*o*-DPO) from peaches treated 21 d after bloom displayed a catechol band absent in untreated peaches or peaches treated 46 d after

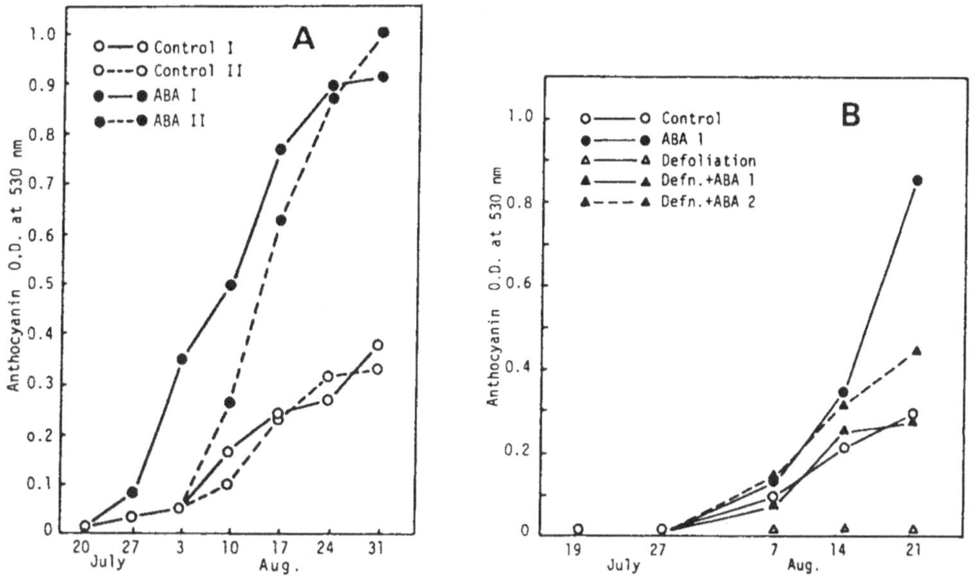

FIGURE 82. Effect of ABA (1000 ppm) (A) and defoliation (B) on anthocyanin concentration of Kyoho grapes. (From Kataoka, I., Sugiura, A., Utsunomiya, N., and Tomana, T., *Vitis*, 21, 325, 1982. With permission.)

bloom. Treatment with growth regulators had no effect on the qualitative distribution of phenolic compounds.[304] A combination of gibberellic acid and ethephon also reduced browning in plums.[323] The mechanism by which browning decreased was not consistent. Multiple regression analysis shows that 81% of variation in browning can be accounted for by differences in treatment, dates of treatment, *o*-diphenol content, *o*-DPO activity, and variations in fresh weight.[303] It has, nevertheless, been shown that treatment with GA_{4+7} has no effect on the browning rate and on skin color of ripe peaches.[360] In Bing cherries, gibberellic acid is more effective than ethephon in reducing the anthocyanin content of the fresh fruit.[78] There are thus few data on the relation between exogenous gibberellin and phenolic compounds in fruits. Relations between gibberellic acid and the polyphenol metabolism are complex; there is sometimes inhibition and sometimes stimulation, depending on the plant material.[301]

4. Abscisic Acid (ABA)

Application of exogenous abscisic acid stimulates accumulation of anthocyanins in grape berries without having an effect on the total soluble solids or titratable acids content (Figure 82A).[211] Since considerable increase in the endogenous ABA content has been observed in skin after veraison,[56,211] it can be considered that the synthesis of anthocyanins in the skin of Kyoho grape berry may be controlled by ABA accumulation. Total defoliation at veraison totally inhibits synthesis of anthocyanins and results in a low sugar level in the skin. The stimulating effect of ABA on anthocyanin synthesis after defoliation (Figure 82B) does not appear to be related to an increase in the flux of sugar to berry skin.[211] It has also been shown that sucrose and ABA operate independently to induce accumulation of anthocyanins and other phenolic compounds in isolated berries of Cabernet Sauvignon.[308] There is little information on the mechanisms by which ABA stimulates anthocyanin production in grape berries. Abscisic acid may act by activating the enzymes — PAL in particular — involved in the biosynthesis of phenolic compounds.[439]

5. Growth Retardants

Succinic acid-2,2-dimethylhydrazide (SADH or Alar), is a growth retardant whose effects on the development and maturation of fruits are well known.[83] In particular, the effect on the anthocyanin metabolism has been studied in cherries,[70,78,79,361] cranberries,[91,94] and apples.[24,96] On McIntosh apples, the red color of peel was enhanced by Alar applications combined with ethephon and auxins. At 2000 ppm, Alar induced early production of anthocyanins in several sweet and sour cherry cultivars.[70,361] The most significant effect of Alar on cherry fruit color is the modification of pigment ratios. In cherries, it was concluded that Alar enhanced the biosynthesis of anthocyanins, but that it did not measurably advance the physiological maturity of fruits. Color development in SADH-treated peaches was 10 to 20% more than in controls, and color developed not only in the skin but also around the pit.[103] In contrast, preharvest sprays of SADH on Early Black cranberries have no effect on their anthocyanin content.[94] Chlorocholine chloride, another growth retardant, stimulates rutin synthesis in tomato fruits.[198]

F. OTHER FACTORS

1. Anaerobiosis

Abnormal browning frequently occurs when fruits are stored in very low-oxygen atmospheres. Extended treatment in pure nitrogen enhances the appearance of brown surfaces in fruits, which then rot rapidly when they are returned to air. These observations are probably the result of cell disorganization under anaerobiosis, but may also be related to variations in phenolic metabolism. Whereas hydroxycinnamic acids are never detected in a free state in fruits, they accumulate in tomato stored in a nitrogen atmosphere or under CO_2 in the first few hours of treatment. The compounds include caffeic acid in particular and also *p*-coumaric and ferulic acids.[111] This feature is accompanied by a decrease in certain corresponding combined forms, especially chlorogenic acid (Figure 83). Free acids are probably derived from hydrolysis of native compounds and not from activation of biosynthesis pathways since during treatment there is a simultaneous decrease in PAL which reaches zero after the first 24 h.[112] The most sensitive fruits are those picked at the beginning of growth, and the phenomenon is not reversible. The much greater toxicity of free acids in comparison with the combined forms may cause the acceleration of metabolic diversion in the fruit.

The effect of anaerobiosis on the phenolic metabolism is also of considerable technological importance in carbonic maceration winemaking techniques whose first stages consist of placing grape berries rapidly under anaerobiosis in a CO_2-enriched atmosphere. There is a decrease in all phenolic compounds (e.g., anthocyanins, flavonols, and caffeoyl tartaric and *p*-coumaroyl tartaric acids) in both skins and pulp.[326] However, only very little free hydroxycinnamic acids appear, and skin anthocyanins move towards the pulp before disappearing during the treatment. Similar observations have been made during classic maceration of grapes,[284] and in this case, the authors considered that yeast plays an active role in the disappearance of native molecules and also in the nonappearance of the corresponding free acids. Anaerobiosis generally appears to be harmful for the fruit products formed, with the frequent appearance of unwanted browning or loss of anthocyanins. In contrast, this treatment becomes necessary in the case of removal of astringency from persimmon fruit by means of CO_2 or N_2 atmospheres. These treatments — particularly CO_2 — result in the production of acetaldehyde, and deastringency is due to the insolubilization of kaki-tannin by reaction with the acetaldehyde.[126,262,307] This example is examined in more detail in Chapter 4.

2. Sugars

Synthesis of anthocyanins requires the presence of free sugars which, added to the medium, can stimulate or trigger accumulation of anthocyanins.[60,308] In addition, this accumulation is often correlated with that of soluble sugars during maturation. Thus, in certain grape cultivars, good correlation has been established between the skin sugar content and

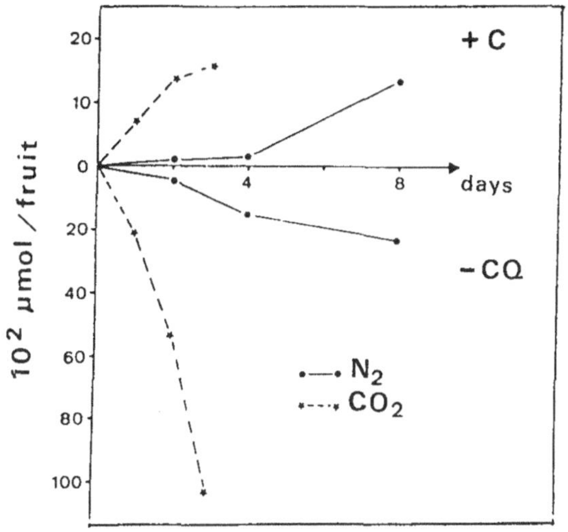

FIGURE 83. Changes in chlorogenic (CQ) and caffeic (C) acids levels in cherry tomato submitted to anaerobiosis (N_2: nitrogen or CO_2). (+): appearance; (−) disappearance. (Adapted from Fleuriet, A. and Macheix, J. J., *Physiol. Vég.*, 14, 407, 1976.)

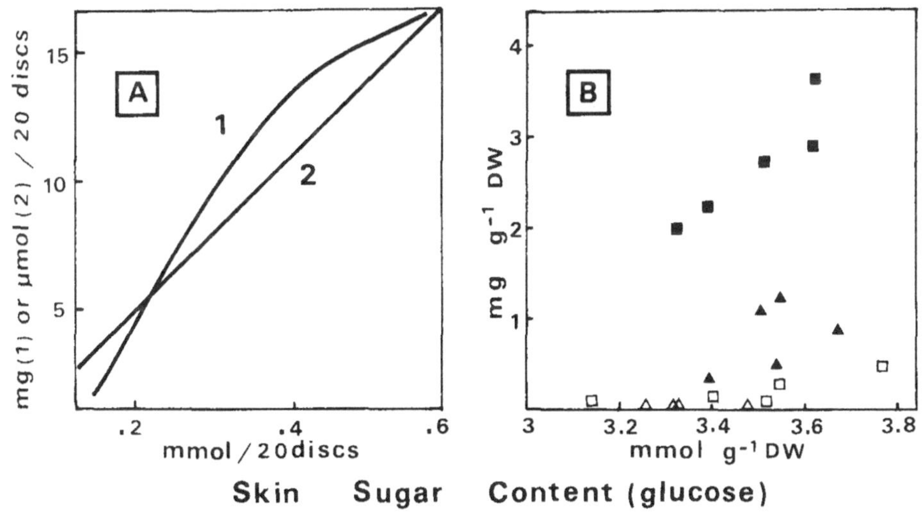

FIGURE 84. Relationship between anthocyanin (A1 and B) or total phenolics (A2) and concentration of sugar in grape skin. A: cvs. Cabernet Sauvignon and Shiraz. B: cv. Emperor. (A: Adapted from Pirie, A. and Mullins, M. G., *Am. J. Enol. Vitic.*, 28, 204, 1977; B: adapted from Wicks, A. S. and Kliewer, W. M., *Am. J. Enol. Vitic.*, 34, 114, 1983.)

the total phenolic or anthocyanin content (see Figure 84) in the 5-week period after veraison.[183,309,310,352] Total blockage of anthocyanin synthesis is observed when the supply of sugar to berries is cut off by defoliation at veraison (see Figure 82).[211] Nevertheless, the sugar content continues to increase during the final maturation stages whereas the anthocyanin content levels off and may even fall slightly.[309,310] In addition, there is no satisfactory correlation between the soluble sugar content of berry pulp and the accumulation of anthocyanins in the skin.[310] In olives, nearly stoicheiometric relations between reduction in sugar content and increase in anthocyanin synthesis have been reported.[375]

Even in cases where there are good correlations during fruit maturation, this does not necessarily imply that the soluble sugar content is responsible for the accumulation of anthocyanins. Indeed, in certain cases, the two parameters may vary independently, for example after light and/or ethephon treatments (Figure 84B).[448] The sugar content may then appear as only one of the factors which limit the accumulation of anthocyanins by interactive effect with other factors during maturation.[357]

In **apple** skin, fructose, glucose, lactose, maltose, and sucrose all stimulate anthocyanin synthesis.[433] On isolated epidermal fruit disks incubated in the light, stimulation by sucrose is more pronounced in dark pretreated disks (229%) than in disks which have not been pretreated (53%).[127] Anthocyanin production by apple skin disks was limited by carbon availability, and sucrose stimulates this production by providing increased substrate for the requisite synthetic pathways. Catabolism of glucose via the pentose phosphate pathway (PPP) has been associated with anthocyanin production in apple fruit.[102,127] Indeed, PPP is at the origin of the formation of erythrose-4-phosphate which, by condensation with phosphoenolpyruvate, enables entry to the shikimate pathway. It might be interesting, *a priori* to seek the regulation of phenolic biosynthesis at the level of PPP activity and its variations in relation to physiological stages. However, the results are generally fairly contradictory. Thus, whereas the natural development of red color in apple is associated with an increasing role of PPP,[102] a shift of PPP to the glycolytic degradation of glucose has been reported fairly frequently during the maturation of climacteric fruits.[447] In addition, chemical diversion of carbon flow to PPP does not always result in increased anthocyanin synthesis in apple.[127]

In comparison with anthocyanins, which generally accumulate during maturation when the soluble sugar content increases in the fruit, most of the other phenolic compounds accumulate at an earlier stage. Peaks in contents are often observed, as has been mentioned many times, in young fruits where the soluble sugar contents are very low.[447] It is, therefore, totally impossible to find a correlation between the levels of the two parameters concerned and their evolution in time, which does not mean that sugars are not involved in the biosynthesis of phenols. On the contrary, it is even probable that the low soluble sugar contents are truly limiting (together with other factors) with regard to the accumulation of phenols.

3. Calcium

Calcium carbonate increases red color formation in McIntosh apples, and an additive effect on calcium carbonate and sugars (sucrose and galactose) on anthocyanin synthesis has been shown.[388,389] Ca^{2+} ions thus appear to enhance synthesis of anthocyanins in disks of apple peel, but the effects depend on the stage of maturity of the fruit and above all on the form in which the calcium is supplied.[433] Anthocyanins expressed in percentage of control (100) were 152, 145, and 187 for calcium carbonate 0.00 1 *M*, calcium-chelate (tetrasodium EDTA, 0.0005 to 0.001 *M*) and calcium succinate 0.002 *M*, respectively. Among the other calcium compounds tested, Ca-lactate, Ca-nitrate, Ca-acetate, Ca-gluconate, and Ca-glucuronate did not have a significant effect in the amounts applied (0.0005 to 0.001 *M*).[433] Curiously, the same stimulation was observed with Ca-chelate (Na_2Ca-EDTA) and tetrasodium EDTA (Na_2-EDTA), which would tend to prove that the stimulating effect was not specifically related to calcium.[433]

Calcium delays the appearance of senescence in fruits, particularly by protecting cell membrane from disorganization.[104,311] Brown discoloration of fruit tissues frequently occurs in Ca-deficiency disorders due to a leakage of phenolic compounds from the vacuole into the cytoplasm.[12] This may account for the effect observed on the browning and the color of apples[387] and avocado.[430]

4. Nitrogen and Fertilizers

As in other parts of plants, excess nitrogen reduces the accumulation of anthocyanins in fruits. In **cranberries** (cv. Early Black) treated with five levels of fertilizer (N, P, K) in three ratios, anthocyanin formation was inversely related to increasing applications of nitrogen and phosphorus.[122]

In **grape** (cv. Emperor), treatment with large amounts of nitrogen in NH_4NO_3 form 2 weeks before veraison reduced the anthocyanin content of berries by 22% and delayed maturation.[214] This is thought to have been caused by the diversion of photosynthetic assimilates to the synthesis of amino acids and proteins, but it would also be possible to envisage an indirect effect through an increase in vegetative growth slowing down maturation of the fruit. The limiting factor governing polyphenol accumulation in numerous plant tissues might be the availability of phenylalanine rather than the level of PAL activity.[257]

Ciders obtained from **apples** from trees grown with 'fed' and 'unfed' fertilizer regimes (N, P, K) were compared. The ciders from fed trees were less bitter and astringent than those from unfed trees. This can be related to a fall of 17% in total phenol concentration: 350 against 410 mg/100 ml.[229]

5. Miscellaneous Compounds

Various compounds used in the treatment of trees or fruits may have repercussions on the coloring of the latter and on their phenolic compound contents. In fact, very few precise studies have been carried out on the effect of these compounds at phenolic level, and it is probable that the effects observed are indirect. In **cranberries,** application of malathion (organophosphorus insecticide) as foliar spray 2 to 3 weeks before harvesting enhances the synthesis and accumulation of anthocyanins in the cultivars Early Black, McFarlin, and Searles.[72,86,91] The anthocyanin content of fruits treated with malathion can increase by 30% in comparison with control fruits.[94] In the same fruits, treatment with different fungicides, particularly in the presence of fertilizer, leads either to a markedly lower anthocyanin content or on the other hand to an increase in red coloration.[122,277]

Concentrations of less than 10^{-5} *M* of 3-(3,4-dichlorophenyl)1,1-dimethylurea (DCMU or diuron), a photosynthesis inhibitor used as an herbicide, is effective in increasing the red color of apples. Several other substitute ureas have the same effect.[103] Unlike other compounds that induce color in fruits, DCMU has no effect on the ripening of apples.

It would in fact be interesting to specify the consequences on the phenolic metabolism of fruits of treatments with compounds such as aminoethoxyvinylglycine, a classic inhibitor of ethylene synthesis in particular in fruits,[98,353] or of glyphosate, a nonselective herbicide which acts as an inhibitor of the shikimic pathway.[180,201]

6. Miscellaneous Factors

Numerous other agronomic or climatic factors may also affect the phenolic compound contents of fruits. They include, for example, water supply or modified atmospheres which are frequently used in storage. Thus, high concentrations of CO_2 cause the disappearance of astringency in persimmon; this is related to a reduction in the tannin content.[126,264] The latter point is discussed in Chapter 4.

Water supply to fruit is also responsible for modifications in the phenolics content. In the case of grape, limited water supply results in berries being smaller but they contain more anthocyanins and tannins than the control.[33,84] However, the response is not uniform for all phenolic compounds since the acid-phenol and catechin contents are reduced.[33] Mineral nutrition may also be involved.[313] However, it remains difficult in all cases to determine whether the variations are the result of specific effect on the phenolic metabolism or whether they are the indirect result of more general physiological disturbance at the level of the fruit or the plant as a whole. The latter hypothesis remains the most probable.

IV. CONCLUSION

A. GENERAL REMARKS CONCERNING VARIATIONS IN THE PHENOLIC METABOLISM

In spite of several noteworthy exceptions, a number of general features emerge from the experimental data which has been presented and discussed. (1) With the exception of anthocyanins and of some particular compounds (capsaicin, dopamine, . . .), high concentrations of phenolic compounds are found in young fruits, the maximum generally being observed during an early growth phase associated with extensive cell division, as opposed to the subsequent cell enlargement period. Concentrations then fall until ripening. (2) The amount per fruit increases over a much longer period and may then stabilize until ripeness or in contrast fall during the final phases of growth. (3) Anthocyanins generally accumulate during the maturation of red fruits. (4) Noteworthy variations in the concentration ratios of the different compounds are observed during growth and ripening, leading to the situation described in the ripe fruit. (5) External factors may modify these evolutions, but genetic control appears to be the primary factor in all cases. (6) There are generally fairly good correlations between the accumulations observed and activity of phenolic anabolism enzymes, some of which (PAL, CoAL, etc.) emerge as key parameters in the quantitative expression or the orientation of the metabolism.

B. ORIGIN OF PHENOLIC COMPOUNDS IN FRUITS

The observations above, together with the results of experiments on biosynthesis carried out on fruits lead to several clear conclusions: (1) There is a very active *in situ* biosynthesis phase in very young fruits; this then slows during the most active growth period and, in the case of anthocyanin synthesis, increases again after veraison. The balance of accumulation rate and growth rate accounts for the evolution of concentrations. (2) The turnover of certain phenolic compounds in fruits leads either to the formation of other phenolic molecules (e.g., increased polymerization of tannins, formation of lignin from monomers, possible formation of the B-ring of anthocyanins from the phenylpropanoid skeleton during maturation, etc.) or their more or less advanced degradation in one of the classic manners,[15] in particular the oxidative pathways which lead to the browning that is common in fruits. (3) Very different orientation of the phenolic metabolism is noted according to the part of the fruit (skin, pulp, stone, seeds) which leads to very distinct tissue compartmentation. (4) Inward and outward migration of either immediate precursors of phenolic compounds or of these compounds themselves cannot be ruled out. Data on possible movement in phloem remained unconvincing for a long time[456] in spite of the intensity of phloem sap movement which feeds fruit sinks. However, recent work[421-423] revealed the presence of several groups of flavonoids (flavanones, isoflavones, flavones) in *Prunus* phloem. Concentrations varied with the season and the position of the twig on the tree. In addition, flavonoid contents increased above the union in senescing interspecific grafts as compared to healthy homospecific ones,[422] the level of prunin being a possible marker for incompatible heterograph.[421,423] These data obtained in a fruit species suggest probable migration of certain phenolic compounds in the phloem, although no direct experimental proof has been put forward. These observations could be broadened to cover supply of the fruit itself. Several organic acids which are more or less remote precursors of phenolic compounds (quinic, shikimic, and tartaric acids, etc.) and phenylalanine are also found in phloem but at very small concentrations; it is probable that fruits that need large amounts of shikimic acid in the formation of phenylpropanoid derivatives use transported sugars as basic skeletons.[456] In this case, accumulation of phenolic compounds

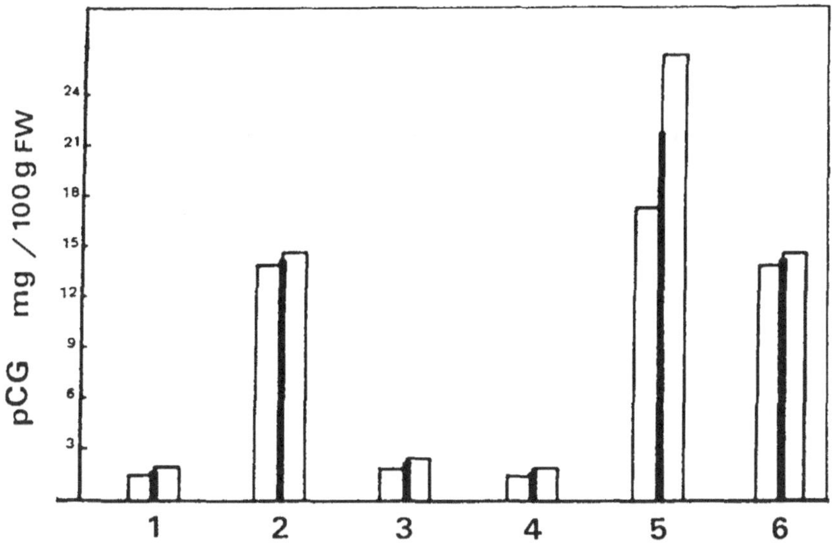

FIGURE 85. Accumulation of *p*-coumaroylglucose (pCG) in apple pulp discs supplied with phenylalanine (2), glucose (3), quinic acid (4), phenylalanine + glucose (5), phenylalanine + quinic acid (6), (1): control. (Adapted from Macheix, J. J., *Physiol. Vég.*, 13, 591, 1975.)

probably depends both on the supply of precursors and on biosynthesis activity which, as we have mentioned, is particularly high in young fruits.

C. RELATIONS BETWEEN VARIATIONS IN CERTAIN SUBSTRATES AND THE ACCUMULATION OF PHENOLIC COMPOUNDS

Insofar as the levels of certain substrates may be limiting factors for the accumulation of phenolic compounds,[65,257,310] it is interesting to see whether there are correlations between the two parameters during the life of fruits. Unfortunately, experimental data on these aspects are very fragmentary or contradictory. In the case of intermediates in the shikimate pathway or of closely related molecules, such as quinic acid (which can be both used through this pathway or directly by the esterification of hydroxycinnamic acids), no in-depth research of such correlations appears to have been carried out in fruits, and the grouping of data from different authors does not reveal any interesting phenomena.[242] The same applies to phenylalanine, whose content in fruits is always fairly low in comparison with other amino acids,[44] but whose concentration has been proposed as being one of the limiting factors for entry to the phenylpropanoid pathway.[65,257] Indeed, there are many data on the effectiveness of supplying phenylalanine, combined with other metabolites or not, on the accumulation of phenolic compounds, particularly in certain fruits[164,244] (Figure 85). However, although the stimulation observed can be interpreted as the result of the removal of a limited factor at the physiological stage at which the experiments were carried out, these conclusions cannot be transposed simply when fruits at different stages of development are compared with each other.

The case of soluble sugars has received a great deal of attention. Indeed, they may serve both as more or less direct precursors (either through the shikimate pathway or directly as activated forms for glycosylation) and as a respiratory substrate giving cells enough energy for biosyntheses. However, the results reported are somewhat contradictory, as discussed above (see Section IV.F). Nevertheless, even when there are correlations between phenolic accumulation and soluble sugars, it must not be forgotten that the phenolic metabolism is compartmentalized at a subcellular scale and that it is not the overall concentrations of a possible precursor that should be taken into account (e.g., vacuolar sugars) but those present in the compartment in which it is used. In addition, numerous examples discussed above have drawn attention to the possible importance of variations in anabolism and catabolism enzymes in the regulation of phenolic accumulation.

D. RELATIVE VARIATIONS OF THE DIFFERENT CLASSES OF PHENOLIC COMPOUNDS

The variations of the different classes of phenolic compounds described above during the life of a fruit reveal several extremely noteworthy variations which may suggest metabolic relations between certain compounds and possibly some interconversions. Two of these are particularly distinct:

1. In most red fruits, proanthocyanidins diminish or disappear whereas anthocyanidins accumulate.[136] Although there is no direct metabolic relationship between these two groups of compounds, they both derive from common intermediate substances which are probably dihydroflavonols; orientation towards procyanidins takes place through several reduction steps (Figure 55). The fall in photosynthesis activity and the progressive disorganization of chloroplasts which characterize maturation probably lead to a very large diminution in available reducing potential in the fruit, whence the possible shift to the formation of anthocyanidins at the expense of proanthocyanidins.
2. The appearance of anthocyanins is frequently combined with possible prior decrease of hydroxycinnamic esters. No research has been carried out on fruits to identify possible links between these two contrasting types of evolution. These links are nonetheless probable in the light of the importance of *p*-coumaroyl-CoA in the formation of the flavonoid B-ring as a whole on the one hand and in the acylation of C_{15} skeletons on the other.[207]

It is curious to observe that these two orientations have been rarely approached in fruits, which are, nevertheless, particularly suitable material given the large variations observed according to physiological stage. Other comparisons concerning the various compounds belonging to the same class of phenolic compounds have already been mentioned (more or less substituted or glycosylated anthocyanins, quinic or glycosylated hydroxycinnamic esters, etc.). These comparisons can now be developed using modern analytical techniques and should make it possible to extend to fruits the observations already made for other parts of plants and which suggest interconversions between the different compounds and certain phylogenetic remarks.

E. PHENOLIC COMPOUNDS AND THE BIOCHEMICAL CHARACTERIZATION OF MATURATION

In comparison with other parts of plants, fruits possess the original feature of ripening, whose principal biochemical parameters have been studied extensively.[34] Is it also possible to characterize the phenomenon through phenolic compounds and their metabolism? In addition, the phenomena of growth and maturation are generally partially superposed with the growth of the fruit continuing during maturation. Is it possible, therefore, to use phenolic markers to achieve more distinct differentiation? Some of the numerous examples discussed above provide partial answers to these questions. Thus, maturation can be characterized as follows, depending on the case:

1. The appearance of new compounds: anthocyanins in red fruits, demethyloleuropein in certain olive cultivars, neochlorogenic acid in tomato. The anthocyanin content increases gradually during maturation and can be considered as an index of maturity.
2. The evolution of ratios between certain compounds (e.g., glucosylated derivatives of hydroxycinnamic acids:quinic derivatives in tomato or the ratio of various forms of anthocyanins in grape or blackcurrant).

3. The disappearance of certain compounds, such as tannins in certain persimmon cultivars.
4. Finally, the evolution of the activity of numerous enzymes leading to the biosynthesis or degradation of phenolic compounds (PAL, chalcone synthase, glucosyltransferase, esterases, etc.). All these characteristic variations lead to the notion of biochemical markers of maturation applied to phenolic compounds and their metabolism. The examples studied with precision and the results obtained are still not sufficient for this notion to be generalized but are sufficiently promising to lead to the undertaking of research on the subject.

F. PHYSIOLOGICAL AND ECOLOGICAL ASPECTS

Variation can occur in phenolic compounds as a more or less direct result of certain modifications which take place during the life of fruits: possible regulation of growth and maturation, modifications of resistance of fruits to their biological and physicochemical environment, acquisition of certain properties which determine their qualities (color, flavor, etc.) and which may play a substantial role in utilization by man or animals (consumption, processing, dispersal of the species, etc.). All these physiological, ecological, and technological aspects are discussed at length in subsequent chapters.

REFERENCES

1. **Afek, U., Szteinberg, A., and Carmely, S.,** 6,7-Dimethoxycoumarin, a citrus phytoalexin conferring resistance against *Phytophthora gummosis, Phytochemistry*, 25, 1855, 1986.
2. **Albach, R. F., Redman, G. H., Cruse, R. R., and Petersen, H. D.,** Seasonal variations of bitterness components, pulp and Vitamin C in Texas commercial citrus juices, *J. Agric. Food Chem.*, 29, 805, 1981.
3. **Alibert, G. and Ranjeva, R.,** Recherches sur les enzymes catalysant la biosynthèse des acides phénoliques chez *Quercus pedonculata* Ehrh. I. Formation des premiers termes des séries cinnamique et benzoique, *FEBS Lett.*, 19, 11, 1971.
4. **Amiot, M. J., Fleuriet, A., and Macheix, J. J.,** Importance and evolution of phenolic compounds in olive during growth and maturation, *J. Agric. Food Chem.*, 34, 823, 1986.
5. **Amiot, M. J., Fleuriet, A., and Macheix, J. J.,** Accumulation of oleuropein derivatives during olive maturation, *Phytochemistry*, 28, 67, 1989.
6. **Aoki, S., Yahagi, Y., and Tamura, T.,** The use of polyvinylpryrrolidone column for the quantitative determination of chlorogenic acid in apple fruit, *J. Jpn. Soc. Food Sci. Technol.*, 31, 110, 1984.
7. **Aoki, S., Araki, C., Kaneko, K., and Katayama, O.,** Occurrence of L-phenylalanine ammonia-lyase activity in peach fruit during growth, *Agric. Biol. Chem.*, 35, 784, 1971.
8. **Arakawa, O., Hori, Y., and Ogata, R.,** Relative effectiveness and interaction of ultraviolet-B, red, and blue light in anthocyanin synthesis of apple fruit, *Physiol. Plant.*, 64, 323, 1985.
9. **Aubert, S.,** L'aubergine (*Solanum melongena* L.) I. Composition et facteurs de qualité, *Ann. Technol. Agric.*, 20, 241, 1971.
10. **Avigdori-Avidov, H.,** Strawberry, in *Handbook of Fruit Set and Development*, Monselise, S. P., Ed., CRC Press, Boca Raton, FL, 1986, 419.
11. **Bajaj, Y. P. S. and Collins, W. B.,** Some factors affecting the *in vitro* development of strawberry fruits, *Proc. Am. Soc. Hortic. Sci.*, 93, 326, 1968.
12. **Bangerth, F.,** Calcium related physiological disorders of plants, *Annu. Rev. Phytopath.*, 17, 97, 1979.
13. **Barritt, B. H. and Torre, L. C.,** Fruit anthocyanin pigments of red raspberry cultivars, *J. Am. Soc. Hortic. Sci.*, 100, 98, 1975.
14. **Bartholomew, D. P. and Paull, R. E.,** Pineapple, in *Handbook of Fruit Set and Development*, Monselise, S. P., Ed., CRC Press, Boca Raton, FL, 1986, 371.
15. **Barz, W.,** Degradation of polyphenols in plants and plant cell suspension cultures, *Physiol. Vég.*, 15, 261, 1977.
16. **Barz, W. and Wierman, R.,** Recent aspects of flavonoid biosynthesis and degradation in plants, in *Proc. Int. Bioflavonoid Symp.*, Munich, 1981, 185.

17. **Barz, W., Köster, J., Weltring, K. M., and Strack, D.,** Recent advances in the metabolism and degradation of phenolic compounds in plants and animals, in *Ann. Proc. Phytochem. Soc. Europe,* Vol. 25, Van Sumere, C. F. and Lea, P. J., Eds., Clarendon Press, Oxford, 1985, 307.
18. **Ben-Arie, R. and Lurie, S.,** Prolongation of fruit life after harvest, in *Handbook of Fruit Set and Development,* Monselise, S. P., Ed., CRC Press, Boca Raton, FL, 1986, 493.
19. **Bergmann, L., Grosse, W., and Ruppel, H. G.,** Die Bildung von Serotonin in *Juglans regia* L., *Planta,* 94, 47, 1970.
20. **Beyrich, T.,** Die Bildungsstätte der furocumarine von *Pastinaca sativa* L., *Planta Med.,* 15, 306, 1967.
21. **Billot, J., Hartmann, C., Macheix, J. J., and Rateau, J.,** Les composés phénoliques au cours de la croissance de la poire Passe-Crassane, *Physiol. Vég.,* 16, 693, 1978.
22. **Bishop, R. C. and Klein, R. M.,** Photo-promotion of anthocyanin synthesis in harvested apples, *HortScience,* 10, 126, 1975.
23. **Blankenship, S. M. and Richardson, D. G.,** Changes in phenolic acids and internal ethylene during long-term cold storage of pears, *J. Am. Soc. Hortic. Sci.,* 110, 336, 1985.
24. **Blanpied, G. D., Forshey, C. G., Styles, W. C., Green, D. W., Lord, W. J., and Bramlage, W. J.,** Use of Ethephon to simulate red color without hastening ripening of 'McIntosh' Apples, *J. Am. Soc. Hortic. Sci.,* 100, 379, 1975.
25. **Blommaert, K. L. J., Theron, T., and Steenkamp, J.,** Earlier and more uniform ripening of Santa Rosa plums using etephon, *Dec. Fruit Grower,* 25, 267, 1975.
26. **Bollard, E. C.,** The physiology and nutrition of developing fruits, in *The Biochemistry of Fruits and Their Products,* Vol. 2, Hulme, A. C., Ed., Academic Press, London, 1970, 387.
27. **Bolwell, G. P.,** Synthesis of cell wall components: aspects of control, *Phytochemistry,* 27, 1235, 1988.
28. **Bompeix, G. and Bousquet, J. F.,** Les composés phénoliques dans la résistance des pommes à l'infection par les *Pezicula malicortis* et *P. alba, Fruits,* 29, 693, 1974.
29. **Bostock, R. M., Thomas, C. S., Ogawa, J. M., Rice, R. E., and Uyemoto, J. K.,** Relationship of wound-induced peroxidase activity to epicarp lesion development in maturing Pistachio fruit, *Phytopathology,* 77, 275, 1987.
30. **Boudet, A. M., Graziana, A., and Ranjeva, R.,** Recent advances in the regulation of the prearomatic pathway, in *Ann. Proc. Phytochem. Soc. Europe,* Vol. 25, Van Sumere, C. F., and Lea, P. J., Eds., Clarendon Press, Oxford, 1985, 135.
31. **Boudet, A., Ranjeva, R., and Gadal, P.,** Propriétés allostériques spécifiques des deux iso-enzymes de la phénylalanine-ammoniaque lyase chez *Quercus pedonculata, Phytochemistry,* 10, 997, 1971.
32. **Boursiquot, J. M.,** Contribution à l'étude des esters hydroxycinnamoyltartriques chez le genre *Vitis.* Recherche d'application taxonomique, Thèse Doct. Ing., ENSA, Montpellier, 1987.
33. **Bourzeix, M., Heredia, N., Meriaux, S., Rollin, H., and Rutten, P.,** De l'influence de l'alimentation hydrique de la Vigne sur les caractéristiques anatomiques des baies de raisins et leur richesse en couleur, tannins et autres constituants phénoliques, *C. R. Acad. Sci.,* Ser. D, 284, 365, 1977.
34. **Brady, C. J.,** Fruit ripening, *Annu. Rev. Plant Physiol.,* 38, 155, 1987.
35. **Bramlage, W. J., Devlin, R. M., and Smagula, J. M.,** Effects of preharvest application of ethephon on 'Early Black' cranberries, *J. Am. Soc. Hortic. Sci.,* 97, 625, 1972.
36. **Brown, S. A.,** Coumarins, in *The Biochemistry of Plants,* Vol. 7, Conn, E. E., Ed., Academic Press, New York, 1981, 269.
37. **Brown, S. A.,** Recent advances in the biosynthesis of coumarins, in *Ann. Proc. Phytochem. Soc. Europe,* Vol. 25, Van Sumere, C. F. and Lea, P. J., Eds., Clarendon Press, Oxford, 1985, 257.
38. **Brunet, G. and Ibrahim, R. K.,** Tissue culture of citrus peel and its potential for flavonoid synthesis, *Z. Pflanzenphysiol.,* 69, 152, 1973.
39. **Brunet, G. and Ibrahim, R. K.,** *O*-Methylation of flavonoids by cell-free extracts of calmondin Orange, *Phytochemistry,* 19, 741, 1980.
40. **Brunet, G., Saleh, N. A. M., and Ibrahim, R. K.,** Stepwise methylation of quercetin by cell-free extracts of citrus tissues, *Z. Naturforsch.,* C, 33, 786, 1978.
41. **Buchanan, D. W., Hall, C. B., Biggs, R. H., and Knapp, F. W.,** Influence of Alar, Ethrel and gibberellic acid on browning of peaches, *HortScience,* 4, 302, 1969.
42. **Bulatović, M. S.,** Analysis of the anthocyanin pigment content in the plum cultivars with the different time of maturity, *Acta Hortic.,* 74, 163, 1978.
43. **Bureau, D., Macheix, J. J., and Rouet-Mayer, M. A.,** Relations entre la teneur en *o*-diphenols, l'activité polyphénoloxydasique et l'aptitude au brunissement de quelques variétés de pêches, *Lebensm. Wiss. Technol.,* 10, 211, 1977.
44. **Burroughs, L. F.,** Amino acids, in *The Biochemistry of Fruits and Their Products,* Vol. 1, Hulme, A. C., Ed., Academic Press, New York, 1970, 119.
45. **Butt, V. S.,** Oxidases in aromatic metabolism, in *Recent Advances in Phytochemistry,* Vol. 12, Swain, T., Harborne, J. B., and Van Sumere, C. F., Eds., Plenum Press, New York, 1979, 433.

46. **Butt, V. S.,** Oxygenation and oxidation in the metabolism of aromatic compounds, in *Ann. Proc. Phytochem. Soc. Europe,* Vol. 25, Van Sumere, C. F. and Lea, P. J., Eds., Clarendon Press, Oxford, 1985, 349.
47. **Butt, V. S. and Lamb, C. J.,** Oxygenases and the metabolism of plants products, in *The Biochemistry of Plants,* Vol. 7, Conn, E. E., Ed., Academic Press, New York, 1981, 627.
48. **Butt, V. S. and Wilkinson, E. M.,** Enzyme changes accompanying lignification and flavonoid synthesis in illuminated pea shoots, *Proc. 12th FEBS Meet.,* Dresden, 55, 147, 1979.
49. **Buttrose, M. S., Hale, C. R., and Kliewer, W. M.,** Effect of temperature on the composition of "Cabernet Sauvignon" berries, *Am. J. Enol. Vitic,* 22, 71, 1971.
50. **Camm, E. L. and Towers, G. H. N.,** Phenylalanine ammonialyse, *Phytochemistry,* 12, 961, 1973.
51. **Ceska, O., Chaudhary, S. K., Warrington, P. J., and Ashwood-Smith, M. J.,** Photoactive furocoumarins in fruits of some Umbellifers, *Phytochemistry,* 26, 165, 1987.
52. **Chalmers, D. J. and Faragher, J. D.,** Regulation of anthocyanin synthesis in apple skin. I. Comparison of the effects of cycloheximide, ultraviolet light, wounding and maturity, *Aust. J. Plant Physiol.,* 4, 111, 1977.
53. **Chalmers, D. J. and Faragher, J. D.,** Regulation of anthocyanin synthesis in apple skin. II. Involvement of ethylene, *Aust. J. Plant Physiol.,* 4, 123, 1977.
54. **Chalmers, D. J., Faragher, J. D., and Raff, J. W.,** Changes in anthocyanin synthesis as an index of maturity in red apple varieties, *J. Hortic. Sci.,* 48, 387, 1973.
55. **Chalutz, E.,** Ethylene-induced phenylalanine ammonia-lyase activity in carrot roots, *Plant Physiol.,* 51, 1033, 1973.
56. **Coombe, B. G.,** The development of fleshy fruits, *Annu. Rev. Plant Physiol.,* 27, 207, 1976.
57. **Cotter, D. J.,** A review of studies on chili, *Bull. N.M. State Univ. Agric. Exp. Stn.,* 673, 1980.
58. **Craker, L. E.,** Post-harvest color promotion in cranberry with ethylene, *HortScience,* 6, 137, 1971.
59. **Crane, J. C., Marei, N., and Nelson, M. M.,** Growth and maturation of fig fruits stimulated by 2-chloroethylphosphonic acid, *J. Am. Soc. Hortic. Sci.,* 95, 367, 1970.
60. **Creasy, L. L.,** The significance of carbohydrate metabolism in flavonoid synthesis in strawberry leaf disks, *Phytochemistry,* 7, 1743, 1968.
61. **Creasy, L. L.,** The role of low temperature in anthocyanin synthesis in McIntosh apples, *Proc. Am. Soc. Hortic. Sci.,* 93, 716, 1968.
62. **Creasy, L. L., Zucker, M., and Wong, P. P.,** Anomalous effects of cycloheximide on phenylalanine ammonia-lyase: role of synthesis and inactivation in leaf disks of *Helianthus annuus, Phytochemistry,* 13, 2117, 1974.
63. **Crippen, D. D. and Morrison, J. C.,** The effects of sun exposure on the phenolic content of Cabernet Sauvignon berries during development, *Am. J. Enol. Vitic.,* 37, 243, 1986.
64. **Czochanska, Z., Foo, L. Y., and Porter, L. J.,** Compositional changes in lower molecular weight flavans during grape maturation, *Phytochemistry,* 18, 1819, 1979.
65. **Da Cunha, A.,** The estimation of L-phenylalanine ammonia-lyase shows phenylpropanoid biosynthesis to be regulated by L-phenylalanine supply and availability, *Phytochemistry,* 26, 2723, 1987.
66. **Daigle, D. J., Conkerton, E. J., Sanders, T. H., and Rixon, A. C.,** Peanut hull flavonoids: their relationship with peanut maturity, *J. Agric. Food Chem.,* 36, 1179, 1988.
67. **Darné, G.,** Evolution des différentes anthocyanes des pellicules de Cabernet-Sauvignon au cours du développement des baies, *Connaiss. Vigne Vin,* 22, 225, 1988.
68. **Deacon, W. and Marsh, H. V.,** Properties of an enzyme from bananas *(Musa sapientum)* which hydroxylates tyramine to dopamine, *Phytochemistry,* 10, 2915, 1971.
69. **Dekazos, E.,** Quantitative determination of anthocyanin pigments during the maturation and ripening of red tart cherries, *J. Food Sci.,* 35, 242, 1970.
70. **Dekazos, E. D. and Worley, J. F.,** Effect of succinic acid 2,2-dimethyl hydrazide on anthocyanin metabolism and cell wall carbohydrates in red tart cherries, *Prunus Cerasus* L., Cv. Montmorency, *J. Am. Soc. Hortic. Sci.,* 95, 703, 1970.
71. **Devlin, R. M. and Demoranville, I. E.,** Influence of 2-chloroethylphosphonic acid on anthocyanin formation, size, and yield in *Vaccinium macrocarpon* cv. Early Black, *Physiol. Plant.,* 23, 1139, 1970.
72. **Devlin, R. M., Zuckerman, B. M., and Demoranville, I. E.,** Influence of preharvest applications of malathion and indole-3-acetic acid on anthocyanin development in *Vaccinium macrocarpon,* var. 'Early Black', *J. Am. Soc. Hortic. Sci.,* 94, 52, 1969.
73. **Dick, A. J.,** Apple fruit polyphenols and fruit softening, *Bull. Liaison Groupe Polyphénols,* 13, 358, 1986.
74. **Diener, H. A. and Naumann, W. D.,** Influence of dry and night temperatures on an thocyanin synthesis in the apple skin, *Gartenbauwissenschraft,* 46, 125, 1981.
75. **Dieudonné, G.,** Les dérivés hydroxycinnamiques de l'ananas Comosus variété Cayenne Lisse (Cote d'Ivoire) et leurs relations avec le brunissement interne, Thèse Doct. Ing., Université Pierre et Marie Curie, Paris, 1977.
76. **Dilley, D. R.,** Enzymes, in *The Biochemistry of Fruits and Their Products,* Vol. 1, Hulme, A. C., Ed., Academic Press, London, 1970, 179.

77. **Downs, R. J., Siegelman, H. W., Butler, W. L., and Hendricks, S. B.,** Photoreceptive pigments for anthocyanin synthesis in apple skin, *Nature (London),* 205, 909, 1965.
78. **Drake, S. R., Proebsting, E. L., Jr., and Nelson, J. W.,** Influence of growth regulators on the quality of fresh and processed 'Bing' cherries, *J. Food Sci.,* 43, 1695, 1978.
79. **Drake, S. R., Proebsting, E. L., Jr., Thompson, J. B., and Nelson, J. W.,** Effects of daminozide, maturity, and cultivar on the color grade and character of sweet cherries, *J. Am. Soc. Hortic. Sci.,* 105, 668, 1980.
80. **Duke, S. O. and Vaughn, K. C.,** Lack of involvment of polyphenoloxidases in ortho-hydroxylation of phenolic compounds in mung bean seedlings, *Physiol. Plant.,* 59, 381, 1982.
81. **Dumazert, G.,** Les substances phénoliques des raisins et des vins blancs de Mauzac, Thèse Doct. Ing., Université Paul Sabatier, Toulouse, 1971.
82. **Dumazert, G., Margulis, H., and Montreau, F. R.,** Evolution of the phenolic compounds during maturation of a white *Vitis vinifera* Mauzac, *Ann. Technol. Agric.,* 22, 137, 1973.
83. **Dumont, L. C., Judge, F. D., Longeon, R., and Rognon, J.,** Les effets de l'Alar (S.A.D.H. — Aminozide) sur le développement et la maturation des fruits, in *Facteurs et Régulation de la Maturation des Fruits,* Coll. Int. CNRS, no. 238, 1975, 133.
84. **Duteau, J., Guilloux, M., Glories, Y., and Seguin, G.,** Influence de l'alimentation en eau de la vigne sur la teneur en sucre réducteurs, acides organiques et composés phénoliques des raisins, *C.R. Acad. Sci.,* Ser. III, 292, 965, 1981.
85. **Eastmond, R. and Gardner, R. J.,** (^{14}C)-epicatechin and (^{14}C)-procyanidins from seed shells of *Aesculus hippocastanum, Phytochemistry,* 13, 1477, 1974.
86. **Eaton, G. W., Zuckerman, B. M., Shawa, A. Y., Eck, P., Dana, M. N., Garren, R., and Lockhart, C. L.,** The effect of preharvest Malathion sprays upon cranberry fruit color, *J. Am. Soc. Hortic. Sci.,* 94, 590, 1969.
87. **Ebel, J. and Grisebach, H.,** Reduction of cinnamic acids to cinnamyl alcohols with an enzyme preparation from cell suspension cultures of soybean *(Glycin max.), FEBS Lett.,* 30, 141, 1973.
88. **Ebel, J. and Hahlbrock, K.,** Enzymes of flavone and flavonol glycoside biosynthesis. Coordinated and selective induction in cell suspension culture of *Petroselinum hortense, Eur. J. Biochem.,* 75, 201, 1977.
89. **Ebel, J. and Hahlbrock, K.,** Biosynthesis, in *The Flavonoids: Advances in Research,* Harborne, J. B. and Mabry, T. J., Eds., Chapman and Hall, New York, 1982, 641.
90. **Ebel, J., Schaller-Hekeler, B., Knobloch, K. H., Wellman, E., Grisebach, H., and Halhbrock, K.,** Coordinated changes in enzyme activities of phenylpropanoid metabolism during the growth of soybean cell suspension cultures, *Biochim. Biophys. Acta,* 362, 417, 1974.
91. **Eck, P.,** Chemical color enhancement of cranberry fruit, *HortScience,* 3, 70, 1968.
92. **Eck, P.,** Effect of preharvest sprays of Ethrel, Alar, Malathion on anthocyanin content of early black Cranberry (*Vaccinium macrocarpon* Ait.), *HortScience,* 4, 224, 1969.
93. **Eck, P.,** Influence of Ethrel upon highbush blueberry fruit ripening, *HortScience,* 5, 23, 1970.
94. **Eck, P.,** Cranberry yield and anthocyanin content as influenced by ethephon, SADH, and Malathion, *J. Am. Soc. Hortic. Sci.,* 97, 213, 1972.
95. **Eck, P.,** Cranberry, in *Handbook of Fruit Set and Development,* Monselise, S. P., Ed., CRC Press, Boca Raton, FL, 1986, 109.
96. **Edgerton, L. J. and Blanpied, G. D.,** Interaction of succinic acid 2,2,-dimethylhydrazide, 2-chlorophosphonic acid and auxins on maturity, quality and abscission of apples, *J. Am. Soc. Hortic. Sci.,* 95, 664, 1970.
97. **Erez, A. and Flore, J. A.,** The quantitative effect of solar radiation on "Redhaven" peach fruit skin color, *HortScience,* 21, 1424, 1986.
98. **Even-Chen, Z., Mattoo, A. K., and Goren, R.,** Inhibition of ethylene biosynthesis by aminoethoxyvinylglycine and by polyamines shunts label from 3,4-(^{14}C)methionine into spermidine in aged orange peel discs, *Plant Physiol.,* 69, 385, 1982.
99. **Faragher, J. D.,** Temperature regulation of anthocyanin accumulation in apple skins, *J. Exp. Bot.,* 34, 1291, 1983.
100. **Faragher, J. D. and Brohier, R. L.,** Anthocyanin accumulation in apple skin during ripening: regulation by ethylene and phenylalanine ammonialyase, *Sci. Hortic.,* 22, 89, 1984.
101. **Faragher, J. D. and Chalmers, D. J.,** Regulation of anthocyanin synthesis in apple skin. III. Involvement of phenylalanine ammonia-lyase, *Aust. J. Plant Physiol.,* 4, 133, 1977.
102. **Faust, M.,** Physiology of anthocyanin development in McIntosh apple. I. Participation of pentose phosphate pathway in anthocyanin development, *Proc. Am. Soc. Hortic. Sci.,* 87, 1, 1965.
103. **Faust, M.,** Effect on growth regulators on firmness and red color of fruit, *Acta Hortic.,* 34, 407, 1973.
104. **Faust, M.,** The role of calcium in the respiratory mechanism and senescence of apples, in *Facteurs et Regulation de la Maturation des Fruits,* Coll. Int. CNRS, no. 238, 1975, 87.
105. **Finkle, B. J. and Nelson, R. F.,** Enzyme reactions with phenolic compounds: Effect of *O*-methyltransferase on a natural substrate of fruit polyphenol oxidase, *Nature (London),* 197, 902, 1963.

106. **Finkle, B. J. and Nelson, R. F.**, Enzyme reactions with phenolic compounds: a meta-*O*-methyltransferase in plants, *Biochim. Biophys. Acta,* 78, 747, 1963.
107. **Fleurence, J. and Negrel, J.**, Partial purification of tyramine feruloyl transferase from TMV inoculated tobacco leaves, *Phytochemistry,* 28, 733, 1989.
108. **Fleuriet, A.**, Evolution des composés phénoliques au cours de la croissance et de la maturation des fruits de tomatoes "cerise" (*Lycopersicum esculentum* var. cerasiforme), *Fruits,* 31, 117, 1976.
109. **Fleuriet, A.**, Expression et régulation du métabolisme des dérivés hydroxycinnamiques au cours de la croissance, de la maturation et de la réaction aux blessures du fruit de la tomate "cerise" (*Lycopersicum esculentum,* var. cerasiforme), These Doct. Etat, Université de Montpellier, France, 1982.
110. **Fleuriet, A. and Deloire, A.**, Aspects histochimiques et biochimiques de la cicatrisation des fruits de Tomate blessés, *Z. Pflanzenphysiol.,* 107, 259, 1982.
111. **Fleuriet, A. and Macheix, J. J.**, Effet de conditions anaérobies sur les composés phénoliques des fruits de tomates "cerise" (*Lycopersicum esculentum,* var. cerasiforme), *Physiol. Vég.,* 14, 407, 1976.
112. **Fleuriet, A. and Macheix, J. J.**, Effets de traitements par l'azote et l'éthylène sur l'activité de la phénylalanine ammoniaque-lyase (PAL) des fruits de la tomate 'cerise' (*Lycopersicum esculentum,* var. cerasiforme), *C.R. Acad. Sci. Paris,* Ser. D, 285, 1103, 1977.
113. **Fleuriet, A. and Macheix, J. J.**, Dérivés glucosés des acides hydroxycinnamiques de la tomate: évolution au cours de la vie du fruit et formation *in vitro, Phytochemistry,* 19, 1955, 1980.
114. **Fleuriet, A. and Macheix, J. L.**, Quinyl esters and glucose derivatives of hydroxycinnamic acids during growth and ripening of tomato fruit, *Phytochemistry,* 20, 667, 1981.
115. **Fleuriet, A. and Macheix, J. J.**, Orientation nouvelle du métabolisme des acides hydroxycinnamiques dans les fruits de tomate blessés *(Lycopersicum esculentum), Physiol. Plant,* 61, 64, 1984.
116. **Fleuriet, A. and Macheix, J. J.**, Tissue compartmentation of phenylpropanoid metabolism in tomatoes during growth and maturation, *Phytochemistry,* 24, 929, 1985.
117. **Fleuriet, A., Macheix, J. J., Suen, R., and Ibrahim, R. K.**, Partial purification and some properties of a hydroxycinnamoyl glucosyltransferase from tomato fruits, *Z. Naturf. C.,* 35, 967, 1980.
118. **Floss, H. G.**, The shikimate pathway, in *Recent Advances in Phytochemistry,* Vol. 12, Swain, T., Harborne, J. B., and Van Sumere, C. F., Eds., Plenum Press, New York, 1979, 59.
119. **Foo, L. Y. and Porter, L. J.**, The phytochemistry of proanthocyanidin polymers, *Phytochemistry,* 19, 1747, 1980.
120. **Foo, L. Y. and Porter, L. J.**, The structure of tannins of some edible fruits, *J. Sci. Food Agric.,* 32, 711, 1981.
121. **Fordebras, I. and Macheix, J. J.**, Evolution de la teneur en *o*-diphénols et an acide chlorogénique des poires Passe-Crassane et de leurs potentialités de brunissement au cours de la conversation, *C.R. Acad. Agric. France,* 65, 502, 1979.
122. **Francis, F. J. and Atwood, W. M.**, The effect of fertilizer treatments on the pigment content of cranberries, *Proc. Am. Soc. Hortic. Sci.,* 77, 351, 1961.
123. **Frylinck, L., Dubery, I. A., and Schabort, J. C.**, Biochemical changes involved in stress response and ripening behaviour of γ-irradiated mango fruits, *Phytochemistry,* 26, 681, 1987.
124. **Fuke, Y. and Matsuoka, H.**, Changes in content of pectic substances, ascorbid acid and polyphenols and activity of pectinesterase in kiwi fruit during growth and ripening after harvest, *J. Jpn. Soc. Food Sci. Technol.,* 31, 31, 1984.
125. **Gaspar, Th.**, Integrated relationships of biochemical and physiological peroxidase activities, in *Molecular and Physiological Aspects of Plant Peroxidases,* Greppin, H., Penel, C., and Gaspar, Th., Eds., University of Geneva, 1986, 455.
126. **Gazit, S. and Adato, I.**, Effect of carbon dioxide atmosphere on the course of astringency disappearance of persimmon fruits, *J. Food Sci.,* 37, 815, 1972.
127. **Gianfagna, T. J. and Berkowitz, G. A.**, Glucose catabolism and anthocyanin production in apple fruit, *Phytochemistry,* 25, 607, 1986.
128. **Given, N. K., Venis, M. A., and Grierson, D.**, Phenylalanine ammonia-lyase activity and anthocyanin synthesis in ripening strawberry fruit, *J. Plant Physiol.,* 133, 25, 1988.
129. **Given, N. K., Venis, M. A., and Grierson, D.**, Purification and properties of phenylalanine ammonialyase from strawberry fruit and its synthesis during ripening, *J. Plant Physiol.,* 133, 31, 1988.
130. **Glazener, J. A.**, Accumulation of phenolic compounds in cells and formation of lignin like polymers in cell walls of young tomato fruits after inoculation with *Botrytis cinerea, Physiol. Plant Pathol.,* 20, 11, 1982.
131. **Goldstein, J. L. and Swain, T.**, Changes in tannins in ripening fruit, *Phytochemistry,* 2, 371, 1963.
132. **Goldstein, J. L. and Swain, T.**, The inhibition of enzymes by tannins, *Phytochemistry,* 4, 185, 1965.
133. **Goodenough, P. W., Kessel, S., Lea, A. G. H., and Loeffler, T.**, Mono- and diphenolase activity from fruit of *Malus pumila, Phytochemistry,* 22, 359, 1983.
134. **Graham, D. and Patterson, B. D.**, Responses of plants to low, non-freezing temperatures: proteins, metabolism, and acclimation, *Annu. Rev. Plant Physiol.,* 33, 347, 1982.

135. **Grand, C.,** Ferulic acid 5-hydroxylase: a new cytochrome P-450-dependent enzyme from higher plant microsomes involves in lignin synthesis, *FEBS Letters,* 169, 7, 1984.
136. **Green, A.,** Soft fruits, in *The Biochemistry of Fruits and Their Products,* Vol. 2, Hulme, A. C., Ed., Academic Press, London, 1971, 375.
137. **Greene, D. W. and Lord, W. J.,** Influence of light and spray coverage on red color development of Ethephon-treated 'McIntosh' apples, *HortScience,* 10, 254, 1975.
138. **Grisebach, H.,** Biochemistry of lignification, *Naturwissenschaften,* 64, 619, 1977.
139. **Grisebach, H.,** Selected topics in flavonoid biosynthesis, in *Recent Advances in Phytochemistry,* Vol. 12, Swain, T., Harborne, J. B., and Van Sumere, C. F., Eds., Plenum Press, New York, 1979, 221.
140. **Grisebach, H.,** Lignins, in *The Biochemistry of Plants,* Vol. 7, Conn, E. E., Ed., Academic Press, New York, 1981, 457.
141. **Grisebach, H.,** Biosynthesis of anthocyanins, in *Anthocyanins as Food Colors,* Markakis, P., Ed., Academic Press, New York, 1982, 69.
142. **Grisebach, H.,** Topics in flavonoid biosynthesis, in *Ann. Proc. Phytochem. Soc. Europe,* Vol. 25, Van Sumere, C. F. and Lea, P. J., Eds., Clarendon Press, Oxford, 1985, 183.
143. **Gross, G. G.,** The chemistry and biosynthesis of lignin, in *Biochemistry of Plant Phenolics,* Swain, T., Harborne, J. B., and Van Sumere, C. F., Eds., Plenum Press, New York, 1979, 177.
144. **Gross, G. G.,** Phenolic acids, in *The Biochemistry of Plants,* Vol. 7, Conn. E. E., Ed., Academic Press, New York, 1981, 301.
145. **Gross, G. G. and Zenk, M. H.,** Isolation and properties of hydroxycinnamate: CoA ligase from lignifying tissue of Forsythia, *Eur. J. Biochem.,* 42, 453, 1974.
146. **Gross, J.,** *Pigments in Fruits,* Academic Press, London, 1987.
147. **Guern, J., Renaudin, J. P., and Brown, S. C.,** The compartmentation of secondary metabolites in plant cell cultures, in *Cell Culture in Phytochemistry,* Vol. 4, Constabel, F. and Vasil, I. K., Eds., Academic Press, London, 1987, 43.
148. **Guiotto, A., Rodighiero, P., Pastorini, G., and Celon, E.,** Coumarins from unripe fruits of *Poncirus trifoliata; Phytochemistry,* 16, 1257, 1977.
149. **Gupta, R. K., Al-Shafi, S. M. K., Layden, K., and Haslam, E.,** The metabolism of gallic acid and hexahydroxydiphenic acid in higher plants. Part 2. Esters of (S)-hexahydroxydiphenic acid with D-glucopyranose ($^{14}C1$), *J. Chem. Soc. Perkin Trans. 1,* 2525, 1982.
150. **Gur, A.,** Plum, in *Handbook of Fruit Set and Development,* Monselise, S. P., Ed., CRC Press, Boca Raton, FL, 1986, 401.
151. **Haddock, E. A., Gupta, R. K., Al-Shafi, S. M. K., Layden, K., Haslam, E., and Magnolato, D.,** The metabolism of gallic acid and hexahydroxydiphenic acid in plants: biogenetic and molecular taxonomic considerations, *Phytochemistry,* 21, 1049, 1982.
152. **Hahlbrock, K.,** Flavonoids, in *The Biochemistry of Plants,* Vol. 7, Conn, E. E., Ed., Academic Press, New York, 1981, 425.
153. **Hahlbrock, K. and Grisebach, H.,** Biosynthesis of flavonoids, in *The Flavonoids,* Harborne, J. B., Mabry, T. J., and Mabry, H., Eds., Chapman and Hall, London, 1975, 866.
154. **Hahlbrock, K. and Grisebach, H.,** Enzymic controls in the biosynthesis of lignin and flavonoids, *Annu. Rev. Plant Physiol.,* 30, 105, 1979.
155. **Halbrock, K. and Schroeder, J.,** Light-induced changes of enzyme activities in Parsley cell suspension cultures. Increased rate of synthesis of phenylalanine ammonia-lyase, *Arch. Biochem. Biophys.,* 166, 47, 1975.
156. **Hale, C. R., Coombe, B. G., and Hawker, J. S.,** Effects of ethylene and 2-chloroethylphosphonic acid on the ripening of grapes, *Plant Physiol.,* 45, 620, 1970.
157. **Hall, R. D., Holden, M. A., and Yeoman, M. M.,** Studies on the regulation of capsaicin biosynthesis in immobilized cell cultures and the developing fruit of the chili pepper, *Capsicum frutescens,* in *Secondary Metabolism in Plant Cell Cultures,* Morris, P., Scragg, A. H., Stafford, A., and Fowler, M. W., Eds., Cambridge University Press, Cambridge, 1986, 121.
158. **Hall, R. D., Holden, M. A., and Yeoman, M. M.,** The accumulation of phenylpropanoid and capsaicinoid compounds in cell cultures and whole fruit of the chili pepper *Capsicum frutescens, Plant Cell Tissue Organ Culture,* 8, 163, 1987.
159. **Hansen, E.,** Proteins, in *The Biochemistry of Fruits and Their Products,* Vol. 1, Hulme, A. C., Ed., Academic Press, London, 1970, 147.
160. **Hanson, K. R. and Havir, E. A.,** An introduction to the enzymology of phenylpropanoid biosynthesis, in *Recent Advances in Phytochemistry,* Vol. 12, Swain, T., Harborne, J. B. and Van Sumere, C. F., Eds., Plenum Press, New York, 1979, 91.
161. **Hanson, K. R. and Havir, E. A.,** Phenylalanine ammonia-lyase, in *The Biochemistry of Plants,* Vol. 7, Conn, E. E., Ed., Academic Press, New York, 1981, 577.
162. **Harborne, J. B.,** Plant phenolics, in *Encyclopedia of Plant Physiology,* Vol. 8, Bell, E. A. and Charlwood, B. V., Eds., Springer-Verlag, Berlin, 1980, 329.

163. **Harborne, J. B.,** The flavonoids of the Rutales, in *Chemistry and Chemical Taxonomy of the Rutales,* Waterman, P. G. and Grundon, M. F., Eds., Academic Press, London, 1983, 147.
164. **Harborne, J. B. and Corner, J. J.,** Plant polyphenols. 4-Hydroxycinnamic acid-sugar derivatives, *Biochem. J.,* 81, 242, 1961.
165. **Harel, E. and Mayer, A. M.,** Partial purification and properties of catechol oxidases in grapes, *Phytochemistry,* 10, 17, 1971.
166. **Harkin, J. M. and Obst, J. R.,** Lignification in trees: indication of exclusive peroxydase participation, *Science,* 180, 296, 1973.
167. **Hasegawa, S. and Maier, V. P.,** Cinnamate hydroxylation and the enzyme loading from phenylpyruvate to *p*-coumarate synthesis in grapefruit tissues, *Phytochemistry,* 11, 1365, 1972.
168. **Haslam, E.,** Symmetry and promiscuity in procyanidin biochemistry, *Phytochemistry,* 16, 1625, 1977.
169. **Haslam, E.,** Vegetable tannins, in *Recent Advances in Phytochemistry,* Vol. 12, Swain, T., Harborne, J. B., and Van Sumere, C. F., Eds., Plenum Press, New York, 1979, 475.
170. **Haslam, E.,** Vegetable tannins, in *The Biochemistry of Plants,* Vol. 7, Conn, E. E., Ed., Academic Press, New York, 1981, 527.
171. **Haslam, E.,** Proanthocyanidins, in *The Flavonoids: Advances in Research,* Harborne, J. B. and Mabry, T. J., Eds., Chapman and Hall, London, 1982, 417.
172. **Haslam, E. and Lilley, T. H.,** New polyphenols for old tannins, in *Proc. Phytochem. Soc. Eur.,* Vol. 25, Sumere, C. F. and Lea, P. J., Eds., Clarendon Press, Oxford, 1985, 237.
173. **Haslam, E. and Lilley, T. H.,** Natural astringency in foodstuffs. A molecular interpretation, *CRC Crit. Rev. Food Sci. Nutr.,* 27, 1, 1988.
174. **Haslam, E., Lilley, T. H., and Ozawa, T.,** Polyphenol complexation. Astringency in fruit and beverages, *Bull. Liaison Groupe Polyphénols,* 13, 352, 1986.
175. **Heinicke, D. R.,** Characteristics of Mc Intosh and red Delicious apples as influenced by exposure to sunlight during the growing season, *Proc. Am. Soc. Hortic. Sci.,* 89, 10, 1966.
176. **Heller, W. and Hahlbrock, K.,** Highly purified "Flavanone synthase" from parsley catalyzes the formation of naringenin chalcone, *Arch. Biochem. Biophys.,* 200, 617, 1980.
177. **Heller, W., Britsch, L., Forkmann, G., and Grisebach, H.,** Leucoanthocyanidins as intermediates in anthocyanidin biosynthesis in flowers of *Matthiola incana* R. Br., *Planta,* 163, 191, 1985.
178. **Herrmann, K.,** The phenolics of fruits. I. Our knowledge of occurrence and concentrations of fruit phenolics and of their variations in the growing fruit, *Z. Lebensm. Unters. Forsch.,* 151, 41, 1973.
179. **Ho, L. C.,** Regulation of assimilate translocation between leaves and fruits in the tomato, *Ann. Bot.,* 43, 437, 1979.
180. **Hollander, H. and Amrhein, N.,** The site of the inhibition of the shikimate pathway by glyphosate. I. Inhibition by glyphosate of phenylpropanoid synthesis in buckwheat (*Fagopyrum esculentum* Moench.), *Plant Physiol.,* 66, 823, 1980.
181. **Howell, G. S., Stergios, B. G., Stackhouse, S. S., Bittenbender, H. C., and Burton, C. L.,** Ethephon as a mechanical harvesting aid for highbush blueberries (*Vaccinium australie* Small), *J. Am. Soc. Hortic. Sci.,* 101, 111, 1976.
182. **Hrazdina, G. and Wagner, G. J.,** Compartmentation of plant phenolic compounds: sites of synthesis and accumulation, in *Ann. Proc. Phytochem. Soc. Eur.,* Vol. 25, Van Sumere, C. F. and Lea, P. J., Eds., Clarendon Press, Oxford, 1985, 119.
183. **Hrazdina, G., Parsons, G. F., and Mattick, L. R.,** Physiological and biochemical events during development and maturation of grape berries, *Am. J. Enol. Vitic.,* 35, 220, 1984.
184. **Hrazdina, G., Kreuzaler, F., Hahlbrock, K., and Grisebach, H.,** Substrate specificity of flavonone synthase from cell suspension cultures of parsley and structure of release products *in vitro, Arch. Biochem. Biophys.,* 175, 392, 1976.
185. **Hulme, A. C.,** Transformations des constituants phénoliques de la "peau" des pommes et des poires pendant la croissance et la sénescence, *Bull. Soc. Fcse. Phys. Vég.,* 4, 43, 1958.
186. **Hulme, A. C. and Rhodes, M. J. C.,** Pome fruits, in *The Biochemistry of Fruits and Their Products,* Vol. 2, Hulme, A. C., Ed., Academic Press, London, 1971, 363.
187. **Hwang, B. K.,** Contents of sugars, fruit acids, aminoacids and phenolic compounds of apple fruits in relation to their susceptibility to *Botryosphaeria ribis, Phytopath. Z.,* 108, 1, 1983.
188. **Hyodo, H.,** Phenylalanine ammonia-lyase in strawberry fruits, *Plant Cell Physiol.,* 12, 989, 1971.
189. **Ibrahim, R. K.,** Regulation of synthesis of phenolics, in *Cell Culture and Somatic Cell Genetics of Plants,* Vol. 4, Constabel, F. and Vasil, I. K., Eds., Academic Press, San Diego, 1987, 77.
190. **Ibrahim, R. K. and Boulay, B.,** Purification and some properties of UDP-glucose: *O*-dihydroxycoumarin 7-*O*-glycosyltranferase from tobacco cell cultures, *Plant Sci. Lett.,* 18, 177, 1980.
191. **Ibrahim, R. K. and Grisebach, H.,** Purification and properties of UDP-glucose; Coniferyl alcohol glucosyltransferase from suspension cultures of Paul's Scarlet Rose, *Arch. Biochem. Biophys.,* 176, 700, 1976.

192. **Ishikura, N., Murakami, H., and Fujii, Y.,** Conversion of (+)-Dihydroquercetin to 3,4-cis-leucocyanidin by a reductase extracted from cell suspension cultures of *Cryptomeria japonica*, *Plant Cell Physiol.*, 29, 795, 1988.
193. **Ismail, A. A.,** Preharvest application of ethephon and SADH on ripening and quality of lowbush blueberry fruit, *HortScience*, 9, 205, 1974.
194. **Ito, S.,** The persimmon, in *The Biochemistry of Fruits and Their Products*, Vol. 2, Hulme, A. C., Ed., Academic Press, London, 1971, 281.
195. **Itoo, S.,** Persimmon, in *Handbook of Fruit Set and Development*, Monselise, S. P., Ed., CRC Press, Boca Raton, FL, 1986, 355.
196. **Itoo, S., Matsuo, T., Ibushi, Y., and Tamari, N.,** Seasonal changes in the levels of polyphenols in guava fruit and leaves and some their properties (Japanese), *J. Jpn. Soc. Hortic. Sci.*, 56, 107, 1987.
197. **Iwai, K., Lee, K. R., Kobashi, M., and Suzuki, T.,** Formation of pungent principles in fruits of sweet pepper, *Capsicum annum* L. var. grossum, during post-harvest ripening under continuous light, *Agric. Biol. Chem.*, 41, 1873, 1977.
198. **Jablonowski, W.,** Effect of chlorocholine chloride (CCC) on choline and rutin content in tomato fruits, *Bromatol. Chem. Toksykol.*, 5, 27, 1972.
199. **Jacques, D., Opie, C. T., Porter, L. J., and Haslam, E.,** Plant proanthocyanidins. Part 4. Biosynthesis of procyanidins and observations on the metabolism of cyanidin in plants, *J. Chem. Soc., Perkin I*, 1637, 1977.
200. **Jensen, F. L., Kissler, J. J., Peacock, W. L., and Leavitt, G. M.,** Effect of ethephon on color and fruit characteristics of 'Tokay' and 'Emperor' table grapes, *Am. J. Enol. Vitic.*, 26, 79, 1975.
201. **Jensen, R. A.,** The shikimate/arogenate pathway: link between carbohydrate metabolism and secondary metabolism, *Physiol. Plant.*, 66, 164, 1986.
202. **Joslyn, M. A. and Goldstein, J. L.,** Astringency of fruits and fruit products in relation to phenolic content, *Adv. Food Res.*, 13, 179, 1964.
203. **Jourdan, P. S., Weiler, E. W., and Mansell, R. L.,** Naringin levels in Citrus tissues. I. Comparison of different antibodies and tracers for the radioimmunoassay of naringen, *Plant Physiol.*, 77, 896, 1985.
204. **Kahan, R. S., Monselise, S. P., Riov, J., Van Kooy, J., and Chadwick, K.,** Comparison of the effect of radiation of various penetrating powers on the damage to Citrus fruit peel, *Rad. Bot.*, 8, 415, 1968.
205. **Kahn, V. and Pomerantz, S. H.,** Monophenolase activity of avocado polyphenol oxidase, *Phytochemistry*, 19, 379, 1980.
206. **Kamsteeg, J., Van Brederode, J., and Van Nigtevecht, G.,** Identification, properties, and genetic control of UDP-glucose:cyanidin-3-rhamnosyl-(1-6)-glucoside-5-*O*-glucosyltransferase isolated from petals of the red Campion *(Silene dioica)*, *Biochem. Genet.*, 16, 1059, 1978.
207. **Kamsteeg, J., Van Brederode, J., and Van Nigtevecht, G.,** The pH-dependent substrate specificity of UDP-glucose:anthocyanidin 3-rhamnosylglucoside, 5-*O*-glucosyltransferase in petals of *Silene dioica*: the formation of anthocyanidin 3,5-diglucosides, *Z. Pflanzenphysiol.*, 96, 87, 1980.
208. **Kamsteeg, J., Van Brederode, J., Verschuren, P. M., and Van Nigtevecht, G.,** Identification, properties and genetic control of para-coumaroyl coenzyme A, 3-hydroxylase isolated from petals of *Silene dioica*, *Z. Pflanzenphysiol.*, 102, 435, 1981.
209. **Kärppä, J.,** Analysis of anthocyanins during ripening of crowberry, *Empetrum nigrum* coll., *Lebensm. Wiss. Technol.*, 17, 175, 1984.
210. **Kataoka, I., Kubo, Y., Sugiura, A., and Tomana, T.,** Changes in L-phenylalanine ammonia-lyase activity and anthocyanin synthesis during berry ripening of 3 grape cultivars, *J. Jpn. Soc. Hortic. Sci.*, 52, 273, 1983.
211. **Kataoka, I., Sugiura, A., Utsunomiya, N., and Tomana, T.,** Effect of abcissic acid and defoliation on anthocyanin accumulation in Kyoho grapes (*Vitis vinifera* L. × *V. labruscana* Bailey), *Vitis*, 21, 325, 1982.
212. **Kato, K.,** Factors related to lusterless and reddish brown coloring of eggplant fruits in greenhouse cultivation, *J. Jpn. Soc. Hortic. Sci.*, 56, 431, 1988.
213. **Kliewer, W. M.,** Effect of day temperature and light intensity on coloration of *Vitis vinifera* L. grapes, *J. Am. Soc. Hortic. Sci.*, 95, 693, 1970.
214. **Kliewer, W. M.,** Influence of temperature, solar radiation and nitrogen on coloration and composition of Emperor grapes, *Am. J. Enol. Vitic.*, 28, 96, 1977.
215. **Kliewer, W. M. and Schultz, H. B.,** Effect of sprinkler cooling of grapevines on fruit growth and composition, *Am. J. Enol. Vitic.*, 24, 17, 1973.
216. **Kliewer, W. M. and Torres, R. E.,** Effect of controlled day and night temperature on grape coloration, *Am. J. Enol. Vitic.*, 22, 71, 1972.
217. **Knapp, F. W., Hall, C. B., Buchanan, D. W., and Biggs, R. H.,** Reduction of polyphenoloxidase activity in Peaches sprayed with alar, ethrel or gibberellic acid, *Phytochemistry*, 9, 1453, 1970.
218. **Knobloch, K. H. and Hahlbrock, K.,** Isoenzyme der *p*-coumarat: CoA ligase aus Zell suspensionkulturen von *Glycine max*, *Planta Med.*, 102, 1975.

219. **Kojima, M. and Uritani, I.**, Studies on chlorogenic acid biosynthesis in sweet potato root tissue using *trans*-cinnamic acid-2-^{14}C and quinic acid-G-^{3}H^{1}, *Plant Cell Physiol.*, 13, 311, 1972.
220. **Kojima, M. and Uritani, I.**, Studies on chlorogenic acid biosynthesis in sweet potato root tissue in special reference to the isolation of a chlorogenic acid intermediate, *Plant Physiol.*, 51, 768, 1973.
221. **Kojima, M. and Villegas, R. J. A.**, Detection of the enzyme potato root wich catalyzes *trans*-esterification between 1-*O*-*p*-coumaroyl-D-glucose and D-quinic acid, *Agric. Biol. Chem.*, 48, 2397, 1984.
222. **Koukol, J. and Conn, E. E.**, The metabolism of aromatic compounds in higher plants. IV. Purification and properties of the phenylalanine deaminase of *Hordeum vulgare*, *J. Biol. Chem.*, 236, 2692, 1961.
223. **Koumba-Koumba, D. and Macheix, J. J.**, Biosynthesis of hydroxycinnamic derivatives in apple fruit cell suspension culture, *Physiol. Vég.*, 20, 137, 1982.
224. **Kozukue, N. and Ogata, K.**, Physiological and chemical studies of chilling injury in pepper fruits. III. The changes of phenylalanine ammonia-lyase and tyrosine ammonia-lyase activities in pepper seeds stored at 6°C and 20°C, *J. Jpn. Soc. Hortic. Sci.*, 40, 80, 1971.
225. **Lakshminarayana, S., Mathew, A. G., and Parpia, H. A. B.**, Changes in polyphenols of Sapota fruit (*Achras zapota* L.) during maturation, *J. Sci. Food Agric.*, 20, 651, 1969.
226. **Lal, R. K., Gary, M., and Krishnan, P. S.**, Biochemical aspects of developing and ripening bananas, *Phytochemistry*, 13, 2365, 1974.
227. **Lamikanra, O.**, Development of anthocyanin pigments in Muscadine grapes, *HortScience*, 23, 597, 1988.
228. **Lavee, S.**, Abscission studies of olive fruit. Physiological and horticultural aspects, *Olea*, 35, 1976.
229. **Lea, A. G. H. and Beech, F. W.**, The phenolics of ciders: effect of cultural conditions, *J. Sci. Food Agric.*, 29, 493, 1978.
230. **Lee, C. Y. and Jaworski, A.**, Phenolic compounds in white grapes grown in New York, *Am. J. Enol. Vitic.*, 38, 277, 1987.
231. **Lee, J. C. and Lee, V. B.**, Physiological study on coloration of plum fruits. 1. Effects of etephon on fruit composition and anthocyanin development in Santa Rosa plum *(Prunus salicina)*, *J. Kor. Soc. Hortic. Sci.*, 21, 36, 1980.
232. **Lees, D. H. and Francis, F. J.**, Effect of gamma radiation on anthocyanin and flavonol pigments in cranberries (*Vaccinium macrocarpon* Ait.), *J. Am. Soc. Hortic. Sci.*, 97, 128, 1972.
233. **Levitt, J.**, *Responses of Plants to Environmental Stresses*, Vol. 1, *Chilling, Freezing and High Temperature Stresses*, Academic Press, New York, 1980.
234. **Lewinsohn, E., Berman, E., Mazur, Y., and Gressel, J.**, Glucosylation of exogenous flavanones by grapefruit *(Citrus paradisi)* cell cultures, *Phytochemistry*, 25, 2531, 1986.
235. **Lindsey, K.**, Incorporation of (^{14}C) phenylalanine and (^{14}C)cinnamic acid into capsaicin in cultured cells of *Capsicum frutescens*, *Phytochemistry*, 25, 2793, 1986.
236. **Linko, R. R., Kärppä, J., Kallio, H., and Ahtonen, S.**, Anthocyanin contents of Crowberry and Crowberry juice, *Lebensm. Wiss. Technol.*, 16, 343, 1983.
237. **Lotfy, S., Fleuriet, A., Ramos, T., and Macheix, J. J.**, Biosynthesis of phenolic compounds in *Vitis vinifera* cv. Gamay Freau cell suspension cultures: study on hydroxycinnamoyl CoA:ligase, *Plant Cell Rep.*, 8, 93, 1989.
238. **Luckner, M.**, Expression and control of secondary metabolism, in *Encyclopedia of Plant Physiol.*, Vol. 8, Bell, E. A. and Charlwood, B. V., Eds., Springer-Verlag, Berlin, 1980, 23.
239. **Lyons, J. M.**, Chilling injury in plants, *Annu. Rev. Plant Physiol.*, 24, 445, 1973.
240. **Macheix, J. J.**, Sur l'acide chlorogénique des pommes: méthode de dosage et variations au cours de la croissance, de la maturation et de la conservation, *C.R. Acad. Sci.*, Paris, Ser. D, 264, 3010, 1967.
241. **Macheix, J. J.**, Role de divers facteurs intervenant dans le brunissement enzymatique des pommes pendant la croissance, *Physiol. Vég.*, 8, 585, 1970.
242. **Macheix, J. J.**, Les esters hydroxycinnamiques de la pomme: identification, variations au cours de la croissance du fruit et métabolisme, Thèse Doct. Etat, Université Pierre et Marie Curie, Paris, 1974.
243. **Macheix, J. J.**, Evolution de la teneur du fruit de *Pirus Malus* L. en esters des acides *p*-coumarique et cafeique au cours de la croissance, *Physiol. Vég.*, 12, 25, 1974.
244. **Macheix, J. J.**, Etude de quelques caractéristiques de l'accumulation du *p*-coumarylglucose dans le parenchyme de la Pomme (*Pirus Malus* L.), *Physiol. Veg.*, 13, 591, 1975.
245. **Macheix, J. J.**, Biosynthèse des esters glucosés des acides hydroxycinnamiques à partir d'uridine diphosphoglucose et des acides libres chez la pomme, *C.R. Acad. Sci.*, Paris Ser. D., 284, 33, 1977.
246. **Macheix, J. J. and Ibrahim, R. K.**, The *O*-methyltransferase system of apple fruit cell suspension culture, *Biochem. Physiol. Pflanzen.*, 179, 659, 1984.
247. **Macheix, J. J., Fleuriet, A., and Quessada, M. P.**, Involvement of phenols and peroxidases in wound healing and grafting, in *Molecular and Physiological Aspects of Plant Peroxidases*, Greppin, H., Penel, C., and Gaspar, Th., Eds., University of Geneva, Switzerland, 1986, 267.
248. **Macheix, J. J., Suen, R., and Ibrahim, R. K.**, Metabolism of phenylpropanoid compounds in apple fruit cell suspension culture, *Biochem. Physiol. Pflanzen*, 176, 195, 1981.

249. **Mäder, M., Nessel, A., and Bopp, M.,** On the physiological significance of the isoenzyme groups of peroxidase from tobacco demonstrated by biochemical properties. II. pH optima, Michaelis constants, maximal oxidation, *Z. Pflanzenphysiol.*, 82, 247, 1977.
250. **Mäder, M., Nessel, A., and Schloss, P.,** Cell compartimentation and specific roles of isoenzymes, in *Molecular and Physiological Aspects of Plants Peroxidases*, Greppin, H., Penel, C., and Gaspar, Th., Eds., University of Geneva, Switzerland, 1986, 247.
251. **Maestro Duran, R. and Vasquez Roncero, A.,** Colorantes antocianicos de las aceitunas manzanillas maduras, *Grasas Aceites*, 27, 237, 1976.
252. **Magness, J. R.,** Observation on color development in apple, *Proc. Am. Soc. Hortic. Sci.*, 25, 289, 1928.
253. **Maier, V. P. and Hasegawa, S.,** L-Phenylalanine ammonia-lyase activity and naringenin glycoside accumulation in developing grapefruit, *Phytochemistry*, 9, 139, 1970.
254. **Makus, D. J. and Ballinger, W. E.,** Characterization of anthocyanins during ripening of fruit of *Vaccinium corymbosum* L. cv. Wolcoot, *J. Am. Soc. Hortic. Sci.*, 98, 99, 1973.
255. **Mancinelli, A. L.,** Light-dependence of anthocyanin synthesis: a model system for the study of plant photomorphogenesis, *Bot. Rev.*, 51, 107, 1985.
256. **Mansell, R. L., Stöckigt, J., and Zenk, M. H.,** Reduction of ferulic acid to coniferyl alcohol in a cell free system from a higher plant, *Z. Pflanzenphysiol.*, 68, 286, 1972.
257. **Margna, U.,** Control at the level of substrate supply. An alternative in the regulation of phenylpropanoid accumulation in plant cells, *Phytochemistry*, 16, 419, 1977.
258. **Martinez-Cayuela, M., Faus, M. J., and Gil, A.,** Effects of some reductants on the activity of Cherimoya polyphenol oxidase, *Phytochemistry*, 27, 1589, 1988.
259. **Mathew, A. G. and Govindarajan, V. S.,** Polyphenolic substances of Arecanut. II. Changes during maturation and ripening, *Phytochemistry*, 3, 657, 1964.
260. **Matsuo, T. and Itoo, S.,** On mechanisms of removing astringency in persimmon fruit by carbon dioxide treatment. I. Some properties of the two processes in the de-astringency, *Plant Cell Physiol.*, 18, 17, 1977.
261. **Matsuo, T. and Ito, S.,** The chemical structure of Kaki-tannin from immature fruit of the persimmon (*Diospyros Kaki* L.), *Agric. Biol. Chem.*, 42, 1637, 1978.
262. **Matsuo, T. and Itoo, S.,** A model experiment for de-astringency of persimmon fruit with high carbon dioxide treatment: *in vitro* gelation of Kaki-tannin by reacting with acetaldehyde, *Agric. Biol. Chem.*, 46, 683, 1982.
263. **Matsuo, T., Kawazoe, H., and Itoo, S.,** Seasonal changes in the polyphenol content of Persimmon leaves, calyxes and young fruit, *Bull. Facult. Agr., Kagoshima Univ.*, 31, 1, 1981.
264. **Matsuo, T., Shinohara, J., and Itoo, S.,** An improvement on removing astringency in persimmon fruits by carbon dioxide gas, *Agric. Biol. Chem.*, 40, 215, 1976.
265. **Mayer, A. M.,** Polyphenol oxidases in plants, in *Recent Advances in Phytochemistry*, Vol. 26, Harborne, J. B., Smith, T. A., and Swain, T., Eds., Pergammon Press, New York, 1977, 11.
266. **Mayer, A. M.,** Polyphenol oxidases in plants. Recent progress, *Phytochemistry*, 26, 11, 1987.
267. **Mayer, A. M. and Harel, E.,** Polyphenol oxydases in plants, *Phytochemistry*, 18, 193, 1979.
268. **Mayer, A. M. and Harel, E.,** Polyphenol oxidases in fruits. Changes during ripening, in *Recent Advances in the Biochemistry of Fruits and Vegetables*, Friend, J. and Rhodes, M. J. C., Eds., Academic Press, London, 1981, 161.
269. **McBride, R. L. and Faragher, J. D.,** The sensory evaluation of Jonathan and Delicious apples treated with the growth regulator ethephon, *J. Sci. Food Agric.*, 29, 465, 1978.
270. **McCloskey, L. P. and Yengoyan, L. S.,** Analysis of anthocyanins in *Vitis vinifera* wines and red color versus aging of HPLC and spectrophotometry, *Am. J. Enol. Vitic.*, 32, 257, 1981.
271. **McClure, J. W.,** The physiology of phenolic compounds in plants, in *Biochemistry of Plant Phenolics*, Vol. 12, Swain, T., Harborne, J. B., and Van Sumere, C. F., Eds., Plenum Press, New York, 525, 1979.
272. **Melin, C., Billot, J., and Dupin, J. F.,** Les composés phénoliques de la Cerise Bigareau Napoléon. II. Les flavonoides, *Physiol. Vég.*, 17, 573, 1979.
273. **Melin, C., Moulet, A. M., Dupin, J. F., and Hartmann, C.,** Phenylalanine ammoniaque lyase et composés phenoliques au cours de la maturation de la cerise, *Phytochemistry*, 16, 75, 1977.
274. **Melin-Moulet, A. M. and Hartmann, C.,** La phénylalanine ammoniaque-lyase (PAL) dans le fruit de deux Rosacées, la cerise et la poire, *Physiol. Vég.*, 16, 563, 1978.
275. **Mohr, H. and Drumm-Herrel, H.,** Coaction between phytochrome and blue/UV light in antocyanin synthesis in seedlings, *Physiol. Plant.*, 58, 408, 1983.
276. **Monselise, S. P.,** Citrus, in *Handbook of Fruit Set and Development*, Monselise, S. P., Ed., CRC Press, Boca Raton, FL, 1986, 87.
276a. **Monselise, S. P.,** *Handbook of Fruit Set and Development*, CRC Press, Boca Raton, FL, 1986.
277. **Moore, D. H.,** New formulation restores interest in Ferbam as a means of enhancing berry coloration while combatting fruit rots, cranberries, *Natl. Cranberry Mag.*, 26, 10, 1962.
278. **Moriguchi, T., Villegas, R. J. A., Kondo, T., and Kojima, M.,** Isolation of 1-*O-trans*-para-coumaroyl-beta-deuterium-glucopyranose from sweet potato roots and examination of its role in chlorogenic acid biosynthesis, *Plant Cell Physiol.*, 29, 221, 1988.

279. **Mosel, H. D. and Herrmann, K.,** The phenolics of fruits. III. The content of catechins and hydroxycinnamic acids in pome and stone fruits, *Z. Lebensm. Unters. Forsch.*, 154, 6, 1974.
280. **Mosel, H. D. and Herrmann, K.,** Changes in catechins and hydroxycinnamic acid derivatives during development of apples and pears, *J. Sci. Food Agric.*, 25, 251, 1974.
281. **Mosel, H. D. and Herrmann,** Die phenolischen inhaltsstoffe des Obstes. IV. Die phenolischen inhalsstoffe der Brombeeren und Himbeeren und deren veränderungen während wachstum und reife der fruchte, *Z. Lebens. Unters. Forsch.*, 154, 324, 1974.
282. **Murphey, A. S. and Dilley, D. R.,** Anthocyanin biosynthesis and maturity of McIntosh apples as influenced by ethylene-releasing compounds, *J. Am. Soc. Hortic. Sci.*, 113, 718, 1988.
283. **Murray, R. D. H., Mendez, J., and Brown, S. A.,** *The Natural Coumarins. Occurrence, Chemistry and Biochemistry*, J. Wiley-Interscience, Chicester, 1982.
284. **Nagel, C. W. and Wulf, L. W.,** Changes in the anthocyanins, flavonoids, and hydroxycinnamic acid esters during fermentation and aging of Merlot and Cabernet Sauvignon, *Am. J. Enol. Vitic.*, 30, 111, 1979.
285. **Nagels, L. and Parmentier, F.,** Kinetic study of possible intermediates in the biosynthesis of chlorogenic acid in *Cestrum poeppigii*, *Phytochemistry*, 15, 703, 1976.
286. **Nakamura, K., Amano, Y., and Kagami, M.,** Purification and some properties of a polyphenol oxidase from Koshu grapes, *Am. J. Enol. Vitic.*, 34, 122, 1983.
287. **Ndubizu, T. O. C.,** Relations of phenolic inhibitors to resistance of immature apple fruits to rot, *J. Hortic. Sci.*, 51, 311, 1976.
288. **Nelson, R. F. and Finkle, B. J.,** Enzyme reactions with phenolic compounds: effects of *O*-methyltransferase and high pH on the polyphenol oxidase substrates in apple, *Phytochemistry*, 3, 321, 1964.
289. **Nothmann, J.,** Eggplant, in *Handbook of Fruit Set and Development*, Monselise, S. P., Ed., CRC Press, Boca Raton, FL, 1986, 145.
290. **Nothmann, J., Rylski, I., and Spigelman, M.,** Color and variations in color intensity of fruit of different eggplant cultivars, *Sci. Hortic.*, 4, 191, 1976.
291. **Nothmann, J., Rylski, I., and Spigelman, M.,** Effects of air and soil temperatures on colour of eggplant fruits (*Solanum melongena* L.), *Expl. Agric.*, 14, 189, 1978.
292. **Nothmann, J., Rylski, I., and Spigelman, M.,** Flowering pattern, fruit growth and color development of eggplant during the coal season in a subtropical climate, *Sci. Hortic.*, 11, 217, 1979.
293. **Noé, W., Langebartels, C., and Seitz, H. U.,** Anthocyanin accumulation and PAL activity in a suspension culture of *Daucus carota* L. Inhibition by L-alpha aminoxy-beta-phenylpropionic acid and t-cinnamic acid, *Planta*, 149, 283, 1980.
294. **Ojha, N. L., Roy, A. K., and Shrivastava, A. K.,** Variation in total phenols in bottle Gourd fruits during maturation and storage, *Natl. Acad. Sci. Lett.*, 6, 43, 1983.
295. **Okombi, G.,** Les pigments de la cerise, *Prunus avia* (L.), variété "Bigarreau Napoléon": variations au cours de la croissance, de la maturation et de la conservation, Thèse Doctorat 3ème cycle, Université d'Orléans, France, 1979.
296. **Ong, B. Y. and Nagel, C. W.,** Hydroxycinnamic acid tartaric acid ester content in mature grapes and during maturation of white riesling grapes, *Am. J. Enol. Vitic.*, 29, 277, 1978.
297. **Oydvin, J.,** Inheritance of four cyanidin-3-glycosides in red Currant, *Hortic. Res.*, 14, 1, 1974.
298. **Ozawa, T., Lilley, T. H., and Haslam, E.,** Polyphenol interactions: astringency and the loss of astringency in ripening fruit, *Phytochemistry*, 26, 2937, 1987.
299. **Ozeki, Y., Komamine, A., Noguchi, H., and Sankawa, U.,** Changes in activities of enzymes involved in flavonoid metabolism during the initiation and suppression of anthocyanin synthesis in carrot suspension cultures regulated by 2,4-dichlorophenoxyacetic acid, *Physiol. Plant*, 69, 123, 1987.
300. **Palmer, J. K.,** The banana, in *The Biochemistry of Fruits and Their Products*, Vol. 2, Hulme, A. C., Ed., Academic Press, London, 1971, 65.
301. **Panisset, B. and Tissut, M.,** Métabolisme des flavonols dans les tuniques isolées de huit variétés d'oignon: effet d'un traitement par l'acide gibbérellique et l'aminozide, *Physiol. Vég.*, 21, 49, 1983.
302. **Park, E. Y. and Luh, B. S.,** Polyphenol oxidase of Kiwi fruit, *J. Food Sci.*, 50, 678, 1985.
303. **Paulson, A. T., Vanderstoep, J., and Eaton, G. N.,** Effects of gibberellic acid and Ethephon on enzymatic browning of 'Redhaven' peaches, *HortScience*, 14, 711, 1979.
304. **Paulson, A. T., Vanderstoep, J., and Porritt, S. W.,** Enzymatic browning of peaches: effect of gibberellic acid and ethephon on phenolic compounds and polyphenoloxidase activity, *J. Food Sci.*, 45, 341, 1980.
305. **Pecheur, J. and Ribaillier, D.,** La régulation de la maturation des fruits par l'éthéphon, in *Facteurs et Régulation de la Maturation des Fruits*, Coll. Int. CNRS, no. 238, 1975, 121.
306. **Pecket, R. C. and Small, C. J.,** Occurrence, location and development of anthocyanoplasts, *Phytochemistry*, 19, 2571, 1980.
307. **Pesis, E. and Ben-Arie, R.,** Carbon dioxide assimilation during postharvest removal of astringency from Persimmon fruit, *Physiol. Plant.*, 67, 644, 1986.

308. **Pirie, A. and Mullins, M. G.,** Changes in anthocyanin and phenolics content of grapevine leaf and fruit tissues treated with sucrose, nitrate and abscisic acid, *Plant Physiol.*, 58, 468, 1976.
309. **Pirie, A. and Mullins, M. G.,** Interrelationships of sugars, anthocyanins, total phenols and dry weight in the skin of grape berries during ripening, *Am. J. Enol. Vitic.*, 28, 204, 1977.
310. **Pirie, A. J. G. and Mullins, M. G.,** Concentration of phenolics in the skin of grape berries during fruit development and ripening, *Am. J. Enol. Vitic.*, 31, 34, 1980.
311. **Poovaiah, B. W.,** The role of calcium and calmodulin in senescence, in *Plant Senescence: Its Biochemistry and Physiology*, Thomson, W. W., Nothnagel, E. A., and Huffaker, R. C., Eds., *Am. Soc. Plant Physiol.*, 1987, 182.
312. **Porter, L. J., Foo, L. Y., and Furneaux, R. H.,** Isolation of three naturally occurring O-beta-glycopyranosides of procyanidin polymers, *Phytochemistry*, 24, 567, 1985.
313. **Pouget, R.,** Action de la concentration de la solution nutritive sur quelques caractéristiques physiologiques et technologiques chez *Vitis vinifera* L. cv. "Cabernet Sauvignon". I. Vigueur, rendement, qualité du moût et du vin, *Agronomie*, 4, 437, 1984.
314. **Poulton, J. E.,** Transmethylation and demethylation reaction in the metabolisme of secondary plant products, in *Biochemistry of Plants*, Vol. 7, Conn, E. E., Ed., Academic Press, New York, 1981, 668.
315. **Poulton, J. E. and Kauer, M.,** Identification of an UDP-glucose:flavonol 3-*O*-glucosyltransferase from cell suspension cultures of soybean (*Glycine max* L.), *Planta*, 136, 53, 1977.
316. **Poulton, J., Halhbrock, K., and Grisebach, H.,** Purification and properties of the *S*-adenosyl L-methlonine:caffeic acid-3-*O*-methyltransferase from soybean cell suspension cultures, *Arch. Biochem. Biophys.*, 176, 449, 1976.
317. **Poulton, J., Grisebach, H., Ebel, J., Schaller-Hekeler, B., and Hahlbrock, K.,** Two distinct *S*-adenosyl-L-methionine, 3-4-dihydric phenol 3-*O*-methlythansferases of phenyl propanoid metabolism in soybean cell suspension cultures, *Arch. Biochem. Biophys.*, 173, 301, 1976.
318. **Prabha, T. N. and Patwardhan, M. V.,** Polyphenols of avocado (*Persea americana*) and their endogenous oxidation, *J. Food Sci. Technol.*, 17, 215, 1980.
319. **Prasad, U. S. and Jha, O. P.,** Changes in pigmentation patterns during Litchi ripening: flavonoid production, *Plant Biochem. J.*, 5, 44, 1978.
320. **Proctor, J. T. A.,** Color stimulation in attached apples with supplementary light, *Can. J. Plant Sci.*, 54, 499, 1974.
321. **Proctor, J. T. A. and Creasy, L. L.,** Effect of supplementary light to anthocyanin synthesis in "McIntosh" apples, *J. Am. Soc. Hortic. Sci.*, 96, 523, 1971.
322. **Proctor, J. T. A. and Lougheed, E. C.,** The effect of covering apples during development, *HortScience*, 11, 108, 1976.
323. **Proebsting, E. L. Jr., and Mills, H. H.,** Effects of 2-chloroethane phosphonic acid and its interaction with gibberellic acid on quality of Early Italian prunes, *J. Am. Soc. Hortic. Sci.*, 94, 443, 1969.
324. **Puech, A. A., Rebeiz, C. A., and Crane, J. C.,** Pigment changes associated with application of ethephon ((2-chloroethyl) phosphonic acid) to fig (*Ficus carica* L.) fruits, *Plant Physiol.*, 57, 504, 1976.
325. **Rabino, I. and Mancinelli, A. L.,** Light, temperature, and anthocyanin production, *Plant Physiol.*, 81, 922, 1986.
326. **Ramos, T. and Macheix, J. J.,** Evolution des anthocyanols et des esters hydroxycinnamoyltartriques des baies de raisin Carignan au cours d'un traitement anaérobie sous CO_2, 4e Symp. Int. Oenol., Bordeaux, June 15 to 17, 1989.
327. **Ranadive, A. S. and Haard, N. F.,** Changes in polyphenolics on ripening of selected pear varieties, *J. Sci. Food Agric.*, 22, 86, 1971.
328. **Ranadive, A. S., and Haard, N. F.,** Peroxidase localization and lignin formation in developing pear fruit, *J. Food Sci.*, 37, 381, 1972.
329. **Ranjeva, R., Boudet, A. M., and Faggion, R.,** Phenolic metabolism in Petunia tissues: IV. Properties of *p*-coumarate:coenzyme A ligase isoenzymes, *Biochimie*, 58, 1255, 1976.
330. **Ranjeva, R., Faggion, R., and Boudet, A. M.,** Métabolisme de composés phénoliques des tissus de Pétunia. II. Etude du système enzymatique d'activation des acides cinnamiques; mise en évidence de deux formes de la "cinnamoyl-coenzyme A ligase", *Physiol. Vég.*, 13, 725, 1975.
331. **Ranjeva, R., Boudet, A. M., Harada, H., and Marigo, G.,** Phenolic metabolism in Petunia tissues. I. Characteristic responses of enzymes involved in different steps of polyphenol synthesis to different hormonal influences, *Bioch. Biophys. Acta*, 399, 23, 1975.
332. **Raymond, W. R. and Maier, V. P.,** Chalcone cyclase and flavonoid biosynthesis in grapefruit, *Phytochemistry*, 16, 1535, 1977.
333. **Rhodes, M. J. C.,** The maturation and ripening of fruits, in *Senescence in Plants*, Thimann, K. V., Ed., CRC Press, Boca Raton, FL, 1980, 157.
334. **Rhodes, M. J. C. and Wooltorton, L. S. C.,** The effect of ethylene on the respiration and on the activity of phenylalanine ammonia-lyase in Swede and parsnip root tissue, *Phytochemistry*, 10, 1989, 1971.

335. **Rhodes, M. J. C. and Wooltorton, L. S. C.,** Stimulation of phenolic acids and lignin biosynthesis in Swede root tissue by ethylene, *Phytochemistry,* 12, 107, 1973.
336. **Rhodes, M. J. C. and Wooltorton, L. S. C.,** The *p*-coumaryl CoA ligase of potato tubers, *Phytochemistry,* 14, 2161, 1975.
337. **Rhodes, M. J. C. and Wooltorton, L. S. C.,** The enzymic conversion of hydrocinnamic acids to *p*-coumarylquinic and chlorogenic acids in tomato fruit, *Phytochemistry,* 16, 655, 1977.
338. **Rhodes, M. J. C. and Wooltorton, L. S. C.,** Changes in the activity of enzymes of phenylpropanoid metabolism in tomatoes stored at low temperatures, *Phytochemistry,* 15, 947, 1976.
339. **Rhodes, M. J. C. and Wooltorton, L. S. C.,** The biosynthesis of phenolic compounds in wounded plant storage tissues, in *Biochemistry of Wounded Plant Tissues,* Kahl, G., Ed., Walter de Gruyter, Berlin, 1978, 243.
340. **Rhodes, M. J. C., Wooltorton, L. S. C., and Hill, A. C.,** Changes in phenolic metabolism in fruit and vegetable tissues under stress, in *Recent Advances in the Biochemistry of Fruits and Vegetables,* Friend, J. and Rhodes, M. J. C., Eds., Academic Press, London, 1981, 193.
341. **Ribereau-Gayon, P.,** Evolution des composés phénoliques au cours de la maturation du raisin. I. Expérimentation 1969, *Connaiss. Vigne Vin,* 2, 247, 1971.
342. **Ribereau-Gayon, P.,** The anthocyanins of grapes and wines, in *Anthocyanins as Food Colors,* Markakis, P., Ed., Academic Press, New York, 1982, 209.
343. **Ribereau-Gayon, P. and Glories, Y.,** Phenolics in grape and wines, *Proc. Sixth Austr. Wine Ind. Techn. Conf.,* Australian Industrial Publishers, Adelaide, 1987, 247.
344. **Rigby, B., Dana, M. N., and Binning, L. K.,** Ethephon sprays and cranberry fruit color, *HortScience,* 7, 82, 1972.
345. **Riov, J.,** 6,7-Dimethoxycoumarin in the peel of gamma-irradiated grapefruit, *Phytochemistry,* 10, 1923, 1971.
246. **Riov, J. and Goren, R.,** Effects of gamma radiation and ethylene on protein synthesis in peel of mature grapefruit, *Rad. Bot.,* 10, 155, 1970.
347. **Riov, J., Monselise, S. P., and Goren, R.,** Stimulation of phenolic biosynthesis in citrus fruit peel by gamma radiation, *Rad. Res. Rev.,* 3, 417, 1972.
348. **Riov, J., Monselise, S. P., and Kahan, R. S.,** Effect of gamma radiation on phenylalanine ammonia-lyase activity and accumulation of phenolic compounds in Citrus fruit peel, *Radiat. Bot.,* 8, 463, 1968.
349. **Riov, J., Monselise, S. P., and Kahan, R. S.,** Ethylene-controlled induction of phenylalanine ammonia-lyase in Citrus fruit peel, *Plant Physiol.,* 44, 631, 1969.
350. **Riov, J., Monselise, S. P., and Kahan, R. S.,** Radiation damage to grapefruit in relation to ethylene production and phenylalanine ammonia-lyase activity, *Radiat. Bot.,* 10, 281, 1970.
351. **Riov, J., Monselise, S. P., and Kahan, R. S.,** Effect of gamma radiation on the synthesis of scopoletin and scopolin in grapefruit peel in relation to radiation damage, *Radiat. Res.,* 45, 326, 1971.
352. **Roggero, J. P., Coen, S., and Ragonnet, B.,** High performance liquid chromatography survey on changes in pigment content in ripening grapes of Syrah. An approach to anthocyanin metabolism, *Am. J. Enol. Vitic.,* 37, 77, 1986.
353. **Romani, R., Labavitch, J., Yamashita, T., Hess, B., and Rae, H.,** Preharvest AVG treatment of 'Bartlett' pear fruits: effects on ripening, color change, and volatiles, *J. Am. Soc. Hortic. Sci.,* 108, 1046, 1983.
354. **Romeyer, F. M., Macheix, J. J., and Sapis, J. C.,** Changes and importance of oligomeric procyanidins during maturation of grape seeds, *Phytochemistry,* 25, 219, 1986.
355. **Romeyer, F. M., Macheix, J. J., Goiffon, J. P., Reminiac, C. C., and Sapis, J. C.,** The browning capacity of grapes. 3. Changes and importance of hydroxycinnamic acid-tartaric acid esters during development and maturation of the fruit, *J. Agric. Food Chem.,* 31, 346, 1983.
356. **Roubelakis-Angelakis, K. A. and Kliewer, W. W.,** Phenylalanine ammonia-lyase in berries of *Vitis vinifera* L.: extraction and possible sources of error during assay, *Am. J. Enol. Vitic,* 36, 314, 1985.
357. **Roubelakis-Angelakis, K. A., and Kliewer, W. M.,** Effects of exogenous factors on phenylalanine ammonialyase activity and accumulation of anthocyanins and total phenolics in grape berries, *Am. J. Enol. Vitic.,* 37, 275, 1986.
358. **Russel, D. W. and Conn, E. E.,** The cinnamic-acid 4-hydroxylase of pea seedlings, *Arch. Biochem. Biophys.,* 122, 256, 1967.
359. **Rylski, I.,** Pepper *(Capsicum)* in *Handbook of Fruit Set and Development,* Monselise, S. P., Ed., CRC Press, Boca Raton, FL, 1986, 341.
360. **Rymal, K. S., Dozier, W. A., Jr., Satjawatcharaphong, C., Knowles, J. W., and Smith, D. A.,** Effects of growth regulators on quality of rip peach fruit: discoloration of canned peaches, *J. Food Sci.,* 51, 1065, 1986.
361. **Ryugo, K.,** Persistence and mobility of Alar (B-995) and its effect on anthocyanin metabolism in sweet cherries, *Prunus avium, Proc. Am. Soc. Hortic. Sci.,* 88, 160, 1966.
362. **Sagi, F., Kollanyi, L., and Simon, I.,** Changes in the colour and anthocyanin content of raspberry fruit during ripening, *Acta Aliment.,* 3, 397, 1974.

363. **Sakamura, S. and Obata, Y.,** Anthocyanase and anthocyanine occurring in eggplant (*Solanum melongena* L.), *Agric. Biol. Chem.*, 25, 750, 1961.
364. **Sakamura, F. and Suga, T.,** Changes in chemical components of ripening oleaster fruits, *Phytochemistry*, 26, 2481, 1987.
365. **Sal'Kova, E. G., Zuyagintseva, Y. V., Kuliev, A. A., and Akhundov, R. M.,** Effect of Ethrel on biochemical processes during fruit ripening, *Appl. Biochem. Microbiol.*, 13, 71, 1977.
366. **Saleh, N. A. M., Fritsch, H., Kreuzaler, F., and Grisebach, H.,** Flavanone synthase from cell suspension cultures of *Haplopappus gracilis* and comparison with the synthase from parsley, *Phytochemistry*, 17, 183, 1978.
367. **Sapers, G. M., Graff, G. R., Phillips, J. G., and Deubert, K. H.,** Factors affecting the anthocyanin content of cranberry, *J. Am. Soc. Hortic. Sci.*, 111, 612, 1986.
368. **Sapers, G. M., Hicks, K. B., Burgher, A. M., Hargrave, D. L., Sondey, S. M., and Billyk, A.,** Anthocyanin patterns in ripening thornless blackberries, *J. Am. Soc. Hortic. Sci.*, 111, 945, 1986.
369. **Sarkar, S. K. and Phan, C. T.,** Effect of ethylene on the phenylalanine ammonia-lyase activity of carrot tissues, *Physiol. Plant.*, 32, 318, 1974.
370. **Schmidtlein, H. and Herrmann, K.,** Phenolic acids of vegetables. II. Hydroxycinnamic acids and hydroxybenzoic acids of fruit and seed vegetables, *Z. Lebensm. Unters. Forsch.*, 159, 213, 1975.
371. **Schröeder, J., Betz, B., and Hahlbrock, K.,** Messenger RNA-controlled increase of phenylalanine ammonia-lyase activity in parsley. Light independent induction by dilution of cell suspension cultures into water, *Plant Physiol.*, 60, 440, 1977.
372. **Sharma, V. and Strack, D.,** Vacuolar localization of 1-sinapoylglucose:L-malate sinapoyltransferase in protoplasts from cotyledons of *Raphanus sativus*, *Planta*, 163, 563, 1985.
373. **Shawa, A. Y.,** Effect of ethephon on color, abscission, and keeping quality of 'McFarlin' cranberry, *HortScience*, 14, 168, 1979.
374. **Shen, Z., Haslam, E., Falshaw, C. P., and Begley, M. J.,** Procyanidins and polyphenols of *Larix gmelini* bark, *Phytochemistry*, 25, 2629, 1986.
375. **Shulman, Y.,** Physiology of ripening and coloration of olive fruits *(Olea europaea)* Ph.D. thesis, The Hebrew University of Jerusalem, Rehovot, Israel, 1976.
376. **Shulman, Y. and Lavee, S.,** The effect of kinetin on anthocyanin formation in green harvested olive fruits, *J. Am. Soc. Hortic. Sci.*, 96, 808, 1971.
377. **Shulman, Y. and Lavee, S.,** The effect of cytokinins and auxins on anthocyanin accumulation in green Manzanillo olives, *J. Exp. Bot.*, 24, 655, 1973.
378. **Shulman, Y. and Lavee, S.,** Fruit development and maturation of olives as affected by treatments with auxin, *Riv. Ortoflorofruttic. Ital.*, 63, 31, 1979.
379. **Shulman, Y., Erez, A., and Lavee, S.,** Delay in ripening of picked olives due to ethylene treatments, *Sci. Hortic.*, 2, 21, 1974.
380. **Siegelman, H. W. and Hendricks, S. B.,** Photocontrol of anthocyanin synthesis in apple skin, *Plant Physiol.*, 33, 185, 1958.
381. **Singh, N. S.,** Influence of UV light on the biosynthesis of anthocyanin-like pigments in ripening bananas, *Phytochemistry*, 11, 163, 1972.
382. **Singleton, V. L.,** Grape and wine phenolics, background and prospects, *C.R. Coll. Centenaire*, Paris, 215, 1980.
383. **Singleton, V. L. and Esau, P.,** *Phenolic Substances in Grapes and Wine, and Their Significance*, Academic Press, New York, 1969.
384. **Singleton, V. L., Zaya, J., and Trousdale, E.,** Compositional changes in ripening grapes: caftaric and coutaric acids, *Vitis*, 25, 107, 1986.
385. **Slinkard, K. W. and Singleton, V. L.,** Phenol content of grape skins and the loss of ability to make anthocyanins by mutation, *Vitis*, 23, 175, 1984.
386. **Smith, H., Billett, E. E., and Giles, A. B.,** The photocontrol of gene expression in higher plants, in *Regulation of Enzyme Synthesis and Activity in Higher Plants*, Smith, H., Ed., Academic Press, London, 1977, 93.
387. **Smith, R. B. and Cline, R. A.,** A laboratory and field study of the relationship between calcium sources and browning in Apple juice, *J. Food Sci.*, 49, 1419, 1984.
388. **Smock, R. M.,** Laboratory studies of anthocyanin development in McIntosh apples, *Proc. Am. Soc. Hortic. Sci.*, 88, 80, 1966.
389. **Smock, R. M.,** Laboratory studies on the effects of chemicals on the coloration of apples, *Proc. Am. Soc. Hortic. Sci.*, 94, 49, 1969.
390. **Somers, T. C.,** Pigment development during ripening of the grape, *Vitis*, 14, 269, 1976.
391. **Spencer, C. M., Cai, Y., Martin, R., Gaffney, S. H., Goulding, P. N., Magnolato, D., Lilley, T. H., and Haslam, E.,** Polyphenol complexation. Some thoughts and observations, *Phytochemistry*, 27, 2397, 1988.

392. **Sponga, F., Deitzer, G. F., and Mancinelli, A.**, Cryptochrome, phytochrome, and the photoregulation of anthocyanin production under blue light, *Plant Physiol.*, 82, 952, 1986.
393. **Stafford, H. A.**, Compartmentation in natural product biosynthesis by multienzyme complexes, in *The Biochemistry of Plants*, Vol. 7, Conn, E. E., Ed., Academic Press, New York, 1981, 118.
394. **Stafford, H. A.**, Enzymatic regulation of procyanidin biosynthesis, lack of a flav-3-en-ol intermediate, *Phytochemistry*, 22, 2643, 1983.
395. **Stafford, H. A., Kreitlow, K. S., and Lester, H. H.**, Comparison of proanthocyanidins and related compounds in leaves and leaf-derived cell cultures of *Ginkgo biloba* L., *Pseudotsuga menziesii* Franco, and *Ribes sanguineum* Pursh, *Plant Physiol.*, 82, 1132, 1986.
396. **Stafford, H. A., Shimamoto, M., and Lester, H. H.**, Incorporation of ^{14}C phenylalanine into flavan-3-ols and procyanidins in cell suspension cultures of Douglas Fir, *Plant Physiol.*, 69, 1055, 1982.
397. **Starke, H. and Herrmann, K.**, Phenolic of fruits. 8. Changes in flavonol concentrations during fruit development, *Z. Lebensm. Unters. Forsch.*, 161, 131, 1976.
398. **Steer, B. T. and Darbyshire, B.**, Phenylalanine ammonia-lyase activity in developing orange fruit, *J. Aust. Inst. Agric. Sci.*, 39, 201, 1973.
399. **Stockigt, J. and Zenk, M. H.**, Enzymatic synthesis of chlorogenic acid from caffeoyl coenzyme A and quinic acid, *FEBS Lett.*, 42, 131, 1974.
400. **Stohr, H. and Herrmann, K.**, On the occurrence of derivatives of hydroxycinnamic acids, hydroxybenzoic acids and hydroxycommarins in Citrus Fruits, *Z. Lebensm. Unters. Forsch.*, 159, 305, 1975.
401. **Stohr, H. and Herrmann, K.**, Die phenolischen Inhaltsstoffe des Obstes V. Die phenolischen Inhaltsstoffe der Erdbeeren und deren Veranderungen wahrend wachstrum und reife der Fruchte, *Z. Lebensm. Unters. Forsch.*, 159, 341, 1975.
402. **Stohr, H. and Herrmann, K.**, Die phenolischen Inhaltsstoffe des Obstes. VI. Die phenolischen Inhaltsstoffe der Johannisbeeren, Stachelbeeren und Kulturheidelbeeren. Veranderungen der Phenolsauren und Catechine wahrend wachstum und reife von Swarzen Johannisbeeren, *Z. Lebensm. Unters. Forsch.*, 159, 31, 1975.
403. **Stohr, H., Mosel, H. D., and Herrmann, K.**, The phenolics of fruits. VII. The phenolics of cherries and plums and the changes in catechins and hydroxycinnamic acid derivatives during the development of fruits, *Z. Lebensm. Unters. Forsch.*, 159, 85, 1975.
404. **Strack, D., Ruhoff, R., and Grawe, W.**, Hydroxycinnamoyl-coenzyme-A:tartronate hydroxycinnamoyl-transferase in protein preparation from mung bean, *Phytochemistry*, 25, 833, 1986.
405. **Sugiura, A. and Tomana, T.**, Relationships of ethanol production by seeds of different types of Japanese persimmons and their tannin content, *HortScience*, 18, 319, 1983.
406. **Sugiura, A., Yonemori, K., Harada, H., and Tomana, T.**, Changes of ethanol and acetaldehyde contents in Japanese Persimmon fruits and their relation to natural deastringency, *Studies Inst. Hortic. Kyoto Univ.*, 9, 41, 1979.
407. **Sutfeld, R. and Wiermann, R.**, Untersuchungen zum Vorkommen zweier O-methyltransferase in Tulipa-antheren, *Biochem. Physiol. Pflanzen*, 172, 111, 1978.
408. **Swain, T. and Hillis, W. E.**, The phenolic constituents of *Prunus domestica*. I. The quantitative analysis of phenolic constituents, *J. Sci. Food Agric.*, 10, 63, 1959.
409. **Takeda, F., Saunders, M. S., and Saunders, J. A.**, Physical changes in muscadine grapes during postharvest storage, *Am. J. Enol. Vitic.*, 34, 180, 1983.
410. **Tan, S. C.**, Relationships and interactions between phenylalanine ammonialyase, phenylalanine ammonia-lyase inactivating system, and anthocyanin in Apples, *J. Am. Soc. Hortic. Sci.*, 104, 581, 1979.
411. **Tan, S. C.**, Phenylalanine ammonialyase and the phenylalanine ammonialyase inactivating system. Effects of light, temperature and mineral deficiencies, *Aust. J. Plant Physiol.*, 7, 159, 1980.
412. **Tan, S. C. and Lam, P. F.**, Effect of gamma irradiation on PAL activity and phenolic compounds in Papaya (*Carica papaya* L.) and Mango (*Mangifera indica* L.) Fruits, *Asean Food J.*, 1, 134, 1985.
413. **Teisson, C.**, Le brunissement interne de l'ananas, *Fruits*, 34, 245, 1979.
414. **Teisson, C., Martin-Prevel, P., Combres, J. C., and Py, C.**, A propos du brunissement interne de l'ananas accident de la réfrigération, *Fruits*, 33, 48, 1978.
415. **Teusch, M., Forkmann, G., and Seyffert, W.**, Genetic control of hydroxycinnamoyl-coenzyme A:anthocyanidin 3-glucoside-hydroxycinnamoyltransferase from petals of *Mattiola incana*, *Phytochemistry*, 26, 991, 1987.
416. **Thomas, P. and Janave, M. T.**, Polyphenol oxidase activity and browning of mango fruits induced by gamma irradiation, *J. Food Sci.*, 38, 1149, 1973.
417. **Thomas, P. and Nair, P. M.**, Effect of gamma irradiation on polyphenol oxidase activity and its relation to skin browning in bananas, *Phytochemistry*, 10, 771, 1971.
418. **Thorpe, T. A., Maier, V. P., and Hasegawa, S.**, Phenylalanine ammonialyase activity in Citrus fruit tissue cultured *in vitro*, *Phytochemistry*, 10, 711, 1971.
419. **Timberlake, C. F.**, Anthocyanins in fruits and vegetables, in *Recent Advances in the Biochemistry of Fruits and Vegetables*, Friend, J. and Rhodes, M. J. C., Eds., Academic Press, London, 1981, 221.

420. **Towers, G. H. N. and Yamamoto, E.**, Interactions of cinnamic acid and its derivatives with light, in *The Biochemistry of Plant Phenolics*, Van Sumere, C. F. and Lea, P. J., Eds., Clarendon Press, Oxford, 1985, 271.
421. **Treutter, D., Feucht, W., and Schmid, P. P. S.**, Polyphenole des Phloems in Beziehung zur inkompatibilität von interspezifischen Prunus-Veredlungen (*P. avium* L., *P. cerasus* L.) I. Flavanone und Flavonole uber der veredlungsstelle, *Gartenbauwisenchaft*, 51, 77, 1986.
422. **Treutter, D., Feucht, W., and Schmid, P. P. S.**, Ageing-depending responses of phloem flavonoids of *Prunus avium* graftings: flavanone, flavone and isoflavone-glucosides, *Sci. Hortic.*, 32, 183, 1987.
423. **Treutter, D., Galensa, R., Feucht, W., and Schmid, P. P. S.**, Flavanone glucosides in callus and phloem of *Prunus avium*: identification and stimulation of their synthesis, *Physiol. Plant.*, 65, 95, 1985.
424. **Tsuji, M., Harakawa, M., and Komiyama, Y.**, Inhibition of increase of pulp color and phenylalanine ammonia-lyase activity in plum fruit at high temperature (30°C), *J. Jpn. Soc. Food Sci. Technol.*, 30, 688, 1983.
425. **Ulbrich, B., and Zenk, M. H.**, Partial purification and properties of hydroxycinnamoyl-CoA:quinate hydroxycinnamoyltransferase from higher plants, *Phytochemistry*, 18, 929, 1979.
426. **Ulbrich, B. and Zenk, M. H.**, Partial purification and properties of *p*-hydroxycinnamoyl-CoA:shikimate-*p*-hydroxycinnamoyl transferase from higher plants, *Phytochemistry*, 11, 1625, 1980.
427. **Valero, E., Varon, R., and Garcia-Carmona, F.**, Characterization of polyphenoloxidase from Airen grapes, *J. Food Sci.*, 53, 1482, 1988.
428. **Vamos-Vigyazo, L., Gajzago, I., Nadudvari-Markus, V., and Magdolna-Kerek, M.**, Changes in some enzymes activities and browning rate of apricots during storage, *Qual. Plant. Plant Foods Hum. Nutr.*, 35, 139, 1985.
429. **Van Buren, J.**, Fruit phenolics, in *The Biochemistry of Fruits and Their Products*, Vol. 1, Hulme, A. C., Ed., Academic Press, London, 1970, 269.
430. **Van Rensburg, E. and Engelbrecht, A. H. P.**, Effect of calcium salts on susceptibility to browning of avocado fruit, *J. Food Sci.*, 51, 1067, 1986.
431. **Vandercook, C. E. and Tisserat, B.**, Flavonoid changes in developing lemons grown *in vivo* and *in vitro*, *Phytochemistry*, 28, 799, 1989.
432. **Vaughn, K. C., Lax, A. R., and Duke, S. O.**, Polyphenol oxidase: the chloroplast oxidase with no established function, *Physiol. Plant*, 72, 659, 1988.
433. **Vestrheim, S.**, Effects of chemical compounds on anthocyanin formation in 'McIntosh' apple skin, *J. Am. Soc. Hortic. Sci.*, 95, 712, 1970.
434. **Villegas, R. J. A. and Kojima, M.**, Purification and characterization of hydroxycinnamoyl D-glycose quinate hydroxycinnamoyl transferase in the root of sweet potato, *Ipomoea batatas* Lam, *J. Biol. Chem.*, 261, 8729, 1986.
435. **Volk, R., Harel, E., Mayer, A. M., and Gan-Zvi, E.**, Catechol oxidase in tissue culture of apple fruit, *J. Exp. Bot.*, 28, 820, 1977.
436. **Vshivkova, G. F., Nilov, G. I., Sivtsev, M. V., and Bagleva, L. Y.**, Juglone and ascorbate in fruit and leaves of various forms, *Fiziol. Biochim. Kul't Rast.*, 11, 181, 1979.
437. **Wagner, G. J. and Hrazdina, G.**, Endoplasmic reticulum as a site of phenylpropanoid and flavonoid metabolism in *Hippeastrum*, *Plant Physiol.*, 74, 901, 1984.
438. **Wallis, P. J. and Rhodes, M. J. C.**, Multiple forms of hydroxycinnamate:CoA ligase in etiolated pea seedlings, *Phytochemistry*, 16, 1891, 1977.
439. **Walton, D. C. and Sondheimer, E.**, Effect of abscisin II on phenylalanine ammonia-lyase activity in excised bean axes, *Plant Physiol.*, 43, 467, 1968.
440. **Wang, C. Y. and Mellenthin, W. M.**, Chlorogenic acid levels, ethylene production and respiration of d'Anjou pears affected with cork spot, *HortScience*, 8, 180, 1973.
441. **Wang, C. Y. and Mellenthin, W. M.**, Relationship of friction discoloration to phenolic compounds in d'Anjou Pears, *HortScience*, 8, 321, 1973.
442. **Wardale, D. A.**, Effect of phenolic compounds in *Lycopersicum esculentum* on the synthesis of ethylene, *Phytochemistry*, 12, 1523, 1973.
443. **Warren, J. M., Ballinger, W. E., and Mainland, C. M.**, Effects of ethephon upon fruit development and ripening of highbush blueberries in the greenhouse, *HortScience*, 8, 504, 1973.
444. **Weaver, R. J. and Montgomery, R.**, Effect of Ethephon on coloration and maturation of wine grapes, *Am. J. Enol. Vitic.*, 25, 39, 1974.
445. **Weaver, R. J. and Pool, R. M.**, Effect of (2-chloroethyl) phosphonic acid (ethephon) on maturation of *Vitis vinifera* L., *J. Am. Soc. Hortic. Sci.*, 96, 725, 1971.
446. **Wengenmayer, H., Ebel, J., and Grisebach, H.**, Enzymatic synthesis of lignin presursors. Purification and properties of a cinnamoyl-CoA:NADPH reductase from cell suspension cultures of soybean *(Glycine max)*, *Europ. J. Biochem.*, 65, 529, 1976.
447. **Whiting, G. C.**, Sugars, in *The Biochemistry of Fruits and Their Products*, Vol. 1, Hulme, A. C., Ed., Academic Press, London, 1970, 1.

448. **Wicks, A. S. and Kliewer, W. M.,** Further investigations into the relationship between anthocyanins, phenolics and soluble carbohydrates in grape berry skins, *Am. J. Enol. Vitic.*, 34, 114, 1983.
449. **Woodward, J. R.,** Physical and chemical changes in developing strawberry fruits, *J. Sci. Food Agric.*, 23, 465, 1972.
450. **Workman, M.,** Color and pigment changes in Golden Delicious and Grimes Golden apples, *Proc. Am. Soc. Hortic. Sci.*, 83, 149, 1963.
451. **Yonemori, K. and Matsushima, J.,** Chemical characteristics of tannins from non-astringent and astringent type fruits of Japanese persimmon *(Diospyros kaki)* with particular reference to ultracentrifugal behavior, *J. Jpn. Soc. Hortic. Sci.*, 53, 121, 1984.
452. **Yonemori, K. and Matsushima, J.,** Morphological characteristics of tannin cells in Japanese persimmon fruit, *J. Am. Soc. Hortic. Sci.*, 112, 812, 1987.
453. **Yonemori, K. and Matsushima, J.,** Changes in tannin cell morphology with growth and development of Japanese persimmon fruit, *J. Am. Soc. Hortic. Sci.*, 112, 818, 1987.
454. **Yonemori, K., Matsushima, J., and Sugiura, A.,** Differences in tannins of non-astringent and astringent type fruits of Japanese persimmon *(Diospyros kaki* Thunb.), *J. Jpn. Soc. Hortic. Sci.*, 53, 121, 1983.
455. **Zenk, M. H.,** Recent work on cinnamoyl CoA derivatives, in *Recent Advances in Phytochemistry*, Vol. 12, Swain, T., Harborne, J. B., and Van Sumere, C. F., Eds., Plenum Press, New York, 1979, 139.
456. **Ziegler, H.,** Nature of transported substances, in *Encylopedia of Plant Physiology*, Vol. 1, Zimmermann, M. H. and Milburn, J. A., Eds., Springer-Verlag, Berlin, 1975, 59.
457. **Zucker, M.,** Light and enzymes, *Annu. Rev. Plant Physiol.*, 23, 133, 1972.

Chapter 4

IMPORTANCE AND ROLES OF PHENOLIC COMPOUNDS IN FRUITS

I. INTRODUCTION

The study of the qualitative and quantitative variations of phenolic compounds in the last chapter showed that they possess two interesting characteristics: first, some of them are excellent indicators of the physiological stages of fruit development, and second, they are particularly sensitive to environment factors and especially to light and temperature. The question which naturally arises is that of knowing whether the variations observed are the cause or simply the consequence of the physiological responses of the fruit. It is very difficult to give a definitive answer to this, and two aspects are discussed: the physiological role during the normal development of the fruit on the one hand and the role in resistance to attacks and stress on the other. The second point (resistance) is one of the most interesting aspects of research on phenolic compounds over the past 10 years.

Phenolic compounds in fruits are also of great interest to man in two different domains. They first contribute to sensory qualities of fruits: color, astringency, bitterness, and aroma. Modifications of these parameters in connection with the technological processes used to manufacture fruit-based products are discussed in Chapter 5. In addition, some compounds possess pharmacological properties and are used for therapeutic purposes.

These different aspects, which justify the increasing interest in phenolic compounds in fruits, are examined in turn in the first four parts of this chapter. This is followed by discussion of the ecological role of these compounds and their chemotaxonomic interest.

II. PHYSIOLOGICAL ROLE IN FRUITS

A. PRELIMINARY REMARKS

The problem of the possible role played by polyphenols in the physiology of plants is far from having been solved and, in spite of numerous works and reviews over the past 20 years, knowledge of the physiological functions of these compounds is still imprecise.[123,133,142,196,244,245,320] Integration of the phenolic metabolism in the general program of development of a plant organ, such as fruit, as described in Chapter 3, raises the question of the possible role of these substances in the regulation of the various physiological stages. The hypotheses which have been put forward are based on three types of experimental argument:

- The effects of exogenous supply of phenolic compounds (*in vitro* studies)
- Observation of variations in the phenolic metabolism preceding or accompanying the physiological events in the life of the plant, e.g,. the growth and ripening of fruits
- The revealing of modifications in the levels of certain phenolic compounds during the application of various stimuli leading to responses in growth and development.

Only a small proportion of all the work carried out concerns fruits. In addition, although research has resulted in a large body of presumptions in favor of the potential involvement of phenolic compounds in the development process, neither observation of correlation *in vivo* nor the obtaining of *in vitro* inhibitions or stimulation by means of exogenous supply form absolute proof of the participation of these compounds in the normal functioning of cells. Indeed, these approaches pay little attention to the problems of cellular or metabolic compartmentation, which is nevertheless a fundamental aspect. On this point, the decisive progress made in the past few years concerning the microlocalization of both phenolic compounds and enzymes[8,31,38,168,169,175,192,417] should lead to data which are closer to physiological reality.

These restrictive remarks should be borne in mind in the interpretation of the data presented below. Examined successively is the intervention of phenolic compounds in the growth of fruits, in maturation, and finally in various other physiological phenomena connected with the life of fruits.

B. ROLE IN THE GROWTH OF FRUITS

In contrast with pip fruits, such as apple and pear which display simple sigmoid growth, the development of stone fruits can be divided into several stages referred to as I, II, and III[50,257] which show the evolution of diameter and fresh matter in double sigmoid form. However, the cell division growth phase in all fruits is situated right at the beginning of the life of the fruit and most of the growth of the fruit is by auxesis.[50]

As in all parts of plants, growth of fruits after fertilization is controlled by endogenous hormones, and in particular by those produced by the developing seeds.[127,275] It can be put forward that phenolic compounds can act on the growth of fruits through interactions with phytohormones. Most of the works published over the past two decades have concentrated mainly on relations between polyphenols and the auxin metabolism. The main hypotheses today concerning the regulation of cell growth by phenolic compounds are grouped in Figure 86. They mainly concern auxin degradation, but also cover polar auxin transport[188,236] and the balance between free and bound forms. In addition, phenolic compounds can have a direct effect by bonding to certain constituents of the primary wall, reducing its extensibility.[120,149,212] Photoregulation of growth might take place by means of changes in conformation connected with *cis-trans* type isomerization of the hydroxycinnamic acids of the cell wall.[389,432]

Numerous works have shown the *in vitro* antagonist or stimulating effect of phenolic compounds on the degradation of auxin by IAA-oxidase.[124,195,216,276-278,372,387,388] In spite of a number of contradictions caused in particular by the different concentrations used, the results obtained generally show that *o*-diphenols (e.g., rutin, quercetin and caffeic acid derivatives) are inhibitors of IAA-oxidase activity, whereas, in contrast, monophenols (derivatives of *p*-coumaric, *p*-hydroxybenzoic, and vanillic acids, etc.) are activators.[216] In addition, certain *o*-diphenols, such as chlorogenic acid are better substrates for IAA-oxidase than auxin and can thus play an effective role of auxin protector.[94,363] These results have generally been obtained *in vitro* and, as mentioned above, transposition to *in vivo* conditions is always delicate. The hypothesis can nevertheless be put forward that the growth of fruits is linked with their relative mono- and *o*-diphenolic contents.[47,111,225,275]

Experimental data on variation in the monophenols/*o*-diphenols (M/D) ratio during the growth of fruits are in fact scarce. In cherry tomato, the IAA content is high in young fruits and falls about 20 d after anthesis.[1] At the same time, the M/D ratio increases first rapidly and then more slowly.[106] The most rapid increase in this ratio precedes a fall in IAA content and an increase in peroxidase activity,[106] and particularly IAA-oxidase,[382] which occurs in tomato at the end of growth and during maturation.

Distinct variations in the M/D ratio have also been observed during growth of cv. Passe Crassane pear (Figure 87A).[47] During the first 4 weeks when the fruits are very small and growth fairly slow, the *p*-coumaric derivatives/chlorogenic acid (curve 1) and monophenols/total *o*-diphenols (curve 2) ratios are low and practically constant. However, at the start of exponential growth of the fruit, which coincides approximately with the beginning of cell growth and the end of cell division, there is a considerable increase in the ratios in question with a peak on June 15th. Similar results have been reported for Calville blanc apple[225] (Figure 38). The low M/D ratio during the first few weeks of the life of the fruit may

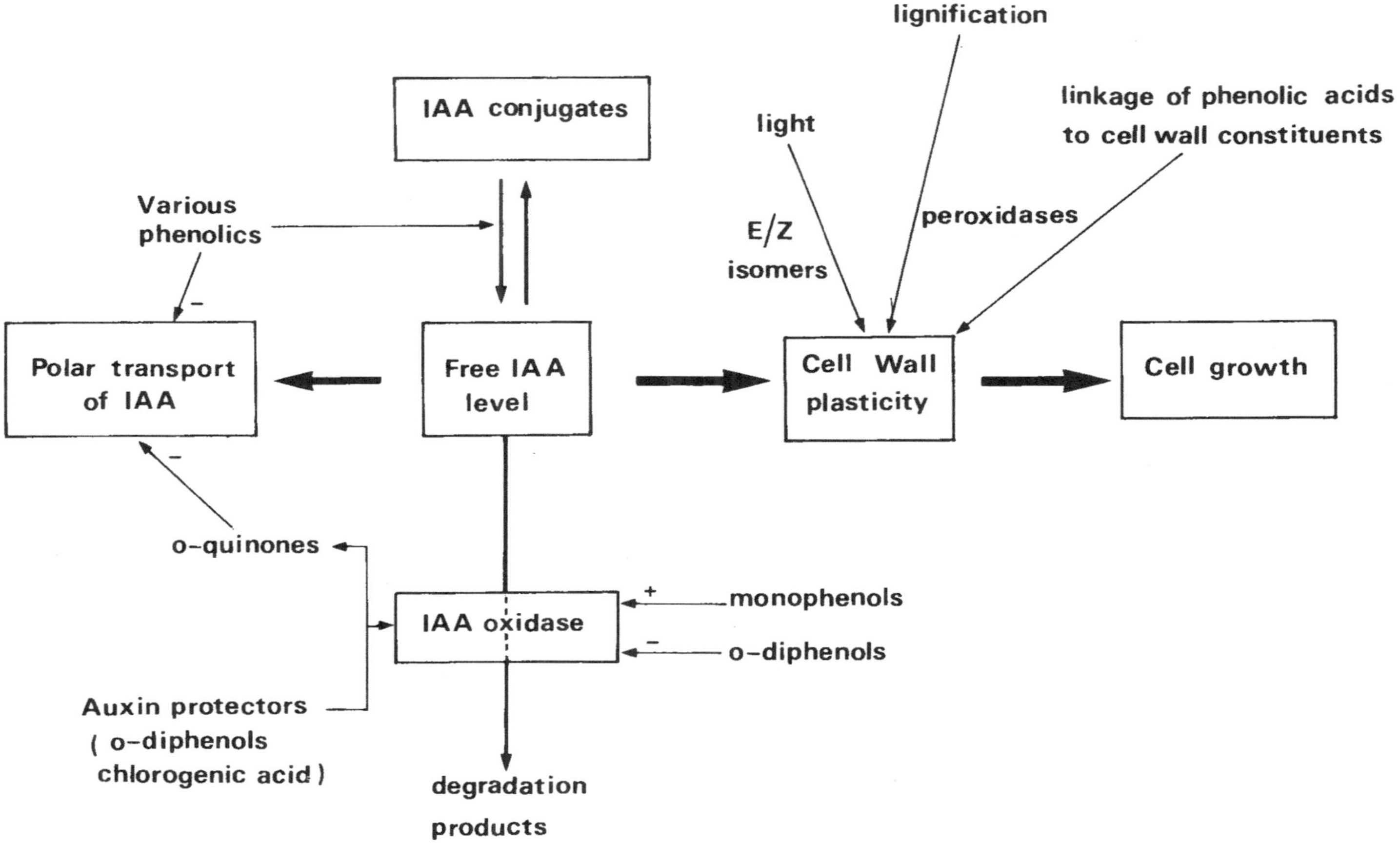

FIGURE 86. Roles of phenolic compounds in the regulation of growth cell: main hypothesis.

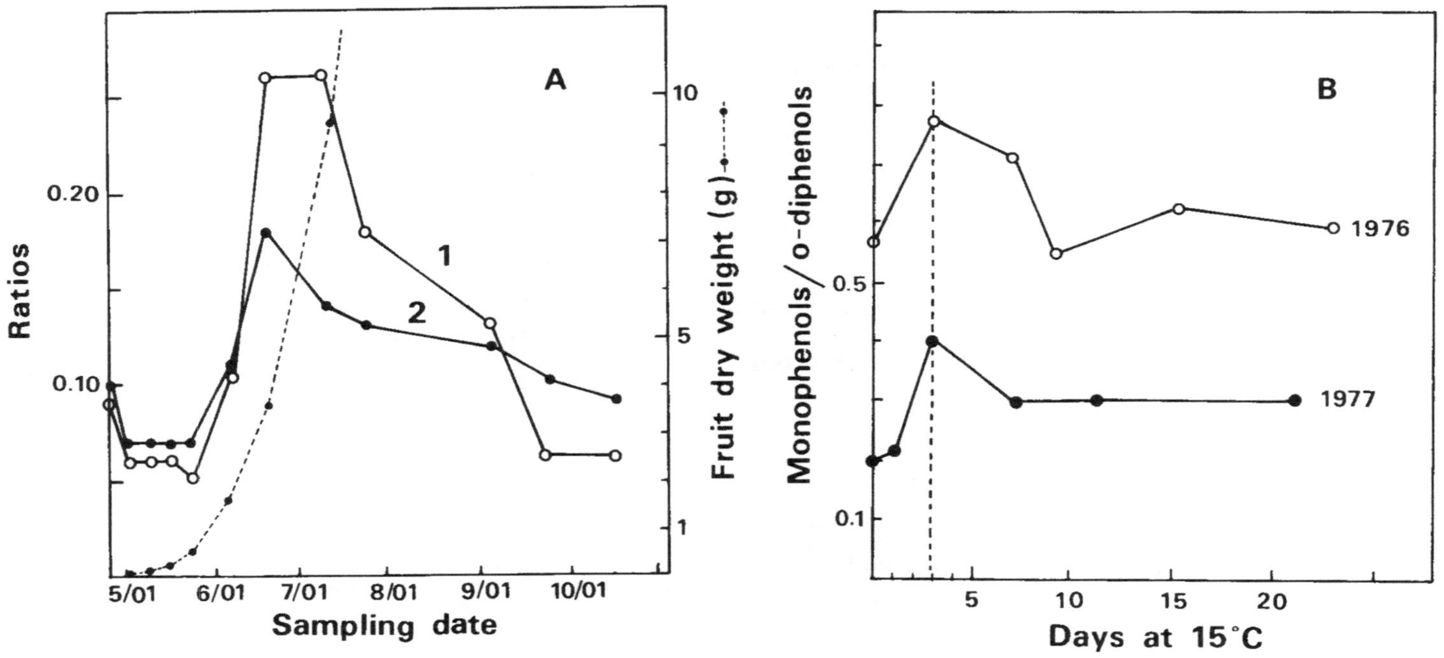

FIGURE 87. Changes in different ratios during growth (A) and maturation (B) of Passe-Crassane pear. A.1: *p*-coumaric derivatives/chlorogenic acid; A.2: monophenols/total *o*-diphenols; B: monophenols/*o*-diphenols in peel during normal ripening at 15°C. After harvest, fruits have been stored at 0°C during twelve weeks before subsequent ripening at 15°C. (Adapted from References 46 and 47.)

enhance auxin accumulation, whereas the value subsequently observed may increase its degradation. This agrees with the variations in stone fruits, where auxin accumulates during the first growth stages and then falls rapidly during growth of the cells of the fleshy part.[80]

C. ROLE IN THE RIPENING OF FRUITS

Ripening, which begins at veraison, is an important stage in the development of fleshy fruits. The biochemical and physiological phenomena which characterize ripening have been studied closely:[56,77,150,234,317,319] release of ethylene, respiratory peak in climacteric fruits, decrease in chlorophylls, etc. The problem which has not yet been solved is that of determinism and regulation of ripening. All the phytohormones, such as auxin, giberellins, cytokinins, abscisic acid, and ethylene affect the progress of ripening in fruits,[56,77,248] but it has now been accepted that ethylene or products of its metabolism play a fundamental role in the expression of maturation.[152,163,220,246,248,433]

Characteristic changes in phenolic compounds and of the enzymes of their metabolism are associated with the ripening of fruits, and in particular, with the triggering of ripening (Chapter 3). It can be seen that there are no spectacular differences in the evolution of phenolic compounds during maturation between climacteric and nonclimacteric fruits. The question raised is that of determining whether these variations are the result of a general trend in ripening or whether they can act directly to control the phenomenon. In the first case, phenolic compounds could possibly play the role of early biochemical markers of ripening[106] (see Chapter 3). In the second case, the evolution of phenolic compounds may result in variations in the monophenols/*o*-diphenols (M/D) ratio which may participate in the control of ripening.[46,106,108,421] In cherry tomato, the chlorogenic acid/caffeic acid glucoside and quinyl esters/glucose derivatives ratios may play a role in growth and maturation.[111]

The normal progress of ripening may be disturbed by stress, caused in particular by wounding. This causes (see below) simultaneous variations in phenolic compounds, ethylene, and maturation.[106,108,110] Although the question remains of whether the variations in phenolic compounds and the release of ethylene are the cause or the result of the acceleration of the maturation of wounded fruits, observation of these concomitant variations has led to seeking the possibility of the intervention of phenolic compounds in the biosynthesis of ethylene.

The first hypothesis put forward was that of intervention of phenolics as direct cofactors of the biosynthesis of ethylene.[233] However, knowledge of the biosynthesis of ethylene has made decisive progress over the past 10 years and the steps from methionine are now known.[2,220,434] They make it possible to suppose that phenolic compounds may play an indirect role through auxin. Indeed, IAA is involved in both the activity and *de novo* synthesis of ACC-synthase[7,103,126,435] and the conversion of ACC into ethylene, which is dependent on the oxidative degradation of IAA via IAA-oxidase.[116,330,408] *In vitro,* the phenolic compounds in avocado tissues inhibit oxidation of ACC into ethylene by hypochlorite, and a negative correlation was found between the amount of plant material in the assay mixture and the conversion efficiency of ACC to ethylene.[348] Although the oxidation process is chemical and different from that of physiological oxidation by EFE, these results suggest that phenolic compounds may be involved in *in vivo* conversion of ACC into ethylene. As has already been discussed, phenolic compounds may act at the level of the IAA-oxidase system through the M/D ratio and thus regulate auxin contents and control ethylene synthesis.

Such an involvement of the M/D ratio in the control of the ripening of fruits was first suggested for the tomato.[421] More recent results on the same fruit[106] and on pear[46] lead to the hypothesis that the M/D ratio may be involved in regulation of the triggering of ripening. Very significant variations of the M/D ratio were observed from the beginning of normal ripening at 15°C in Passe-Crassane pear previously stored at 0°C for 12 weeks (Figure 87B).[46] The M/D ratio hardly varied at all during cold storage, but increased when the fruits were returned to 15°C, with a maximum on the third day (Figure 87B). This very early evolution of the M/D ratio precedes synthesis of mRNA in the pulp, release of ethylene (peak on 9th day) and the respiratory crisis (peak on 11th day).[96,97,151,153,263] Research on other climacteric and nonclimacteric fruits is required before concluding that the results above can be applied in a general manner and in order to check the contribution of the M/D ratio to the triggering and control of ripening. However, the hypothesis appears to be plausible and is used in recently published representations of the regulation of the ripening of climacteric fruit.[152,262]

Finally, a delay in fruit softening is observed when apples are treated with an extract containing quercetin glycosides, other flavonoids, and chlorogenic acid. The firmness retention benefit may be as much as 50%.[90,92,219] Flavonol glycosides and phloridzin are the most effective phenolics for inhibition of apple softening. Loss of firmness of the fruit during ripening involves several enzymes including β-galactosidase, which has been isolated and purified from apples. Polyphenols exert reversible inhibition on β-galactosidase *in vitro*.[90-92] However, inhibition of the enzyme and apple softening by polyphenols are not well correlated because phenolic acids, which are good inhibitors of the enzyme, are not good inhibitors of apple softening and because phloridzin, a good inhibitor of softening, does not inhibit the enzyme. It thus appears that prevention of softening after application of flavonoids is not caused directly by inhibition of apple β-galactosidase. In addition, these results raise a problem of tissue compartmentation: even after treatment with rutin, quercetin glycosides are located in fruit skin whereas β-galactosidase activity and loss of firmness are located in the pulp. It is suggested that inhibition of softening by polyphenols is caused by a previously unrecognized regulatory mechanism.[90]

D. OTHER ROLES

In addition to the possible roles mentioned above in the growth and ripening of fruits, phenolic compounds may make contributions to numerous physiological processes. It should nevertheless be noted that few works have been carried out on these physiological aspects in fruits. Examined successively are the role of phenolics in abscission, lignification, and several fundamental metabolic processes together with various other roles which may concern fruits.

1. Role in Abscission

Several reviews have been devoted to abscission of plant organs, leaves, and fruits.[4,5,203,343] This physiological phenomenon may cause substantial loss of fruits and can occur during several periods: bloom drop, June drop, preharvest drop, and maturity abscission. Factors and chemical treatments controlling drop during these different periods have been studied closely in various fruits including apple, peach, and citrus fruit.[257] The formation of an abscission zone is controlled by phytohormones and in particular auxin and abscisic acid, but ethylene certainly plays a determinant role.[129,261,288,316] The relations discussed above between phenolic compounds, degradation of auxin via IAA-oxidase and synthesis of ethylene lead to considering that these compounds may perhaps play regulatory roles in abscission of fruits, although there are very few experimental results. In cv. Andross peach, the mesocarp of fruits with a high abscission potential is richer in free *p*-coumaric acid and poorer in the bound form of this acid than the mesocarp of fruits with a low abscission potential. The differences between the two types of fruit are very distinct when the ratio of bound form to free form and its evolution are observed during fruit development (Figure 88).[306] *p*-coumaric acid has also been extracted from pericarp of sour cherry.[331] Although free *p*-coumaric acid displays inhibitory activity on the wheat coleoptile growth test, no correlation was found between levels of bound *p*-coumaric acid and bioassay response. The relations between the abscission potential and the bound/free *p*-coumaric acid ratio may thus be interpreted as part of interactions with auxin.[306]

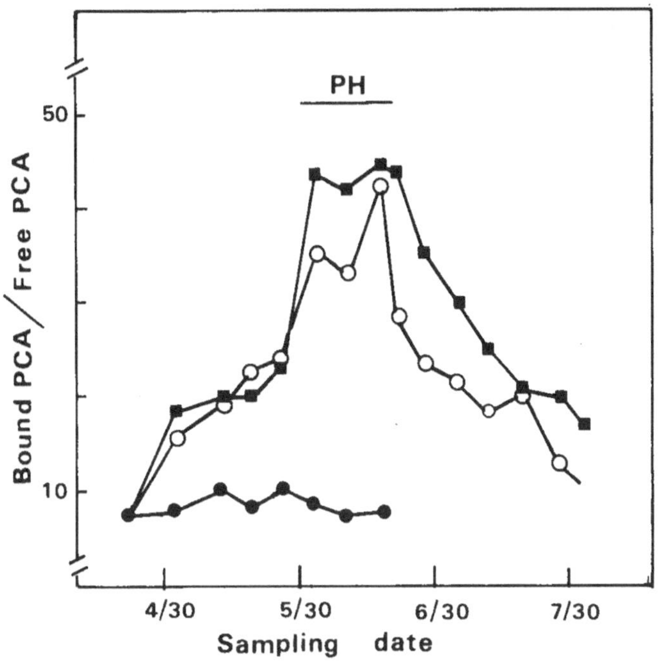

FIGURE 88. Changes in the ratio bound/free para-coumaric acid (PCA) in mesocarp of peach fruit with different abscission potential. ■—■ Fruit with low abscission potential; ○—○ fruit with intermediate abscission potential; ●—● fruit with high abscission potential (PH = pit hardening). (Adapted from Ramina, A. and Masia, A., *Sci. Hortic.*, 16, 375, 1982.)

2. Role in Lignification

Phenolic compounds are involved in the synthesis of lignins, which are complex polymers, by providing monomeric precursors from hydroxycinnamic acids.[132,260] In fruits, this lignification relation to the phenolic metabolism is enhanced after wounding or attack by parasites. This question is examined in detail below. In addition, considerable lignification takes place during the formation of the endocarp during the growth of stone fruits. However, a complete study of this phenomenon remains to be carried out, particularly with regard to the origin of monomeric precursors and the localization of the enzymes involved. Stage II lignification of endocarp requires hydroxycinnamic acids. In peach, localization of phenylalanine ammonia-lyase in the mesocarp and its evolution during fruit growth have been described.[16] In the same fruit, the bound *p*-coumaric acid level in mesocarp appears to be related to pit hardening (Figure 88) since the highest concentrations are observed in stage II of fruit development.[306] Likewise, stage II of cherry, when growth is limited to the pit, is marked by increased PAL activity and concentration of *p*-coumaric esters in the pericarp.[250,331] This raises the question of metabolic relations between various parts of the fruit, e.g., between flesh and pit.

Some fruits contain aggregates of strongly lignified cells (sclereids) which give the pulp a gritty texture. Certain pears may in particular display a physiological disorder characterized by excessive development of sclereids ("Yuzuhada disorder"). A survey of some 40 selections of pear fruit shows an inverse correlation between phenolic content — particularly chlorogenic acid — and the amount of sclereids.[137,308] In addition, cell wall and membrane-bound peroxidase activity is distinctly higher in fruits which display Yuzuhada disorder.[137,308] The revelation of hormonal control of lignification in tomato plants, which can operate among other ways through modulation of parietal peroxidase activities,[235,237] leads to envisaging a regulatory role of phenolic compounds. However, this role remains to be determined in the case of lignification of fruit endocarp.

3. Role as Metabolic Effectors

Very strong inhibition of enzymes is observed when they are extracted from fruits which are rich in phenols, making it necessary to use various protective agents.[318,401] Since they can bond to proteins, phenolic compounds are able to act *in vitro* as inhibitors of the activity of numerous enzymes. However, transposition of these results to *in vivo* conditions remains doubtful in the light of the subcellular compartmentation of enzymes and phenolic compounds.

Numerous data, generally carried out *in vitro*, show that phenolic compounds can be involved in fundamental metabolic processes, usually as inhibitors: respiration and oxidative phosphorylation in mitochondria.[198,244,310,311,386] photosynthesis and photophosphorylation in chloroplasts,[19,79,374,386,424] photorespiration,[82] opening of stomata, and transpiration.[210,254]

With their reversible quinone oxidation, phenols constitute redox systems whose standard potentials at pH 7.0 (Eo) vary from 0.28 V (paradiphenol/hydroquinone) to 0.47 V (3,5,7-trimethyl-quercetin).[258] These compounds can thus act as electron transporters and interact with redox chains in fruit cells. Flavonols are involved as antioxidants at chloroplast level with a scavenger effect on superoxide ion $O_2^{\cdot -}$ and the free radicals formed during lipid peroxidation.[374,375] Carotenoid photobleaching was suppressed by flavonols.[373] Flavonols are also oxidized by chloroplasts in the presence of H_2O_2, suggesting that these compounds can function as effective H_2O_2 scavengers in these organelles.[376] The antioxidant properties of flavonoids, phenolic acids, and other phenols were reviewed recently.[214] In avocado fruits, lipoxygenase activity is regulated by the concentration of flavan-3-ol inhibitor, epicatechin.[304] Infiltration of fruits with a 0.1 m*M* solution of epicatechin delayed the appearance of anthracnose symptoms for 1 to 2 d, in relation with lipoxygenase inhibition and with no decrease in antifungal diene concentration.[302] Relations between the reduction of the superoxide ion, oxidation of *o*-diphenols into semi-quinones and then quinones and the browning phenomenon are examined in Chapter 5.

4. Miscellaneous Roles

Absorption of ultraviolet light is a general feature of phenolic compounds. Some of them can be considered as filters which protect certain fragile cell structures (e.g., chloroplasts) from UV radiation.[67,259,326,344,392] These filters consist mainly of flavonols and are located in the skins of fruits[392] (see Chapter 1). In addition, because of their antioxidant properties discussed above, polyphenols can serve as protection against photooxidation caused by UV light.[214,259]

Phenolic compounds, hydroxycinnamic derivatives and coumarins in particular, may act as *in situ* inhibitors of seed germination in berries or other fleshy fruits. The embryo may be prevented from germinating by inhibitory substances impregnating albumen, seed coat or carpel tissue.[44,66,218,400,428] However, the inhibitory effect may also be due to oxygen fixation by phenolic compounds, in particular in the seed coat, in such a way that the embryo is unable to germinate because of lack of oxygen.[76,78]

By accumulating as glycosides in vacuoles, a number of phenolic compounds probably play a detoxification role for fruit cells as glycosylation reduces toxic effects.[142] In addition, demonstration of a turnover of vacuolar phenolic compounds[32,33,75,291] shows that they are not only waste substances but may also serve as reserves, in particular in certain seeds.[75]

III. THE ROLE OF PHENOLIC COMPOUNDS IN RESISTANCE OF FRUITS TO MECHANICAL AND BIOLOGICAL STRESSES

The involvement of the phenolic metabolism in resistance of plants to different stresses has been described at length over the past 10 years.[41,42,118,397,402] These stresses may be mechanical (wounding), chemical (various types of treatment), or microbiological (infection by fungi, bacteria, or viruses). The role of a great number of factors on the regulation of the phenolic metabolism in fruits was discussed in the last chapter, but we now examine the way in which these changes at phenolic level can play a role in the protection of the plant and more specifically of the fruit. Phenolic compounds are involved in resistance in two ways: first, by contributing to the healing of wounds by lignification of cell walls around wounded zones,[397,402] and second, through the antimicrobial properties that have been shown for many of them.[41,118,232,397] The compounds involved in resistance can be classified in three groups:[143] (1) some are already present in the plant and their level generally increases following stress; (2) others are formed only after injury but are derived from existing substances by hydrolysis or oxidation; (3) finally, others are biosynthesized *de novo* and can be classified as phytoalexins if they possess antimicrobial properties.

It is not possible in most cases to draw a distinct line between resistance to wounding and resistance to microbial attack since pathogenic agents frequently first penetrate the host through tissue and cell wall lesions. Thus, needle-puncture of the pericarp of maturing pistachio fruit induced symptoms similar to those observed after insect feeding.[54] Furthermore, solely mechanical wounding may be followed by infection if healing is not rapid.

Whatever the type of stress, one of the most common responses is the increase in the total phenolic content and of chlorogenic acid in particular.[118,321,398] It may be involved either directly through its antifungal properties[51,172,206,358] in particular when it is oxidized[172,205] and, in addition, as an intermediate substance partially used in the biosynthesis of other phenylpropanoid derivatives such as scopolin,[360] lignin,[119,437] and suberin.[324]

A. MECHANICAL STRESS

Mechanical stress in fruits, resulting for example from insect bites or wounding during mechanized harvesting, causes lesions of varying depths in the skin and pericarp. Physiological disturbance results, in addition to consequences concerning sale of the fruits. Thus, wounding may cause early triggering of maturation[108,110] whether the fruit continues its development on the mother plant or is picked and stored. Numerous aspects of the metabolism of wounded fruits have been studied for a long time; they concern for example respiration,[395] sugar content,[396] and production of ethylene.[247,293,325] In contrast, variations in the metabolism of phenolic compounds have only been approached in rare cases,[108,110,390,391] although the possible contribution of these compounds to the browning of wounded areas has been suggested for a long time.[393]

The most immediate response to wounding is the oxidation of pre-existing phenolic compounds and hence their degradation.[321] Thus, the chlorogenic acid content of tomato fruit pericarp falls for 6 h following wounding (Figure 89). Indeed, after wounding, lesion of membranes results in brutal decompartmentalization which brings into contact enzymes and substrates, particularly phenolic compounds which normally accumulate in the vacuole, and browning enzymes, such as *o*-diphenoloxidase and peroxidase. The brown discoloration or necrosis which appears during maturation in litchi pericarp is thus the combined effect of oxidation of ascorbic acid and activation of browning enzymes.[191] One of the most-studied results of this phenomenon is the appearance of melanic type brown substances caused by oxidative polymerization of quinones. Only the first stage enabling quinone formation is enzymatic, but the sequence of chemical reactions which follows is still very little known; furthermore, the relative contribution of peroxidase and *o*-diphenoloxidase to the browning obtained have never been clearly defined. The *o*-quinones are inhibitors of microbial growth and polymers can precipitate proteins as tannins; such reactions can slow or halt infection of wounded tissue. *o*-Diphenoloxidase substrates are frequently present *in vivo* as glycosides and, on wounding, browning reactions are initiated by the release of glycosidases which act on phenolic glycosides to produce the aglycones which are the substrates for phenolase activity. An example of this is the phloridzin found in apple leaves and fruit skin which,

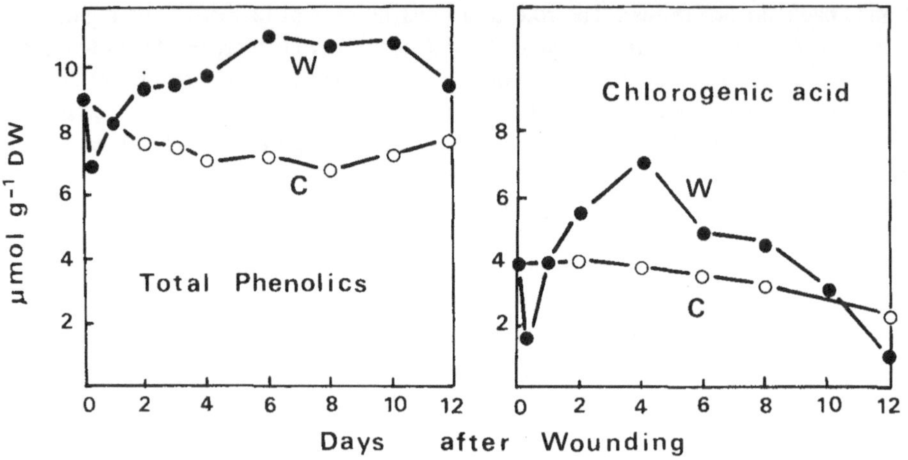

FIGURE 89. Changes in total phenolics and chlorogenic acid content in cherry tomato fruits after wounding C and W: control and wounded fruits. (Adapted from Fleuriet, A., Thèse Doct. d'Etat, Montpellier, France, 1982.)

on wounding, is hydrolyzed by a β-glucosidase to the aglycone, phloretin, which is hydroxylated and oxidized by catechol oxidase to yield 3-hydroxyphloretin and hence the corresponding quinone which possesses antimicrobial activity.[202,427]

After the immediate response of phenolic metabolism there is then an increase in total phenolic content (Figure 89) caused by an accumulation of compounds present before the stress and/or the appearance of compounds which had not previously been detected. Thus, in experimentally wounded tomato, the total increase in phenolic compounds and in particular in hydroxycinnamic derivatives reveals two possibilities: accumulation of chlorogenic acid, feruloylglucose, and *p*-coumaroylglucose in the wounded pericarp, and also the appearance of sinapolyglucose not detected in the control fruit.[112] In *Citrus sinensis,* the phenolic compound content increased more than twofold in injured peel after 48 h.[183] Wounding of apple fruit tissue increased the level of anthocyanin in the skin and replaced the stimulating effect of UV light on anthocyanin formation.[70]

Finally, the third aspect of response to wounding is the formation of healing tissue which protects the fruit from water loss and also forms a physical barrier which prevents possible penetration by pathogens.[402] Wound healing is generally much more rapid and complete in young, growing fruits than in ripe fruits. This was demonstrated in particular for tomato[108] and apple.[349] Healing consists of the formation of polymers, such as lignin and suberin by cells close to the wound. In orange peel, wound healing consists of lignin,[183] but the data on the nature of the polymer involved are sometimes contradictory for the same plant material. For example, in wounded or infected tomato fruit, it may be suberin[59,83] or lignin.[107,125] In fact, both groups of compound are involved simultaneously. Old observations already reported this duality by describing the formation of a ligno-suberous layer.[393] Phenolic compounds are more or less involved in the formation of these polymers (Chapter 1) and it is extremely remarkable that their biosynthesis is observed in wounded fruits whereas this does not occur under normal conditions.

The last two aspects of the response to wounding result from changes in the phenolic metabolism and particularly at the level of enzymes involved in the biosynthesis of simple molecules and their integration in more complex chemical structures like lignins.

Healing of wounding in 'Valencia' orange involved the induction of phenylalanine ammonialyase (PAL) whose activity increases in flavedo but not in the underlying albedo tissue (Figure 90).[184] Such increase in enzymatic activities is responsible for the increase in

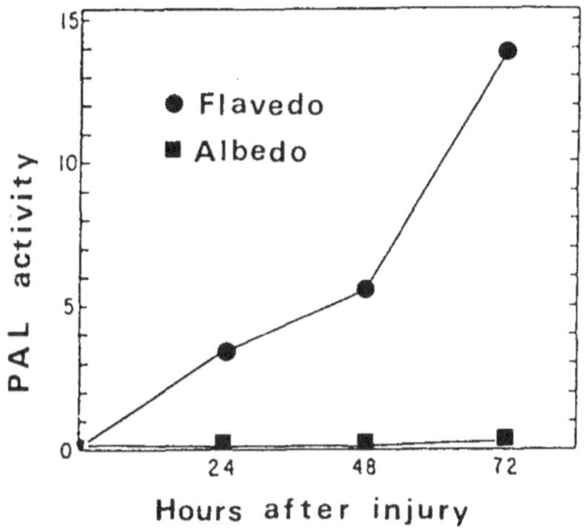

FIGURE 90. Phenylalanine ammonialyase activity (moles cinnamic acid/gDW/h) in "Valencia" orange flavedo and albedo tissues after injury and healing at 30°C. (Adapted from Ismail, M. A. and Brown, G. E., *J. Am. Soc. Hortic. Sci.*, 104, 126, 1979.)

lignin and free phenolic constituents in injured cirtus flavedo[183] or anthocyanins in apple.[100] However, other enzymes are also involved. Thus, besides PAL, *O*-methyltransferase (OMT), hydroxycinnamoyl CoA ligase (CoAL), and glucosyltransferase (GT) activities increase strongly in tomato fruits after wounding (Figure 91); this occurred immediately for most activities (PAL, CoAL, and GT) and later for OMT.[112] The fall or increase in activity subsequently observed for all these enzymes is directly related to the accelerated maturation of the fruits and is fully comparable with the behavior observed during natural maturation (Chapter 3).

In tomato, some of the variations observed may be involved in the lignification of wounded zones which occurs from the 4th d onwards, although lignin itself is only detectable by histochemistry on the 8th d.[107] The first sign is shown by an increase in PAL activity, which occurs early on and leads to accumulation of all the hydroxycinnamic derivatives, and in particular glycosylated derivatives and chlorogenic acid which can supply the precursors required for the biosynthesis of lignin.[119] However, it is CoAL and OMT activities which probably play a determinant role in the orientation of the metabolism of wounded fruits towards biosynthesis of lignins, first, because of the very strong increase in their activity and second, because of their different specificity compared with control fruits. Thus, the relative increase in CoAL activity is much greater for *p*-coumaric (+320% on the 4th day) and ferulic acids (+290%) as substrates than for caffeic acid (+120%). The situation after wounding is thus very different from that in the pericarp of control fruits where activity is essentially oriented towards the utilization of caffeic acid. In wounded fruits, CoAL activity leads to more intense biosynthesis of *p*-coumaroyl CoA and feruloyl CoA, which are precursors of lignin.[338] These variations in specificity might suggest the presence of multiple isoenzymatic forms which may orientate the phenolic metabolism qualitatively.[159,309] The effect of wounding on one of these isoenzymatic forms may result in either neosynthesis or activation of pre-existing forms, as has been demonstrated for example for peroxidases.[240] However, although the lignins formed in wounded fruits react to Maüle and Wiesner's reagents and thus, like all Angiosperms, contain both guaiacyl and syringyl units, we have been unable to characterize the formation of sinapoyl CoA at any stage. Absence of the formation of this derivative has frequently been reported[134] and it is likely to involve a different biosynthetic pathway.[209]

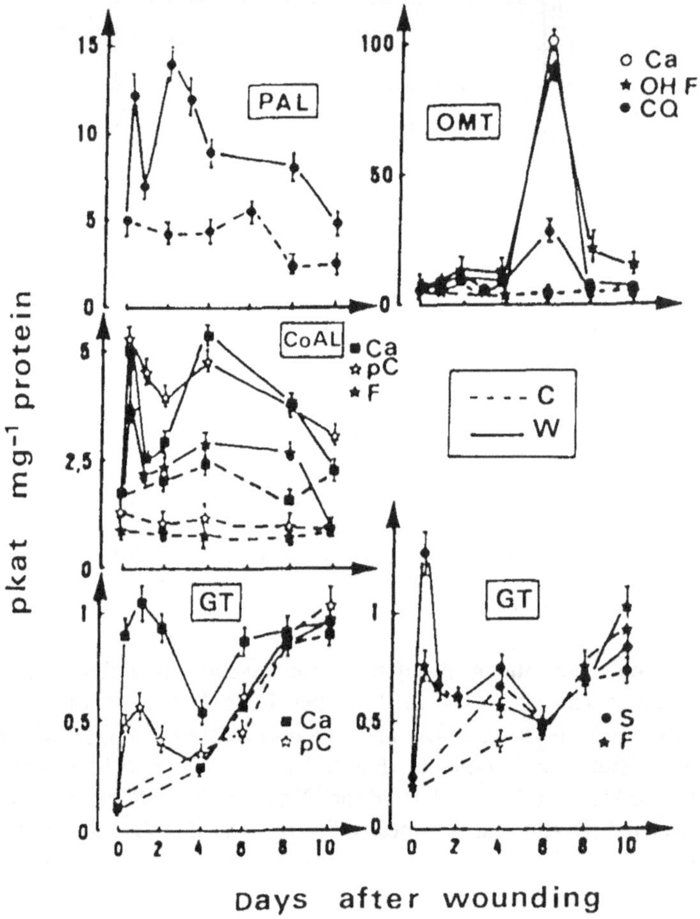

FIGURE 91. Changes in PAL, OMT, CoAL, and GT activities in cherry tomato fruit after wounding. Substrates of enzymes: *p*C: *p*-coumaric; Ca: caffeic; F: ferulic; OHF: hydroxyferulic; S: sinapic; CQ: chlorogenic acids. C and W: control and wounded fruits. (From Fleuriet, A. and Macheix, J. J., *Physiol. Plant.*, 61, 64, 1984. With permission.)

The specificity of OMT activity is another example of a special qualitative orientation of the phenolic metabolism following wounding. After the strong increase from the 4th to the 6th d, relative activity of OMT with 5-OH ferulic and caffeic acids (leading to sinapic and ferulic acids) evolves considerably. Thus, the ratio of activities changes from an average of 1.2 (with nonsignificant variations in wounded fruits during the first 6 d and in control fruits) to 4.2 on the 8th d.[112,207] From this stage, lignin accumulates for about 10 d[107] and it is probable that OMT activity plays both a qualitative and quantitative role in the regulation of lignification of wounded zones. It may in particular orientate the metabolism towards the formation of "wound lignin"; it is known that the composition of this may be different to that of natural lignins.[21]

Ethylene may be one of the determinant factors in the later increase in certain enzymatic activities. Indeed, the two successive peaks of PAL activity in wounded tomato fruits are affected differently by treatment with ethylene, which causes a strong increase of the second only.[109] It is, therefore, probable that the increase in ethylene released immediately after wounding[110] does not participate at all, or only a little, in the early inducing of enzymatic activities, but may be responsible for subsequent evolution.

These observations make it possible to account for the sequence of events leading to healing. First of all, accumulation of monomers due to the increase in the enzymatic activities involved in their biosynthesis (PAL, CoAL, OMT); this is followed by the polymerization of these monomers in the cell wall as a result of lignification-specific peroxidasic activities which appear during healing. This is the case in tomato fruit, where the increase in peroxidase activity observed after wounding is closely associated with the appearance of an isoperoxidase with strong affinity for syringaldazin.[107] Wounding thus causes new orientation of the metabolism of hydroxycinnamic derivatives. Tomato fruit pericarp is never lignified under natural conditions as its phenolic metabolism is normally orientated towards the formation of flavonol glycosides (mainly rutin) and simple hydroxycinnamic derivatives. This new orientation towards lignification is the sign of cell differentiation induced by an exogenous factor. It occurs secondarily and is determined by cell potentialities which are not expressed under normal conditions.

At wounding, there is, therefore, a set of complex and still inadequately studied interactions involving phenol metabolites, peroxidases and ethylene but also other factors, such as plant hormones.[226] Normal development of the wound leading to healing results from the harmonious operation of these interactions, controlled by environmental factors.

What occurs with phenol metabolism enzymes after the wounding of whole fruits is probably comparable to the phenomena observed during aging of tissue slices: genome derepression, activation or switching-on of metabolic pathways, biosynthesis of new enzymes.[194] All these aspects contribute to intensifying metabolic features, such as respiratory activity and biosynthesis of ethylene. Numerous data have been obtained on PAL variations during the ageing of different tissues.[379,437] Study of skin disks of apple fruit shows that both the rate of anthocyanin accumulation and PAL activity were greater than in whole fruit[100] in close relation with the increase in tissue ethylene levels.[71] These data are in conformity with the results previously obtained for PAL in other plant materials.[379,437] However, in most cases and in particular in fruits, the results have not been discussed in reference to the wounding effect[377,378] caused by working with slices and which, as we have just seen, profoundly disturbs the phenolic metabolism.

The tissue slice system was also used as a model for studies of host-parasite relationships[398] which also involve phenolic components whose role in the resistance of fruit to biological stress is examined below.

B. BIOLOGICAL STRESS

Almost all pathogenic bacteria and a few fungi enter plant tissue passively through natural openings or wounds, while most of the pathogenic fungi have evolved sophisticated means for active penetration.[256] Response of the phenolic metabolism is very similar to that described above for wounds. Phenols frequently increase after infection and in particular in fruits such as the tomato (Figure 92). Lignin is thought to be formed commonly as a response to microbial penetration, and is discussed as a mechanism in disease resistance.[117,118,323,402] This is the case of tomatoes inoculated with *Botrytis cinerea* where lignin formation, increased PAL activity, and accumulation of *p*-coumaric and ferulic acids are observed.[125] Likewise, in papaya fruit, resistance to *Colletotrichum gloeosporioides,* which is responsible for anthracnose, has been attributed to the formation of a lignified periderm and callose deposition.[359] In 'Valencia' orange, resistance to *Penicillium digitatum* was reported to be caused by the formation of lignin in injured flavedo tissue which prevented hyphal penetration.[65]

The other types of response concern a different property of phenolic compounds; this is the inhibition of microbial growth or the germination of certain spores. Numerous examples have been studied but few concern the mechanisms of action. However, it has been established that there are relations between the structure of several groups of natural or synthetic compounds and their ability to inhibit pathogen enzymes.[312-315] Control of infection of avocado fruit by *C. gloesporioides* is through lipoxygenase activity which is regulated by

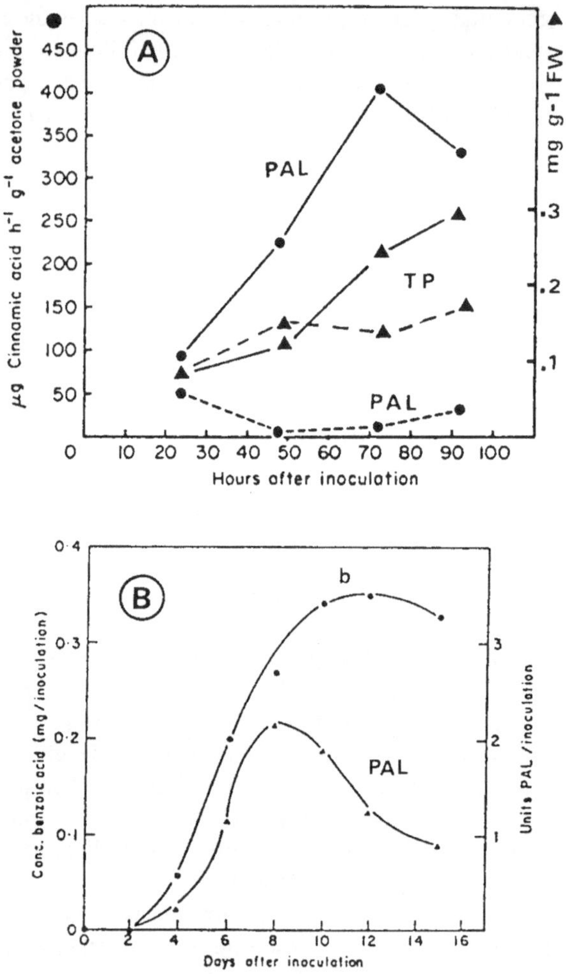

FIGURE 92. PAL activity and total phenolics (TP) or benzoic acid (b) content in tomato (A) and apple (B) at varying times after inoculation with *Botrytis cinerea* (tomato) or *Nectria galligena* (apple). (——) infected tissue; (----) uninoculated control. (Adapted from Glazener, J. A., *Physiol. Plant Pathol.*, 20, 11, 1982; and Swinburne, T. R. and Brown, A. E., *Physiol. Plant Pathol.*, 6, 259, 1975.)

the natural antioxidant epicatechin.[304] Consequently, infiltration of avocado fruits delayed the appearance of anthracnose symptoms.[302]

Three different cases may then occur:[143]

1. The phenolic compounds present in fruits possess antimicrobial properties and their content may increase after infection. Thus chlorogenic (CQ) and *p*-coumaroylquinic (pCQ) acids in apple are both inhibitors of *Botrytis* spore germination and also of mycelial growth.[272] If their effects on the growth of certain fungi liable to infect apple are compared, pCQ is more inhibitory than CQ at the same concentrations (Table 48) for *B. cinerea* and *Alternaria* sp. whereas *P. expansum* is less sensitive. pCQ and CQ are stimulatory at low concentration for *Botrytis* and *Penicillium*. Anthocyanins can be fungitoxic, and cyanidin itself has been shown to inhibit germination of *Gloesporium perennans*, a fungus causing apple rot.[172] The 3-4 dihydrobenzaldehyde extracted from banana skin inhibited the growth and germination of *G. musarum* at respective concentrations of 250 μg/ml and 500 μg/l.[267] The 7-hydrox-

TABLE 48
Inhibitory Effects of *p*-Coumaroylquinic (pCQ) and Caffeoylquinic (CQ) Acids on Growth of *Botrytis cinerea* (BC), *Penicillium expansum* (PE) and *Alternaria* sp. (A), in Liquid Medium after 9 d

Concentration (μg/ml)	Fungal growth expressed as mg N					
	BC		PE		A	
	pCQ	CQ	pCQ	CQ	pCQ	CQ
0	1.10	5.85	10.55	10.5	1.36	4.6
100	0.01	4.20	10.5	8.3	0.16	2.45
200	—	3.70	8.19	5.0	—	1.45

Data from Reference 272.

ycoumarin from grapefuit flavedo was also found to inhibit spore germination and germ tube elongation of *P. digitatum*.[185] Pterostilbene is very active *in vitro* on several fungal microorganisms.[213] This compound was recently found in grape berries and may contribute to the natural resistance of grapes to grey rot.[292] Seselin, a chromeno coumarin identified in citrus tissues, inhibits the growth of *Phytophtora*.[406] Caffeic acid heterosides, and in particular verbascoside revealed in olive, also inhibit growth of a number of fungi.[15] In contrast, the antimicrobial properties of other compounds, such as oleuropein, the main phenolic compound in olive, are contested. Thus, although oleuropein inhibited the growth of *Geotrichum candidum*[193] and slightly delayed onset of the growth and sporulation of *Aspergillus parasiticus*,[130] it would appear that it is the products of hydrolysis which possess antimicrobial properties.[105] Likewise, the antimicrobial properties of phloridzin or the products of its degradation (after hydrolysis and oxidation) are strongly questioned: no direct connection with apple scab resistance was demonstrated.[173] Tannins are also able to block the development of most microorganisms. Gallotannins and condensed proanthocyanidins have long been used in the leather industry to denature animal protein in skin and also to protect it from microbial degradation. Likewise, these substances can inactivate extracellular enzymes produced by microorganisms. Certain flavones, and in particular nobiletin isolated from fruit peel, are the natural compounds involved in resistance to *Deuterophoma tracheiphila* in some varieties of *Citrus*.[43]

Thus the presence, content, and location of phenolic compounds is frequently considered to be an important factor in resistance: on the one hand, the greater quantities of these compounds are concentrated predominantly in the epidermal tissue (Chapter 1) and might, therefore, function as a natural barrier to microbial infection.[272,342] The epicatechin concentration in the peel of resistant cultivars of avocado is hence higher than that in susceptible cultivars.[303] On the other hand — in close connection with their high phenolics content — young fruits are much more resistant than ripe fruits.[125,174,272] Nevertheless, there is not always a very distinct relation between the phenolic compounds content and resistance. Thus, qualitative and quantitative differences were not detected between cultivars susceptible and resistant to leaf rust with regard to the composition and content of the most abundant phenolic compounds in fresh coffee pulp.[307]

2. The second case is the acquisition of antimicrobial properties by phenolic compounds after conversion by oxidation or hydrolysis. Firstly, *o*-quinones are generally more active than *o*-diphenols:[172] browning intensity is often greatest in highly resistant plants suggesting that black and brown pigments contribute to resistance.[41,42] For example, in *Pyrus*, oxidation products of arbutin were toxic to *Erwinia amylovora*.[300] In addition, after infection, an increase of polyphenol oxidase was observed in a few cases[381] which may have resulted

from activation of a latent form.[88] However, the problem is different from that encountered for wounding insofar that when phenolic compounds have been released from the vacuoles, they may be oxided no longer by the host plant's own polyphenoloxidase but by that of the microorganism. Secondly, hydrolysis of certain phenolic compounds can lead to the formation of products with antimicrobial properties. As already mentioned, this is the case of oleuropein and phloridzin. The same applied to arbutin, which is involved in the resistance of pear shoots to *E. amylovora,* which might be the result of a complex interaction mechanism involving β-glucosidases, polyphenoloxidases, and peroxidases.[300] Hydrolysis may be effected by fungal pectic enzymes, as suggested by the appearance of free *p*-coumaric, caffeic, and ferulic acids in apple infected with *P. expansum.*[419] In this case, the damage does not cause much browning around the site of infection; the author accounts for this by inhibition of the phenolase system by acids released after hydrolysis of chlorogenic and *p*-coumaroyl quinic acids in the fruit. In apple again, antifungal compounds, such as 4-hydroxybenzoic acid produced after infection with *Sclerotinia fructigena* is thought to be derived from the transformation of chlorogenic acid by the fungus.[102]

3. In contrast with the preceding cases, certain phenolic compounds may be biosynthesized *de novo* after infection. These compounds, which do not exist before infection and which have antimicrobial properties, are called phytoalexins. They are produced by plants as a defense mechanism in response to microbial infection.[232] However, they are also produced after chemical (herbicide, fungicide) and physical (UV, temperature shock, wounding) treatments. Their chemical nature may vary but many of them are phenolic.

Very few phenolic phytoalexins have been identified to date in fleshy fruits. The appearance of 6-7-dimethoxycoumarin (scoparone) in citrus, conferring resistance to *Phytophtora citrophtora*[6] and a stilbene (resveratrol) in grape berries, with antimicrobial activities with regard to *B. cinerea,* have been reported.[81] The same compound also appears after UV irradiation, and this may be useful in assessing sensitivity to berry disease. The potential biosynthetic capacity of this phytoalexin decreases during maturation (Figure 93) and accounts for the greater sensitivity of berries during this period. In apple, benzoic acid (not phenolic) accumulates after infection by *Nectria galligena.*[369] It appears to be the main phytoalexin in this fruit.

In contrast, numerous works have been devoted to phytoalexins in the pods of Leguminosae[179] where the nature of the phenolic compounds involved is very varied: isoflavones, isoflavanones, pterocarpans, isoflavans, chromones, coumestans, and stilbenes. All these compounds thus possess antimicrobial activities, but pathogens are sometimes able to detoxify the phytoalexins produced by their hosts. Some pathogens may also become tolerant to high concentrations of phytoalexins or may be able to secrete molecules to specifically inhibit the synthesis or phytoalexins in their hosts.

The biosynthetic pathways of phenolic phytoalexins are identical to those of naturally accumulated phenolic compounds.[361,362] Thus, radioactivity from ^{14}C-phenylalanine was incorporated in pea pod pisatin and cinnamic acid was also an excellent precursor.[138] Accumulation of phytoalexins is accompanied by much greater enzyme activities than those in untreated tissues. In particular, PAL activity increased during accumulation of isoflavonoid phytoalexin.[211] In apple, PAL activity was induced by infection by a fungal pathogen (Figure 92)[369] and maximum benzoic acid concentration was attained approximately 3 d after maximum PAL level had been reached. Similarly, stimulation of PAL activity occurred before phaseollin could be detected in pod tissue of *Phaseolus vulgaris.*[160]

CoA ligase, flavanone synthase, and cinnamic acid 4-hydroxylase activies increase during accumulation of phytoalexins.[211] The same effect can be obtained by the sole action of elicitors such as the glucan elicitor of *Phytophtora megasperma* which caused coordinated increases in PAL and 4-coumarate:CoA ligase activities in parsley cells.[139] It appears that healthy, untreated cells contain material capable of initiating synthesis of the appropriate

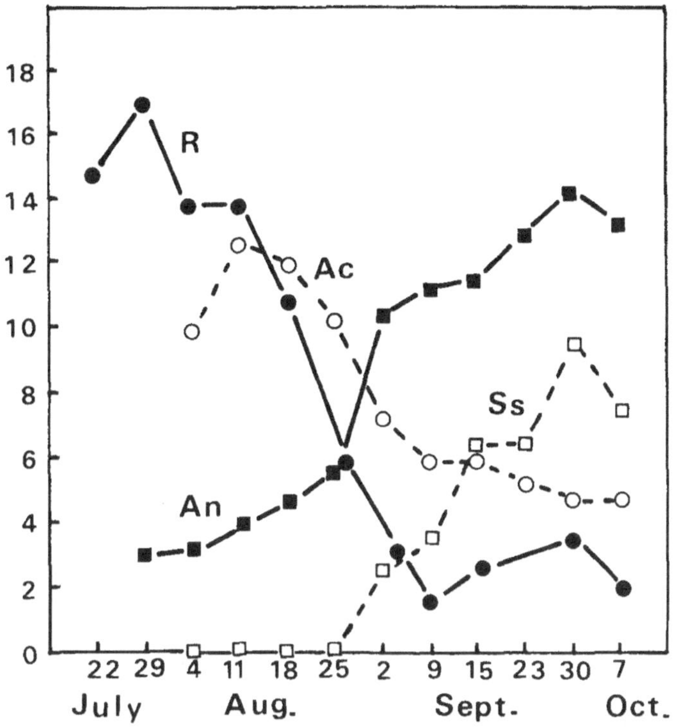

FIGURE 93. Resveratrol potential and maturity parameters for grape (cv. Concord) collected from the vineyard on different dates. R: resveratrol (μg·cm^{-2}); Ac: acidity (meq 2.5·ml^{-1}); An: anthocyanin (μg·cm^{-2}); Ss: percent soluble solids. (Adapted from Creasy, L. L. and Coffee, M., *J. Am. Soc. Hortic. Sci.*, 113, 230, 1988.)

enzymes, but they remain inactive. When cells are injured, these materials become active and thus enzyme synthesis and subsequently phytoalexin synthesis occur in the injured cells and in adjacent cells which remain healthy. It has been suggested that the action of these materials is facilitated by their release as a direct consequence of membrane dysfunction.[25] However, it is not clear whether phytoalexin formation is merely the result of mechanical injury caused by the entry of pathogen into plant tissue. Indeed, there is sometimes no difference in the levels of fungistatic phenolic compounds in sensitive and resistant cultivars.[295] In addition, it appears for example in 'Valencia' orange tissue that lignin formation during healing provides a mechanical barrier which retards or inhibits penetration of injured tissues by *Penicillium digitatum.*[183]

IV. ROLE IN THE QUALITY OF FRUITS

Maturation of fruits is characterized by a complex set of coordinated biochemical changes which lead to the progressive acquisition of the specific properties of the ripe fruit. Several points, such as the acquiring of color or contribution to the forming of the characteristic aroma of a species or a variety, are associated with the presence of phenolic compounds. These parameters are of particular importance insofar as they are taken into account in the sensorial evaluation of the fruit by the consumer — whether human or animal.[228] They then form criteria of quality which may lead to the choosing or refusal of a ripe fruit for consumption. These aspects are discussed here; modifications of the parameters involved in fruit processing are examined in the next chapter.

A. IMPORTANCE OF PHENOLIC COMPOUNDS IN THE COLOR OF FRUITS

Color is an important feature of fruits both from the ecological point of view and for identification or determination of their commercial value. Indeed, as with all human foods,[249] consumer choice is influenced by color, both in the case of direct use of fruit and in that of processed products (fruit juices, fermented beverages, jellies, jams, preserves, etc.).

1. Methods of Measuring Color

The first appraisal of fruit color is provided by the eye, but this must be complemented by quantitative analysis, which involves two types of method: nondestructive and destructive.[74]

a. Nondestructive Methods

Nondestructive methods are based essentially on determination of the characteristics of light transmitted or reflected by the fruit. The results obtained thus depend on the nature and content of fruit pigments and also on their location in the tissues and on all the physical and geometrical characteristics of the fruit cells. These methods have been widely used to assess for example color caused by anthocyanins in grape,[422] strawberry,[355] or cranberry.[333] In the case of sweet cherry, highly significant correlations were obtained for reflectance color with either anthocyanin content or subjective USDA color determination.[95]

Likewise, the anthocyanidin content of blueberries can be assessed directly by light transmittance measurements, leading to the definition of a maturity index correlated with certain other conventional parameters of fruit maturity (pH, total acidity, soluble solids, etc.)[85,208] These principles have been applied to the sorting of small red fruits and the separation of red grapes in relation to their degree of maturity.[28,329]

Nevertheless, insofar as anthocyanins are frequently located in the outer tissues of fruits, correlation of reflectance measurements and anthocyanin content of juice can lead to ambiguous results since the size of the fruit comes into play. The problem has been studied in particular in the case of cranberry[333,334,409] where variability in sample anthocyanin content depended more on berry size differences than on differences in surface coloration. These data lead to orienting selection procedures for breeding programs with the aim of increasing the anthocyanin content of cranberries.[335,336,409]

b. Destructive Methods

Destructive methods are based, first, on the extraction of pigments followed by assaying frequently combined with the separation of the different molecules. Analysis of anthocyanins has been the subject of much work, and a number of significant examples have been given in preceding chapters. Reference can usefully be made to the very complete updates published on the subject.[115,167,186] One of the difficulties in interpretation nevertheless resides in the complex relations that may exist between the overall color of the fruit, such as perceived directly by the eye, and the quantitative analysis data for individual pigments. Indeed, extraction can even lead to profound modifications in color, in particular of anthocyanins. There can be several reasons for this: (1) the mixture during extraction of tissues rich in anthocyanins with tissues which contain none; (2) mixing on a subcellular scale of anthocyanins from the vacuole with other cell compartments where the physicochemical conditions may be very different; (3) modification of the immediate environment of the anthocyanins (pH in particular) and of bonds with other molecules (copigmentation phenomena); (4) instability of anthocyanins under the action of different environmental factors responsible for their degradation, enzymatic or otherwise; (5) disappearance of interference between light absorption by anthocyanins and other pigments (phenolic or not).

In the light of these modifications as a whole, it is always very important to seek in all fruits relations between the data obtained by nondestructive methods and by analysis of the anthocyanins present in the extracts or fruit juices; nevertheless, the results sometimes remain somewhat ambiguous.[333]

2. The Main Pigments in Fruits

Fruit color is caused by pigments which specifically absorb certain radiation in the visible spectrum. Several types of pigment, whose evolution is characteristic of physiological stage, are frequently found in fruits.[128,135]

Chlorophyll and carotenoid pigments — Located in plastids (chloroplasts and chromoplasts) are not discussed in this work. Chlorophylls cause the green color of young fruits; they are abundant during growth and then generally diminish during maturation. Carotenoid pigments are usually yellow-orange and sometimes bright red. They are frequently masked by chlorophyll and become visible during maturation; however, they may also accumulate, an example being lycopene which accounts for the red color of ripe tomato.

Phenolic pigments located in vacuoles — The most abundant are anthocyanins which give their color to red, blue, mauve, and black fruits, etc. The color of the fruit may depend on the relative proportions of anthocyanins and chlorophylls. Thus, in eggplant, the color variations between the different cultivars which have reached maximum coloration are caused partly by the anthocyanin content of the skin and partly by the chlorophyll content which may have a darkening effect on purple as dark purple fruits contain more chlorophyll than dark green ones.[279]

Flavonol and flavone derivatives are generally found in addition to anthocyanins. Although very broadly distributed in fruits, they make only a very small contribution to their yellow coloration.[128] An exception is the isoflavanone osajin which contains two isoprenoid residues and participates in the color of the osaje orange *(Maclura pomifera)*.[128] However, flavonol glycosides may be important isofar as they enhance anthocyanin coloration by copigmentation (see below).

Brown pigments — The color of fruits may be strongly modified by the appearance of brown compounds which generally result from the oxidation and polymerization of phenolic compounds under the effect of phenoloxidases or peroxidases. These melanic type brown pigments may appear naturally during the maturation of certain fruits (date, medlar, sapodilla, etc.) and in these cases, the brown color is a feature of the quality of the fruit. Nevertheless, browning generally occurs accidentally after wounding or crushing of fruits or in physiological disorders during postharvest storage. It results in discoloration of fruit which generally affects commercial quality. Browning can also occur when fruits are processed and in this case, the change in color is due to the combined effect of browning and anthocyanin loss.[356] These phenomena are examined in detail and control methods discussed in Chapter 5. The present paragraph covers only the relations between fruit color and anthocyanin content.

3. Anthocyanin Composition and Fruit Color

Comparative study of different species or different cultivars of red fruits has already revealed very large variations in the nature and level of anthocyanins (Chapters 1 and 2). In addition, very significant changes take place during the life of the fruit, one of the most characteristic being the accumulation of anthocyanins during the maturation of numerous red fruits. In addition, the proportions of the different anthocyanins may evolve with time, as has been shown by several examples (Chapter 3). These variations have important consequences on fruit color. Thus, in blueberry, cyanidin 3-glucoside and cyanidin 3-galactoside are predominant in ripe fruit whereas malvidin derivatives are more abundant in overripe fruits, accounting for the color changes observed.[231] Thornless blackberry changes steadily from red to black during maturation and a certain reversion towards red may nevertheless be observed during frozen storage.[190] Here, all the anthocyanins are derivatives of cyanidin whose amounts and proportions varied greatly at the beginning of maturation. It is, therefore, unlikely that in this case the color differences were caused by differences in the proportions of individual anthocyanins.[337]

The nature, concentration, and relative proportions of anthocyanins are mainly governed by genetic factors which lead to more or less clearcut distinction between cultivars and species. Coloration depends on several genes and in a number of cases it has been made possible to use crosses or mutation to obtain varieties with low anthocyanin contents (e.g., the "pink-fruited" hybrid of highbush blueberry) or no anthocyanins at all (eggplant). These aspects are discussed in Chapter 1. The distribution of anthocyanins in fruit (either only in a few external cell layers or throughout the fruit) is also under genetic control and the consequences with regard to coloration may be important, as for example in the case of holly where the number of colored cell layers plays a large part in determining the intensity of fruit color.[180]

External factors also play a major role in fruit color by acting on the rate of accumulation or degradation of anthocyanins (Chapter 3) and hence on the final content in the fruit. The most colored berries (Rubus, Ribes, Empetrum, Vaccinium, Vitis, Sambucus, etc.) are generally those which are very rich in anthocyanins. Total anthocyanin contents in some vacuoles can reach 0.1 *M*[264] with mixtures of anthocyanin which are often complex and in which it is difficult to discern the contribution of each pigment to the color obtained. The situation in fact varies in complexity according to the fruit. As is explained in Chapters 1 and 2, cases range from the fairly simple situation of apple where all the anthocyanins derive from cyanidin alone to complex situations, such as *Vitis* or *Vaccinium* in which several anthocyanidins are present.

The relationships between the position of the absorption spectrum and the structure have been discussed extensively for pigments in the flavylium form in acidic media.[22,140,365,384] Given the general structure of the flavylium cation, anthocyanidins appear highly colored in a strongly acid medium with two absorption zones, one between 500 and 550 nm which is responsible for pigment color and a second less intense zone in the UV range towards 280 nm. Color depends partly on the absorption maximum.[135] For example, absorption maxima are respectively 520, 535, and 546 nm (in 0.01% HCl in methanol) for pelargonidin (scarlet), cyanidin (crimson), and delphinidin (blue-mauve) although the figures can vary with the solvent. Sugars, acylated sugars, methoxyl and hydroxyl groups have a marked effect on the color and reactivity of anthocyanins.[242,322]

Additional hydroxylation on the B-cycle results in a large bathochromic shift. Methylation has a small effect whereas glycosylation causes modifications: that at C-3 produces a hypsochromic shift (about 15 nm in C-3 glycosides compared to the corresponding anthocyanidins) and a shoulder at 440 nm.[115] This shoulder is less intense in 5 and 3,5-diglucosides. At a given pH, the latter are less colored than 3-glucosides.[242] Acylation of sugar by a hydroxycinnamic acid leads to an additional peak in ultraviolet at between 310 and 330 nm depending on the nature of the acylating group. In contrast, this peak is absent in derivatives acylated by nonaromatic acids (acetic, malonic etc.) and although this property does not have a direct influence on color, it enables the rapid identification of the two groups of compounds after HPLC separation, with hydrolysis techniques used to complete identification.[115]

These spectral differences between pure pigments lead to variations in coloration; this applies, first, to differences between cultivars as regards the characteristic pigment composition, and in particular the characteristic proportions of anthocyanins at maturity, and second, to a single cultivar in relation to degree of maturity insofar as the proportions of individual anthocyanins vary during maturation (see Figure 42).

4. Factors in the Expression of Color

The color obtained with a given anthocyanin content also depends on the immediate physicochemical environment of the molecules (pH, metal ions, other phenolic molecules, etc.). The color may vary considerably if this environment is modified, either naturally

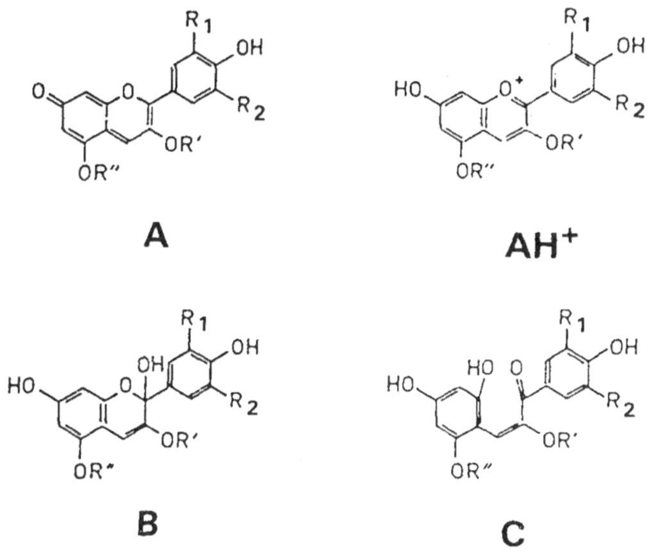

FIGURE 94. The four anthocyanin structures present in aqueous acidic solutions at room temperature. R_1 and R_2 are usually H, OH, or OCH_3. R′ is glycosyl and R″ is H or glycosyl. (Adapted from Brouillard, R., in *Anthocyanins as Food Colors,* Academic Press, New York, 1982, 1.)

during normal physiological evolution of the fruit or artificially during extraction of anthocyanins. These aspects are discussed briefly below, after a reminder of some of the physicochemical properties of anthocyanins.

a. Importance of pH

One of the most striking features of flavylium chemistry is the relative stability of this organic cation in aqueous solutions. Anthocyanins can be in four different forms, whose pH-related interconversions have been studied intensively (see References 60,62,63,166 and references therein): neutral and ionized quinonoidal A bases (blue or purple), flavylium cation AH^+ (red or orange), pseudobase or carbinol B (colorless), and chalcone C (yellow) (Figure 94). The interconversions between these different forms according to pH are shown schematically in Figure 95.[59,62]

As a general rule, anthocyanins in an aqueous medium at a temperature of around 25°C, and with slight acidity close to natural conditions, are rapidly transformed into colorless forms[59] with the exception of a few pigments acylated by hydroxycinnamic acids.[436] The main factor in the fading process is the great reactivity of flavylium towards nucleophilic reagents, in particular the water molecule and the hydroxyl ion: covalent addition of water takes place according to the hydration equilibrium (Figure 95). Possible intermediate mechanisms have been discussed.[59,64]

Addition of water takes place principally in position 2 because of unfavorable kinetic and thermodynamic conditions.[60] At equilibrium, the B/A ratio is independent of the value of the medium and is equal to the equilibrium constants ratio K′h/K′a (Figure 95).[59] It is 43 for cyanidin 3-glucoside at 25°C but may be very small for certain acylated anthocyanins which then give colored solutions unrelated to the pH.

The pH of the medium plays a particularly important role in the equilibrium between the different forms (Figure 95), the the more important color changes will occur at pHs close to pK′a values (Figure 96). Anthocyanins exhibit intense red coloration only in a very limited acidic pH range (between 1 and 3);[166] this corresponds to an equilibrium between the flavylium cation and the colorless carbinol base. Increase in pH causes reversible struc-

$$AH^+ \underset{}{\overset{k'_a}{\rightleftharpoons}} A + H^+ \quad \text{Acid-base equilibrium}$$

$$AH^+ + H_2O \underset{}{\overset{k'_h}{\rightleftharpoons}} B + H^+ \quad \text{Hydration equilibrium}$$

$$B \underset{}{\overset{k'_t}{\rightleftharpoons}} C \quad \text{Ring-chain tautomeric equilibrium}$$

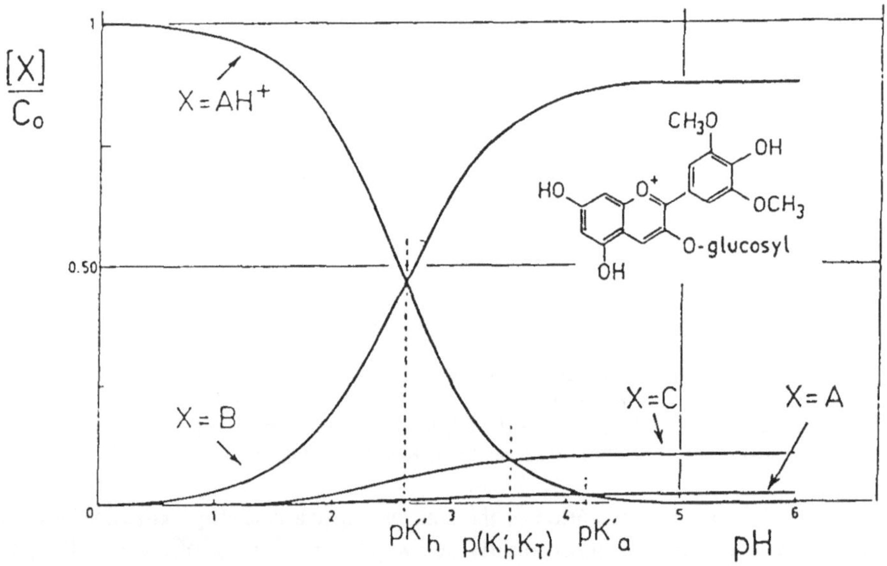

FIGURE 95. Equilibrium distribution at 25°C of AH^+, A, B, and C for malvidin 3-glucoside as a function of pH. (From Brouillard, R.; in *Anthocyanins as Food Colors*, Academic Press, New York, 1982, 1. With permission.)

tural transformations with loss of red color or formation of undesired color characteristics. In the pH range 4 to 6.5, the flavylium salt is converted immediately to the two purple tautomeric neutral quinonoidal bases which slowly hydrate to the carbinol base, with progressive disappearance of the color. Ionized quinonoidal bases also appear at neutrality and at a higher pH ring fission in the heterocycle produces a yellow chalcone progressively. The appearance of this chalcone is enhanced by a rise in temperature; this can occur from pH 4 onwards. However, the chalcone is not stable and the yellow coloring disappears at a rate depending on pH. In addition, quantitative reversibility to the flavylium cation only occurs in freshly prepared alkaline solutions; thus, alkaline conditions — in particular at high temperatures — should be avoided in food color processing.[59]

Unlike anthocyanins, 3-deoxyanthocyanins are very rare in fruits and exist mainly in flavylium form at pH $\leq$4 and as a mixture of the three neutral forms at a higher pH.[243] Unlike the common anthocyanins, 3-deoxyflavylium salts could be used to color foods and beverages at neutral pH.

Vacuolar contents in fruits are distinctly acid because of the sometimes considerable accumulation of organic acids.[394] Acidity generally falls during maturation: for example, the pH of apple changes from 2.8 to 3.5. Measurement of the vacuolar pH has in fact rarely been carried out on vacuoles alone. Nevertheless, a method for the determination of pH in intact vacuoles has been reported for grape.[264] It is based on the linear relation between pH and the logarithm of the ratio AH^+/B, and enables accurate determination of vacuole pH in living tissue. The vacuolar pH of ripe grape is about 2.8 in subepidermal cells which contain anthocyanins, whereas the values determined by measurements on grape skin extracts

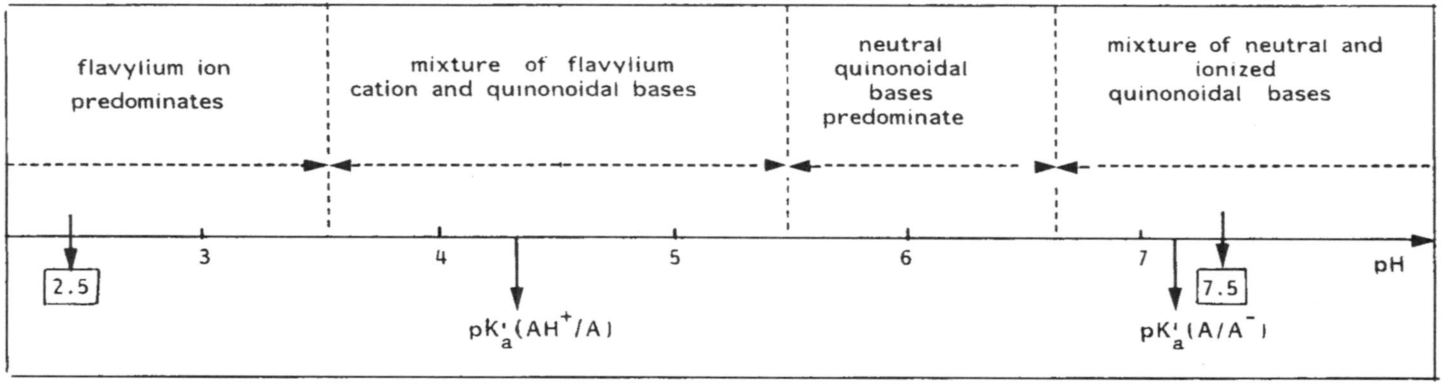

FIGURE 96. Predominant anthocyanin chromophores in different pH ranges. (From Brouillard, R., *Phytochemistry*, 22, 1311, 1983. With permission.)

were approximately 1 unit higher, probably because of the mixing of cytoplasm with vacuoles.[264] It would appear in all cases that the small variations in pH during maturation occur in a zone where they could cause considerable modification to the intensity of fruit color. Nevertheless, there are no experimental data today which can demonstrate such modifications, which may be masked by steady accumulation of anthocyanins during maturation and by regulatory phenomena (copigmentation, self association, complexes with metals) discussed below. In addition, it would be extremely important to confirm that pH variations measured in fruit extracts correspond to the same variations in vacuolar contents during maturation.

b. **In Vivo** ***Stabilization of Color***

In a slightly acid aqueous solution similar to the vacuole content, anthocyanins rapidly lose their color through the hydration phenomena mentioned above. If this does not take place under natural conditions, there must be protection phenomena which permit the stabilization of anthocyanins in the colored form of the flavylium cation. These questions have been studied by many authors, and an exhaustive bibliography on the subject can be found in the reviews by Brouillard[60,61] and Osawa.[287] In agreement with Brouillard,[60] we make a distinction between three main processes which can lead to stabilization of anthocyanin color in vacuoles.

The self-association phenomenon — When the anthocyanic pigment content increases, there is deviation from the Beer-Lambert law for pH values greater than 1. The increase in absorbance at λmax was greater than expected.[24,383] This feature can be interpreted as a self-association phenomenon whose mechanism has not yet been fully described.[24,60] Its importance in *in vivo* expression of fruit color remains to be specified, but it is probably involved, given the high anthocyanin concentrations found in certain fruit vacuoles (e.g., around 0.1 *M* in De Chaunac grape).[264] There may also be competition between self-association and copigmentation.[341]

The copigmentation phenomenon — This is involved when chemical bonds occur either between the anthocyanin molecules themselves (intramolecular copigmentation) (Figure 97) or by the formation of complexes with other colorless phenolic molecules which play the role of copigment (intermolecular copigmentation). In addition to the stabilization of the pigment in its colored form, there is in both cases a bathochromic shift and an increase in the absorbance of the visible band, with intramolecular copigmentation being more effective in color enhancement than the intermolecular type.[59,60,287]

Numerous factors affect *intermolecular copigmentation,*[287] and in particular the pH[425] and the type of anthocyanin and copigment.[24,72,328] Numerous flavonoids and hydroxycinnamic derivatives can play the role of copigment (Table 49), and the number and position of the hydroxyl groups on the molecule are important in the effect obtained.[24,72] Four anthocyanins from berries of *Vitis* species were investigated for complex formation in the presence of the flavonol glycoside rutin.[341] Under pH conditions similar to those in grape juice, deeply colored complexes were formed in the following order: malvidin 3,5-diglucoside (1000% intensification color at λmax), malvidin 3-*p*-coumarylglucoside, 5-glucoside (500%), malvidin 3-*p*-coumarylglucoside (490%), and malvidin 3-glucoside (145%). The copigment effect reduces the production of the colorless carbinol pseudobase.[60] Several mechanisms have been proposed to account for copigmentation based either on the formation of weak complexes or related to hydrogen bonding.[287] Color stabilization of acylated anthocyanins probably results from hydrophobic interactions between the anthocyanin pyrilium ring and the copigment molecule or the acyl groups of the anthocyanin itself, leading to a flavylium structure which is much less reactive towards water.[59] The relative concentrations of anthocyanin and copigment are important to consider: the effect of copigmentation becomes very weak when the two concentrations are weak or when anthocyanin concentration itself is weak,[22,24] e.g., less than 3.5×10^{-5} *M* for cyanidin 3-glucoside.

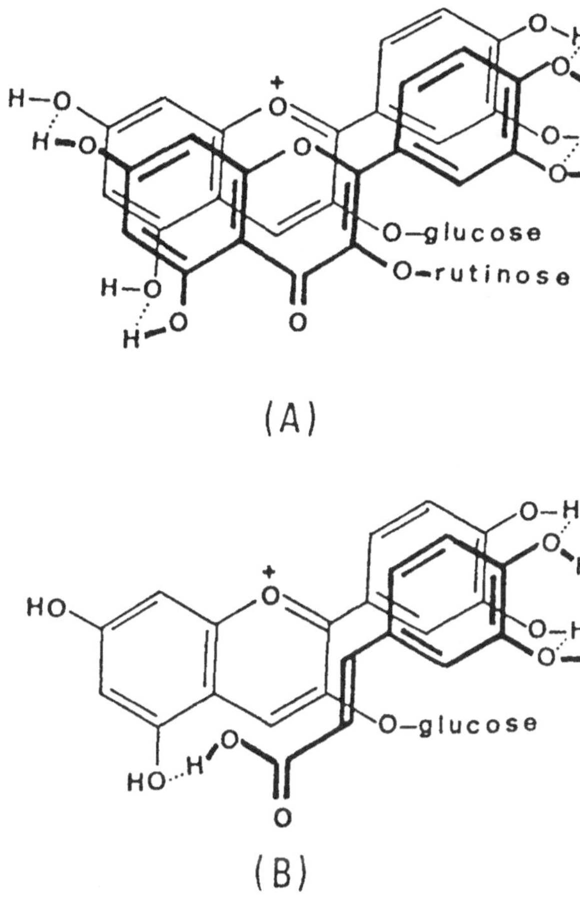

FIGURE 97. (A) Anthocyanin-rutin complex; (B) anthocyanin-caffeic acid complex. (From Maccarone, E., Maccarrone, A., and Rapisarda, P., *J. Food Sci.*, 50, 901, 1985. With permission.)

What may be the importance of *in vivo* intermolecular copigmentation in fruits? Whereas numerous examples studied in flowers stress the role of this phenomenon,[23,60,181,287] the results appear to be more questionable for fruits. Since considerable quantities of flavonoids (flavonol glycosides, tannins, etc.) are found in many blue fruits (cf. Chapters 1 and 2), it is likely that copigmentation occurs[399] although the examples have generally not been the subject of studies as accurate as those on flowers.[287] Furthermore, accurate determination of vacuolar pigment concentrations has only been carried out in rare cases. Reinforcement of anthocyanic pigmentation has long been proposed to account for the strong coloration of certain fruits, such as blackcurrant[328] and black grapes in which complexes with tannins are involved.[351] It was also suggested that copigmentation occurs on the coloration of the blue seed coat of *Ophiopogon jaburan*[182] and of the fruits of *Ilex crenata* and *Ilex rotunda*, which contain ilicicyanin (cyanidin 3-xylosylglucoside).[180] The two latter fruits contain flavonols and exhibit a 12-nm bathochromic shift in λmax over the fruit skins of *Ilex sinensis* which do not contain flavonols, suggesting copigmentation. In the case of grape cv. De Chaunac (a complex interspecies hybrid of *Vitis* in which the anthocyanins are essentially 3,5-diglucosides), concentrations of individual anthocyanins vary from 37.5 to 0.9 m*M* whereas the total anthocyanin content may reach 97 m*M* in the vacuoles.[264] In the same vacuoles, the concentration of hydroxycinnamoyltartaric esters is on the same scale (16.5 m*M* for caffeoyltartrate), whereas the total flavonol glycosides content is only of the order of 0.5

TABLE 49
Copigmentation of Cyanidin 3,5-Diglucoside and Malvidin 3,5-Diglucoside

Copigments	Cyanidin 3,5-diglucoside[a] (2×10^{-3} *M*) (pH 3.32) Copigments: 6×10^{-3}*M*		Malvidin 3,5-diglucoside[b] (10^{-3} *M*) (pH 3.2) Copigments: 4×10^{-3} *M*	
	$\Delta\lambda_{max}$ (nm)	% A increase at λ_{max}	$\Delta\lambda_{max}$ (nm)	% A increase at λ_{max}
Benzoic acid	1	18		
p-Hydroxybenzoic acid	2	19		
Esculin	6	66		
p-coumaric acid	5	32		
Caffeic acid	7	56		
Chlorogenic acid	5	75		
Phloridzin	9	101	7	106
(+)-Catechin	6	78	6	137
[c]Apigenin 7-glucoside	9	68	31	506
8-C-Glucosylapigenin (vitexin)	9	238		
Hesperidin	13	119		
Naringin	10	97		
Quercetin			28	654
Myricetin			34	811
Kaempferol 3-glucoside	22	239		
Robinin	16	185	27	459
Isoquercitrin	19	188	30	580
Quercitrin	19	217	29	584
Hyperin	23	282		
Rutin	20	228	30	596

[a] From Asen et al.[24]
[b] From Chen and Hrazdina.[72]
[c] Formed a slight precipitate.

m*M*. It thus seems that in this particular case, the copigmentation effect of flavonol glycosides is very weak or nil since significant effects were found with equimolar or higher flavonol glycoside contents.[24,264,341] The pronounced color of this fruit may, therefore, be caused by anthocyanins alone, and self-association effects are most likely to be involved in color expression.[264] Such studies should be broadened to other fruits, and other molecules likely to be involved in copigmentation — in particular catechins and proanthocyanidins — should be taken into account.

Intramolecular copigmentation appears to be the most effective mechanism for the stabilization of anthocyanin color. In this case, the copigment is part of the anthocyanin molecule, being bound by covalent bonds. Greater stability is achieved for polyacylated anthocyanins, several examples of which have been studied in flower petals.[60] It is probable that aromatic residues of acyl groups stack with the pyrilium ring of flavylium and decrease the aptitude of water to add to the C2 position.[59,60] The extent of hydration is thus reduced and chromophore stability increases considerably. Different parameters may be involved in the effectiveness of the protection provided, and in particular the nature of the sugar, the position of the acyl groups on the sugar, the nature of these groups (benzoyl being much less effective than cinnamoyl), and finally the extent of acylation as monoacylated anthocyanins are less stable than polyacylated forms.[60] However, intermolecular pigmentation seems more effective with monoacylated anthocyanin than with unacylated pigments.

Interactions between anthocyanins and metal ions — In general, the *in vivo* absorption maxima of anthocyanin-colored fruits tend to be at higher wavelengths than those of the corresponding juice.[287] The factors responsible for this difference are partly the modifications in pH and in copigmentation involved in extraction and partly the possible destruction of anthocyanin-metal complexes. Indeed, anthocyanins with an *o*-diphenolic group are known to be able to form complexes with Al^{3+} and Fe^{3+}[61,141] and are thought to account for the blue color of many flowers; the best-studied metallo-anthocyanins[60] are all blue pigments extracted from flowers. Complexation mechanisms are still not well known.[61] It is probable that Al^{3+} and other metallic ions — often abundant in fruits — play an important role in the expression of blue coloration in numerous fruits.[287,399] Indirect demonstration is provided by comparison of the fruits of *Vaccinium ashei* and *V. corymbosum*.[271] Both fruits are blue, but only the juice of *V. ashei* turns red when the fruit is crushed. Addition of Al salts to the red juice makes it comparable to that of *V. corymbosum*.

It is probable that the different mechanisms mentioned above in the stabilization of the color of anthocyanin fruits operate simultaneously, but it is currently very difficult, given the small number of experimental data available, to assess their relative importance. Nevertheless, the combination of the two types of method has made it possible to determine the importance of copigmentation and the possible presence of metallo-anthocyanins in several fruits and in numerous flowers.[181] The principle is based on comparison of absorption maxima on the one hand of intact cell spectra measured with a microspectrophotometer and on the other of an acid methanolic extract of the same cells. The results are still very fragmentary for fruits (only three species have been studied), but this type of approach merits widespread use, in particular for fruits of economic importance. In addition, the effects would be reinforced by anthocyanin accumulation phenomena, in particular, structures called anthocyanoplasts[290] which would provide protection against hydration. Their presence would seem to be confirmed in the vacuoles of internal cells in cv. Pinot noir grape skin after *in vivo* analyses by resonance Raman spectroscopy.[251,252] Whereas anthocyanins in the vacuoles of the outer cells of the skin (pH less than 3 to 4) are essentially in flavylium form, those at anthocyanoplast level are mainly in the neutral quinonoidal base form at a pH close to neutrality. Extended to other plant materials, and in particular to other fruits, these observations might prove very interesting with regard to both in *in vivo* stabilization of anthocyanins and their biosynthesis.

B. PHENOLS AND BITTERNESS

Numerous distinct taste sensations can be elicited from different parts of the oral cavity by distinct chemical compounds which are frequently very common constituents of foods.[55] Their intensity depends, on the one hand, on the response and on the number of taste buds in fungiform papillae,[20] and on the other, on the nature of the molecules concerned. Whereas astringency uniformly affects the whole of the tongue, bitterness is perceived at the back and sides.[215] This sensation is due to an interaction between polar molecules and the lipid portions of taste papillae membrane and is, therefore, highly dependent on relative lipid solubilities.[215] It might be caused by a conformational change of membrane lipids. The effect of bitter substances on a model membrane also suggests neutralization of the membrane lipid charge by penetration, resulting in a reduction of the effective charge and a decrease in electric potential.[176]

Bitter taste can be elicited by many compounds which may be present in certain fruits, such as heavy metals, amino acids, peptides, alkaloids, terpenes, and different phenolics. For example, the bitterness of wine and cider is caused mainly by phenolic compounds and in particular by oligomeric proanthocyanidins.[18,215,346] Nevertheless, there may be interference between bitterness and astringency in relation to the degree of polymerization of the proanthocyanidins.[215,347] In addition, the sensation of bitterness may be modified by the physicochemical environment of the active molecules, and in particular by the °Brix/acid ratio of fruit juices.[230,380] It can even be completely removed with buffer strength or in the presence of high concentrations of sugars.[215]

Besides proanthocyanidins, certain hydroxycinnamic derivatives are also bitter, with the

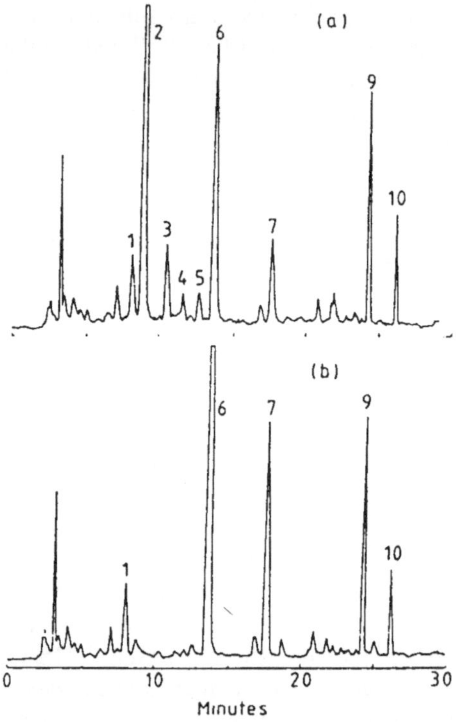

FIGURE 98. Effect of enzyme treatment on the hydroxycinnamate composition of a Chardonnay wine. (a) the intact wine; (b) analysis 6 d after addition of 1.25 $mg \cdot l^{-1}$ hydroxycinnamate ester hydrolase, 20°C storage. (From Somers, T. C., Verette, E., and Pocock, F., *J. Sci. Food Agric.*, 40, 67, 1987. With permission.)

threshold of perception being lowered when several are present simultaneously.[229] In 5% ethanolic solution, caffeic, *p*-coumaric, and chlorogenic acids are considered to be bitter from concentrations varying from 10 to 20 mg/l. In wine, hydroxycinnamoyltartaric derivatives appear to be perceived as being both bitter and astringent[281,286] although recent data showed no significant contribution to sensory properties.[405] Flavor thresholds for caffeoyltartaric acid in aqueous solutions vary from 50 to 80 mg/l, depending on the author,[281,286] whereas the threshold of *p*-coumaroyltartaric acid is clearly lower (10 mg/l). These concentrations may be greatly exceeded in certain wines,[281] but in this case, clear distinction must be made between astringency and bitterness. Recent experimentation made it possible to evaluate the importance of the oxidation of musts in the perception of bitterness of the wines made from them.[268] Although the caffeic derivatives contents of wines made from oxided musts are distinctly lower than those of controls, oxidation significantly increased perception of bitterness. These results appear to be particularly interesting, but their interpretation still raises many questions. In particular, increased bitterness might result either from variation in the proportions of the different phenolic compounds during oxidation or from improved perception of the bitter substances related to a reduction of astringency.

In addition, the importance of hydroxycinnamic derivatives in the sensation of bitterness is still a subject of discussion[269] and the data are fairly contradictory. Indeed, a hydrolase from *Aspergillus japonicus* is capable of specifically hydrolyzing hydroxycinnamic esters[282,283] and may lead to a loss of bitterness in fermented drinks made from fruits.[284] However, although the action of this enzyme may lead to almost complete disappearance of hydroxycinnamoyltartaric esters from Chardonnay wine (Figure 98), the sensory characteristics do not appear to be affected by the treatment.[352] In addition, it was concluded recently that hydroxycinnamates do not play any role at all in the taste of wines, even at concentrations of up to 100 mg/l of caffeoyltartaric acid or 50 mg/l of its oxidation derivative, glutathionyl caffeoyltartaric acid.[405] It remains to be demonstrated more clearly in wines with very different phenol contents how manipulation of their hydroxycinnamic derivative contents, which could be carried out by more or less prolonged oxidation of must and by hydrolysis of esters, can lead to controlling the sensory characteristics of the vintage.

Various other hydroxycinnamic derivatives have also been reported as being bitter, e.g., sinapine, which is abundant in the seeds of certain Cruciferae, and phenylpropanoid glycosides derived directly from caffeic acid and dihydroxyphenylethanol and found in the leaves of *Campsis chinensis*.[178] The latter compounds have not been reported in fruits, but they are very close to verbascoside which is abundant in olive[113] and is bitter itself.[11]

Numerous fruits display slight bitterness which frequently interferes with astringency.[350] However, the two most classic cases of strongly bitter fruits of economic importance are citrus and olive.

The bitterness of some citrus, particularly grapefruit, sour orange, and *Ponderosa lemon*, is due to two chemical classes of compounds.[230,385] First, flavanones, the only type discussed here and which are the cause of the bitterness which exists instantaneously during the preparation of fruit juices, and second, a class of terpenoids called limonoids, which are responsible for delayed bitterness in citrus products. Subthreshold concentrations of these different bitter compounds may have an additive effect.[136] The flavonoids mainly responsible for bitterness are naringin, neohesperidin, and poncirin, the first being the most important. Quantitative data concerning the presence of these compounds in citrus fruits have already been discussed (see Table 29). Numerous studies have been carried out on relations between the intensity of bitterness and the structure of the molecules responsible and many syntheses have been published, in particular by Horowitz and Gentili,[164] Blundstone et al.,[48] and Maier.[230]

Only the principal features of the main results obtained are covered. According to data long considered as "classic," naringenin 7-rutinoside and naringenin itself are not bitter whereas naringin (naringenin 7-neohesperidoside) is very bitter and can be detected in a 10^{-4} or 10^{-5} *M* solution.[164] It thus appears that the bond between the two hexose molecules which make up the disaccharide (1-2 or 1-6 links) (cf. Chapter 1) is determinant with regard to bitterness. The neohesperidosyl group present in naringin, neohesperidin, and poncirin hence appears to be associated with bitterness. Complementary data have been found in more recent studies on the taste of many isomers and analogs of naringin and their structure-taste relationship.[99,200] It was shown that the attachment of rhamnose to the 6-position of D-glucoside or D-galactoside of naringenin does not completely eliminate the taste of the resulting glycosides but elicits a moderately bitter taste. Furthermore, the L-rhamnose methyl group is an essential factor in intense bitterness.

Olive fruit possesses very intense natural bitterness which makes it unfit for consumption without prior treatment. The major bitter principle is a phenolic secoiridoid known as oleuropein[104] which plays an important role in the organoleptic qualities of the fruit both because of its bitter character and its involvement in browning reactions. The oleuropein aglycone is probably also bitter, whereas the two other products of degradation, elenolic acid and 3,4-dihydroxyphenylethanol, were not judged to be bitter.[420] It remains to be clearly specified what might be the degree of bitterness of demethyloleuropein and elenolic acid glycoside, a compound which corresponds to the secoiridoid part of oleuropein only. Verbascoside may also contribute to bitterness in olive,[11] but its concentration is always small in comparison to that of oleuropein.

A natural decrease in bitterness is generally observed during the development of olive; in certain varieties this can go as far as complete disappearance during the final stages of

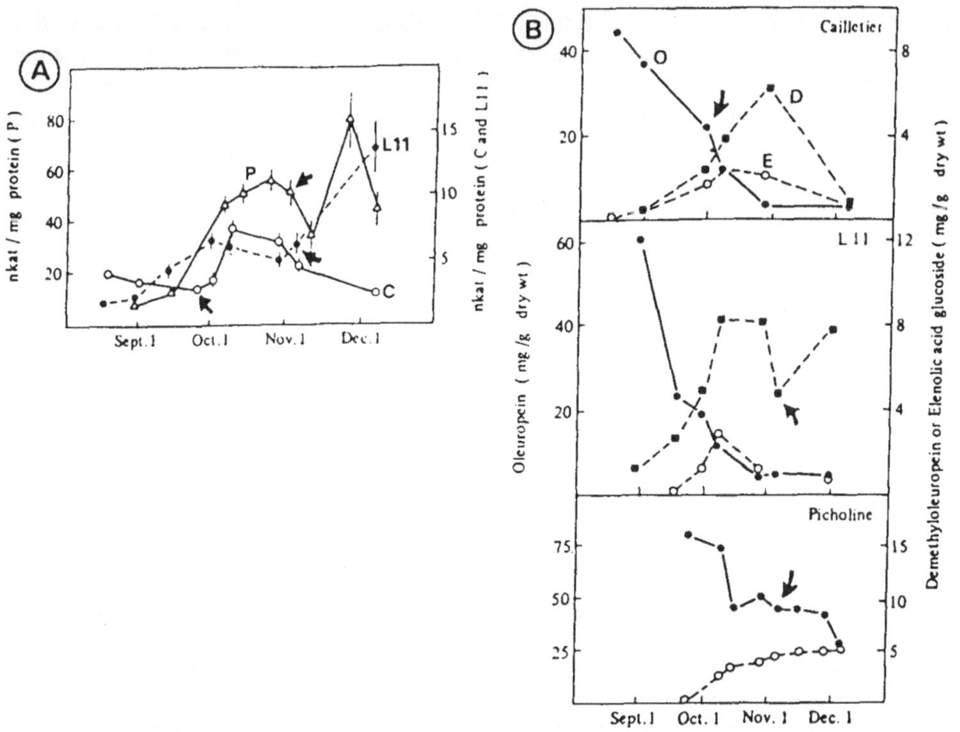

FIGURE 99. Biochemical changes in three olive cultivars (P: Picholine; C: Cailletier and L11) during maturation. A: esterase activity; B: oleuropein (O), demethyloleuropein (D), elenolic acid glucoside (E). Arrows indicate the beginning of black maturation. (From Amiot, M. J., Fleuriet, A., and Macheix, J. J., *Phytochemistry*, 28, 67, 1989. With permission.)

maturation. This decrease in bitterness corresponds to the fall in oleuropein level during maturation accompanied by the appearance of elenolic acid glucoside and also demethyloleuropein in some cultivars (Figure 99).[9,10] Degradation of oleuropein is probably caused by action of the esterases in the fruit.

C. ASTRINGENCY

The taste of fruits and beverages derived from fruits (fruit juice, wine, cider, etc.) is dependent on the presence of tannins. These compounds cause a drying and puckering sensation[215] called astringency which occurs over the whole surface of the tongue and the buccal mucosa following interactions with proteins and glycoproteins in the mouth. Although numerous fruits are astringent, especially before maturation,[350] the most significant examples are cider apples, perry pears, Japanese persimmons, and sloes. The astringency of these fruits is generally considered to be too intense to allow direct consumption. However, a certain degree of astringency, even when not directly perceptible by the consumer since it may be masked by sourness or bitterness, appears to be necessary to ensure high quality in fruits and derived products. Thus, as stressed by Haslam,[154] a red wine with two low a tannin content appears flat and insipid whereas too high a content may result in an aggressive wine.

Astringency is the result of two successive phenomena: first the ability of tannins to bond to proteins and then the physiological response of the eater. Astringency is generally evaluated in model systems using standard testing procedures.[17,18,87,215] In the case of animals, tannins are added to food and water and behavior is recorded in terms of preference and the amounts consumed.[368] Intensity of astringency can also be assessed by the ability of tannins to precipitate certain proteins *in vitro*.[35]

Taste thresholds of astringency have been determined for different polyphenolics in deionized water: tannic acid, (+)-catechin, procyanidin B-3, and a mixture of trimeric and tetrameric proanthocyanidins.[87] The two latter groups are considered as being both astringent and bitter; the greater the molecular size of these substances, the lower their threshold values (from 4.1 to 46.1 mg/ml). These values are considerably increased in the presence of the bitter agents quinine sulfate and caffeine, probably because of competition between these compounds and buccal proteins for complexing with tannins.[357] Interference between astringency and bitterness has also been found in fruit extracts and juice.[17,215,429] Thus, all phenolic fractions from grape seed were found to be astringent and bitter, but relative astringency increased with molecular weight. Likewise, astringency in ciders is characteristic of higher molecular weight procyanidins whereas bitterness is associated with oligomeric procyanidins (up to tetramers); a lower ratio of oligomers/polymers (i.e., 1.85) being necessary for the most astringent cider, and a high ratio (2.50) for the most bitter cider.[215] The presence of ethanol lowers the threshold value of (+)-catechin[87] and decreases astringency by reducing the degree of hydrogen bonding between phenolic groups and proteins.[215]

In addition to their unpleasant taste, tannins also inhibit protein digestion and numerous enzymatic activities. This can even lead to various physiological disorders in the case of excessive consumption.[366] In addition, tannins also appear to be one of the chemical means of resistance of plant and fruit to its biological environment and in particular with regard to pathogenic microorganisms.[37,143,146] These observations partly account for the coevolution phenomena between the higher plants and pathogenic microorganisms or certain animals (insects, reptiles, etc.).[41,143,368] The content and nature of tannins and their bonds with proteins thus appear to be one of the biochemical bases of interactions between plants and other living organisms[143,146,368] and may be at least partly responsible for the choice by man and animals of the fruits that they eat.[143,296,366,368] This might result in better dispersion of seeds of species or varieties whose fleshy fruits are preferred by animals. It would be interesting here to develop research on comparison of varieties of the same species whose fruits have different tannin contents. In addition, insofar as astringency is perceived by man as a fruit quality parameter, it has sometimes been a selection criterion, generally used empirically, in the improvement of fruit-bearing species. High tannin contents have been sought in certain relatively rare cases (perry pears, cider apples, certain red grapes, etc.). Nevertheless, in most other cases, improved varieties for direct consumption by man are usually less astringent than the wild varieties, at least at mature stage. This fall in the tannin content during breeding may, nevertheless, have adverse effects on the resistance of fruits to pathogens, either during growth and maturation on the plant or during postharvest storage. Inspiration should perhaps be drawn in this domain from the work carried out on sorghum which led to the selection of varieties which are rich in tannin during growth (preventing consumption by birds) but whose tannin contents fall very steeply during maturation, increasing their digestibility.[301]

The chemical reactions forming the basis of astringency consist of links between tannins and proteins. As summarized recently,[155,156,357] polyphenol-protein complexation depends partly on the characteristics of the protein and pH and partly on characteristics of the phenolic molecule (molecular size, conformational flexibility, and water solubility). Complexation takes place via the intermediary of hydrogen bonds between phenolic groups and peptide links and via hydrophobic interactions.[156,357,370] However, the relative importance of these two types of interaction remains uncertain, but recent experimentation has made it possible to point to the importance of hydrophobic interactions.[357] The combining of polyphenols and proteins is thought to take place in two successive phases (Figure 100), first, by hydrophobic interactions at protein sites which are particularly rich in aromatic rings, and then by the establishment of hydrogen bonds between the polar groups present at the surface of the protein at these sites. Complexation between tannins and proteins may be inhibited by other macromolecules (polysaccharides, pectic compounds, etc.) able to develop hydrophobic

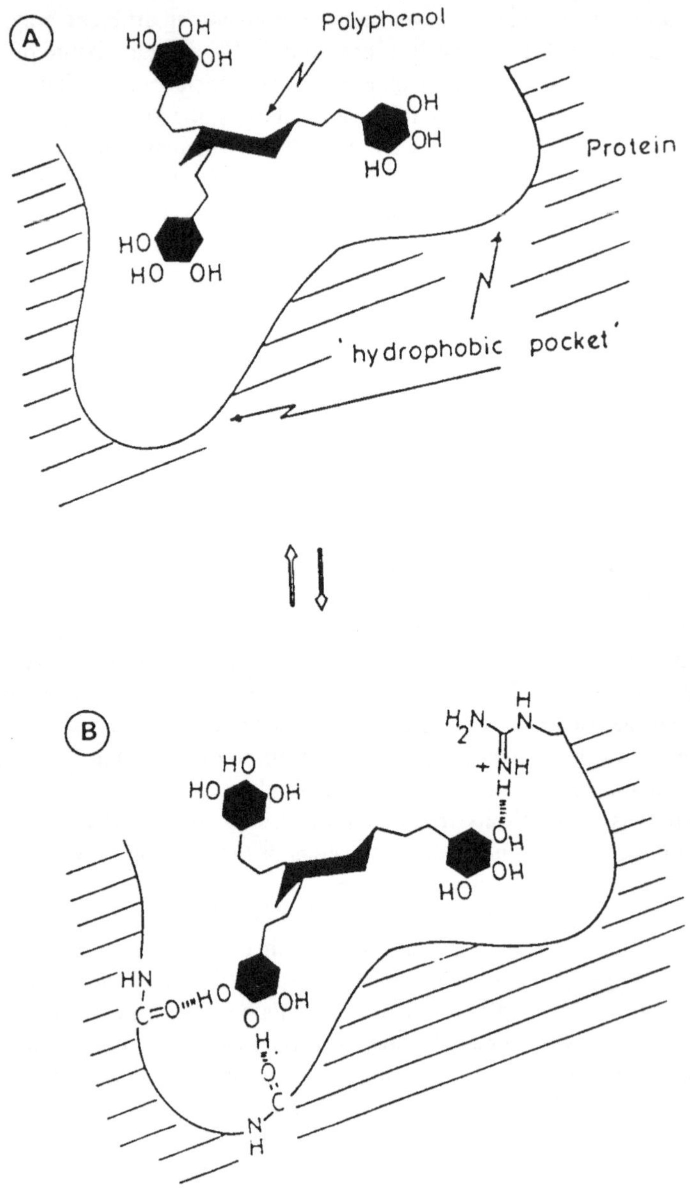

FIGURE 100. Polyphenol-protein complexation: (A) docking of the polyphenol; (B) hydrogen bonding to the protein surface. (From Spencer, C. M., Cai, Y., Martin, R., Gaffney, S. H., Goulding, P. N., Magnolato, D., Lilley, T. H., and Haslam, E., *Phytochemistry,* 27, 2397, 1988. With permission.)

cavities in their secondary structures.[156,289] This competition throws into relief the interference of the chemical environment of tannins in the response of proteins and, hence, probably in astringency. As mentioned above, this competitive complexation may be responsible in certain fruits for the delayed perception of astringency in the mouth and the fall in astringency generally observed during maturation (Chapter 3). Phenolic compounds other than tannins are capable of establishing links with proteins and can even cause them to precipitate,[156,357] but their importance in astringency phenomena is certainly very small in comparison with tannins. Thus, coniferin has been reported as being an astringent constituent in citrus juice.[380] In this case, taste tests in water adjusted to pH 3.5 showed that coniferin could be detected

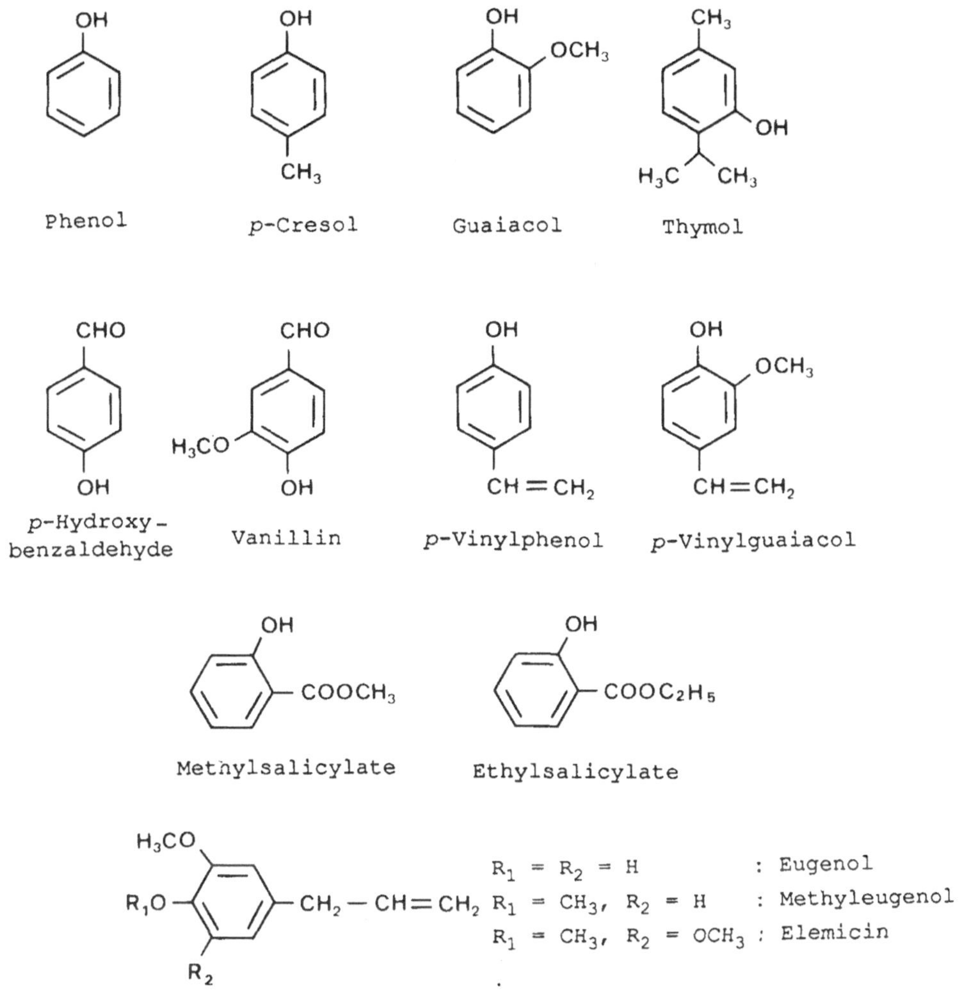

FIGURE 101. Chemical structures of some volatile phenols.

at 100 ppm. Although the conferin contents of citrus fruits appear to be much lower (approximately 3 to 11.5 ppm), it is possible that there are additive effects with the other compounds present in the juice.[380] Nevertheless, given the small quantities present, it is perhaps difficult to distinguish between their respective contributions to the two phenomena of astringency and bitterness. The same applies to the hydroxycinnamoyltartaric derivatives present in wine.[281]

D. PHENOLIC COMPOUNDS AND AROMA

Fruit aromas are formed by a very large number of compounds which belong to various chemical families,[280] and each fruit has a specific aromagram. All possess the property of being partly volatile at room temperature and display both appreciable lipid solubility and a certain degree of water solubility.[347] In the case of phenolic compounds, only a few simple phenols display these properties and their contribution to the aroma of the fruit is generally small. However, it is frequently more marked in beverages and other fruit products. In these cases, volatile phenols of some interest in foods range from phenol, cresols, and guaiacol (Figure 101) with "medicinal", "phenolic" or "smoky" odors to more pleasant odorants, such as vanillin and methyl salicylate.[347]

In the vanilla pod,[305] the three most abundant volatile constituents were found to be vanillin (about 2%), *p*-hydroxybenzaldehyde (0.2%), and *p*-hydroxybenzyl methyl ester (0.02%) (Figure 101). Furthermore, additional volatile phenolics have been identified in vanilla pods and vanilla extracts, including aromatic alcohols (anisyl, coniferyl, *p*-hydroxybenzyl, phenylethyl, and vanillyl alcohols), aromatic acids (including salicylic and vanillic acids) and simple phenols (*p*-cresol, eugenol, guaiacol, *p*-vinyl guaiacol, *p*-vinyl phenol, etc.) The abundance of *p*-hydroxybenzoic acid can make it possible to distinguish between species of vanilla. Green vanilla pods contain very little vanillin, and the development of aroma is associated with the change in pod color from green to brown during maturation or processing. Vanillin itself is formed from glucovanillin by the action of a β-glucosidase whose activity increases as the pod matures. After hydrolysis of the glucosides, vanillin is transformed by oxidizing enzymes to give quinones and stable pigments. However, the final transformations are nonenzymatic reactions.[305]

Some phenols may also contribute to the aroma of other ripe fruits,[280] for example eugenol, methyleugenol, elimicin, and vanillin in banana (Figure 101). In citrus, several coumarins are also natural constituents of aroma, e.g., citral, nootketone, and citropten (see Table 12) in lime. In mandarin, thymol may form up to 0.08% of peel oil. Strawberry contains esters of certain phenolic acids (ethylsalicylate) or related acids (methyl cinnamate, ethyl benzoate, etc.) In plum, only methyl salicylate is found in the greengage whereas Mirabelle contains eugenol, thymol and carvacrol.[407]

V. PHARMACOLOGICAL INTEREST OF PHENOLIC COMPOUNDS IN FRUITS

In addition to alkaloids, which are of fundamental interest in medicine, phenolic compounds — flavonoids in particular — have a non-negligible position among the natural substances used by man for therapeutic purposes. In traditional medicine, fruits are used in the preparation of raw extracts generally used as infusions. It is noticeable that the pharmacopoeia contains fewer preparations based on fruits than on other parts of plants: leaves, bark, and roots. However, the isolation of active principles of the phenolic type from certain fruits is of great interest to the pharmaceutical industry.

There is no question of discussing here all the medicinal and pharmacological utilizations of fruit extracts which contain phenolic compounds. The main applications are described briefly when they have been established scientifically and are related to active principles which have been well characterized chemically. Numerous reviews on the pharmacological interest of phenolic compounds have been published.[49,121,122,223,414,415]

A. ANTI-INFLAMMATORY ACTIVITY

This was the first known pharmacological effect of the flavonoids. It was discovered some 50 years ago by Szent-Gyorgyi who observed that lemon extracts had an effect on capillary permeability and resistance.[171,244] These flavonoids were given the name "vitamin P" and later "bioflavonoids". The term "vitamin P" is no longer used, presumably because of a failure to demonstrate that flavonoids satisfy the "essential growth factor" criterion for vitamins.[171] Numerous works over the past 20 years have shown that flavonoids may represent an interesting new therapeutical approach because they can interfere with different steps involved in the development of vascular diseases. This is illustrated by a number of examples involving fruits and substances produced from them.

After concentration and fermentation, the phenolic compounds extracted from bilberry juice *(Vaccinium myrtillus)* form the active principle of a drug used for its vascular properties.[299] Here, the anthocyanins in the extract of *V. myrtillus* (VMA) act on capillary permeability. Applied orally or by intravenal or intramuscular injection, VMA is more active than *o*-(β-hydroxyethyl)-rutin in the reduction of capillary permeability and fragility.[415] The anti-inflammatory activity of VMA accounts for their significant anti-edema properties[121,415] and their action on diabetic microangiopathy.[52] Anthocyanins of *Rubus occidentalis, Sambucus nigra,* and *Ribes nigrum* also have vascular protection properties.[241,415]

Procyanidins are physiologically very active substances which in particular combat the formation of atheromas which cause myocardial infarction.[238] Grape is very rich in procyanidins, whence the great interest in this fruit and in derived products (fruit juice, grape jelly, red wines and wine coolers, etc.). The research carried out by Masquelier and coworkers led to the conclusion that procyanidins were the wine constituents responsible for this protective effect with regard to atherosclerosis. They also showed that the compounds acted at several levels: stabilization of the collagen fibers supporting the various vessel tunicae, inhibition of histidine decarboxylase, which makes it prevent over-production of histamine which would increase wall permeability and lead to an atheromatous process, and finally intervention as cofactors of vitamin C, whose contribution to the cleansing of cholesterol is reinforced.[238] Today, catechins and procyanidins are proposed for the control of gastric ulcers caused by excess histamine in stomach mucous membranes.[238]

It has been shown that citrus peel extracts inhibit the increase in capillary permeability which is the initial sign of inflammation.[223] The biologically active flavonoids in these extracts (polymethoxylated flavones and flavanones, Chapter 1) are sometimes referred to as "citroflavonoids".[170] They act in a similar way to that described above for anthocyanins and procyanidins, and they appear to play an important role in protection of the vascular system.[170,171] For example, severe brain edema and subpleural hemorrhages in rat are significantly reduced by rutin or hesperidin derivatives (50 mg/kg).[223] The citrus flavonoid complex displayed broader inflammation than either cortisone or ACTH.[223] Polymethoxylated flavones, and particularly sinensetin and nobiletin, reduce erythrocyte aggregation and blood sedimentation rate very effectively. This suggests that they play a role in the control of the high blood viscosity syndrome.[327]

3,6-diC-glucosylapigenin (a C-glycosylflavonoid) and several phenyl propanoids isolated respectively from *Citrus unshui* and *C. hassaku* peels have a hypotensive effect in rat.[239,339] Citrus extracts also possess an antiallergic activity which may be caused first by the polymethoxylated flavones nobiletin, tangeretin, and methoxynobiletin.[415] Release of histamine by rat peritoneal mast cells was inhibited by 99% by nobiletin, 84% by 3-methoxynobiletin, and only 35% by tangeretin.[415]

In addition to the examples above, it can be noted that the fruit hull of *Garcinia mangostana* is used in India as an anti-inflammatory agent in the treatment of diarrhea. This action is thought to be caused by the xanthones present in the fruit.[27] Schizandrin and its derivative isoschizandrin, two lignans isolated from fruits of *Schizandra chinensis*, have an inhibitory effect on stress-induced gastric ulceration.[177]

B. ANTIHEPATOTOXIC ACTIVITY

The fruits of milk thistle (*Silybum marianum*, Compositae) have long been known for their medicinal properties. Study of the principal active principles has been the subject of several publications and reviews.[371,411-415,418] The plant contains a mixture of flavonoligans, silybin, silydianin, and silychristin which all possess antihepatotoxic activity. They are important since they are nonimmunosuppressant drugs which can be used to treat liver diseases and prevent the action of several liver poisons. The activity of these substances is due to stabilization of the liver membrane and cell-regenerating activity, promoting ribosomal RNA and protein synthesis.[353,354,415] Several other flavonolignans have recently been revealed in the same material.[371,416] Relations between antihepatotoxic activity and the structure of the molecules have been described using a model of liver damage in hepatocyte cultures.[162]

The medicinal properties of the dibenzocyclooctadiene lignans of fruits of *Schizandra chinensis*,[270] an oriental medicinal plant, are particularly interesting for their antihepatotoxic properties.[29,227] These compounds have an effect on chronic viral hepatitis and protect the liver from damage caused by pharmaceuticals.[415]

Fruits are an important source of tannins, and it is interesting to note that a study of 25 tannins of various origins showed that they inhibit lipid peroxidation in liver mitochondria and microsomes, each tannin being more effective that α-tocopherol.[285]

C. ANTITUMOR ACTIVITY

Few fruit extracts or phenolic compounds isolated from them have been found to be effective on various tumors. The following can be mentioned: juice of the fruit of *Merrillia caloxylon*, a monotypic species of Rutaceae from Malaysia, which after mild acid hydrolysis contains eupatorin, a cytotoxic flavone.[3] Several phenolic compounds isolated from the fruits of *Mammea americana* and *Kielmeyera pumila* (Guttiferae) display antitumoral activity for sarcoma 180 in an *in vivo* test system.[223] They are 4-phenylcoumarins: mammeisin, mammeigin, and isomammeigin (Chapter 1). A number of lignans possess cytotoxic activity and can be used as antimitotic agents,[223] but the results obtained do not appear to concern lignans extracted from fruits. In some plant extracts, antitumoral activity appears to be related to the presence of tannins.[298] Extracts of green walnut pericarp possess such activity; this was first attributed to juglone, but ellagic acid is also thought to be involved.[298] Quercetin pentamethylether and rutin — the latter being very common in fruits — act as inhibitors of chemically induced tumors.[244,423]

In spite of several examples of flavonoids with antitumoral activity, no flavonoids are currently used in cancer treatment.[223] This is most frequently for reasons of high toxicity resulting in mutagenic and cardiotoxic effects which make use in therapy delicate.

D. LEUCODERMIC AND ANTIPSORIASIS ACTIVITY

The fruits of some species of Umbelliferae and *Citrus* contain furanocoumarins (Chapter 1). These substances are psoralen derivatives and are of great therapeutic importance since they can induce photosensitization and are used in treatment of vitiligo and, above all, psoriasis.[114,197] The fruits of *Ammi majus* (Umbelliferae) contain bergapten (5-methoxypsoralen or 5-MOP) and xanthotoxin (8-methoxypsoralen or 8-MOP).[199] Bergapten is the active principle of *Citrus* essences, and particularly essence of bergamote *(Citrus bergamia)*. Psoralens are used in cosmetics (pigment synthesis effect) and in therapy. PUVA therapy is used in treatment of psoriasis; this consists of a combination of 8-methoxypsoralen (8-MOP) and irradiation with longwave ultraviolet light (UV A, 320 to 400 nm, maximum 365 nm). Clinical studies using PUVA therapy have demonstrated that 77 to 85% of patients treated displayed 90 to 99% clearing of psoriatic lesions over the treatment period.[197] PUVA is thought to have produced therapeutic results by the formation of psoralen-DNA photo adducts which inhibit DNA synthesis and decrease cell division in the epidermis.[114]

E. ANTIBACTERIAL AND ANTIVIRAL ACTIVITY

Phenolic compounds belonging to the main groups represented in fruits frequently possess antibacterial or antiviral activities.[101,223,244,298] Certain caffeic acid derivatives, such as chlorogenic acid, rosmarinic acid, and heterosidic esters, such as verbascoside and forsythiaside display antimicrobial activities which have been studied on numerous pathogenic strains of bacteria (*Staphylococcus aureus* and *Pseudomonas aeruginosa* in particular) and microscopic fungi.[12-14] These esters are more effective at pH 7.4 than at pH 6, which appears to be related to partial oxidation leading to very labile and reactive semi-quinonic forms.[13] The pharmacological activity of most of these molecules is related to caffeic acid, but the nature of the sugars and their position in the molecule are often as important. Thus, forsythiaside inhibits AMPc phosphodiesterase, whereas verbascoside, which is a position isomer of forsythiaside is inactive.[13] Caffeic and chlorogenic acids possess antimicrobial properties and their use as stabilizers in cosmetology can be envisaged.[217] Phenol acids in the benzoic series also have an antibacterial effect.[26,98] Study on the microbial activity of phenolic compounds in olive has shown that oleuropein and the products of its hydrolysis, aglycone and elenolic acids, are active.[105,193] Likewise, numerous naphtoquinones and anthraquinones, such as juglone in walnut, possess antibacterial and antifungal properties.[86]

The antifungal properties of tannins have long been known and used in the form of plant extracts; the great majority are from roots and rhizomes and not from fruits.[298] Nevertheless, extracts of ripe caroub pods *(Ceratonia silica)* are used to treat infectious diarrhea, and their antibacterial properties are attributed to their high tannin content.[431] Tannins of the fruits of *Diospyros kaki* inhibit various microbial toxins and act on different viruses.[298]

Most of the phenolic compounds tested for antimicrobial properties display low activity (10 to 50 μg/ml) compared with antibiotics derived from microorganisms (5 μg/ml) and, according to Wagner, without optimization of their antibacterial activity by structural modification, these compounds will not gain any pharmaceutical importance.[415]

A number of flavonoids also possess antiviral activity.[39,266,410] Strawberry tannins inactivate polioviruses, enteric viruses, and a herpes virus.[201] There have also been many reports of antiviral and antibacterial activities of lignans.[226a] It has been demonstrated that quercetin has an antiviral effect *in vitro* and *in vivo,* in particular on poliovirus type 1, herpes simplex virus type 1, parainfluenza virus type 3, and respiratory synctial virus.[255,265,403] Compared to other flavonoids like catechin or naringin, quercetin is the only compound which possesses both anti-infective and antireplicative activities against each virus.[255] As an additive antiviral effect is observed when quercetin and interferon are combined, the hypothesis has been put forward that flavonoids may act like interferon by increasing the amount of cyclic AMP in cells.[265]

F. MISCELLANEOUS ACTIVITIES

We only give a few examples of other pharmacological activities in which phenolic compounds in fruits are involved.

- Olive contains oleuropein and ligstroside, two secoiridoids with molluscicide properties.[204] Anacardic acid in cashew nut shell *(Anacardium occidentale)* also displays molluscicide activity at a concentration of 0.35 ppm, whereas vitamin K3, 7-methyljuglone and plumbagin are less active (2 to 5 ppm).[364] Plants containing tannins may also possess molluscicidal activity, thought to be due to a higher condensation product.[340]
- Fairly old work showed that the anthocyanins extracted from *Vaccinium myrtillus* enhanced the regeneration of visual purple during the first few minutes of adaptation of the eye to darkness.[34,299] This property led to their use in improving night vision. The mechanism was thought to be different from that above (Section A), with anthocyanins acting at the level of retina enzymes and in particular those involved in glycolytic utilization of glucose.[73]
- Certain caffeic derivatives, and in particular rosmarinic acid and verbascoside, possess an antihormonal activity which essentially takes the form of inhibition of hypophyseal gonadostimulines.[12] This gives them indirect contraceptive properties. Reference to classic reviews[223,415] should be consulted on the subject of the estrogenic and contraceptive activities of flavonoids. The fruits of *Vitex agnus-castus* (chaste tree, Verbenaceae), which are reputed to be antiaphrodisiac, contain polymethoxylated flavonols[430] (Chapter 1).

In conclusion, it should be noted that the mechanisms of the action of phenolic compounds at the biochemical level[157,158] have not been approached here. We mention only that these compounds act by inhibiting enzymes of great physiological importance: aldose-reductase,[93,161] c-AMP-phosphodiesterase,[131,273,274] aryl-hydroxylase,[157] histidine-decarboxylase,[415] cyclooxygenase, and lipoxygenase.[36,89,253] In addition, phenolic compounds have scavenging, antiperoxidant, and antioxidant properties which are clearly demonstrated, for example, by the anthocyanins extracts from bilberry and grape.[57,58,297,332,404]

VI. CONCLUSIONS

Phenolic compounds are thus seen as markers of the physiological stages of growth and maturation during fruit life. Their contribution to mechanisms of resistance of fruits to parasite attack appears to be well established, although their physiological functions and modes of action are still being discussed. These compounds are also very good markers of fruit quality which gives them great practical interest. Two other aspects are examined briefly: the chemotaxonomic interest and the ecological role of fruit phenols.

A. CHEMOTAXONOMIC INTEREST OF FRUIT PHENOLICS

The distribution of phenolic compounds, and, in particular, that of flavonoids, is of great chemotaxonomic interest,[145] for example in the Rutales[144] and the Rosaceae.[426] Material from the vegetative parts of plants are analyzed most frequently, such as leaves and bark (e.g., in the genera *Malus* and *Pyrus*,[69,426]) or flowers and flower bracts (e.g., in varieties of banana[165]). Fleshy fruits are less commonly used for chemotaxonomic study, possibly for practical reasons of picking and storage. However, in certain cases, chemotaxonomic interest was noted during study of the phenolic compounds found in fruits (Chapter 1): anthocyanins in particular in the genera *Ribes, Rubus, Vaccinium,* and *Vitis*; polymethoxylated flavones and flavanones in *Citrus*; coumarins in Umbelliferae and *Citrus*, etc. Reference should be made to Chapter 1 for the corresponding details and bibliographical references as only a few examples are mentioned again here:

- Fruit flavonoids have been surveyed in some 100 species representing all the major tribes of the Umbelliferae.[148] The distribution of flavones and flavonol glycosides (25 different compounds) is extremely interesting from the chemotaxonomic point of view. For example, the revelation of the presence of apigenin and luteolin 7-glucuronosylglucosides in *Cuminum cyminum* supports its transfer from the Apieae to the Caucalideae.
- Coffee cherry would be excellent material for chemotaxonomic studies, and flavonoid analysis was carried out in 10 species of the genus *Coffea;*[221] 75 flavonoids were detected, only 5 of which were common to all 10 species. This result shows that chromatographic studies could be used to improve the systematics of the genus *Coffea.*
- The power of HPLC as a tool for quantitative analysis (Chapter 1) has found its applications in the identification of cultivars and the characterization of hybrids in *Citrus*. For example, five polymethoxylated flavones were separated by HPLC in crude extract of two *Citrus*, tangerine and orange. Although the same compounds were found in both cases, concentrations were very different, clearly distinguishing between the two cultivars.[45]
- Specific anthocyanin patterns characterize different species and cultivars (Chapter 1), e.g., in soft fruits, such as red raspberries.[30] Only one example is examined here — that of the Rosaceae family which has been particularly closely investigated. Table 50 shows the glycosidic patterns of the anthocyanins in fruits of this family and in Saxifragaceae. The branched triglycosides cyanidin 3-(2^G-xylosylrutinoside) and Cy and Pg 3-(2^G-glucosylrutinoside) are of very limited distribution (*Rubus, Ribes,* and

TABLE 50
Distribution of Anthocyanins in Fruits of Rosaceae and Saxifragaceae

	Glycosidic Pattern							Anthocyanidin			
	3-glu	3-gal	3-ara	3-rut	3-soph	2G xyl-rut	2G glu-rut	Pg	Cy	Pn	Dp
Rubus fructicosus (blackberry)	+ +	–	–	+	–	–	–	–	+	–	–
R. idaeus (red raspberry)	+	–	–	+	+ +	–	+	+	+ +	–	–
R. occidentalis (black raspberry)	+	–	–	+	–	+	–	–	+	–	–
Prunus cerasus (sour cherry)	+	–	–	+	+	+[a]	+ +	–	+ +	+[a]	–
P. avium (sweet cherry)	+	–	–	+ +	–	–	–	–	+ +	+	–
P. domestica (plum)	+ +	–	–	+	–	–	–	–	+ +	+	–
P. persica (peach)	+ +	–	–	+	–	–	–	–	+	–	–
Malus pumila (apple)	+	+ +	+	–	–	–	–	–	+	–	–
Pyrus communis (pear)	–	+ +	+	–	–	–	–	–	+	–	–
Fragaria ananassia (strawberry)	+	–	–	–	–	–	–	+ +	+	–	–
Aronia melanocarpa (chokeberry)	+	+ +	+	–	–	–	–	–	+	–	–
Ribes nigrum (blackcurrant)	+	–	–	+ +	–	–	–	–	+	–	+
R. rubrum (redcurrant)	+	–	–	+	+	+	+	–	+	–	–

Note: 3-glu; 3-glucoside; 3-gal: 3-galactoside; 3-ara: 3-arabinoside; 3-rut: 3-rutinoside; 3-soph: 3-sophoroside; 2^G-xyl-rut: 3-(2^G-xylosylrutinoside); 2^G-glu-rut: 3-(2^G-glucosylrutinoside); Pg: pelargonidin; Cy: cyanidin; Pn: peonidin; Dp: delphinidin.

[a] In very small quantity (Dekazos;[84] Shrikhande and Francis[345]).

Prunus cerasus) and are thus of taxonomic interest at generic and specific level.[147] In *Rubus*, these compounds occur only in raspberry-type species (*Ideobatus* section), and all blackberry types (*Rubus* section) contain only rutinoside and glucoside. The distribution of anthocyanins in *Ribes* is remarkably similar to that in *Rubus* since the Saxifragaceae family is closely related to the Rosaceae. Branched triglycosides are present in sour cherry but uniformly absent in sweet cherries.

The fruits of *Opuntia ficus-indica* and those of other Cactaceae do not contain anthocyanins but betacyanins and betaxanthins, compounds which belong to the betalain group.[222,294] These nitrogenous pigments are completely different to flavonoids and are found only in the order Centrospermae. Chemotaxonomic importance here is due to the fact that two classes of pigments, betalains and anthocyanins, are mutually exclusive, never having been detected together in the same species or even in different species of the same family.

B. ECOLOGICAL SIGNIFICANCE OF FRUIT PHENOLICS

Numerous authors have put forward the hypothesis that the products of secondary metabolism are involved in interactions between plants and their environment. The possible roles of phenolic compounds in these interactions have been the subject of several reviews.[40,41,143,366] There is increasing evidence that a considerable number of phenolic compounds play an ecological role in plants.[142] In fruits, phenolic compounds are involved in the following areas:

- **Plant/pathogen interactions.** This point is discussed above (Section III).
- **Plant/animal interactions.** Ripe fleshy fruits are attractive to fruit-eating animals. Their bright colors, caused by anthocyanins and flavonols (together with carotenoids), are probably an essential attraction signal for birds. This role of feeding attractants is involved in seed dispersal, as recorded by Darwin who, in *The Origin of Species* wrote: "But this (a fruit's) beauty serves merely as a guide to birds and beasts in

order that the fruit may be devoured and the matured seeds disseminated." Fruits whose seeds are dispersed by birds are generally reddish-orange, which can be related to the fact that birds are particularly sensitive to these colors, which contrast strongly with the green of foliage. Red is the only color that is at once inconspicuous to most insects and an excellent visual signal to birds.[244]

Phenolic compounds in fruits may also act as feeding repellents; immature fruits are rich in phenols, and particularly tannins, and these compounds are probably involved in protection against fruit-eating animals. For example, the immature fruit pulp of banana is very rich in tannins (up to 17% of dry weight) which probably accounts to a great extent for the relative freedom of immature bananas from herbivores.[189] Likewise, the extreme sourness and bitterness of immature citrus fruits appears to play a protective role. In agreement with Janzen,[189] it can be considered that the loss of an immature fruit is a much greater strategic and material loss than the loss of a leaf, because seed dispersal is no longer ensured.

According to Harborne,[146] in order to understand the interactions that occur between plant and animal in the natural environment, it is useful to consider them in the context of an evolutionary theory. Co-evolutionary trends in phenolic metabolisms have been suggested in several cases,[146,367] but accurate study in fruits has not yet been carried out.

REFERENCES

1. **Abdel-Rahman, M.,** Patterns of hormones, respiration and ripening enzymes during development, maturation and ripening of cherry tomato fruits, *Physiol. Plant.*, 39, 115, 1977.
2. **Adams, D. O. and Yang, S. F.,** Ethylene biosynthesis: identification of 1-amino cyclopropane-1-carboxylic acid as an intermediate in the conversion of methionine to ethylene, *Proc. Natl. Acad. Sci. U.S.A.*, 76, 170, 1979.
3. **Adams, J. H. and Lewis, J. R.,** Eupatorin a constituent of Merrillia caloxylon, *Planta Med.*, 32, 86, 1977.
4. **Addicott, F. T.,** *Abscission*, University of California Press, Berkeley, 1982.
5. **Addicott, F. T. and Lyon, J. L.,** Physiological ecology of abscission, in *Shedding of Plant Parts*, Kozlowski, T. T., Ed., Academic Press, New York, 1973, 85.
6. **Afek, U., Szteinberg, A., and Carmely, S.,** 6,7-Dimethoxycoumarin, a citrus phytoalexin conferring resistance against *Phytophothora gummosis*, *Phytochemistry*, 25, 1855, 1986.
7. **Aharoni, N. and Yang, S. F.,** Auxin-induced ethylene production as related to auxin metabolism in leaf discs of tobacco and sugar beet, *Plant Physiol.*, 73, 598, 1983.
8. **Alibert, G. and Boudet, A. M.,** Compartimentation tissulaire et intracellulaire des polyphénols, *Bull. Liaison Groupe Polyphénols*, 10, 129, 1980.
9. **Amiot, M. J., Fleuriet, A., and Macheix, J. J.,** Importance and evolution of phenolic compounds in olive during growth and maturation, *J. Agric. Food Chem.*, 34, 823, 1986.
10. **Amiot, M. J., Fleuriet, A., and Macheix, J. J.,** Accumulation of oleuropein derivatives during olive maturation, *Phytochemistry*, 28, 67, 1989.
11. **Andary, C.,** Contribution à l'étude botanique, chimique et pharmacodynamique d'Orobanche rapum-genistae Thuill. (Orobanchacées), Thèse Doct. Pharm., Montpellier, France, 1975.
12. **Andary, C.,** Activité pharmacologique de certains esters de l'acide 3,4-dihydroxycinnamique, *Bull. Liaison Groupe Polyphénols*, 11, 544, 1982.
13. **Andary, C.,** Esters hétérosidiques de l'acide caféique: connaissances actuelles, *Bull. Liaison Groupe Polyphénols*, 13, 15, 1986.
14. **Andary, C., Personne, C., and Privat, G.,** Activité antibactérienne et antifongique de certains dérivés de l'acide caféique, *Bull. Liaison Groupe Polyphénols*, 10, 279, 1980.
15. **Andary, C., Roussel, J. L., Motte, M. E., Rascol, J. P., and Privat, G.,** Activité antifongique comparée de divers esters de l'acide dihydroxy-3,4 cinnamique, *Cryptog. Mycol.*, 2, 119, 1981.

16. **Aoki, S., Araki, C., Kaneko, K., and Katayama, O.,** Occurrence of L-phenylalanine ammonia-lyase activity in peach fruit during growth, *Agric. Biol. Chem.*, 35, 784, 1971.
17. **Arnold, R. A. and Noble, A. C.,** Bitterness and astringency of grape seed phenolics in a model wine solution, *Am. J. Enol. Vitic.*, 29, 150, 1978.
18. **Arnold, R. A., Noble, A. C., and Singleton, V. L.,** Bitterness and astringency of phenolic fractions in wine, *J. Agric. Food Chem.*, 28, 675, 1980.
19. **Arntzen, C. F., Falkenthal, S. V., and Bobick, S.,** Inhibition of photophosphorylation by kaempferol, *Plant Physiol.*, 53, 304, 1974.
20. **Arvidson, K. and Friberg, U.,** Human taste: response and taste bud number in fungiform papillae, *Science*, 209, 807, 1980.
21. **Asada, Y. and Matsumoto, I.,** The nature of lignin obtained from downy mildew-infected Japanese radish root, *Phytopathol. Z.*, 73, 208.
22. **Asen, S.,** Known factors responsible for infinite flower color variations, *Acta Hortic.*, 63, 217, 1976.
23. **Asen, S., Norris, K. H., and Stewart, R. N.,** Effect of pH and concentration of the anthocyanin-flavonol co-pigment complex on the color "Better Tones Roses, *J. Am. Soc. Hortic. Sci.*, 96, 770, 1971.
24. **Asen, S., Steward, R. N., and Norris, K. H.,** Co-pigmentation of anthocyanins in plant tissues and its effect on color, *Phytochemistry*, 11, 1139, 1972.
25. **Bailey, J. A.,** Mechanisms of phytoalexin accumulation in *Phytoalexins*, Bailey, J. A. and Mansfield, J. W., Eds., Blackie, Glasgow, 1982, 289.
26. **Balansard, G., Zamble, D., Dumenil, G., and Cremieux, A.,** Mise en évidence des propriétés antimicrobiennes du latex obtenu par incision du tronc de *Alafia multiflora* Stapf. Identification de l'acide vanillique, *Plant. Med. Phytother.*, 2, 99, 1980.
27. **Balasubramanian, K. and Rajagopalan, K.,** Novel xanthones from Garcinia mangostana, structures of BR-xanthone-A and BR-xanthone-B, *Phytochemistry*, 27, 1552, 1988.
28. **Ballinger, W. E., McClure, W. F., Nesbitt, W. B., and Maners, E. P.,** Light-sorting muscadine grapes (*Vitis rotundifolia* Mich x) for ripeness, *J. Am. Soc. Hortic. Sci.*, 103, 629, 1978.
29. **Bao, T., Liu, G., Song, Z., Xu, G., and Sun, R.,** A comparison of the pharmacologic actions of 7 constituents isolated from fructus *Schizandrae*, *Chim. Med. J.*, 93, 41, 1980.
30. **Barrit, B. H. and Torre, L. C.,** Fruit anthocyanin pigments of red raspberry cultivars, *J. Am. Hortic. Sci.*, 100, 98, 1975.
31. **Barthe, G. A., Jourdan, P. S., Mc Intosh, C. A., and Mansell, R. L.,** Radioimmunoassay for the quantitative determination of hesperidin and analysis of its distribution in *Citrus sinensis*, *Phytochemistry*, 27, 249, 1988.
32. **Barz, W.,** Degradation of polyphenols in plants and plant cell suspension cultures, *Physiol. Vég.*, 15, 261, 1977.
33. **Barz, W. and Koster, J.,** Turnover and degradation of secondary (natural) products, in *The Biochemistry of Plants*, Vol. 7, *Secondary Plant Products*, Conn, E. E., Ed., Academic Press, New York, 1981, 35.
34. **Bastide, P., Rouher, F., and Tronche, P.,** Rhodopsine et anthocyanosides. A propos de quelques faits expérimentaux, *Bull. Soc. Ophthal. Fr.*, 68, 801, 1968.
35. **Bate-Smith, E. C.,** Haemanalysis of tannins: the concept of relative astringency, *Phytochemistry*, 12, 907, 1973.
36. **Baumann, J. V., Bruchhausen, F., and Wurm, G.,** Flavonoids and arachidonic acid metabolism, in *Flavonoids and Bioflavonoids, 1981*, Farkas, L., Gabor, M., Kallays, F., and Wagner, H., Eds., Elsevier, Amsterdam, 1982, 411.
37. **Beart, J. E., Lilley, T. H., and Haslam, E.,** Plant polyphenols. Secondary metabolism and chemical defence: some observations, *Phytochemistry*, 24, 33, 1985.
38. **Beerhues, L., Robenek, H., and Wiermann, R.,** Chalcone synthases from spinach *(Spinacia oleracea)*.II Immunofluorescence and immunogold localization, *Planta*, 173, 544, 1988.
39. **Beladi, I., Musci, I., Pusztai, R., Bakay, M., Rosztoczy, I., Gabor, M., and Veckenstedt, A.,** *In vitro* and *in vivo* antiviral effects of flavonoids, in *Flavonoids and Bioflavonoids, 1981*, Farkas, L., Gabor, M., Kallay, F., and Wagner, H., Eds., Elsevier, Amsterdam, 1982, 443.
40. **Bell, E. A.,** The possible significance of secondary products in plants, in *Secondary Plants Products, Encyclopedia of Plant Physiology*, New Series, Vol. 8, Bell, E. A. and Charlwood, B. V., Eds., Springer-Verlag, Berlin, 1980, chap. 2.
41. **Bell, E. A.,** The physiological role(s) of secondary (natural) products, in *The Biochemistry of Plants*, Vol. 8, Conn, Ed., Academic Press, New York, 1981, 1.
42. **Bell, E. A.,** Biochemical mechanisms of disease resistance, *Annu. Rev. Plant. Physiol.*, 32, 21, 1981.
43. **Ben-Aziz, A.,** Nobilitin is main fungistat in tangerines resistant to mal secco, *Science*, 155, 1026, 1967.
44. **Bewley, J. D. and Black, M.,** *Physiology and Biochemistry of Seeds in Relation to Germination*, Vol. 2, *Viability, Dormancy, and Environmental Control*, Springer-Verlag, Berlin, 1982.
45. **Bianchini, J. P. and Gaydou, E. M.,** Separation of polymethoxylated flavones by straight-phase high-performance liquid chromatography, *J. Chromatogr.*, 190, 233, 1980.

46. **Billot, J.**, Evolution des composés phénoliques au cours de la maturation de la Poire Passe-crassane, *Physiol. Vég.*, 21, 527, 1983.
47. **Billot, J.**, Evolution des monophénols et des orthodiphénols au cours de la croissance et de la maturation de la Poire Passe-crassane, *Bull. Liaison Groupe Polyphénols*, 13, 428, 1986.
48. **Blundstone, H. A. W., Woodman, J. S., and Adams, J. B.**, Canned citrus products, in *The Biochemistry of Fruits and Their Products*, Vol. 2, Hulme, A. C., Ed., Academic Press, London, 1971, 543.
49. **Bognar, R. and Rakosi, M.**, New biologically active flavonoid derivatives, in *Flavonoids and Bioflavonoids Current Research Trends*, Farkas, L., Gabor, M., and Kallay, F., Eds., Elsevier, Amsterdam, 1977, 117.
50. **Bollard, E. G.**, The physiology and nutrition of developing fruits, in *The Biochemistry of Fruits and Their Products*, Vol. 1, Hulme, A. C., Ed., Academic Press, London, 1970, chap. 14.
51. **Bompeix, G. and Bousquet, J. F.**, Les composés phénoliques dans la résistance des pommes à l'infection par les *Pezicula malicorticis* et *P. alba*, *Fruits*, 10, 693, 1974.
52. **Boniface, R., Miskulin, M., Robert, L., and Robert, A. M.**, Pharmacological properties of Myrtillus anthocyanosides: correlation with results of treatment of diabetic microangiopathy, in *Flavonoids and Bioflavonoids 1985*, Farkas, L., Gabor, M., and Kallay, F., Eds., Elsvier, Amsterdam, 1986, 293.
53. **Borchert, R.**, Time course and spatial distribution of phenylalanine ammonia-lyase and peroxidase activity in wounded potato tuber tissue, *Plant Physiol.*, 62, 789, 1978.
54. **Bostock, R. M., Thomas, C. S., Ogawa, J. M., Rice, R. E., and Uyemoto, J. K.**, Relationship of wound-induced peroxidase activity to epicarp lesion development in maturing pistachio fruit, *Phytopathology*, 77, 275, 1987.
55. **Boudreau, J. C.**, Taste and the taste of foods. A review and a report symposium, *Naturwissenschaften*, 67, 14, 1980.
56. **Brady, C. J.**, Fruit ripening, *Annu. Rev. Plant Physiol.*, 38, 155, 1987.
57. **Braquet, F., Nomboisse, J. C., Salvayre, R., Douste-Blazy, L., and Borel, J. P.**, Protective effect of flavonoids against the non-enzymatic degradation of collagen by oxy-free radicals, *Bull. Liaison Groupe Polyphénols*, 11, 496, 1982.
58. **Brasseur, Th., Angenot, L., Pincemail, and Deby, C.**, Propriétés antiradicalaires, antilipoperoxydantes et antioxydantes de flavonoides, *Bull. Liaison Groupe Polyphénols*, 13, 557, 1986.
59. **Brouillard, R.**, Chemical structure of anthocyanins, in *Anthocyanins as Food Colors*, Markakis, P., Ed., Academic Press, New York, 1982, 1.
60. **Brouillard, R.**, The *in vivo* expression of anthocyanin colour in plants, *Phytochemistry*, 22, 1311, 1983.
61. **Brouillard, R.**, Organisation des anthocyanes dans leurs sites naturels, *Bull. Liaison Groupe Polyphénols*, 13, 76, 1986.
62. **Brouillard, R. and Delaporte, B.**, Chemistry of anthocyanin pigments. 2. Kinetic and thermodynamic study of proton transfer hydration and tautomeric reactions of malvidin 3-glucoside, *J. Am. Chem. Soc.*, 99, 8461, 1977.
63. **Brouillard, R. and Delaporte, B.**, Etude par relaxation des équilibres de transfert de proton, d'hydratation et de tautomérie phototropique de l'anthocyane majoritaire de Vitis vinifera, la glucoside-3-malvidine, in *Protons and Ions Involved in Fast Dynamic Phenomena*, Laszlo, P., Ed., Elsevier, Amsterdam, 1978, 403.
64. **Brouillard, R. and Dubois, J. E.**, Mechanism of the structural transformations of anthocyanin in acidic media, *J. Am. Chem. Soc.*, 99, 1359, 1977.
65. **Brown, G. E.**, Development of green mold in degradated oranges, *Phytopathology*, 63, 1104, 1973.
66. **Buta, J. G. and Lusby, W. R.**, Catechins as germination and growth inhibitors in Lespedeza seeds, *Phytochemistry*, 25, 93, 1986.
67. **Caldwell, M. M., Robberecht, R., and Flint, S. D.**, Internal filters: prospects for UV-acclimatation in higher plants, *Physiol. Plant.*, 58, 445, 1983.
68. **Cazenave, J. P., Beretz, A., and Anton, R.**, The effects of flavonoids on human platelet function, in *Flavonoids and Bioflavonoids 1985*, Farkas, L., Gabor, M., and Kallay, F., Eds., Elsevier, Amsterdam, 1986, 373.
69. **Challice, J. S.**, Phenolics of Pyrus interspecific hybrids, *Phytochemistry*, 11, 3015, 1972.
70. **Chalmers, D. J. and Faragher, J. D.**, Regulation of anthocyanin synthesis in apple skin. I. Comparison of the effects of cycloheximide, ultraviolet light, wounding and maturity, *Aust. J. Plant Physiol.*, 4, 111, 1977.
71. **Chalmers, D. J. and Faragher, J. D.**, Regulation of anthocyanin synthesis in apple skin. II. Involvement of ethylene, *Aust. J. Plant Physiol.*, 4, 123, 1977.
72. **Chen, L. J. and Hrazdina, G.**, Structural aspects of anthocyanin-flavonoid complex formation and its role in plant color, *Phytochemistry*, 20, 297, 1981.
73. **Cluzel, C., Bastide, P., Wegman, R., and Tronche, P.**, Activités enzymatiques de la rétine et anthocyanosides des extraits de *Vaccinium myrtillus*, *Biochem. Pharmacol.*, 19, 2295, 1970.
74. **Clydesdale, F. M.**, Color measurement, in *Current Aspects of Food Colorants*, Furia, T. E., Ed., CRC Press, Boca Raton, FL, 1977, 1.

75. **Colonna, J. P.**, L'acide chlorogénique et les depsides de divers caféiers africains et malgaches: leur participation au métabolisme et leur signification biologique, *Travaux Documents ORSTOM*, 102, 1, 1979.
76. **Come, D.**, *Les Obstacles de la Germination*, Masson, Paris, 1970.
77. **Coombe, B. G.**, The development of fleshy fruits, *Annu. Rev. Plant Physiol.*, 27, 207, 1976.
78. **Corbineau, F., Lecat, S., and Côme, D.**, Dormancy of three cultivars of oat seeds (*Avena sativa* L.), *Seed Sci. Technol.*, 14, 725, 1986.
79. **Cosio, E. G. and McClure, J. W.**, Kaempferol glycosides and enzymes of flavonol biosynthesis in leaves of a soybean strain with a low photosynthetic rates, *Plant Physiol.*, 74, 877, 1984.
80. **Crane, J. C.**, Role of hormones in fruit set and development, *HortScience*, 4, 108, 1969.
81. **Creasy, L. L. and Coffee, M.**, Phytoalexin production potential of grape berries, *J. Am. Soc. Hortic. Sci.*, 113, 230, 1988.
82. **De Jong, D. W.**, Modification of tobacco leaf glycolate oxidase activity by chlorogenic acid and other polyphenols, *Physiol. Plant.*, 29, 150, 1973.
83. **Dean, B. B. and Kolattukudy, P. E.**, Synthesis of suberin during wound-healing in Jade leaves, tomato fruit, and bean pods, *Plant Physiol.*, 58, 411, 1976.
84. **Dekazos, E. D.**, Anthocyanin pigments in red tart cherries, *J. Food Sci.*, 35, 237, 1970.
85. **Dekazos, E. D. and Birth, G. S.**, A maturity index for blueberries using light transmittance, *J. Am. Soc. Hortic. Sci.*, 95, 610, 1970.
86. **Delaveau, F. and Vidal-Tessier, A. M.** Propriétés biologiques et pharmacologiques des quinones, *Bull. Liaison Groupe Polyphénols*, 13, 299, 1986.
87. **Delcour, J. A., Vandenberghe, M. M., Corten, P. F., and Dondeyne, P. C.**, Flavor thresholds of polyphenolics in water, *Am. J. Enol. Vitic.*, 35, 134, 1984.
88. **Deverall, B. J.**, Phenolase and pectic enzyme activity in the chocolate spot disease of beans, *Nature (London)*, 184, 311, 1961.
89. **Dewhirst, F. E.**, Structure-activity relationships for inhibition of prostaglandin cyclooxygenase by phenolic compounds, *Prostaglandin*, 20, 209, 1980.
90. **Dick, A. J.**, Apple fruit polyphenols and fruit softening, *Bull. Liaison Groupe Polyphenols*, 13, 358, 1986.
91. **Dick, A. J. and Bearne, L. L.**, Inhibition of beta-galactosidase of apple by flavonoids and other polyphenols, *J. Food Biochem.*, 12, 97, 1988.
92. **Dick, A. J., Williams, R., Bearne, S. L., and Lidster, P. D.**, Quercetin glycosides and chlorogenic acid: inhibitors of apple beta-galactosidase and of apple softening, *J. Agric. Food Chem.*, 33, 798, 1985.
93. **Dons, R. F. and Doughty, C. C.**, Isolation and characterization of aldose reductase from calf brain, *Biochim. Biophys. Acta*, 452, 1, 1976.
94. **Doumenjou, N. and Marigo, G.**, Relations polyphénols-croissance: role de l'acide chlorogénique dans le catabolisme auxinique chez *Lycopersicum esculentum*, *Physiol. Vég.*, 16, 319, 1978.
95. **Drake, S. R., Proebsting, E. L., Jr., and Spayd, S. E.**, Maturity index for the color grade of canned dark sweet cherries, *J. Am. Soc. Hortic. Sci.*, 107, 180, 1982.
96. **Drouet, A. and Hartman, C.**, Polyribosomes from pear fruit. Changes during ripening and senescence, *Plant Physiol.*, 64, 1104, 1979.
97. **Drouet, A. and Hartmann, C.**, Polyribosomes from pear fruit. II. Changes occurring in pulp tissues during ripening and senescence, *Plant Physiol.*, 69, 885, 1982.
98. **Dumenil, G., Vasquez, M., Cremieux, A., and Balansard, G.**, Action antibactérienne des acides phénols de la série benzoique, *Bull. Liaison Groupe Polyphénols*, 10, 302, 1980.
99. **Esaki, S., Konishi, F., and Kamiya, S.**, Relationship between the structure and taste of some naringin analogs, *Agric. Biol. Chem.*, 43, 231, 1979.
100. **Faragher, J. D. and Chalmers, D. J.**, Regulation of anthocyanin synthesis in apple skin. III. Involvement of phenylalanine ammonia-lyase, *Aust. J. Plant Physiol.*, 4, 133, 1977.
101. **Farkas, L., Gabor, M., and Kallay, F.**, Flavonoids and bioflavonoids, 1985, *Proc. 7th Hungarian Bioflavonoid Symposium*, Elsevier, Amsterdam, 1986.
102. **Fawcett, C. H. and Spencer, D. M.**, Sclerotinia fructigena infection and chlorogenic acid content in relation to antifungal compounds in apple fruits, *Ann. Appl. Biol.*, 61, 245, 1968.
103. **Ferguson, I. B.**, Calcium stimulation of ethylene production induced by 1-aminocylopropane-1-carboxylic acid and indole-3-acetic acid, *J. Plant Growth Regul.*, 2, 205, 1983.
104. **Fernandez Diez, M. J.**, The olive, in *The Biochemistry of Fruits and Their Products*, Vol. 2, Hulme, A. C., Ed., Academic Press, London, 1971, 255.
105. **Fleming, H. P., Walter, W. H., Jr., Etchells, J. L.**, Antimicrobial properties of oleuropein and products of its hydrolysis from green olives, *Appl. Microbiol.*, 26, 777, 1973.
106. **Fleuriet, A.**, Expression et régulation du métabolisme des dérivés hydroxycinnamiques au cours de la croissance, de la maturation et de la réaction aux blessures du fruit de la tomate "cerise", Thèse Doct. d'Etat, Montpellier, 1982.
107. **Fleuriet, A. and Deloire, A.**, Aspects histochimiques et biochimiques de la cicatrisation des fruits de tomate blessés, *Z. Pflanzenphysiol.*, 107, 259, 1982.

108. **Fleuriet, A. and Macheix, J. J.,** Relations entre la maturation accélérée de fruits blessés et leur teneur en composés phénoliques, *Coll. Int. CNRS No. 238,* 1975, 147.
109. **Fleuriet, A. and Macheix, J. J.,** Effets de traitements par l'azote et l'éthylène sur l'activité de la phénylalanine ammonia-lyase (PAL) des fruits de la tomate "cerise" (*Lycopersicum esculentum,* var. cerasiforme), *C.R. Acad. Sci. Paris, Ser. D,* 285, 1103, 1977.
110. **Fleuriet, A. and Macheix, J. J.,** Effet des blessures sur les composés phénoliques des fruits de tomates "cerise" (*Lycopersicum esculentum,* var. cerasiforme), *Physiol. Vég.,* 1977, 15, 239, 1977.
111. **Fleuriet, A. and Macheix, J. J.,** Quinyl esters and glucose derivatives of hydroxycinnamic acids during growth and ripening of tomato fruit, *Phytochemistry,* 20, 667, 1981.
112. **Fleuriet, A. and Macheix, J. J.,** Orientation nouvelle du métabolisme des acides hydroxycinnamiques dans les fruits de tomates blessés *(Lycopersicon esculentum), Physiol. Plant.,* 61, 64, 1984.
113. **Fleuriet, A., Macheix, J. J., Andary, C., and Villemur, P.,** Mise en évidence et dosage par chromatographie liquide à haute performance du verbascoside dans le fruit de six cultivars d'*Olea europaea* L., *C.R. Acad. Sci. Paris, Ser. III,* 299, 253, 1984.
114. **Forlot, P.,** Les furocoumarines: leur utilisation en cosmétique et en thérapeutique; données toxicologiques, pharmacologiques et cliniques, *Bull. Liaison Groupe Polyphénols,* 12, 268, 1984.
115. **Francis, F. J.,** Analysis of anthocyanins, in *Anthocyanins as Food Colors,* Markakis, P. Ed., Academic Press, New York, 1982, 182.
116. **Frenkel, C. and Mukai, M. K.,** Possible role of fruit cell oxidative activity in ethylene evolution, in *Ethylene: Biochemical Physiological and Applied Aspects,* Fuchs, Y., and Chalutz, E., Eds., Martinus Nijhoff, Dr. W. Junk Publishers, The Hague, 1984, 303.
117. **Friend, J.,** Phenolic substances and plant disease, in *Recent Advances in Phytochemistry,* Vol. 12, Swain, T., Harborne, J. B., and Van Sumere, C. F., Eds., Plenum Press, New York, 1979, 557.
118. **Friend, J.,** Plant phenolics, lignification and plant disease, in *Progress in Phytochemistry,* Vol. 7, Reinhold, L., Harborne, J. B., and Swain, T., Eds., Pergamon Press, Oxford, 1981, 197.
119. **Friend, J., Reynolds, S. B., and Aveyard, M. A.,** Phenylanine ammonia lyase, chlorogenic acid, and lignin in Potato tuber inoculated with *Phytophtora infestans, Physiol. Plant Pathol.,* 3, 495, 1973.
120. **Fry, S. C.,** Phenolic components of the primary cell wall and their possible role in the hormonal regulation of growth, *Planta,* 146, 343, 1979.
121. **Gabor, M.,** Advances in the pharmacology of benzopyrone derivatives and related compounds, in *Flavonoids and Bioflavonoids, 1981,* Farkas, L., Gabor, M., Kallay, F., and Wagner, H., Eds., Elsevier, Amsterdam, 1982, 363.
122. **Gabor, M.,** New results in the pharmacology of benzopyrone derivatives and related compounds, in *Flavonoids and Bioflavonoids 1985,* Farkas, L., Gabor, M., and Kallay, F., Eds., Elsevier, Amsterdam, 1986, 303.
123. **Galston, A. W.,** Flavonoids and photomorphogenesis in peas, in *Perspectives in Phytochemistry,* Harborne, J. B. and Swain, T., Eds., Academic Press, London, 1969, 193.
124. **Galston, A. W.,** The relation of flavonoid and peroxidase activity to the control of growth by phytochrome and hormone, *Am. J. Bot.,* 57, 2, 1970.
125. **Glazener, J. A.,** Accumulation of phenolic compounds in cells and formation of lignin-like polymers in cell walls of young tomato fruits after inoculation with *Botrytis cinerea, Physiol. Plant Pathol.,* 20, 11, 1982.
126. **Goodman, T. C., Montoya, A. L., Williams, S., and Chilton, M. D.,** Sustained ethylene production in *Agrobacterium*-transformed carrot disks caused by expression of the t-DNA tms gene products, *J. Bacteriol.,* 167, 387, 1986.
127. **Goodwin, P. B.,** Phytohormones and fruit growth, in *Phytohormones and Related Compounds, a Comprehensive Treatise,* Vol. II, Letham, D. S., Goodwin, P. B., and Higgins, T. S. V., Eds., Elsevier/North-Holland, Amsterdam, 1978, 175.
128. **Goodwin, T. W. and Mercer, E. I.,** *Introduction to Plant Biochemistry,* 2nd ed., Pergamon Press, Oxford, 1983.
129. **Goren, R., Huberman, M., and Zamski, E.,** Anatomical aspects of citrus abscission. Effects of ethylene on leaf and fruits explants, in *Ethylene, Biochemical Physiological and Applied Aspects,* Fuchs, Y. and Chalutz, E., Eds., Martinus Nijhoff, Dr. W. Junk Publishers, the Hague, 1984, 241.
130. **Gourama, H. and Bullerman, L. B.,** Effects of oleuropein on growth and aflatoxin production by *Aspergillus parasiticus, Lebensm. Wiss. Technol.,* 20, 226, 1987.
131. **Graziani, Y. and Chayath, R.,** Elevation of the c-AMP level in ehrlich ascites tumor cells by quercetin, *Biochem. Pharmacol.,* 26, 1259, 1977.
132. **Grisebach, H.,** Lignins, in *The Biochemistry of Plants,* Vol. 7, *Secondary Plant Products,* Conn, E. E., Ed., Academic Press, New York, 1981, 457.
133. **Gross, D.,** Growth regulating substances of plant origin, *Phytochemistry,* 14, 105, 1975.
134. **Gross, G. G.,** The chemistry and biochemistry of lignin, in *Biochemistry of Plant Phenolics,* Swain, T., Ed., Plenum Press, New York, 12, 177, 1979.

135. **Gross, J.**, *Pigments in Fruits*, Academic Press, London, 1987.
136. **Guadagni, D. G., Maier, V. P., and Turnbaugh, J. G.**, Effect of subthreshold concentrations of limonin, naringin and sweeteners on bitterness perception, *J. Sci. Food Agric.*, 25, 1349, 1974.
137. **Haard, N. F.**, Physiological role of peroxidases in postharvest fruits and vegetables, in *Enzymes in Foods and Beverages Processing*, Ory, R. L. and St. Angelo, A. J., Eds., Am. Chem. Soc. Symp. Ser. 47, American Chemical Society, Washington, D.C., 1977, 143.
138. **Hadwiger, L. A.**, The biosynthesis of pisatin, *Phytochemistry*, 5, 523, 1965.
139. **Hahlbrock, K., Lamb, C. F., Purwin, C., Ebel, J., Fautz, E., and Schafer, E.**, Rapid response of suspension-cultured parsley cells to the elicitor from *Phytophtora megasperma* var. sojae, *Plant Physiol.*, 67, 768, 1981.
140. **Harborne, J. B.**, *Comparative Biochemistry of the Flavonoids*, Academic Press, New York, 1967.
141. **Harborne, J. B.**, Function of flavonoids in plants, in *Chemistry and Biochemistry of Plant Pigments*, Vol. 1, 2nd ed., Goodwin, T. W., Ed., Academic Press, New York, 1976, 736.
142. **Harborne, J. B.**, Plant Phenolics, in *Encyclopedia of Plant Physiology*, New series, Vol. 8, *Secondary Products*, Bell, E. A. and Charlwood, B. V., Eds., Springer-Verlag, Berlin, 1980, 329.
143. **Harborne, J. B.**, *Introduction to Ecological Biochemistry*, 2nd ed., Academic Press, London, 1982.
144. **Harborne, J. B.**, The flavonoids of the Rutales, in *Chemistry and Chemical Taxonomy of the Rutales*, Waterman, P. G. and Grundon, M. F., Eds., Academic Press, London, 1983, 147.
145. **Harborne, J. B.**, Chemical data in practical taxonomy, in *Current Concepts in Plant Taxonomy*, Heywood, V. H. and Moore, D. M., Eds., Academic Press, London, 1984, 237.
146. **Harborne, J. B.**, Phenolics and plant defence, in *Ann. Proc. Phytochem. Soc. Eur.*, Vol. 25, Van Sumere, C. F. and Lea, P. J., Eds., Clarendon Press, Oxford, 393, 1985.
147. **Harborne, J. B. and Hall, E.**, Plant polyphenols. XIII. The systematic distribution and origin of anthocyanins containing branched trisaccharides, *Phytochemistry*, 3, 453, 1964.
148. **Harborne, J. B. and Williams, C. A.**, Flavonoid patterns in the fruits of the Umbelliferae, *Phytochemistry*,
149. **Hartley, R. D. and Jones, E. C.**, Diferulic acid as a component of cell walls of *Lolium multiflorum*, *Phytochemistry*, 15, 1157, 1976.
150. **Hartmann, C.**, Quelques aspects de la senescence du fruit, *Physiol. Vég.*, 21, 137, 1983.
151. **Hartmann, C., Boulay, M., and Moulet, A. M.**, Enzyme malique maturation et besoin de froid chez la Poire Passe-crassane, *C.R. Acad. Sci. Paris*, *281*, *Ser. D*, 135, 1975.
152. **Hartmann, C., Drouet, A., and Morin, F.**, Ethylene and ripening of apple, pear and cherry fruit, *Plant Physiol. Biochem.*, 25, 505, 1987.
153. **Hartmann, C., Drouet, A., and Rigault, C.**, Changes in enzymes and polysomes during aging of cylinders excised from pulp tissues of Passe-Crassane pears, *Physiol. Vég.*, 21, 59, 1983.
154. **Haslam, E.**, Vegetable tannins, in *The Biochemistry of Plants*, Vol. 7, Conn, E. E., Eds., Academic Press, New York, 1981, 527.
155. **Haslam, E. and Lilley, T. H.**, Natural astringency in foodstuffs. A molecular interpretation, *CRC Crit. Rev. Food Sci. Nutr.*, 27, 1, 1988.
156. **Haslam, E., Lilley, T. H., and Ozawa, T.**, Polyphenol complexation. Astringency in fruits and beverages, *Bull. Liaison Groupe Polyphénols*, 13, 352, 1986.
157. **Havsteen, B.**, Biochemical effects of flavonoids, *Z. Lebensm. Unters. Forsch.*, 170, 36, 1980.
158. **Havsteen, B.**, Flavonoids, a class of natural products of high pharmacological potency, *Biochem. Pharmacol.*, 32, 1141, 1983.
159. **Heinzmann, Y., Seitz, U., and Seitz, U.**, Purification and substrate specificities of hydroxycinnamate:CoA ligase from anthocyanin-containing and anthocyanin-free carrot cells, *Planta*, 135, 313, 1977.
160. **Hess, S. L., Hadwiger, L. A., and Schwochau, M. E.**, Studies in biosynthesis of phaseollin in excised pods of *Phaseolus vulgaris*, *Phytopathology*, 61, 79, 1971.
161. **Heyman, S. and Kinoshito, J. H.**, Isolation and properties of lens aldose reductase, *J. Biol. Chem.*, 240, 877, 1976.
162. **Hikino, H., Kiso, Y., Wagner, H., and Fiebig, M.**, Antihepatotoxic actions of flavonolignans from *Silybum marianum* fruits, *Planta Med.*, 50, 248, 1984.
163. **Hobson, G. E., Harman, J. E., and Nichols, R.**, Ethylene and the control of tomato fruit ripening, in *Ethylene: Biochemical, Physiological and Applied Aspects*, Fuchs, Y. and Chalutz, E., Eds., Martinus Nijhoff, Dr. W. Junk Publishers, The Hague, 1984, 281.
164. **Horowitz, R. M. and Gentili, B.**, Taste and structure in phenolic glycosides, *J. Sci. Food Agric.*, 17, 696, 1969.
165. **Horry, J. P. and Jay, M.**, Distribution of anthocyanins in wild and cultivated banana varieties, *Phytochemistry*, 27, 2667, 1988.
166. **Hrazdina, G.**, Anthocyanins and their role in food products, *Lebensm. Wissen. Technol.*, 14, 283, 1981.
167. **Hrazdina, G.**, Anthocyanins, in *The Flavonoids, Advances in Research*, Harborne, J. B. and Mabry, T. J., Eds., Chapman and Hall, London, 1982, 135.

168. **Hrazdina, G. and Wagner, G. J.**, Metabolic pathways as enzyme complexes: evidence for the synthesis of phenylpropanoids and flavonoids in membrane associated enzymes complexes, *Arch. Biochem. Biophys.*, 237, 88, 1985.
169. **Hrazdina, G. and Wagner, G. J.**, Compartmentation of plant phenolic compounds: sites in synthesis and accumulation, in *Ann. Proc. Phytochem. Soc. Eur.*, Vol. 25, Van Sumere, C. F. and Lea, P. J., Eds., Clarendon Press, Oxford, 1985, 118.
170. **Huet, R.**, Constituants des agrumes à effet pharmacodynamique; les citroflavonoides, *Fruits*, 37, 267, 1982.
171. **Hugues, R. E.**, Fruit flavonoids: some nutritional implications, *J. Hum. Nutr.*, 32, 47, 1978.
172. **Hulme, A. C. and Edney, K. L.**, Phenolic substances in the peel of Cox's orange Pippin apples with reference to infection by *G. perennans*, in *Phenolics in Plants in Health and Disease*, Pridham, J. B., Ed., Pergamon Press, Oxford, 1960, 87.
173. **Hunter, L. D.**, Phloridzin and apple scab, *Phytochemistry*, 14, 1519, 1975.
174. **Hwang, B. K.**, Contents of sugars, fruit acids, amino acids and phenolic compounds of apple fruits in relation to their susceptibility to *Botryospharia ribis, Phytopathol. Z.*, 108, 1, 1983.
175. **Ibrahim, R. K., de Luca, V., Khouri, H., Latchinian, L., Brisson, L., and Charest, P. M.**, Enzymology and compartmentation of polymethylated flavonol glucoside in *Chrysosplenium americanum, Phytochemistry*, 26, 1237, 1987.
176. **Iiyama, S., Toko, K., and Yamafuji, K.**, Effect of bitter substances on a model membrane system of taste reception, *Agric. Biol. Chem.*, 50, 2709, 1986.
177. **Ikeya, Y., Taguchi, H., Mitsuhashi, H., Takeda, S., Kase, Y., and Aburada, M.**, A lignin from *Schizandra chinensis, Phytochemistry*, 27, 569, 1988.
178. **Imakura, Y., Kobayashi, S., and Mima, A.**, Bitter phenyl propanoid glycosides from *Campsis chinensis, Phytochemistry*, 24, 139, 1985.
179. **Ingham, J. L.**, Phytoalexins from the Leguminosae, in *Phytoalexins*, Bailey, J. A. and Mansfield, J. W., Eds., Blackie, Glasgow, 1982, 21.
180. **Ishikura, N.**, A further survey of anthocyanins and other phenolics in *Ilex* and *Euonymus, Phytochemistry*, 14, 743, 1975.
181. **Ishikura, N.**, Light absorption patterns of anthocyanin containing cells, *Plant Cell Physiol.*, 19, 887, 1978.
182. **Ishikura, N. and Hayashida, S.**, Kaempferol glycosides in the seed-coat of *Ophiopogon jaburan* (Kunth) Lodd, *Agric. Biol. Chem.*, 43, 1923, 1979.
183. **Ismail, M. A. and Brown, G. E.**, Phenolic content during healing of "Valencia" orange peel under high humidity, *J. Am. Soc. Hortic. Sci.*, 100, 249, 1975.
184. **Ismail, M. A. and Brown, G. E.**, Postharvest wound healing in citrus fruit: induction of phenylalanine-ammonia-lyase in injured "Valencia" orange flavedo, *J. Am. Soc. Hortic. Sci.*, 104, 126, 1979.
185. **Ismail, M. A., Rouseff, R. L., and Brown, G. E.**, Wound healing in citrus: isolation and identification of 7-hydroxycoumarin (Umbelliferone) from grapefruit flavedo and its effect on *Penicillium digitatum* Sacc., *HortScience*, 13, 358, 1978.
186. **Jackman, R. L., Yada, R. Y., and Tung, M. A.**, A review: separation and chemical properties of anthocyanins used for their qualitative and quantitative analysis, *J. Food Biochem.*, 11, 279, 1987.
187. **Jackman, R. L., Yada, R. Y., Tung, M. A., and Speers, R. A.**, Anthocyanins as food colorants, *J. Food Biochem.*, 11, 201, 1987.
188. **Jacobs, M. and Rubery, P. H.**, Naturally occurring auxin transport regulators, *Science*, 241, 346, 1988.
189. **Janzen, D. H.**, Physiological ecology of fruits and their seeds, in *Physiological Plant Ecology*. III. *Responses to Chemical and Biological Environment*, Encylopedia of Plant Physiology, New Series, Vol. 12C, Lange, O. L., Novel, P. S., Osmond, C. R., and Ziegler, H., Eds., Springer-Verlag, Berlin, 1983, 625.
190. **Jennings, D. L. and Carmichael, E.**, Colour changes in frozen blackberries, *Hortic. Res.*, 19, 15, 1979.
191. **Joubert, A. J. and Van Lelyveld, L. J.**, An investigation of pre-harvest browning of Litchi peel, *Phytophylactica*, 7, 9, 1975.
192. **Jourdan, P. S., McIntosh, C. A., and Mansell, R. L.**, Naringin levels in citrus tissues. II. Quantitative distribution of naringen in *Citrus paradisi* Macfad., *Plant Physiol.*, 77, 903, 1985.
193. **Juven, B. and Henis, Y.**, Studies on the antimicrobial activity of olive phenolic compounds, *J. Appl. Bacteriol.*, 33, 721, 1970.
194. **Kahl, G.**, Induction and degradation of enzymes in aging plant storage tissues, in *Biochemistry of Wounded Plant Tissues*, Kahl, G., Ed., de Gruyter, W., Berlin, 1978, 347.
195. **Kefeli, V. I.**, Interaction between phytohormones and natural inhibitors during plant growth, *Fiziol. Rast.*, 18, 614, 1971.
196. **Kefeli, V. I. and Kutacek, M.**, Phenolic substances and their possible role in plant growth regulation, in *Plant Growth Regulation*, Pilet, P. E., Ed., Springer, Berlin, 1977, 181.
197. **Knott, R. P., Mikeal, R. L., and Shrader, K. R.**, Drug therapy of psoriasis, *Pharm. Index*, January, 7, 1981.
198. **Koeppe, D. E. and Miller, R. J.**, Kaempferol inhibitions of corn mitochondrial phosphorylation, *Plant Physiol.*, 54, 374, 1974.

199. **Kohlmunzer, S. and Rozkrutowa, B.,** An improved simple procedure for isolation of furocoumarins from *Ammi majus* L. fruits, *Bull. Liaison Groupe Polyphénols*, 12, 293, 1984.
200. **Konishi, F., Esaki, S., and Kamiya, S.,** Synthesis and taste of flavanone and dihydrochalcone glycosides containing 3-*O*-α-L-Rhammnopyranosyl-D-glucopyranose or 4-*O*-α-L-rhamnopyranosyl-D-glyropyranose in the sugar moiety, *Agric. Biol. Chem.*, 47, 265, 1983.
201. **Konowalchuk, J. and Speirs, J.,** Antiviral activity of fruit extracts, *J. Food Sci.*, 41, 1013, 1976.
202. **Kosuge, T.,** The role of phenolics in host response to infection, *Annu. Rev. Phytopathol.*, 7, 195, 1969.
203. **Kozlowski, T. T.,** *Shedding of Plants Parts*, Academic Press, New York, 1973.
204. **Kubo, I. and Matsumoto, A.,** Molluscicides from Olive *Olea europaea* and their efficient isolation by countercurrent chromatographies, *J. Agric. Food Chem.*, 32, 687, 1984.
205. **Kuc, J.,** Phytoalexins, *Annu. Rev. Phytopathol.*, 10, 207, 1972.
206. **Kuc, J., Henze, R. E., Ullstrup, A. J., and Quackenbush, F. W.,** Chlorogenic and caffeic acids as fungistatic agents produced by potatoes in response to inoculation with *Helminthosporium carbonium*, *J. Am. Chem. Soc.*, 78, 3123, 1956.
207. **Kuroda, H., Shimada, M., and Higuchi, T.,** Characterization of a lignin-specific *O*-methyltransferase in Aspen wood, *Phytochemistry*, 20, 2635, 1981.
208. **Kushman, L. J. and Ballinger, W. E.,** Relation of quality indices of individual blueberries to photoelectric measurement of anthocyanin content, *J. Am. Soc. Hortic. Sci.*, 100, 561, 1975.
209. **Kutsuki, H., Shimada, M., and Higughi, T.,** Distribution of roles of *p*-hydroxy cinnamate:CoA ligase in lignin biosynthesis, *Phytochemistry*, 21, 267, 1982.
210. **Laloraya, M. M., Nozzolillo, C., Purohit, S., and Stevenson, L.,** Reversal of abscissic acid-induced stomatal closure by *trans*-cinnamic and *p*-coumaric acid, *Plant Physiol.*, 81, 253, 1986.
211. **Lamb, C. J., Lawton, M. M., Taylor, S. J., and Dixon, R. A.,** Elicitor modulation of phenylalanine ammonia-lyase in *Phaseolus vulgaris*, *Ann. Phytopathol.*, 12, 423, 1980.
212. **Lamport, D. T. A.,** Cell wall carbohydrates in relation to structure and function, in *Frontiers of Plant Tissue Culture*, Thorpe, T. A., Ed., Calgary University Press, 1978, 235.
213. **Langcake, R.,** Disease resistance of *Vitis* spp. and the production of the stress metabolites, ε-viniferin, α-viniferin and pterostilbene, *Physiol. Plant Pathol.*, 18, 213, 1981.
214. **Larson, R. A.,** The antioxidants of higher plants, *Phytochemistry*, 27, 969, 1988.
215. **Lea, A. G. H. and Arnold, G. M.,** The phenolics of ciders: bitterness and astringency, *J. Sci. Food Agric.*, 29, 478, 1978.
216. **Lee, T. T., Starratt, A. N., and Jevnikar, J. J.,** Regulation of enzymic oxidation of indole-3 acetic acid by phenols: structure-activity relationships, *Phytochemistry*, 21, 517, 1982.
217. **Lener, M. and Bittoni, A.,** Utilisation des acides caféique et chlorogénique comme antioxydants alimentaires, *Bull. Liaison Groupe Polyphénols*, 10, 289, 1980.
218. **Lenoir, C., Corbineau, F., and Côme, D.,** Barley *(Hordeum vulgare)* seed dormancy as related to glumella characteristics, *Physiol. Plant.*, 68, 301, 1986.
219. **Lidster, P. D., Dick, A. J., Demarco, A., and MacRae, K. B.,** Application of flavonoid glycosides and phenolic acids to suppress firmness loss in apples, *J. Am. Soc. Hortic. Sci.*, 11, 892, 1986.
220. **Lieberman, M.,** Biosynthesis and action of ethylene, *Annu. Rev. Plant Physiol.*, 30, 533, 1979.
221. **Lopez, C. R. and Monaco, L. Cl.,** Chemotaxonomic studies of some species of the genus *Coffea*, *J. Plant Crops*, 7, 6, 1979.
222. **Mabry, T. J.,** Betalains, in *Secondary Plant Products*, Encyclopedia of Plant Physiology, New Series, Vol. 8, Bell, E. A. and Charlwood, B. V., Eds., Springer-Verlag, Berlin, 1980, chap. 11.
223. **Mabry, T. J. and Ulubelen, A.,** Chemistry and utilization of phenylpropanoids including flavonoids, coumarins and lignans, *J. Agric. Food Chem.*, 28, 188, 1980.
224. **Maccarone, E., Maccarone, A., and Rapisarda, P.,** Stabilization of anthocyanins of blood orange fruits juice, *J. Food Sci.*, 50, 901, 1985.
225. **Macheix, J. J.,** Evolution de la teneur du fruit de *Pirus Malus* L. en esters des acides *p*-coumarique et caféique au cours de la croissance, *Physiol. Vég.*, 12, 25, 1974.
226. **Macheix, J. J., Fleuriet, A., and Quessada, M. P.,** Involvement of phenols and peroxidases in wound healing and grafting, in *Molecular and Physiological Aspects of Plant Peroxidases*, University of Geneva, Switzerland, 1986, 267.
226a. **MacRae, W. D. and Towers, G. H. N.,** Biological activities of lignans, *Phytochemistry*, 23, 1207, 1984.
227. **Maeda, S., Sudo, K., Miyamoto, Y., Taheda, S., Shinbo, M., Harada, M., Ikeya, Y., and Taguchi, H.,** Pharmacological studies on Schizandra fruits I. General pharmacological effects of Gomisin A and Schizandrin, *Yakugaku Zasshi*, 102, 579, 1982.
228. **Maga, J. A.,** Simple phenols and phenolic compounds in food flavor, *CRC Crit. Rev. Food Nutr.*, 10, 323, 1978.
229. **Maga, J. A. and Lorentz, K.,** Taste threshold values for phenolic acids which can influence flavor properties of certain flours grains and oilseeds, *Cereal Sci. Today*, 10, 326, 1973.

230. **Maier, V. P.,** Chemistry and significance of selected citrus limonoids and flavonoids, in *Chemistry and Chemical Taxonomy of the Rutales,* Waterman, P. G. and Grundon, M. F., Eds., Academic Press, London, 1983, 319.
231. **Makus, D. J., and Ballinger, W. E.,** Characterization of anthocyanins during ripening of fruit of *Vaccinium corymbosum* L. cv. Wolcott., *J. Am. Soc. Hortic. Sci.,* 98, 99, 1973.
232. **Mansfield, J. W.,** The role of phytoalexins in disease resistance, in *Phytoalexins,* Bailey, J. A. and Mansfield, J. W., Eds., Blackie, Glasgow, 1982, 253.
233. **Mapson, L. W. and Mead, A.,** Biosynthesis of ethylene. Dual nature of cofactor required for the enzymic production of ethylene from methional, *Biochem. J.,* 108, 875, 1968.
234. **Marcellin, P.,** La respiration des fruits après récolte: la crise climactérique, *Bull. Soc. Bot. Fr.,* 129, 107, 1982.
235. **Marigo, G.,** Polyphénols et croissance végétale. Essai d'évaluation du role joué *in vivo* par les composés phénoliques chez *Lycopersicum esculentum,* Thèse Doct. Etat, Toulouse, 1979.
236. **Marigo, G. and Boudet, A. M.,** Relations polyphénols-croissance: mise en évidence d'un effet inhibiteur des composés phénoliques sur le transport polarisé de l'auxine, *Physiol. Plant.,* 41, 197, 1977.
237. **Marigo, G. and Boudet, A. M.,** Relations polyphénols-croissance: lignification et limitation de croissance chez *Lycopersicum esculentum, Physiol. Plant.,* 49, 425, 1980.
238. **Masquelier, J.,** Vin et athérosclérose; action protectrice des procyanidines, *C.R. Symp. Int. "L'alimentation et la Consommation de Vin",* Vérone, Italie, 147, 1982.
239. **Matsubara, Y., Kumamoto, H., Iizuka, Y., Murakami, T., Okamoto, K., Miyake, H., and Yokoi, K.,** Structure and hypotensive effect of flavonoid glycosides in citrus unshiu peelings, *Agric. Biol. Chem.,* 49, 909, 1985.
240. **Matsuno, H. and Uritani, I.,** Physiological behavior of peroxidase isozymes in sweet potato root tissue injured by cutting or with black rot, *Plant Cell Physiol.,* 13, 1091, 1972.
241. **Mazoyer-Allermoz, C. and Cier, A.,** Sur l'étude pharmacologique d'un extrait de cassis, *Bull. Liaison Groupe Polyphénols,* 5, 1974.
242. **Mazza, G. and Brouillard, R.,** Recent developments in the stabilization of anthocyanins in food products, *Food Chem.,* 25, 207, 1987.
243. **Mazza, G. and Brouillard, R.,** Color stability and structural transformations of cyanidin 3,5-diglucoside and four 3-deoxyanthocyanins in aqueous solutions, *J. Agric. Food Chem.,* 35, 422, 1987.
244. **McClure, J. W.,** Physiology and functions of flavonoids, in *The Flavonoids,* Harborne, J. B., Mabry, T. J., and Mabry, H., Eds., Chapman and Hall, London, 1975, 970.
245. **McClure, W.,** The physiology of phenolic compounds in plants, in *Biochemistry of Plant Phenolics,* Swain, T., Harborne, J. B., and Van Sumere, C. F., Eds., Rec. Adv. Phytochem. 12, Plenum Press, New York, 1979, 525.
246. **McGlasson, W. B.,** Ethylene and fruit ripening, *HortScience,* 20, 51, 1985.
247. **McGlasson, W. B. and Pratt, H. K.,** Effects of wounding on respiration and ethylene production by cantaloupe fruit tissue, *Plant Physiol.,* 39, 128, 1964.
248. **McGlasson, W. B., Wade, N. L., and Adato, I.,** Phytohormones and fruit ripening, in *Phytohormones and Related Compounds: A Comprehensive Treatise,* Vol. II, Letham, D. S., Goodwin, P. B., and Higgins, T. J. V., Ed., Elsevier/North-Holland, Amsterdam, 1978, 447.
249. **Meggos, H. N.,** Color-key food ingredients, *Food Technol.,* 38, 70, 1984.
250. **Melin, C., Moulet, A. M., Dupin, J. F., and Hartmann, C.,** Phénylalanine-ammonialyase et composés phénoliques au cours de la maturation de la cerise, *Phytochemistry,* 16, 75, 1977.
251. **Merlin, J. C. and Brouillard, R.,** Resonance Raman spectrometry of anthocyanins, *Bull. Liaison Groupe Polyphénols,* 13, 145, 1986.
252. **Merlin, J. C., Statoua, A., and Brouillard, R.,** Investigation of the *in vivo* organization of anthocyanins using resonance Raman microspectrometry, *Phytochemistry,* 24, 1575, 1985.
253. **Michel, F., Mercklein, L., Rey, R., and Crastes de Paulet, A.,** Comparative effects of some flavonoids on cyclooxygenase and lipoxygenase activities in different cell systems or subfractions, in *Flavonoids and Bioflavonoids 1985,* Farkas, L., Gabor, M., and Kallay, F., Eds., Elsevier, Amsterdam, 1986, 389.
254. **Michniewicz, M. and Rozej, B.,** Stimulation of transpiration rate in bean plants by ferulic acid, *Naturwissenschaften,* 61, 33, 1974.
255. **Middleton, E. J.,** Some effects of flavonoids on mammalian cell systems, in *Flavonoids and Bioflavonoids 1985,* Farkas, L., Gabor, M., and Kallay, F., Eds., Elsevier, Amsterdam, 1986, 381.
256. **Misaghi, I. J.,** *Physiology and Biochemistry of Plant-Pathogen Interactions,* Plenum Press, New York, 1982.
257. **Monselise, S. P.,** *Handbook of Fruit Set and Development,* CRC Press, Boca Raton, FL, 1986.
258. **Monties, B.,** Etude de quelques relations entre polyphénols et photosynthèse: présence, compartimentation et métabolisme des polyphénols dans les chloroplastes, Thèse Doct. Etat, Orsay, 1975.
259. **Monties, B.,** Rayonnement ultraviolet et photosynthèse, in *Photosynthese et Production Végétale,* Costes, C., Ed., Gauthier-Villars, Paris, 1975, 193.

260. **Monties, B.**, Les lignines, in *Les Polymères Végétaux*, Monties, B., Ed., Gauthier-Villars, 1980, 122.
261. **Morgan, P. W.**, Is ethylene the natural regulator of abscission?, in *Ethylene Biochemical, Physiological and Applied Aspects*, Fuchs, Y. and Chalutz, E., Eds., Martinus Nijhoff, Dr. W. Junk Publishers, The Hague, 1984, 231.
262. **Morin, F.**, Métabolisme de l'éthylène et sénescence après récolte de deux fruits climactériques, la pomme, *Pyrus malus* L., cv Golden Delicious, la poire *Pyrus Communis* L. cv Passe-Crassane, Thèse Doctorat, University Orléans, 1988.
263. **Morin, F., Rigault, C., and Hartmann, C.**, Conséquences d'un séjour au froid sur le métabolisme de l'éthylene au cours de la maturation de la poire Passe-Crass ane après récolte, *Physiol. Vég.*, 23, 353, 1985.
266. **Moskowitz, A. H. and Hrazdina, G.**, Vacuolar contents of fruit subepidermal cells from *Vitis* species, *Plant Physiol.*, 68, 686, 1981.
265. **Mucsi, I.**, Combined antiviral effect of quercetin and interferon on the multiplication of herpes simplex virus in cell cultures, in *Flavonoids and Bioflavonoids 1985*, Farkas, L., Gabor, M., and Kallay, F., Eds., Elsevier, Amsterdam, 1986, 333.
266. **Mucsi, I., Beladi, I., Pusztai, R., Bakay, M., and Gabor, M.**, Antiviral effect of flavonoids, in *Flavonoids and Bioflavonoids Current Research Trends*, Farkas, L., Gabor, M., and Kallay, F., Eds., Elsevier, Amsterdam, 1977, 401.
267. **Mulvena, D., Webb, E. C., and Zerner, B.**, 3-4-dihydroxybenzaldehyde, a fungistatic substance from green Cavendish Bananas, *Phytochemistry*, 8, 393, 1969.
268. **Nagel, C. W. and Graber, W. R.**, Effect of must oxidation on quality of white wines, *Am. J. Enol. Vitic.*, 39, 1, 1988.
269. **Nagel, C. W., Herrick, I. W., and Graber, W. R.**, Is chlorogenic acid bitter?, *J. Food Sci.*, 52, 213, 1987.
270. **Nakajima, K., Taguchi, H., Ikeya, Y., Endo, T., and Yosioka, I.**, The constituents of *schizandre chinensis* Bail. XIII. Quantitative analysis of lignans in the fruit of *Schizandra chinensis*. Bail. by high performance liquid chromatography, *Yakugaku Zasshi*, 103, 743, 1983.
271. **Nakayama, T. O. M. and Powers, J. J.**, in *Chemistry of Plant Pigments*, Chichester, C. O., Ed., Academic Press, New York, 193, 1972.
272. **Ndubizu, T. O. C.**, Relations of phenolic inhibitors to resistance of immature apple fruits to rot, *J. Hortic. Sci.*, 51, 311, 1976.
273. **Nikaiko, T., Ohmoto, T., Kinoshita, T., Sankawa, U., Nishibe, S., and Hisada, S.**, Inhibition of cyclic AMP phospho-diesterase by lignans, *Chem. Pharm. Bull.*, 29, 3586, 1981.
274. **Nikaiko, T., Ohmoto, T., Noguchi, H., Kinoshita, T., Saitoh, H., and Sankawa, U.**, Inhibition of cyclic AMP phosphodiesterase in medicinal plants, *Planta Med.*, 43, 18, 1981.
275. **Nitsch, J. P.**, Hormonal factors in growth and development, in *The Biochemistry of Fruits and Their Products*, Vol. 1, Hulme, A. C., Ed., Academic Press, London, 1970, 427.
276. **Nitsch, J. P. and Nitsch, C.**, Synergie d'action entre les dérivés de l'acide caféique et l'acide 3-indoly-lacétique, *Bull. Soc. Bot. Fr.*, 106, 414, 1959.
277. **Nitsch, J. P. and Nitsch, C.**, Synergistes naturels des auxines et des gibbérel lines, *Bull. Soc. Bot. Fr.*, 108, 349, 1961.
278. **Nitsch, J. P. and Nitsch, C.**, Composés phénoliques et croissance végétale, *Ann. Physiol. Vég.*, 4, 211, 1962.
279. **Nothmann, J., Rylski, I., and Spigelman, M.**, Color and variations in color intensity of fruit of different eggplant cultivars, *Sci. Hortic.*, 4, 191, 1976.
280. **Nurstern, H. E.**, Volatile compounds: the aroma of fruits, in *The Biochemistry of Fruits and Their Products*, Vol. 1, Hulme, A. C., Ed., Academic Press, London, 239, 1970.
281. **Okamura, S. and Watanabe, M.**, Determination of phenolic cinnamates in white wine and their effect on wine quality, *Agric. Biol. Chem.*, 45, 2063, 1981.
282. **Okamura, S. and Watanabe, M.**, Hydroxycinnamic acid ester hydrolase of microorganisms. 1. Properties and application to wine making of hydroxycinnamic acid ester hydrolase from *Aspergillus japonicus*, *J. Agric. Chem. Soc. Jpn.*, 55, 1099, 1981.
283. **Okamura, S. and Watanabe, M.**, Measurement of hydroxycinnamic acid ester hydrolase activity by the change of UV absorption, *Agric. Biol. Chem.*, 46, 297, 1982.
284. **Okamura, S. and Watanabe, M.**, U.S. Patent, 4,370,416, 1983.
285. **Okuda, T., Kimura, Y., Yoshida, T., Hatano, T., Okuda, H., and Arichi, S.**, Studies on the activity of tannins and related compounds from medicinal plants and drugs. Inhibitory effects on lipid peroxidation, *Chem. Pharm. Bull.*, 31, 1625, 1983.
286. **Ong, B. Y. and Nagel, C. W.**, Hydroxycinnamic acid-tartaric acid ester content in mature grapes and during the maturation of white Riesling grapes, *Am. J. Enol. Vitic.*, 29, 277, 1978.
287. **Osawa, Y.**, Copigmentation of anthocyanins, in *Anthocyanins as Food Colors*, Markakis, P., Ed., Academic Press, New York, 1982, 41.

288. **Osborn, D. J. and McManus, M. T.**, Abscission and recognition of zone specific target cells. The role of ethylene, in *Ethylene: Biochemical, Physiological and Applied Aspects*, Fuchs, Y. and Chalutz, E., Ed., Martinus Nijhoff, Dr. W. Junk Publishers, The Hague, 1984, 221.
289. **Ozawa, T., Lilley, T. H., and Haslam, E.**, Polyphenol interactions: astringency and the loss of astringency in ripening fruit, *Phytochemistry*, 26, 2937, 1987.
290. **Pecket, R. C. and Small, C. J.**, Occurrence, location and development of anthocyanoplasts, *Phytochemistry*, 19, 2571, 1980.
291. **Peyron, D. and Tissut, M.**, Le métabolisme des flavonoides de la plantule de pois, *Physiol. Vég.*, 19, 315, 1981.
292. **Pezet, R. and Pont, V.**, Mise en évidence de pterostilbène dans les grappes de *Vitis vinifera, Plant Physiol. Biochem.*, 26, 603, 1988.
293. **Phan, C. T.**, *Ethylène, Métabolisme et Activité Métabolique*, Masson and Cie, Paris, 1971.
294. **Piatelli, P.**, Betalains, in *Chemistry and Biochemistry of Plant Pigments*, 2nd ed., Goodwin, T. W., Ed., Academic Press, London, 1976, 560.
295. **Piatelli, M. and Impellizzeri, G.**, Fungistatic flavons in the leaves of citrus species resistant and susceptible to *Deuterophoma tracheiphila, Phytochemistry*, 10, 2657, 1971.
296. **Pierpoint, W. S.**, Phenolics in food and feedstuffs: the pleasures and perils of vegetarianism, in *The Biochemistry of Plant Phenolics*, Van Sumere, C. F. and Lea, P. J., Eds., Clarendon Press, Oxford, 427, 1985.
297. **Pincemail, J., Deby, C., Lion, Y., Braquet, P., Hans, P., Drieu, K., and Goutier, R.**, Role of flavonoids in lipoperoxidation and radicalar reactions, in *Flavonoids and Bioflavonoids*, Farkas, L., Gabor, M., and Kallay, F., Eds., Elsevier, Amsterdam, 1986, 423.
298. **Pourrat, H.**, Les drogues à tannins dans la pharmacologie moderne, *Bull. Liaison Groupe Polyphénols*, 10, 33, 1980.
299. **Pourrat, H., Bastide, P., Dorier, P., Pourrat, A., and Tronche, P.**, Préparation et activité thérapeutique de quelques glycosides d'anthocyanes, *Chim. Ther.*, 2, 33, 1967.
300. **Powel, C. C. and Hildebrand, D. C.**, Fire-blight resistance in Pyrus: involvement of arbutin oxidation, *Phytopathology*, 60, 337, 1970.
301. **Price, M. L., Stromberg, A. M., and Butler, L. G.**, Tannin content as a function of grain maturity and drying conditions in several varieties of *Sorghum bicolor* (L.) Moench, *J. Agric. Food Chem.*, 27, 1270, 1979.
302. **Prusky, D.**, The use of antioxidants to delay the onset of anthracnose and stem end decay in avocado fruits after harvest, *Plant Dis.*, 72, 381, 1988.
303. **Prusky, D., Kobiler, I., and Jacoby, B.**, Involvement of epicatechin in cultivar susceptibility of avocado fruits to *Colletotrichum gloeosporioides* after harvest, *J. Phytopathol.*, 123, 140, 1988.
304. **Prusky, D., Kobiler, I., Jacoby, B., Sims, J. J., and Midlands, S.**, Inhibitors of avocado lipoxygenase: their possible relationship with the latency of *Colletotrichum gloeosporioides, Physiol. Plant Pathol.*, 27, 269, 1985.
305. **Purseglove, J. W., Brown, E. G., Green, C. L., and Robbins, S. R. J.**, *Spices (Vanilla)*, Vol. 2, Longman, London, 644, 1981.
306. **Ramina, A. and Masia, A.**, Levels of extractable para-coumaric acid in the exomesocarp and seed of persisting and abscising peach fruit, *Sci. Hortic.*, 16, 375, 1982.
307. **Ramirez-Martinez, J. R.**, Phenolic compounds in coffee pulp: quantitative determination by HPLC, *J. Sci. Food Agric.*, 43, 135, 1988.
308. **Ranadive, A. S. and Haard, N. F.**, Chemical nature of stone cells from pear fruit, *J. Food Sci.*, 38, 331, 1973.
309. **Ranjeva, R., Boudet, A., Harada, H., and Marigo, G.**, Phenolic metabolism in Petunia tissues, I. Characteristic responses of enzymes involved in different steps of polyphenol synthesis to different hormonal influences, *Biochem. Biophys. Acta*, 399, 23, 1975.
310. **Ravanel, P., Tissut, M., and Douce, R.**, Effects of flavone on the oxidative properties of intact plant mitrochondria, *Phytochemistry*, 20, 2101, 1981.
311. **Ravenel, P., Tissut, M., and Douce, R.**, Effect of kaempferol on the oxidative properties of intact plant mitochondria, *Plant Physiol.*, 69, 375, 1982.
312. **Ravise, A. and Kirkiacharian, B. S.**, Influence de la structure des composés phénoliques sur l'inhibition du *Phytophthora parasitica* et d'enzymes participant aux processus parasitaires. I. Isoflavonoides et coumestanes, *Phytopathol. Z.*, 85, 74, 1976.
313. **Ravise, A. and Kirkiacharian, B. S.**, Influence de la structure de composés phénoliques sur l'inhibition du *Phytophthora parasitica* et d'enzymes participant aux processus parasitaires. II. Coumarines, *Phytopathol. Z.*, 86, 314, 1976.
314. **Ravise, A. and Kirkiacharian, B. S.**, Influence de la structure de composés phénoliques sur l'inhibition du *Phytophthora parasitica* et d'enzymes participant aux processus parasitaires. III. Homo-isoflavanones, *Phytopathol. Z.*, 92, 36, 1978.

315. **Ravise, A. and Kirkiacharian, B. S.,** Influence of the structure of phenolic compounds on the inhibition of the growth of *Phytophora parasitica* and the activity of parasitogenic enzymes. IV. Neoflavonoids, *Phytopathol. Z.*, 97, 219, 1980.
316. **Reid, M. S.,** Ethylene and abscission, *HortScience*, 20, 45, 1985.
317. **Rhodes, M. J. C.,** The climacteric and ripening of fruits, in *The Biochemistry of Fruits and Their Products*, Vol. 1, Hulme, A. C., Ed., Academic Press, London, 1970, 521.
318. **Rhodes, M. J. C.,** The extraction and purification of enzymes from plant tissues, in *Regulation of Enzyme Synthesis and Activity in Higher Plants*, Smith, H., Ed., Academic Press, London, 1977, 245.
319. **Rhodes, M. J. C.,** The maturation and ripening of fruits, in *Senescence in Plants* Thimann, K. V., Ed., CRC Press, Boca Raton, FL, 1980, 157.
320. **Rhodes, M. J. C.,** The physiological significance of plant phenolic compounds, in *Ann. Proc. Phytochem. Soc. Eur.*, Vol. 25, Van Sumere, C. F. and Lea, P. J., Eds., Clarendon Press, Oxford, 1985, 99.
321. **Rhodes, M. J. C. and Wooltorton, L. S. C.,** The biosynthesis of phenolic compounds in wounded plant storage tissues, in *Biochemistry of Wounded Plant Tissues*, Kahl, G., Ed., De Gruyter, W., 1978, 243.
322. **Ribereau-Gayon, P.,** *Plant Phenolics*, Oliver and Boyd, Edinburgh, 1972.
323. **Ride, J. P.,** The role of cell wall alterations in resistance to fungi, *Ann. Appl. Biol.*, 89, 302, 1978.
324. **Riley, R. G. and Kolattukudy, P. E.,** Evidence for covalently attached *p*-coumaric and ferulic acid in cutins and suberins, *Plant Physiol.*, 56, 650, 1975.
325. **Riov, J., Monselise, S. P., and Kahan, R. S.,** Ethylene controlled induction of phenylalanine ammonia-lyase in citrus fruit peel, *Plant Physiol.*, 44, 631, 1969.
326. **Robberecht, R. and Caldwell, M. M.,** Leaf epidermal transmittance of ultraviolet radiation and its implications for plant sensitivity to ultraviolet-radiation induced injury, *Oecologia*, 32, 277, 1978.
327. **Robbins, R. C.,** Regulatory action of phenylbenzo-γ-pyrone (PBP) derivatives on blood constituents affecting rheology in patients with coronary heart disease (CHD), *Int. J. Vit. Nutr. Res.*, 46, 338, 1976.
328. **Robinson, G. M. and Robinson, R.,** A survey of anthocyanins, *Biochem. J.*, 25, 1687, 1931.
329. **Rohrback, R. P. and McClure, W. E.,** A production capacity conveyor for small fruit sorting: the M-belt, *Trans. Am. Soc. Agric. Eng.*, 21, 1092, 1978.
330. **Rohwer, F. and Mader, M.,** The role of peroxidase in ethylene formation from 1-aminocyclopropane-1-carboxylic acid, *Z. Pflanzenphysiol.*, 104 S., 363, 1981.
331. **Rudnicki, R., Hammond, R. K., and Bukovac, M. J.,** Endogenous plant growth substances in developing fruit of *Prunus cerasus* L. II. Levels of extractable para coumaric acid in the pericarp, *J. Am. Soc. Hortic. Sci.*, 98, 225, 1973.
332. **Salvayre, R., Braquet, P., Perruchot, Th., and Douste-Blazy, L.,** Comparison of the scavenger effect of bilberry anthocyanosides with various flavonoids, in *Flavonoids and Bioflavonoids, 1981*, Farkas, L., Gabor, M., Kallay, F., and Wagner, H., Eds., Elsevier, Amsterdam, 1982, 437.
333. **Sapers, G. M., Jones, S. B., and Maher, G. T.,** Factors affecting the recovery of juice and anthocyanin from cranberries, *J. Am. Soc. Hortic. Sci.*, 108, 246, 1983.
334. **Sapers, G. M., Graff, G. R., Phillips, J. G., and Deubert, K. H.,** Factors affecting the anthocyanin content of cranberry, *J. Am. Soc. Hortic. Sci.*, 111, 612, 1986.
335. **Sapers, G. M., Phillips, J. G., Rudolf, H. M., and Divito, A. M.,** Cranberry quality: selection procedures for breeding programs, *J. Am. Soc. Hortic. Sci.*, 108, 241, 1983.
336. **Sapers, G. M., Jones, S. B., Kelley, M. J., Phillips, J. G., and Stone, E. G.,** Breeding strategies for increasing the anthocyanin content of cranberries, *J. Am. Soc. Hortic. Sci.*, 111, 618, 1986.
337. **Sapers, G. M., Hicks, K. B., Burgher, A. M., Hargrave, D. L., Sondey, S. M., and Bilyk, A.,** Anthocyanin patterns in ripening thornless blackberries, *J. Am. Soc. Hortic. Sci.*, 111, 945, 1986.
338. **Sarkanen, K. V.,** Precursors and their polymerization, In *Lignins, Occurrence, Formation, Structure and Reactions*, Sarkanen, K. V. and Ludwig, C. H., Eds., Wiley Intersciences, New York, 1971, 95.
339. **Sawabe, A., Matsubara, Y., Kumamoto, H., Iizuka, Y., and Okamoto, K.,** Structure and physiological activity of phenyl propanoid glycosides of hassaku (*Citrus Hassaku* Hort.) and orange (*Citrus sinensis* Osbeck.) peelings, *J. Agric. Chem. Soc. Jpn.*, 60, 593, 1986.
340. **Schaufelberger, D. and Hostettmann, K.,** On the molluscicidal activity of tannin containing plants, *Planta Med.*, 48, 105, 1983.
341. **Scheffeldt, P. and Hrazdina, G.,** Co-pigmentation of anthocyanins under physiological conditions, *J. Food. Sci.*, 43, 517, 1978.
342. **Senter, S. D., Horvat, R. J., and Forbus, Jr., W. R.,** Quantitative variation of total phenols in fresh market tomatoes at three stages of maturity, *J. Food Sci.*, 53, 639, 1988.
343. **Sexton, R. and Roberts, J. A.,** Cell biology of abscission, *Annu. Rev. Plant Physiol.*, 33, 133, 1982.
344. **Shimazaki, K., Igarashi, T., and Kondo, N.,** Protection by the epidermis of photosynthesis against UV-C radiation estimated by chlorophyll fluorescence, *Physiol. Plant.*, 74, 34, 1988.
345. **Shrikande, A. J. and Francis, F. J.,** Anthocyanin pigments of sour cherries, *J. Food Sci.*, 38, 649, 1973.
346. **Singleton, V. L. and Esau, P.,** *Phenolic Substances in Grapes and Wine, and Their Significance*, Academic Press, New York, 1969.

347. **Singleton, V. L. and Noble, A. C.,** Wine flavor and phenolic substances, *ACS Symp. Ser.*, 26, 47, 1976.
348. **Sitrit, Y., Riov, J., and Blumenfeld, A.,** Interference of phenolic compounds with the 1-aminocyclopropane-1-carboxylic acid assay, *Plant Physiol.*, 86, 13, 1988.
349. **Skene, D. S.,** Wound healing in apple fruits: the anatomical response of Cox's Orange Pippin at different stages of development, *J. Hortic. Sci.*, 56, 145, 1981.
350. **Skorikova, Yu. G.,** Effect of flavonoids on the taste of fruits and berries. Review, *Chem. Abstr.*, 81, 150, 1974.
351. **Somers, T. C.,** The polymeric nature of wine pigments, *Phytochemistry*, 10, 2175, 1971.
352. **Somers, T. C., Verette, E. and Pocock, F.,** Hydroxycinnamate esters of *Vitis vinifera*: changes during white vinification, and effect of exogenous enzymic hydrolysis, *J. Sci. Food Agric.*, 40, 67, 1987.
353. **Sonnenbichler, J. and Zetl, I.,** Influence of the flavanolignan derivative silybin on nucleic acid and protein synthesis in liver cells, in *Flavonoids and Bioflavonoids 1985*, Farkas, L., Gabor, M., and Kallay, F., Eds., Elsevier, Amsterdam, 1986, 361.
354. **Sonnenbichler, J., Mattersberger, J., Machicao, F., Rosen, H., and Zetl, I.,** Influence of silybin on the synthesis of macromolecules in liver cells, in *Flavonoids and Bioflavonoids, 1981*, Farkas, L., Gabor, M., Kallay, F., and Wagner, H., Eds., Elsevier, Amsterdam, 1982, 475.
355. **Spayd, S. E. and Morris, J. R.,** Prediction of sensory color of strawberry jam, *Ark. Farm. Res.*, 29, 12, 1980.
356. **Spayd, S. E., Nagel, C. W., Hayrynen, L. D., and Drake, S. R.,** Color stability of apple and pear juices blended with fruit juices containing anthocyanins, *J. Food Sci.*, 1984, 49, 411, 1984.
357. **Spencer, C. M., Cai, Y., Martin, R., Gaffney, S. H., Goulding, P. N., Magnolato, D., Lilley, T. H., and Haslam, E.,** Polyphenol complexation. Some thoughts and observations, *Phytochemistry*, 27, 2397, 1988.
358. **Spurr, H. W., Hildebrandt, A. C., Ricker, A. J.,** The integral association of chlorogenic acid to crown-gall, *Phytopathology*, 55, 1004, 1965.
359. **Stanghellini, M. E. and Aragaki, M.,** Relation of periderm formation and callose depositions to anthracnose resistance in papaya fruit, *Phytopathology*, 56, 444, 1966.
360. **Steck, W.,** Metabolism of cinnamic acid in plants, chlorogenic acid formation, *Phytochemistry*, 7, 1711, 1968.
361. **Stoessl, A.,** Phytoalexins. A biogenetic perspective, *Phytopathol. Z.*, 99, 251, 1980.
362. **Stoessl, A.,** Biosynthesis of phytoalexins, in *Phytoalexins*, Bailey, J. A. and Mansfield, J. W., Eds., Blackie, Glasgow, 133, 1982.
363. **Stonier, T., Macgladrie, K., and Shaw, G.,** Studies on auxin protectors. XIV. Chlorogenic acid, a low molecular weight auxin protector in sunflower, *Plant Cell. Environ.*, 2, 79, 1979.
364. **Sullivan, J. T., Richards, C. S., Lloyd, H. A., and Krishna, G.,** Molluscicide in cashew nut shell liquid, *Planta Med.*, 44, 175, 1982.
365. **Swain, T.,** Nature and properties of flavonoids, in *Chemistry and Biochemistry of Plant Pigments*, Vol. 1, 2nd ed., Goodwin, T. W., Ed., Academic Press, New York, 1976, 425.
366. **Swain, T.,** Secondary compounds as protective agents, *Annu. Rev. Plant Physiol.*, 28, 479, 1977.
367. **Swain, T.,** Plant-animal coevolution: a synoptic view of the paleozoic and mesozoic, in *Biochemical Aspects of Plant and Animal Coevolution*, Harborne, J. B., Ed., Academic Press, London, 1978, 3.
368. **Swain, T.,** Phenolics in the environment, in *Biochemistry of Plant Phenolics, Recent Adv. Phytochemistry*, Vol. 12, Swain, T., Harborne, J. B., and Van Sumere, C. F., Eds., Plenum Press, New York, 617, 1979.
369. **Swinburne, T. R. and Brown, A. E.,** Biosynthesis of benzoic acid in Bramley's seedling apples infected by *Nectria galligena*, *Physiol. Plant Pathol.*, 6, 259, 1975.
370. **Synge, R. L. M.,** Polyphenol proteins reactions and their significance for agricultural practices, *Bull. Liaison Groupe Polyphénols*, 8, 13, 1978.
371. **Szilagyi, I., Tetenyi, P., Antus, S., Seligmann, O., Chari, V. M., Seitz, M., and Wagner, H.,** Isolation and structure of silymonin and silandrin, two new flavanolignans from *Silybum marianum* (L.) Gaertn. flore albo, in *Flavonoids and Bioflavonoids*, 1981, Farkas, L., Gabor, M., Kallay, F., and Wagner, H., Eds., Elsevier, Amsterdam, 1982, 345.
372. **Tafuri, F., Businelli, M., and Scarponi, L.,** Effect of caffeic and *p*-coumaric acids on indole-3-acetic acid catabolism, *J. Sci. Food Agric.*, 23, 1417, 1972.
373. **Takahama, U.,** Suppression of carotenoid photobleaching by kaempferol in isolated chloroplasts, *Plant Cell Physiol.*, 23, 859, 1982.
374. **Takahama, U.,** Redox Reactions between Kaempferol and Illuminated Chloroplasts, *Plant Physiol.*, 71, 598, 1983.
375. **Takahama, U.,** Suppression of lipid photoperoxidation by quercetin and its glycosides in spinach chloroplasts, *Photochem. Photobiol.*, 38, 363, 1983.
376. **Takahama, U.,** Hydrogen peroxide-dependent oxidation of flavonols by intact spinach chloroplasts, *Plant Physiol.*, 74, 852, 1984.

377. **Tan, S. C.,** Relationships and interactions between phenylalanine ammonialyase, phenylalanine ammonia-lyase inactivating systeme, and anthocyanin in apples, *J. Am. Soc. Hortic. Sci.*, 104, 581, 1979.
378. **Tan, S. C.,** Phenylalanine ammonialyase and the phenylalanine ammonia lyase inactivating system. Effects of light, temperature and mineral deficiencies, *Aust. J. Plant Physiol.*, 7, 159, 1980.
379. **Tanaka, Y. and Uritani, I.,** Purification and properties of phenylalanine ammonia-lyase in cut-injured sweet potato, *J. Biochem.*, 81, 963, 1977.
380. **Tatum, J. H. and Berry, R. E.,** Coniferin. An astringent glycoside in citrus, *Phytochemistry*, 26, 322, 1987.
381. **Thomas, P. and Janave, M. T.,** Polyphenol oxydase activity and browning of mango fruits induced by gamma irradation, *J. Food Sci.*, 38, 1149, 1973.
382. **Thomas, R. L., Morr, C. V., and Jen, J. J.,** Oxidation of indoleactic acid by a homogeneous tomato fruit peroxidase, *J. Food Biochem.*, 4, 235, 1980.
383. **Timberlake, C. F.,** Anthocyanins—occurrence, extraction and chemistry, *Food Chem.*, 5, 69, 1980.
384. **Timberlake, C. F. and Bridle, P.,** Anthocyanins, in *The Flavonoids*, Harborne, J. B., Mabry, T. J., and Mabry, H., Eds., Chapman and Hall, London, 1975, 214.
385. **Ting, S. V. and Attaway, J. A.,** Citrus fruits, in *The Biochemistry of Fruits and Their Products*, Vol. 2, Hulme, A. C., Ed., Academic Press, London, 1971, 107.
386. **Tissut, M., Chevallier, D., and Douce, R.,** Effet de différents polyphénols sur les mitochondries et les chloroplasts isolés, *Phytochemistry*, 19, 495, 1980.
387. **Tomaszewski, M.,** The mechanism of synergistic effects between auxin and some natural phenolic substances, in *Régulateurs Naturels de la Croissance Végétale, Colloques Int. CNRS*, 123, 335, 1964.
388. **Tomaszewski, M. and Thimann, K. V.,** Interactions of phenolic acids metallic ions and chelating agents on auxin-induced growth, *Plant Physiol.*, 41, 1443, 1966.
389. **Towers, G. H. N. and Yamamoto, E.,** Interactions of cinnamic acid and its derivatives with light, in *Ann. Proc. Phytochem. Soc. Europe*, Vol. 25, Van Sumere, C. F. and Lea, P. J., Eds., Clarendon Press, Oxford, 1985, 271.
390. **Tronchet, J.,** Effets de blessures, scarifications ou abrasion au carborundum sur le contenu flavonoidique de limbes de *Lycopersicum esculentum* Mill, *Ann. Sci. Univ. Besançon, Botanique*, 3e ser., 6, 16, 1969.
391. **Tronchet, J.,** Effets de traumatismes sur la peau des bananes: disparition des flavonols, *Plantes Med. Phytother.*, 5, 147, 1971.
392. **Tronchet, J.,** Les dérivés flavoniques des fruits, localisation, répartition et évolution, *Bull. Soc. Bot. Fr.*, 119, 25, 1972.
393. **Ulrich, R.,** Les réactions des végétaux supérieurs aux excitations traumatiques, *Extraits Rev. Sci.*, 5—6, 293, 1941.
394. **Ulrich, R.,** Organic acids, in *The Biochemistry of Fruits and Their Products*, Vol. 1, Hulme, A. C., Ed., Academic Press, London, 1970, 89.
395. **Ulrich, R. and Lafon, J.,** L'intensité respiratoire des fruits de tomate à plusieurs stades de leur développement et ses variations sous l'effet des blessures, *C.R. Acad. Sci. Paris*, 230, 459, 1950.
396. **Ulrich, R. and Renac, J.,** Le métabolisme des fruits de tomate et son altération sous l'effet des blessures, *C.R. Acad. Sci. Paris*, 230, 567, 1950.
397. **Uritani, I.,** Biochemistry of host responses to infection, *Progr. Phytochem.*, 5, 29, 1977.
398. **Uritani, I. and Oba, K.,** The tissue slice system as a model for studies of host-parasite relationships, in *Biochemistry of Wounded Plant Tissues*, Kahl, G., Ed., Walter De Gruyter, Berlin, 1978, 287.
399. **Van Buren, J.,** Fruit phenolics, in *The Biochemistry of Fruits and Their Products*, Vol. 1, Hulme, A. C., Ed., Academic Press, London, 1970, 269.
400. **Van Sumere, C. F., Cottenie, J., De Greef, J., and Kint, J.,** Biochemical studies in relation to the possible germination regulatory role of naturally occurring coumarin and phenolics, *Recent Adv. Phytochem.*, 4, 165, 1972.
401. **Van Sumere, C. F., Albrecht, J., Dedonder, A., de Pooter, H., and Pe, I.,** Plant proteins and phenolics, in *The Chemistry and Biochemistry of Plant Proteins*, Harborne, J. B. and Van Sumere, C. F., Eds., Academic Press, London, 1975, 211.
402. **Vance, C. P., Kirk, T. K., and Sherwood, R. T.,** Lignification as a mechanism of disease resistance, *Annu. Rev. Phytopathol.*, 18, 259, 1980.
403. **Veckenstedt, A., Guttner, J., Pusztai, R., Heinecke, H., Hartl, A., and Beladi, I.,** Antiviral, immunologic and toxicologic studies with quercetin in mice, in *Flavonoids and Bioflavonoids* 1985, Farkas, L., Gabor, M., and Kallay, F., Eds., Elsevier, Amsterdam, 1986, 339.
404. **Verette, E., Brun, S., and Cabanis, J. C.,** Etude comparative de l'activité antiradicalaire des produits de fractionnement d'un jus fermenté de myrtilles *Vaccinium myrtillus*, *Bull. Liaison Groupe Polyphénols*, 12, 312, 1984.
405. **Verette, E., Noble, A. C., and Somers, T. C.,** Hydroxycinnamates of *Vitis Vinifera*: Sensory assessment in relation to bitterness in white wines, *J. Sci. Food Agric.*, 45, 267, 1988.
406. **Vernenghi, A., Ramiandrasoa, F., Chuilon, S., and Ravise, A.,** Phytoalexines des citrus: séséline. Propriétés inhibitrices et modulation de synthèse, *Fruits*, 42, 103, 1987.

407. **Vernin, G., Bouin, D., and Metzger, J.,** The flavour components of plums and their changes during freezing, fermentation and cooking, in *The Shelf Life of Foods and Beverages,* Charalambous, G., Ed., Elsevier, Amsterdam, 255, 1986.
408. **Vioque, A., Albi, M. A., and Vioque, B.,** Role of IAA-oxidase in the formation of ethylene from 1-aminocyclopropane-1-carboxylic acid, *Phytochemistry,* 20, 1473, 1981.
409. **Vorsa, N. and Welker, W. V.,** Relationship between fruit size and extractable anthocyanin content in cranberry, *HortScience,* 20, 402, 1985.
410. **Wacker, A. and Eilmes, H. G.,** Antivirale wirkung von Planzeninhaltsstoffen. I. Flavonoids, *Arzneim. Forsch. (Drug Res.),* 28, 347, 1978.
411. **Wagner, H.,** The antihepatotoxic principle of Silybum marianum Gaertn. in *Recent Flavonoid Research, Recent Development in the Chemistry of Natural Carbon Compounds,* Vol. 5, Bognar, R., Bruckner, V., and Szantay, Cs. Eds., Akademiai Kiado, Budapest, 1973, 51.
412. **Wagner, H.,** Flavonolignans from *Silybum marianum,* in *Bull. Liaison Groupe Polyphénols,* 5, 1974.
413. **Wagner, H.,** New structures in the flavonoid series, in *Flavonoids and Bioflavonoids Current Research Trends,* Farkas, L., Gabor, M., and Kallay, F., Eds., Elsevier, Amsterdam, 1977, 29.
414. **Wagner, H.,** Phenolic compounds in plants of pharmaceutical interest, in *Biochemistry of Plant Phenolics,* Swain, T., Harborne, J. B., and Van Sumere, C. F., Eds., Rec. Adv. Phytochem., 12, Plenum Press, New York, 1979, chap. 18.
415. **Wagner, H.,** New plant phenolics of pharmaceutical interest, in *Ann. Proc. Phytochem. Soc. Eur.,* Vol. 25, Van Sumere, C. F. and Lea, P. J., Eds., Clarendon Press, Oxford, 1985, 409.
416. **Wagner, H. and Fiebig, M.,** Neue antihepatotoxisch wirksame flavonolignane auseiner weipblühenden *Silybum*-Värietat, *Planta Med.,* 51, 310, 1984.
417. **Wagner, G. J. and Hrazdina, G.,** Endoplasmic reticulum as a site of phenylpropanoid and flavonoid metabolism in *Hippeastrum, Plant Physiol.,* 74, 901, 1984.
418. **Wagner, H., Seligman, O., and Lotter, H.,** Recent advances in the chemistry, stereochemistry synthesis and pharmacology of flavanolignans, in *Flavonoids and Bioflavonoids 1985,* Farkas, L., Gabor, M., and Kallay, F., Eds., Elsevier, Amsterdam, 1986, 139.
419. **Walker, J. R. L.,** Inhibition of the apple phenolase system through infection by *Penicillium expansum, Phytochemistry,* 8, 561, 1969.
420. **Walter, W. M., Fleming, H. P., and Etcheus, J. L.,** Preparation of antimicrobial compounds by hydrolysis of oleuropein from green olives, *Appl. Microbiol.,* 26, 773, 1973.
421. **Wardale, D. A.,** Effect of phenolic compounds in *Lycopersicom esculentum* on the synthesis of ethylene, *Phytochemistry,* 12, 1523, 1973.
422. **Watada, A. E. and Abbott, J. A.,** Objective method of estimating anthocyanin content for determining color grade of grapes, *J. Food Sci.,* 40, 1278, 1975.
423. **Wattenberg, L. W.,** Inhibitors of chemical carcinogens, in *Cancer Achievements, Challenger and Prospects for 1980's,* Vol. 1, Burchenal, J. H. and Oettgen, H. F., Eds., Grune & Stratton, New York, 1981, 517.
424. **Weissenbock, G. and Schneider, V.,** Untersuchungen zur Lokalisation von Flavoiden in Plastenden. III. Das Problem der Kontamination isolierter Plastiden durch Flavonoide, *Z. Pflanzenphysiol.,* 72, 23, 1974.
425. **Williams, A. and Hrazdina, G.,** Anthocyanins as food colorants: effect of pH on the formation of anthocyanin-rutin complexes, *J. Food Sci.,* 44, 66, 1979.
426. **Williams, A. W.,** Chemical evidence from the flavonoids relevant to the classification of *Malus species, Bot. J. Linn. Soc.,* 84, 31, 1982.
427. **Williams, E. B. and Kuc, J.,** Resistance in Malus to *Venturia inaequalis, Annu. Rev. Phytopathol.,* 7, 223, 1969.
428. **Williams, R. D. and Hoagland, R. E.,** The effects of naturally occurring phenolic compounds on seed germination, *Weed Sci.,* 30, 206, 1982.
429. **Wilson, E. L.,** High-pressure liquid chromatography of apple juice phenolic compounds, *J. Sci. Food Agric.,* 32, 257, 1981.
430. **Wollenweber, E. and Mann, K.,** Flavonole aus Fruchten von *Vitex agnus castus, Planta Med.,* 48, 126, 1983.
431. **Wuersch, P., Rosset, J., and Smiley, M.,** Les tanins condensés de la gousse mûre de *Ceratonia silica* L., *Bull. Liaison Groupe Polyphénols,* 10, 328, 1980.
432. **Yamamoto, E. and Towers, G. H. N.,** Cell wall bound ferulic acid in barley seedlings during development and its photoisomerization, *J. Plant Physiol.,* 117, 441, 1985.
433. **Yang, S. F.,** The role of ethylene and ethylene synthesis in fruit ripening, in *Plant Senescence: Its Biochemistry and Physiology,* Thompson, W., Nothnagel, E. A., and Huffaker, R. C., Eds., The American Society of Plant Physiologists, Rockville, MD, 1987, 156.
434. **Yang, S. F. and Hoffman, N. E.,** Ethylene biosynthesis and its regulation in higher plants, *Annu. Rev. Plant Physiol.,* 35, 155, 1984.
435. **Yoshii, H. and Imaseki, H.,** Biosynthesis of auxin-induced ethylene. Effect of indole-3-acetic, benzyladenine and abscissic acid on endogeneous levels of 1-aminocyclopropane-1-carboxylic acid (ACC) and ACC synthetase, *Plant Cell Physiol.,* 22, 369, 1981.

436. **Yoshitama, K.,** Blue and purple anthocyanins isolated from the flowers of *Tradescantia reflexa, Bot. Mag.*, 91, 207, 1978.
437. **Zucker, M.,** Light and enzymes, *Annu. Rev. Plant Physiol.*, 23, 133, 1972.

Chapter 5

PHENOLIC COMPOUNDS IN FRUIT PROCESSING

I. INTRODUCTION

As discussed above, phenolic compounds contribute to the characteristic flavor of ripe fruits. However, involvement is not limited to the whole fruit; they also play a large role in the acquisition of the sensory properties of fruit products. Schematically, the origin of the phenomenon is at two levels. First, the direct participation of phenolic components in fruit qualities is also found in processed products where certain characteristics (color, bitterness, astringency, etc.) are directly dependent on their phenolic contents. In this case, the stability of the molecules concerned is frequently very different from that *in situ* and depends on their immediate physicochemical environment and in particular on the oxygen content of the medium. In addition, the cellular decompartmentation generally associated with processing may cause reactions — enzymic or not — which lead either to the formation or to the transformation and degradation of phenolic compounds. Browning, which occurs frequently, is a very common example.

In all cases, the phenolic composition of the product of processing is different from that of the original fruit. It may display new chemical and organoleptic properties which, depending on the case and the nature of the product, may be considered to be favorable or unfavorable. Some of the biochemical modifications involved may already have started in the whole fruit, for example during cold storage, but they are generally only fully expressed during processing itself; modifications may continue during the storage or aging of the fruit product. In this chapter, examined are the principal modifications related to the presence of phenolic compounds during the processing of fruits, with particular stress on fruit beverages.

II. ENZYMATIC BROWNING OF FRUITS

The organoleptic and biochemical characteristics of fruits may be profoundly modified by the appearance of brown pigments whose color is superposed on the natural colors. With a few exceptions (medlars, prunes, raisins, dates, figs, etc.), these changes are considered as degradation and lower fruit quality both visually and with regard to taste and nutritional characteristics. Some browning may be nonenzymatic in origin and result from the Maillard reaction when mixtures of amino acids and carbohydrates are heated.[4] However, most browning, and particularly the types that occur very rapidly, are caused by enzymatic oxidation of phenolic compounds under the effect of polyphenol oxidase.[165,217,219] The first products of the reaction are quinones, colored substances which then rapidly condense and do or do not combine with amino or sulfhydryl groups of proteins.[218] Relatively insoluble brown polymers then form a substance whose chemical composition has not yet been well defined.[53,269,392,426] The formation of quinones, catalyzed by phenoloxidases, is reversible in the presence of reducing agents, such as ascorbic acid, whereas subsequent polymerization is nonenzymatic and irreversible.

Peroxidases may also be involved in the oxidation of a large number of phenols and are responsible for the development of off-flavors and off-odors in foods.[51] Although these enzymes play an important role in plant and fruit physiology,[206,377] they generally appear to be little involved in the browning reaction. Indeed, with the exception of the role that they play in the internal browning of pineapple,[360,362] their activity is rarely correlated with browning as in olive[309] and apricot.[382]

Three different situations may cause browning in fruits: first, in a few cases, physiological evolution related to maturation (e.g., in dates); second, certain disorders which may occur during cold storage; and third, various technological processes which involve wounding, crushing, extraction etc. Many of these problems have been known for a long time, but they are still topical and have even become more important with the increasing use of mechanical equipment in the harvesting and packaging of fruits and the recent marketing of new, ready-prepared plant products. There is a common feature in all cases: contact between phenolic compounds with predominantly vacuolar localization (Chapter 1) and oxidation enzymes located in the cytoplasm.[225] This modification of compartmentation is needed (in the presence of oxygen) for browning to be triggered.

Although browning is generally considered to be harmful, with the exceptions mentioned above, browning products may contribute to resistance to certain stresses. Thus, the appearance of browning is one of the first signs of response to wounding or to fungal attack (Chapter 4) and the *o*-quinones formed appear to possess more effective antimicrobial properties than the original compounds and participate in resistance mechanisms.[28,147]

The tendency of a fruit to turn brown results from the action of a number of parameters which are naturally involved in limiting browning and its intensity:[217] nature and content of substrates, activity of oxidative enzymes, availability of oxygen, and compartmentation of enzymes and substrates. Several other factors are more secondary but may play an important role in certain cases, e.g., reducing compounds, such as ascorbic acid and glutathione, pH and temperature, existence of protective systems against toxic oxygen (scavengers, superoxide dismutase activity, etc.) and of various cofactors (metal ions, etc.). All these parameters vary with the age of the fruit, its physiological state and its health, and also in relation to the variety and the treatments to which it is subjected.

Excellent reviews on plant phenoloxidases, phenolic compounds involved in browning, and the control of browning by various treatments are available,[217,223,225,226,377,407] and only aspects concerning fruit browning are reported here.

A. *o*-DIPHENOL OXIDASE (*o*-DPO) IN FRUITS

The term "polyphenol oxidases", which is commonly used, refers to several enzymes whose main characteristics is their capacity to use molecular oxygen during the oxidation of phenolic compounds. Numerous reviews have covered work on these enzymes.[52,53,223,225,226,377] They may be divided into two main groups in relation to their substrate specificity and the nomenclature reported here is that of the new International Enzymes Nomenclature:[223] (1) *catechol oxidase* (EC 1.10.3.2) or *o*-diphenol: oxygen oxidoreductase (*o*-DPO) catalyzes two distinct reactions (Figure 102): (a) hydroxylation of a monophenol often referred to as cresolase activity (EC 1.14.18.1), generally followed by oxidation of the diphenol obtained to the corresponding quinone, and (b) the oxidation of the *o*-diphenols referred to as catecholase activity (EC 1.10.3.2), often called phenolase or polyphenol oxidase. To avoid any ambiguity, we avoid the use of these terms and have kept only ***o*-diphenol oxidase** (*o*-DPO). (2) *Laccase* (EC 1.10.3.1) or *p*-diphenol: oxygen oxidoreductase oxidizes numerous substrates but can oxidize *p*-diphenols (Figure 102). Use of the different properties of the enzymes may provide a simple and convenient procedure for the differentiation of *o*- and *p*-diphenol oxidases.[409]

Most of the polyphenol oxidases involved in fruit browning are *o*-diphenol oxidases, except in peach where the simultaneous presence of *o*- and *p*-diphenol oxidases has been reported.[129,224] In this case, catechol oxidase is present in a particulate fraction, whereas laccase is in a soluble fraction. Nevertheless, this laccase activity found in peaches grown in Israel was not detected in other varieties of peaches grown in the U.S.[160,201] Both enzymes are also present in mango fruit, but *o*-DPO is located in the flesh whereas laccase is present in the secretory ducts from cut pedicel.[162] In banana, the presence of laccase[83] is very ambiguous since some authors report only *o*-DPO.[253]

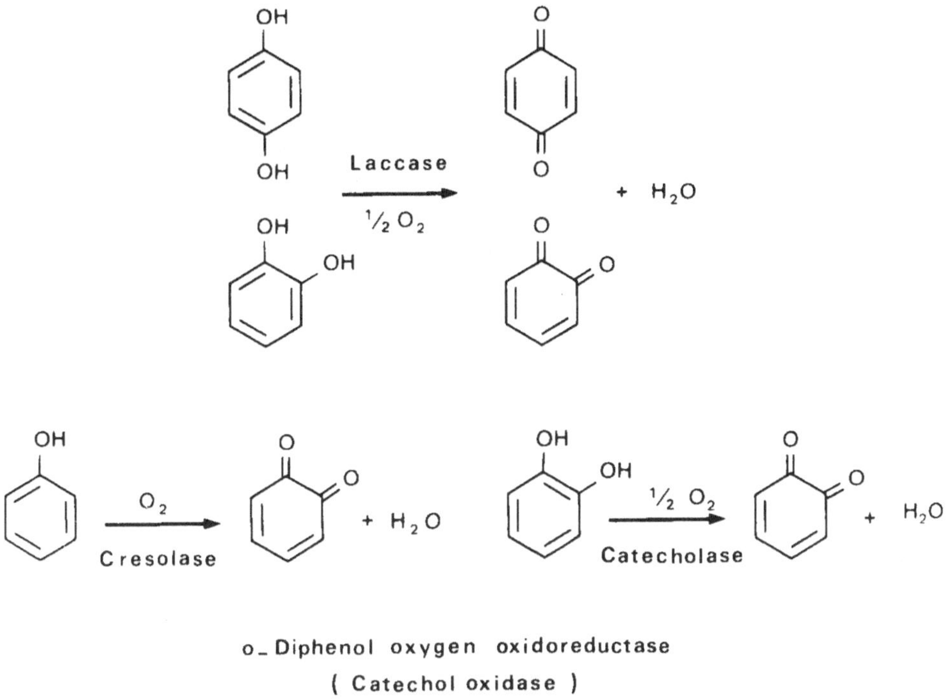

FIGURE 102. Mechanism of enzymatic oxidation of mono and diphenols. (Adapted from Mayer, A. M. and Harel, E., in *Recent Advances in the Biochemistry of Fruits and Vegetables*, Academic Press, London, 1981, 161.)

Cresolase activity has been shown in several fruits, such as apple,[109,204] avocado,[172] banana,[72,365] kiwi,[255] eggplant,[312] grapes,[127,197,376] and cherimoya.[215] The role of this enzymatic activity in the biosynthesis of *o*-diphenolic compounds is controversial and is discussed in Chapter 3.

The physicochemical properties of *o*-DPO and the reactional mechanisms involved are not described here as they are covered at length in the reviews mentioned above. We report only a number of data concerning the location of these enzymes and their variation during the life of fruits insofar as they contribute to browning density.

1. Tissue and Subcellular Localization

One of the main causes of the diversity of *o*-DPO activities lies in their localization. Indeed, measurements of *o*-DPO activity may be totally different depending on the tissue studied or the type of extraction used.

In banana[102,234,250] and in plum[282] *o*-DPO activity (expressed in units per milliliter or in units per milligram protein) is greater in the pulp than in the outer part of the fruit, i.e., plum epicarp or banana peel. In the latter fruit, enzyme preparation from pulp tissue displayed more cresolase activity than skin whereas skin had more catecholase activities.[365] In addition, *o*-DPO activity was two to three times as high in interior pulp than in exterior pulp whatever the physiological stage of the fruit.[234] The opposite applies in apple[130] and in pear[292] where *o*-DPO activity is substantially higher in the peel that in the flesh. In kiwi, the situation differs according to the method of expression used: in units per milliliter, *o*-DPO activity is higher in flesh (52,000 units/ml) than in skin (38,600 units/ml) whereas when expressed in specific activity (units/mg of protein) it is three times higher in the skin (160,900 and 59,100 units/mg of protein, respectively).[255] In mango, distribution in flesh and pulp is related to the physiological stage. The *o*-DPO level is higher in the peel in young fruits and then falls, increasing in the pulp.[164]

o-DPO has been located in a variety of cell fractions in plants: organelles (chloroplasts, mitochondria, peroxisomes) where it may be tightly bound to membranes, and in the soluble fraction of the cell.[225] However, many experiments, including immunocytochemistry in particular, agree that active *o*-DPO is located exclusively in plastids in healthy cells.[398] In fruits, subcellular localization of *o*-DPO has mainly been reported in chloroplasts, as in olive,[32,320] apple,[126,130,131] and grape,[127,196] but it has also been found in mitochondria, e.g., in apple,[131] or in microbodies in avocado.[313]

In most cases, however, this location has not been investigated and only "soluble" and "bound" or "particulate" forms defined by the method of extraction are discussed. Extraction is carried out either by simply using a buffer and the activity obtained is thus from soluble forms, or it is carried out in the presence of detergents which enable the solubilization of bounded forms. However, the increase in enzyme activity in the presence of detergent has also been attributed to activation caused by the detergent[102] and the disruption of chloroplast membranes, releasing more enzyme.[254] Extractions carried out on acetone powders give much greater activity as in peach for example, where it is 20 times as high.[96] This is why this method is often used in the study of *o*-DPO even though it probably leads to changes in enzyme structure.[32]

In most fruits, the insoluble part of the enzymes was found to be dominant and the activity of the soluble part amounted to 10 to 30% of the total value depending on the fruit. The simultaneous presence of "soluble" and "bound" forms has been described in olive,[32] banana,[159] avocado,[313,391] apple,[226] apple tissue culture,[401] peach,[96] grape,[304,305] eggplant,[312] and plum.[384]

There are also modifications which lead to change from "bound" to "soluble" forms. This may happen either during maturation (Section 2) or after stress, e.g., water stress and restricted ventilation for avocado fruit.[42,69] Thus, rapid dessication or deficiency in growth substances cause a change in the subcellular distribution of enzyme in tissue and suspension cultures of apple fruit.[401,402] Cell localization may also depend on the variety studied. In four banana varieties, *o*-DPO activity was partly in the soluble fraction and partly associated with the cell wall, whereas in another variety *o*-DPO was found entirely in the soluble fraction.[159] This may account for the contradictions encountered in this fruit in which *o*-DPO activity has been reported as being either soluble[250] or bound.[102,253]

Finally, the possible existence of latent forms should be added to these causes of variations in *o*-DPO activity. In avocado, for example, much greater activity is obtained with the addition of SDS (sodium dodecyl sulfate), since it is probable that the latent forms of the enzyme are revealed.[171] In this case, four instead of two forms of the enzyme are present: soluble/bound and active/latent. Several types of treatment reveal the latent forms: trypsin or other proteolytic enzymes, fatty acids, detergents, denaturants, and acid treatment.[128] However, it is possible that there is simultaneous alteration of the enzyme pattern.[13]

2. Changes during the Development of Fruits

o-DPO activity is generally more important in young green fruits than in ripe fruits (Figures 103 and 104). This has been observed in apple,[96,129] olive,[32,111] grape,[177,304] plum,[384] banana,[234] mango,[164] tomato,[140] and coffee berries.[117] However, in banana, *o*-DPO activity and concentration in the pulp showed an increase from unripe to ripe and declined only in overripe fruit.[156] In mango, changes in *o*-DPO varied in relation to localization in tissue, as we have mentioned previously, but very little or no activity was detectable in ripe mango, either in pulp or peel tissues.[164] The changes are not the same for the different forms of *o*-DPO and there is generally a change from "bound" to more "soluble" forms during maturation. In olive, changes in enzyme solubility coincide with the appearance of anthocyanins and it is probable that chloroplast membranes are disrupted at this stage, facilitating release of the enzyme.[32] In apple, when particulate *o*-DPO decreases during maturation the

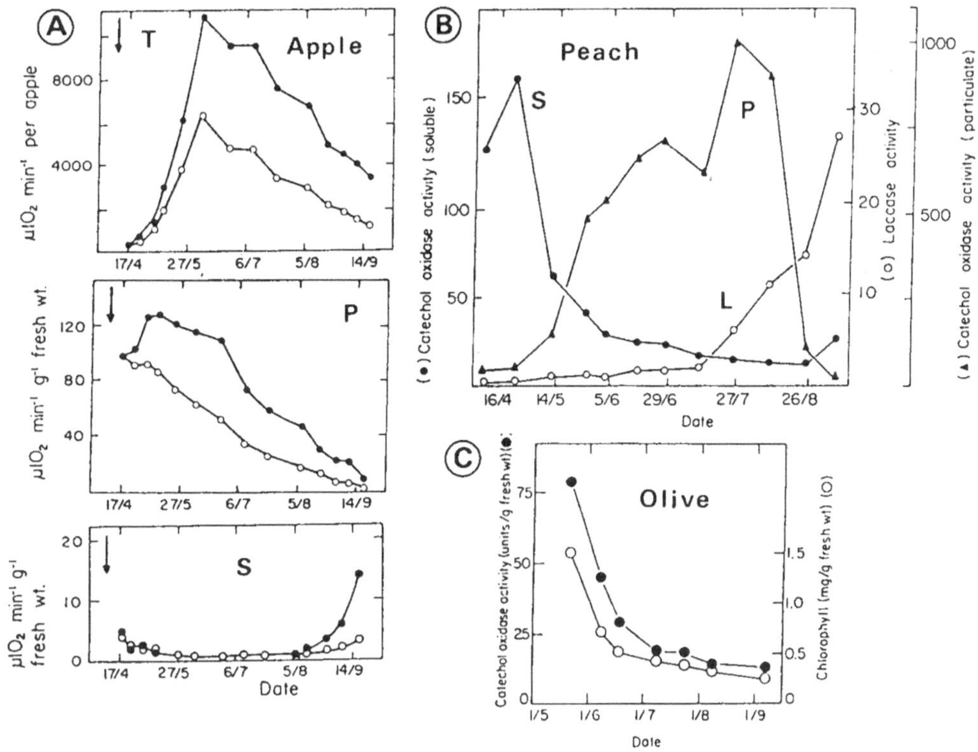

FIGURE 103. Changes in *o*-DPO (apple, peach, olive) and laccase activities (peach) during fruit development; *o*-DPO: T total, P particulate, S soluble. Laccase: L. (From Harel, E., et al., *J. Sci. Food Agric.*, 21, 542, 1970; Harel, E., et al., *J. Sci. Food Agric.*, 17, 389, 1966; Ben Shalom, N., et al., *Phytochemistry*, 16, 1153, 1977. With permission.)

soluble form increases, but its activity in ripe fruits is only about 17% of the maximum activity observed in young fruits (Figure 103).[132] The same applies to grape where the maximum soluble enzyme is attained when the fruit is ripe in both red and white cultivars (Figure 104).[304] This activity forms only 4 to 5% of the bounded activity.

Different evolutions are found in certain fruits. Thus, the activity in the particulate fraction of peach increases during maturation (Figure 103). However, it falls considerably at picking and becomes lower than soluble activity.[129] In avocado, more *o*-DPO activity is associated with the particulate fraction relative to the supernatant fraction as the fruit ripens.[314]

Differences in the variations of *o*-DPO activity during the development of the fruit according to the year (Figure 104) and to the expression used may also be observed.[304] In addition, isoenzyme patterns of oxidative enzymes are reported to change during fruit maturation with the diappearance of some isoenzymes at various stages of ripening, thus suggesting that they play a role in the ripening process.[55,156]

In tissue culture of apple fruit, *o*-DPO activity may be compared to that reported for intact apples with low initial activity which then increases and falls again with time. As in intact apple, the decline in activity is accompanied by a tendency for the enzyme to appear in the soluble fraction of the cell.[401]

B. NATURAL SUBSTRATES OF *o*-DIPHENOL OXIDASE

4-Methylcatechol and catechol are substrates frequently used to measure *o*-DPO activity. However, these phenolics are not natural substrates since they do not accumulate in fruits even though the enzyme often has the strongest affinity for them. Catechin, epicatechin and

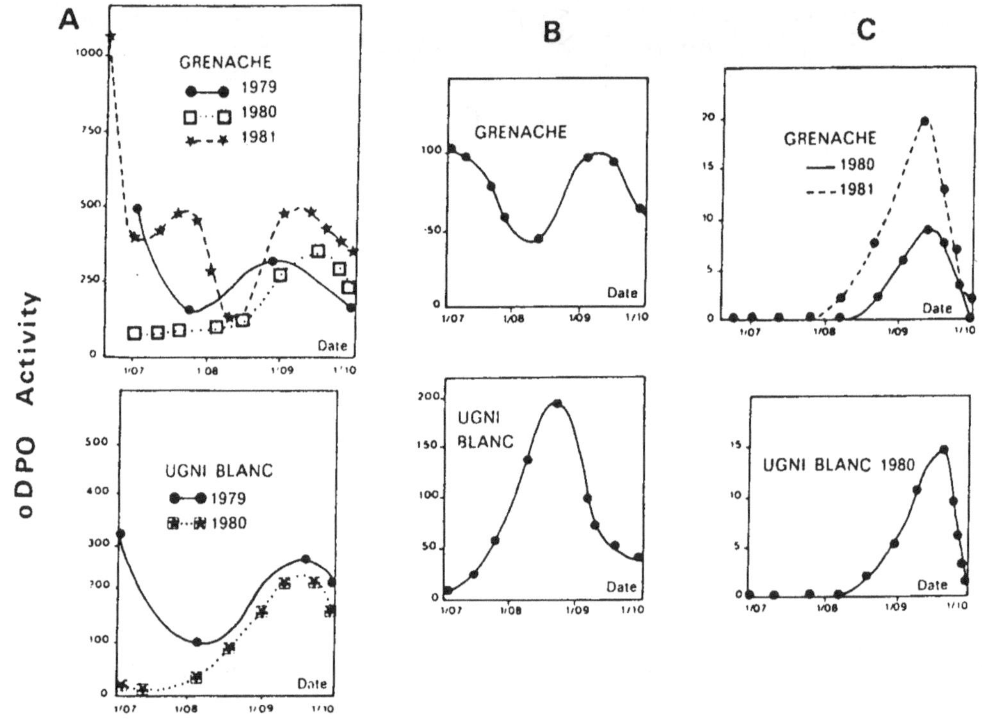

FIGURE 104. Changes in *o*-DPO activity during development and maturation of two grape cultivars. A: solubilized activity as μmol O_2/min/gFW; B: solubilized activity as μmol O_2/min/protein; C: crude soluble activity as μmol O_2/min/g FW. (From Sapis, J. C., Macheix, J. J., and Cordonnier, R. E., *J. Agric. Food Chem.*, 31, 342, 1983. With permission.)

caffeic acid derivatives appear to be the best substrates (Table 51). Chlorogenic acid and its isomers, in numerous fruits, and other caffeic esters (caffeoylshikimic ester in date, caffeoyltartaric ester in grape) are easily degraded by *o*-DPO.[120,134,244,325] It seems that esterification of the carboxyl group of caffeic acids with quinic acids (Table 51) or with tartaric acids[120] leads to an increase in *o*-DPO activities in practically all cases. Chlorogenic acid is not present in grape, but although it is little or not accepted by *o*-DPO extracted from certain cultivars,[55,151] it is in contrast a good substrate *in vitro* fo: others.[127,242] (+)-Catechin is often oxided more rapidly than chlorogenic acid and also appears to be a very good substrate for *o*-DPO as in peach,[201] kiwi,[255] and avocado;[277] (−)-epicatechin is sometimes better accepted than (+)-catechin[151] and contributes more to browning than chlorogenic acid and (+)-catechin. Nevertheless, the browning rates of caftaric and chlorogenic acid added to grape musts are higher than that of catechin.[244]

Substrates with smaller substituent groups on the ring of diphenols were oxidized at a faster rate than those with larger substituent groups by banana[234] and cherry[33] *o*-DPOs. This shows that the nature of the side chain is critical and may, therefore, play a part in the enzyme-substrate interaction. Km were very different when calculated for the same substrate but for *o*-DPOs from a variety of fruits;[377] it is, therefore, very difficult to make a general statement on the affinity of these enzymes for natural substrates. In addition to the question of plant origin, these differences may be due to the degree of purification of the enzymic solution. When enzyme purification is taken as far as separation into several isoenzymes, it has been possible to isolate a specific isoenzyme from one or more substrates. In kiwi, a monophenol-specific cresolase activity was separated from other isoenzymes. Likewise, in banana, several isoenzymes separated by electrophoresis displayed different specificities for substrates: nine were active with catechol, dopamine, DL-DOPA, protocatechuic, and chlo-

TABLE 51
o-DPO Relative Activity (%) with Common Natural Substrates

Fruits	Caffeic acid	Chlorogenic acid	Catechin	Epicatechin	Ref.
Apple	50	100	60		404,405
		89.7[b]	32.6		352
Banana					
Interior	5.9[a]	7.6	19		
Exterior	3.1[c]	4.7	10.3		234
Peel	2[a]	2	19.3		
Cherry	119[a]	142			185
		31[c]			33
Date	2.7		4	100	134
Eggplant	312.5[a]	0.38—19.5[e]	45.9		295,312
Grape	48[a]		58		337
	0[a]	0	4	87	151
	106[a]	154	98		242
	100	52.3—55.4[f]	50.8—82.6		425
	54[b]	134		56	127
	90[a]	3	75		55
	100	51	21		195
Kiwi	24[a]	2	1266	287	255
Mango	85[d]	81			164
Olive		74[b]	8.6		32
Peach	41[a]	67	250		201
	0—39[a,f]	22—81	32—145		427
Pear	43	100	60		404,405
		72[c]	7.8		292

[a] 100% with Catechol.
[b] 100% with 4-Methylcatechol.
[c] 100% with Pyrogallol.
[d] 100% with Dopamine-HCl.
[e] Values from two different authors.
[f] Values for two cultivars.

rogenic acids, seven with caffeic acid and catechin, two with pyrogallol and none with tyrosine.[157] In contrast, it was not possible to distinguish between isoenzymes in several varieties of the same species on the basis of their banding pattern or the relative intensities of the bands with various substrates. This was observed for several cultivars of banana,[157] avocado,[170] pear,[123] and peach.[257]

Other phenolic compounds present in large quantities in fruits do not appear to be direct substrates of *o*-DPO, but their degradation is strongly increased through coupled oxidation phenomena. Apple procyanidins are not themselves substrates for *o*-DPO but are instead oxidized by a chlorogenic acid $\rightleftharpoons$ chlorogenoquinone redox shuttle.[190] In a model system, the procyanidin became oxidized while the amount of chlorogenic acid remained almost constant. Anthocyanins can be destroyed directly by *o*-DPO but remaining fairly poor substrates for these enzymes,[212] but attack is generally more vigorous when other phenolic compounds which are better *o*-DPO substrates than anthocyanin are also present in the system. Thus, enzymatic oxidation of chlorogenic acid may be combined with non-enzymatic degradation of anthocyanins (Figure 105). Destruction of anthocyanins by reaction with the product of enzymatic oxidation of chlorogenic acid has been shown in eggplant,[299] sweet cherry,[270] and d'Ente plum.[282] The same phenomenon has also been observed in the presence of catechin.[270] In cider apple fruit, the product of oxidation of phloridzin is a dimer produced by initial hydroxylation to 3-hydroxyphloridzin followed by multistage oxidation, and the final product is a stable dimeric quinone.[109,190]

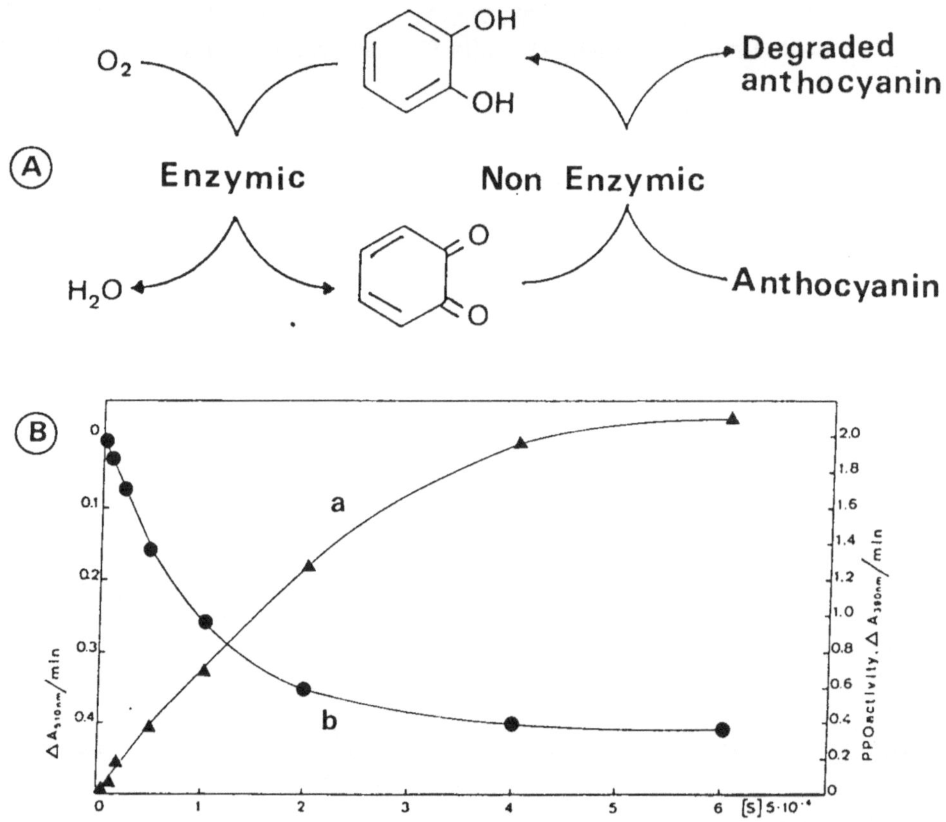

FIGURE 105. Anthocyanin degradation by *o*-DPO. A: possible mechanism. B: saturation graph for chlorogenic acid (a) and cyanidin-3-glucoside degradation (b) as a function of the chlorogenic acid concentration. (From Pifferi, P. G. and Cultrera, R., *J. Food Sci.*, 39, 786, 1974. With permission.)

Although flavonols and their derivatives are abundant in fruits, they are very poor *o*-DPO substrates. In apple, activity of the enzyme in relation to quercetin represents only 4% of that with chlorogenic acid.[352] In orange, relative enzyme activity is 19% with quercetin whereas it is 39 and 12% with catechin and chlorogenic acid, respectively, maximum activity being obtained with L-DOPA.[47] Activity is the same (19%) with quercitrin (quercetin-3-rhamnoside) and weaker (8%) with rutin (quercetin-3-rutinoside). It is, nevertheless, possible that chlorogenic acid and catechin enable and accelerate the oxidation of flavonol glycosides.[138] To this poor affinity of *o*-DPO for flavonols is added the fact that these compounds, together with flavones, are considered to be powerful antioxidants.[186,374] In blackcurrant juice, quercetin, kaempferol, and rutin inhibit oxidation of ascorbic acid and thus exert a protective effect against browning and thus enable color stabilization.[133] Other phenolic compounds found *in vivo* in fruits and which are very close to *o*-DPO substrates do not contribute to browning and even act as inhibitors. These include in particular substituted benzoic and cinnamic acids, such as *p*-coumaric, ferulic, and sinapic acids. These inhibited *o*-DPO from apple,[408,411] pear,[292] sweet cherry,[271] banana pulp,[409] and grape.[120] However, this is not the case of cherimoya *o*-DPO, probably because of the weak concentrations used.[216] Inhibition by these acids is even a characteristic of *o*-DPO which can be used to distinguish them from laccases which are not inhibited by these compounds.[409] Their inhibitory action may operate in case of stress. Thus, after infection of apple by *Penicillium expansum,* the presence of even very small amounts of *p*-coumaric and ferulic acids at the site of penetration may account for the absence of browning around the site of infection.[406]

TABLE 52
Experimental Conditions in Fruit Browning Measurement

	Actual browning	First potential browning	Second potential browning
Fruit powder (g)	1	1	1
Buffer: phosphate-citrate 0.1 *M* pH 4.5 (ml)	50	50	50
Triton X-100 (ml)	0	0.3	0.3
4-methylpyrocatechol 0.2 *M* (ml)	0	0	2
Homogenizing time at 0°C (s)	30	30	30
Shaking time exposed to air at 20°C (min)	30	30	30
Centrifuging time at 40,000 g (min)	10	10	10

However, as we stated in Chapter 1, these acids are not found in a free state in fruits and the role of their esterified forms with acids or sugars with regard to browning is not yet known.

The color of the products of enzymatic oxidation depends on the phenolic compounds used. Few substrates naturally present in fruits have been studied. With catechin, color development is bright orange although caffeic acid was oxidized to a brown compound but of a light shade. Protocatechuic acid also have a light color reminiscent of rosé wine. Dihydroquercetin was initially colorless but oxidized rapidly to a bright yellow compound whereas the color measurement for quercetin, which nevertheless contains *o*-dihydroxy groups, indicated that oxidation had not occurred.[358] The colors produced from mixtures of various compounds suggested that in some cases true interaction had taken rather than two colors combining to make a different shade.

It is clear that the phenolic substrates present in fruits and their quinones formed after oxidation may have a synergic effect (combined reaction) or be antagonistic (*o*-DPO inhibitors, antioxidants). The contribution of a given endogenous substrate depends on its concentration and on the nature of the other substrates present in the tissue. It is thus desirable that future studies concerning browning substrates should investigate combined reactions between natural compounds and not their aglycones.

C. EVALUATION OF BROWNING

In order to evaluate the browning potential of a fruit, the enzymes involved and their substrates should be put into close contact by grinding and the resulting coloration measured. There are two possible types of approach: (1) those based on measurements of reflectance using a differential tristimulus colorimeter (Hunterlab type). These measurements can be carried out either directly on cut surfaces,[30,302] on mixed fruit pulp[71,156-158] or using concentrated juice,[35] but the quantification of browning is always delicate; (2) Those based on measurements of absorbance of solutions after purification by centrifugation and filtration. The wavelengths at which these measurements are made are characteristic of the substrates and their quinones. They range from 300 to 495 nm for seven *o*-diphenolic substrates tested, but those most commonly used are around 400 nm, corresponding to maximum absorption of catechol or 4-methylcatechol quinones.[403] The estimated molar extinction coefficient of these quinones was calculated for different pH values and varied from about 1000 to 9000 M^{-1} cm^{-1}.[403] Quantification is easy in this case but a significant proportion of brown polymers, which are less and less soluble with time and part of which remains bound to particles, is eliminated in the centrifugation pellets.

In order to obtain browning values which are as representative as possible, it is interesting to compare three different experimental conditions (Table 52). (1) The first is simple homogenization of finely ground tissue in an appropriate buffer (whose pH corresponds to maximum browning, which is generally that of maximum activity of the enzyme involved).

The figure obtained is the *actual browning* since only soluble enzymes and endogenous substances are taken into account. (2) A detergent (e.g., Triton X-100) can then be added to the medium to enable the solubilization of bounded *o*-DPO. The browning value obtained then reflects the *first potential browning.* (3) If an exogenous substrate is then added to the medium (2), the *second potential browning* can then be determined. This substrate may be a compound naturally present in the fruit or chosen to cause maximum browning. The latter procedure represents an approximate evaluation of potential *o*-DPO activity insofar as the substrates are no longer limiting and enzymatic activity is expressed in the presence of inhibitors or activators that may be present *in vivo*. These three measurements of browning are thus complementary and make it possible to obtain a satisfactory evaluation of the browning potential of a specific fruit. Thus in peach,[50] grape,[305] olive,[111] and apple,[203] "second potential browning" is always distinctly greater than "actual" and "first potential" browning. It thus appears that at a given physiological stage, the browning capacity of a fruit is always limited by the oxidable substrates content rather than by *o*-DPO activity. However, as will now be seen, the substrate content is not always directly correlated with browning when comparison is made between fruits at different physiological stages or between fruits of different cultivars of the same species.

D. RELATIONS BETWEEN BROWNING, *o*-DPO, AND SUBSTRATES

1. Changes during Growth and Maturation

The time of picking of a fruit may be determinant for the browning intensity that may occur immediately after harvesting or during storage or technological processing. In a general manner, tendency towards browning is particularly high in young fruits and then falls rapidly to a low level during maturation. Changes in phenolic compounds during growth and maturation of fruits are discussed in Chapter 3 and changes in *o*-DPO activities are examined above. However, the relative influence of the different browning parameters may evolve with time.

In apple, (Figure 106), influence of the two main factors in browning reach maximum successively during growth. First, the chlorogenic acid concentration appears to be determinant for about 3 weeks. *o*-DPO activity then become preponderant for a similar length of time. The limiting factor for browning is thus first the oxidable substrates content and then — when the latter is sufficiently high — *o*-DPO activity.[203]

In olive, tendency towards browning reaches a maximum at the end of growth ("green maturation") just when fruit is picked for use as "green table olives".[111] This is a considerable problem with regard to the mechanization of harvesting. Risk of browning then falls strongly during maturation as soon as the fruits change color as a result of accumulation of anthocyanins. As in apple, the two major factors in browning appear to be consecutive. *o*-DPO activity is fairly low and the substrate content very high in young fruits whereas the latter falls considerably at the end of growth and *o*-DPO activity increases.

In banana, pulp browning potential was found to increase from unripe to ripe and then decline or increase steadily depending on the variety.[156,157] A combination of factors involving *o*-DPO, ascorbic acid, and pH influenced the susceptibility of pulp to browning at any one stage of maturation, and it was not possible to assign an exclusive role to a particular factor. Total phenolic content did not appear to affect browning,[158] but these results contrast with other work in which a correlation was found between the dopamine concentration and increase in browning during ripening of banana.[414]

2. Intervarietal Comparisons

In order to orient selection towards obtaining cultivars displaying minimum browning, correlations are frequently sought between browning intensity and the main parameters

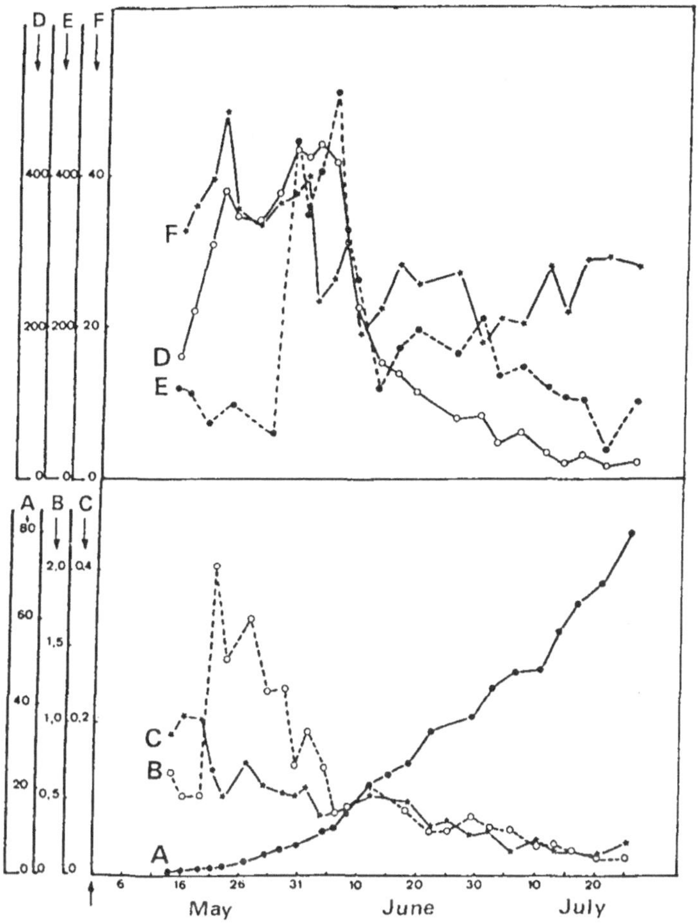

FIGURE 106. Changes during growth of apple fruit. A: fresh weight (g); B: browning (Δ A_{485} g^{-1} FW); C: *o*-DPO activity (Δ A_{420} min^{-1} g^{-1} FW); D: chlorogenic acid level (mg 100 g^{-1} FW); E: catechins level (mg 100 g^{-1} FW); F: ascorbic acid level (mg 100 g^{-1} FW). (Adapted from Macheix, J. J., *Physiol. Vég.*, 8, 585, 1970.)

responsible. The results are often contradictory and there is no simple answer to the problem, as can be seen in Table 53.

There have been few intervarietal comparisons in which there is both good correlation between browning and substrates, on the one hand, and between browning and enzymatic activities on the other. However, this is the case in five olive varieties in which a positive correlation was observed between browning rate and *o*-diphenols content (r = 0.994) and *o*-DPO activity (r = 0.992).[309,310] Nevertheless, in nine other varieties, good correlation was found only between browning and oleuropein content (r = 0.826), the major phenolic compound in olive;[111] no correlation was then observed with *o*-DPO activity. In avocado, a positive correlation was found between the browning tendency of the fruit, total phenols content, and *o*-DPO activity but not with *o*-dihydroxyphenols in two varieties.[108] Avocado fruit with mesocarp discoloration disorder exhibited higher extractable activities of soluble *o*-DPO and peroxidase, as well as higher levels of total phenols, hydroxycinnamic acids, and proanthocyanidins when compared to healthy fruit.[391] Total and specific *o*-DPO activity *in vitro* in three avocado varieties were correlated with the *in vivo* rate of browning.[169] Nevertheless, no correlation was found with the different forms of the enzyme whatever the variety, and the soluble/bound and active/latent form ratios were similar.[171]

TABLE 53
Relations between Browning, *o*-DPO and Phenolic Substrates in Different Fruits

Fruits	Number of cultivars	Correlation[a] with Substrates	Correlation[a] with *o*-DPO	Ref.
Apple	7	0	0	68
	3	+	0	68,203,381
	4	0	+	68,203,383
	6—8	+		278,279
Apricot	2	+	0	101,380
Avocado	3	0	+	169
	2	+	+	108
	6	0	+	315
Banana	3—5	0	+	157,158
Eggplant	3		+	94
Grape	9 (red)	0	+	294
	19 (white)	0	0	
Olive	6	+	+	30
	5		+	310
	5	+	+	309
	9	+		111
Peach	6	+		50
Pear	5	+		378

[a] 0: no correlation; +: positive correlation

In many cases, there is a positive correlation with one or other of the factors but not with both. Thus, in banana, browning of five varieties was influenced mainly by *o*-DPO concentration and ascorbic acid content but not by phenolic level.[158] In apple, relatively low correlation was observed when the degree of browning of seven cultivars was compared with their respective enzyme activity or phenolic levels.[68] However, a better correlation was established when the apple cultivars were divided into two groups (Figure 107). One of the groups shows a direct correlation between *o*-DPO activity and browning ($r = 0.827$), whereas the other displays high correlation with total phenolics ($r = 0.986$). In grape, enzymatic activity did not appear to be a particularly limiting factor in browning, while substrate content appeared to be of capital importance.[305] However, no correlation was found between potential browning and hydroxycinnamic esters content in mature fruit of 28 grape varieties, whereas significant correlation ($r = 0.86$) was observed with enzyme activity in 9 red varieties.[293] Positive correlation may also be observed between browning and *o*-diphenols content in certain fruits, such as peach[50] and apricot.[379,380] An identical relationship was found with catechin contents in pear.[378] In contrast, no correlation was found in other cases between browning and *o*-diphenols content although there is a correlation with total phenolic content.[68,108] The results obtained from intervarietal studies thus appear to be particularly heterogeneous as a whole. Although the phenolic substrates of *o*-DPO are always limiting in the expression of potential browning of the fruit at a specific stage, absence of constant correlation between the two parameters in intervarietal comparisons indicates simultaneous contribution of other factors. They obviously include *o*-DPO activities, but those of other enzymes, such as peroxidases and the presence of natural reducing agents in the fruit are also involved.

Old research on correlation with ascorbic acid contents did not give very clear results in apple.[203,418] However, in banana, there is good agreement between the susceptibility to browning of some varieties and the ascorbic acid content.[158] Likewise, in pineapple, internal browning is accompanied by the disappearance of ascorbic acid.[362] In kiwi, the low poly-

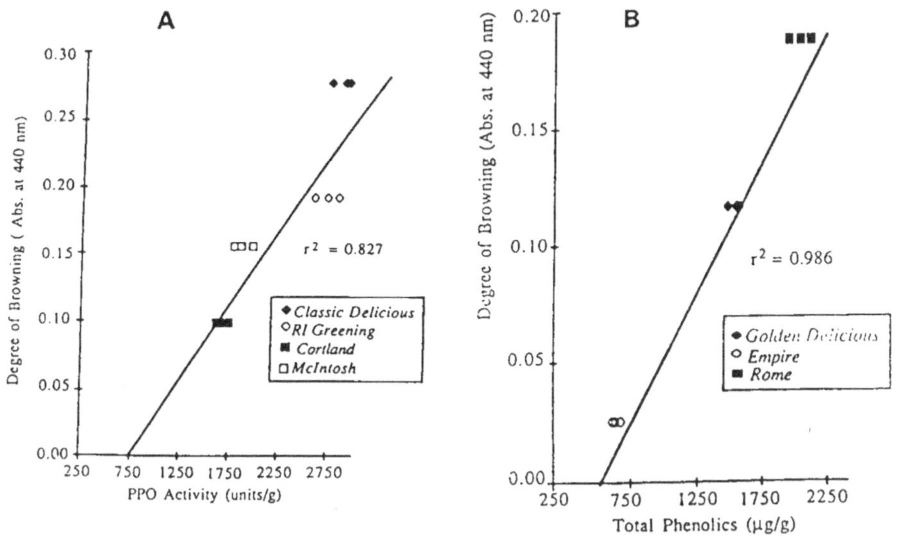

FIGURE 107. Correlation between enzyme activity (A) or total phenolics (B) and degree of browning in some apple cultivars. (From Coseteng, M. Y. and Lee, C. Y., *J. Food Sci.*, 52, 985, 1987. With permission.)

phenolic levels and *o*-DPO activity coupled with the high ascorbic acid content explain, in mature fruit, the lack of browning reaction in tissues of bruised fruits.[245]

Recent data involving other natural reducing agents show the complexity of the mechanisms involved: the browning observed during the process leading from grape to wine is accompanied by a decrease in *trans*-caftaric acid content due to its oxidation. A new caftaric acid derivative, the "grape reaction product" (GRP), forms simultaneously.[335] This compound was the result of a four-component reaction (Figure 108) involving caftaric acid (or coutaric acid), phenoloxidase, oxygen, and glutathione.[333] GRP has been identified as 2-*S*-glutathionyl caftaric acid,[62] and varietal differences in grape susceptibility to browning might depend on the ratios of different substrates.[59] However, GRP is not a substrate for grape *o*-DPO[333] and is believed to be a way of limiting must browning by trapping caftaric acid quinones in the form of the stable glutathione-substituted product and preventing them from proceeding to brown polymers. In contrast, 2-*S* glutathionyl caftaric acid was shown to be oxidized by laccase from *Botrytis cinerea* into 2-5-di-*S* glutathionyl caftaric acid (Figure 108) which may be responsible for the intense browning observed in must and wine made of botrytized grapes.[301]

This study and the results described above show that in comparison of varieties or of different physiological stages numerous parameters are superposed on the two main factors in browning (substrates and *o*-DPO activity). Results thus vary from one fruit to another, and it would probably be necessary to carry out full multifactorial analysis of the data to identify the relative influence of the various parameters. These approaches have already been undertaken in certain cases and should be broadly developed.

E. VARIATIONS IN BROWNING POTENTIAL DURING STORAGE

Many problems of browning of fruits arise during storage. Wounding or bruising, caused by handling at picking and storage, freezing, cold storage and thawing cause marked deterioration in the color and texture of fruits. At a commercial level, the development of browning and sensitivity to bruising are among the factors which limit the storage life of fruits. The various disorders caused by cold (chilling injury) are accompanied by the possibility of fungal attack during cold storage; as described in Chapter 4, this can be responsible for discoloration or browning of stored fruit.

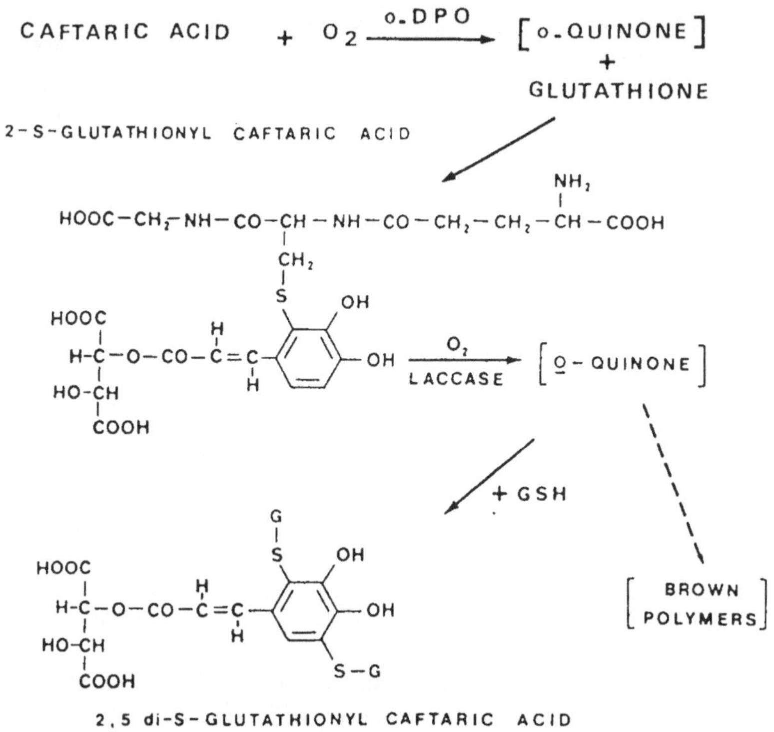

FIGURE 108. The proposed reaction sequence for the oxidation of 2-*S*-glutathionyl caftaric acid by laccase and the formation of 2,5-di-*S*-glutathionyl caftaric acid. GSH = glutathione. (From Salgues, M., Cheynier, V., Gunata, Z., and Wylde, R., *J. Food Sci.*, 51, 1191, 1986. With permission.)

Methods other than cold (e.g., controlled atmosphere, gamma irradiation, and drying) used to preserve fruits may cause browning and discoloration. In all cases, physicochemical modifications of membranes cause subcellular decompartmentation leading to enzyme-substrate contact and cause browning. Disturbance of the phenolic metabolism is also observed in a fair number of fruits. It was seen in Chapter 3 that the qualitative and quantitative changes in phenolic compounds during cold storage vary considerably from one fruit to another and even between the varieties studied. However, the relation between susceptibility to browning and changes in phenolic contents or *o*-DPO activity is not always very distinct.

In 'd'Anjou' pear, friction discoloration tendency increased with storage duration and was related to an increase in total phenol and chlorogenic acid concentrations.[413] In contrast, in Passe-Crassane pears, the appearance of internal browning cannot be accounted for by the small variations in chlorogenic acid and *o*-diphenol contents but is related to an increase in *o*-DPO activity.[97] This increase does not occur when the fruits are stored in controlled atmosphere and no browning occurs in this case. In apple stored at 0°C, browning potential varied little in most cultivars or increased over a period of about 2 months and then decreased. These small changes should be seen in relation to the absence of modifications observed in total phenolic and chlorogenic acid levels and in *o*-DPO activity which remain almost constant during storage.[68,381] The influence of controlled atmosphere on flesh browning of apple varies considerably according to cultivar: it is either increased in the presence of CO_2[275] or no effect is observed.[229] Storage temperature appears to be the determinant factor.[230] The degree of browning decreased in apricot stored at 5°C whereas *o*-DPO activity and *o*-diphenol concentration varied irregularly and peroxidase activity increased.[382] Avocado is usually stored at 4 to 8°C prior to marketing. However, storage at 12°C or below leads to signs of chilling injury, such as pitting, discoloration of flesh, blackening of the skin, etc. If the fruits have been stored at 5°C (chilled fruits), browning is more severe than in fruits stored at 17°C immediately after harvesting. However, *o*-DPO levels in chilled or nonchilled fruits are about the same.[314] However, in another case, *o*-DPO activity was increased by refrigerated storage of fruits for 31 d.[389]

Tropical and subtropical fruits are very sensitive to chilling injury, and this is obviously a serious disadvantage for fruits which are transported for long distances. This is the case of pineapple, which is transported at about 8°C and which displays internal browning which renders the fruits unfit for consumption. The appearance of internal browning in this fruit requires two phases: induction by chilling to 8°C followed by the browning symptoms themselves during subsequent warming to 20°C.[363] The greatest changes in levels in the phenolic metabolism are observed during the latter phase, and in particular a marked increase in derivatives of *p*-coumaric and caffeic acids (see Figure 75).[73] Browning intensity is well correlated with oxidative activities.[362] *o*-DPO activity, which is practically nil throughout storage at 8°C, only appears in sensitive fruits and occurs after the temperature rise. However, it remains low in comparison with peroxidasic activity which increases in parallel with browning intensity and appears to be responsible to a great extent for the oxidation of phenolic compounds in this fruit. Nevertheless, in the same cultivar ('Cayenne'), other authors consider that peroxidase activity had no apparent connection with black heart, whereas *o*-DPO activity is much higher in diseased fruitlets compared to healthy ones.[390] An inverse relationship was found in banana and mango between total phenols content and the degree of blackening during storage at injurious chilling temperature.[2] Another method of storage which delays the ripening process is the use of γ-irradiation. However, this can cause injury which results in browning proportional to radiation dose. In banana[365] and mango,[364] there was a good correlation between browning and *o*-DPO activity in irradiated fruit. The effect of irradiation on these enzymatic activities and the phenolic metabolism is described in detail in Chapter 3. In all cases, irradiation affects cell membrane permeability, bringing enzymes and substrates into closer contact, thus accelerating or even initiating reactions.

The drying of fruits, such as grapes, plums, dates, and figs can be considered as a storage method in addition to being a special preparation technique since the finished products have long shelf lives. The browning which occurs is in most cases an added-value feature, with the exception of certain types of raisins, such as Sultanas. Changes in cell permeability are observed during the process, and structural disturbances promote browning. In d'Ente plum, cell disorganization occurs very rapidly and *o*-DPO displays high resistance to thermal denaturation, in particular at the temperatures used in drying.[237,282] Oxidation of the phenolic compounds thus takes place very rapidly and then stabilizes. Degradation of anthocyanins is also very fast and takes place by means of an oxidation mechanism combined with the enzymatic oxidation of hydroxycinnamic acids, as we have already reported. In Sultana grape berries, browning during drying is also caused by *o*-DPO acting on substances released from cell compartments when cell integrity is lost.[118] Nevertheless, glutathione remains partially compartmentalized during the drying of Thompson seedless grapes in direct sunlight.[331] One of the consequences is more intense browning due to there being no formation of 2-*S*-glutathionyl caftaric acid, a compound which is not brown and does not brown enzymatically.

Total *o*-DPO activity decreases during frozen storage of plums (−20°C).[384] However, browning problems occur during thawing of fruit, in particular in sensitive fruits, such as peach.[296] Browning at thawing is once again the result of decompartmentalization due to freezing of tissue. Thawed, brown fruits become unfit for consumption and unusable in the canning for which they are generally destined. The formation of brown substances can only be prevented by treatment prior to or during freezing.

TABLE 54
Procedures Used to Limit Browning

Mechanism	Experimental conditions
Inhibition of *o*-DPO	NaCl, $CaCl_2$, SO_2, metal chelator (sodium diethyldithiocarbamate), low pH, high temperature, drying, hormones (gibberellic acid, ethephon), NaOH, *p*-coumaric acid, salicylhydroxamic acid
Exclusion of oxygen	Controlled atmospheres, bagging, high concentration of sugars, oil, paraffin, polyvinylpyrrolidone, natural wax
Reduction or trapping of quinones	Ascorbic acid, glutathione, cysteine, SO_2, polyvinylpyrrolidone
Modification of substrates	Methylation by *O*-methyltransferase, oxidative ring-opening by protocatechuate-3,4,-dioxygenase
Reversibility of browning	Ascorbic acid, SO_2
Removal of browning	Ultrafiltration, resins, clarification.

F. CONTROL OF BROWNING

Numerous works on fruit phenolics and their metabolism have originated in investigation of the enzymatic browning which occurs during storage or processing. Simultaneously, more applied work has been carried out to design methods to control this browning. They are generally preventive and consist of slowing the phenomena or blocking them completely.[3,89,272,407] However, they can be completed by other approaches which consist of reducing browning intensity when the quinones formed have not yet been polymerized.

Prevention of browning makes use of a variety of methods whose targets may be very different (Table 54). Action mechanisms concern four main aspects: inhibition or inactivation of *o*-DPO, reduction of oxygen content, reduction or trapping of the quinones formed, and action on the concentration or nature of the substrates. The techniques used require physical or chemical treatments or a combination of these. Physical treatments include filtration on resins and ultrafiltration, which for example make it possible to remove brown color from wine.[10,168] Cold is also an effective method for controlling the development of browning since it slows the metabolism and the rate of chemical reactions. Freezing appears to be even more effective since it blocks most enzymatic activities. However, as we have already reported, abnormal browning of fruits can occur as a result of chilling injury or during thawing.

The browning which occurs during the drying of fruits is often sought after (prunes, certain types of raisins, etc.); it occurs during the first few hours of the process[227] since *o*-DPO activity is very rapidly repressed and subsequently stopped by the increasing concentration of solutes in the plant tissue. Browning during drying is undesirable in certain cases, such as Sultana grapes. In this case, dipping or spraying with oil emulsion prevents browning by increasing the sugar concentration.[118] High temperatures inactivate *o*-DPO and are often used as a means of prevention. As indicated by the results of heat denaturation of *o*-DPO in various sources, short exposure to a temperature of 70 to 90°C is generally sufficient to obtain partial or total inactivation of the enzymes. Thus, 90% destruction of *o*-DPO in cranberry is achieved with a temperature of 50 to 70°C applied for 1 to 15 min.[56] In this case, activation energy for the inactivation of the enzyme was observed to be 27.7 kcal/mol, whereas it was 25.2 for apricot,[394] 18.4 for banana,[250] and 52.4 in grape.[195] Still on the subject of physical treatment, the soaking of fruits in a solution of paraffin wax, polyvinylpyrrolidone, or natural waxes forms a protective film which limits the diffusion of oxygen. Likewise, the use of controlled atmospheres (generally enriched with CO_2 and low in O_2) makes it possible to reduce browning.[97,361] In the case of pineapple, the most effective method of controlling internal browning consists of keeping the fruits in plastic bags in which a CO_2-rich atmosphere can be created.[361]

In addition to or as a complement to physical methods, numerous chemical treatments exist for the control of browning (Table 54), and in particular those which inhibit *o*-DPO activity. Many reagents have this effect, but only a limited number are considered to be acceptable for use in the control of enzymic browning during food processing.[407] NaCl has been found to be effective for various fruits.[33,151,297,427] However, although a concentration of some 0.5 to 1% (w/v) is sufficient to protect fruits from browning, much stronger concentrations are necessary to achieve total inhibition *in vitro*. It has been shown that the chloride anion plays an active role in the inhibition of browning and not the sodium or calcium cations.[71] However, calcium appears to act as a firming agent by limiting both the deterioration of cell structure and hence browning and development of fungal deterioration.[25] This has been shown both in the whole apple[230] and in slices.[268,274] Metal chelators, such as sodium diethyldithiocarbamate, acid pH, and high sugar concentrations also inhibit *o*-DPO activity. Sulfur dioxide is widely used as an additive to prevent browning of fruits during processing. It inhibits both enzymatic and nonenzymatic browning and can be used as an antioxidant, bleaching, or antimicrobial agent. Inhibition has been found to be caused by the formation of quinone-sulfite complexes and the inactivation of *o*-DPO.[122]; Inhibition of *o*-DPO is irreversible and is related to modification of the protein structure.[306]

It has already been seen that certain phenolics, such as ferulic cinnamic, or *p*-coumaric acids inhibit *o*-DPO activity. The addition of such compounds, and in particular cinnamic acid or its more soluble sodium salt, neither of which is expensive, can be used for long-term control of browning and has been tested in fruit juice.[272,408] It has recently been shown that salicylhydroxamic acid (SHAM), which is known to be a selective inhibitor of cyanide-resistant respiration in higher plants, is also a specific inhibitor of *o*-DPO.[8] Possible use of this compound in the limiting of browning remains to be specified.

Ascorbic acid is a very widely used reducing agent whose mode of action is well known. It delays the appearance of browning by reducing quinones as they form. Its role as an inhibitor of *o*-DPO is controversial;[272] indeed, it has been found to inactivate potato, apple, and tobacco enzyme, but no inhibition has been reported in grape or apple by other workers. Numerous studies have led to specifying optimal conditions of utilization of this compound. They differ according to the fruit studied and these aspects are not developed here. In a similar manner, the formation of complexes between grape phenolics and glutathione or cystein[410] by coupled reaction has been considered as a way of limiting must-browning since these complexes are not *o*-DPO substrates and the quinones thus converted cannot form brown polymers.[59,333,335]

A combination of methods is often used to increase the effects of the various chemical treatments. Thus, as an example, slices of apple for refrigeration can be immersed in solutions containing different combinations of sucrose, calcium or sodium chloride, ascorbic acid, and citric acid.[268] Apple slices pretreated in this way can be kept for about 4 weeks at 0°C.

Combinations of treatments are also used to control the browning that appears during thawing: immersion in solutions containing sugar and ascorbic acid and also heat treatment, thawing under a partial vacuum,[86] under nitrogen atmosphere or more recently with micro-waves.[267] However, the results are closely dependent on the treatment received by the fruits prior to freezing, and in particular stone fruit which can be subjected to antioxygen conditioning before freezing.[266]

In olive, different processes, such as blanching, dipping in solutions of *o*-DPO inhibitors or reducers are not very effective in comparison to dipping in NaOH solutions. In addition to facilitating the removal of bitter constituents, the treatment prevents browning.[30] This was apparently due to inactivation of olive *o*-DPO, which displayed almost complete loss of activity after 1 h of incubation in 0.2% NaOH.[12]

Hormone treatments can also be used to control browning in more exceptional cases. Peaches and apples from trees sprayed with giberellic acid and ethephon browned much less than control fruits[49,257] and displayed a decrease in *o*-DPO activity.[49,179,300] This decrease in browning can also be attributed to lower levels of polyphenolic substrates, but no differences in qualitative distributions of these were observed.[257]

Action on *o*-DPO substrates may either affect their content or convert them into non-substrate compounds. A fall in substrate content can only be envisaged during a manufacturing process, such as the production of apple or grape juice. Initial oxidation of the pulp will produce a very pale juice after clarification, or the *o*-DPO substrates can be adsorbed on polyvinylpyrrolidone.[244,387] In addition, two original methods of converting substrates by an enzymatic pathway were proposed a long time ago. One is the methylation of *o*-diphenols under the effect of *O*-methyltransferases. Thus, methylation of chlorogenic acid in apple juice leads to the formation of feruloyquinic acid which is not an *o*-DPO substrate.[90,243] The other consists of the oxidative ring-opening of *o*-DPO substrates through the addition of protocatechuate-3,4-dioxygenase, resulting in non-benzenoid, nonreacting compounds.[176]

Finally, whereas the aim of all these methods is to prevent or limit browning, other techniques can make it possible to lighten a browned product on the condition that action is taken before the quinones formed have polymerized or are still at an early stage of polymerization. The reversibility caused by immersion in ascorbic acid or SO_2 was 65% for peaches which had blackened for half an hour during thawing and only 50% for those darkened for 6 h.[296]

We have only mentioned the most common methods for controlling browning and many others exist. However, if avoiding or reducing these treatments during a process is desired, the long-term solution should be chosen, consisting of selecting and cloning cultivars which have low browning potential because of low phenolic substrate contents or weak *o*-DPO activity. Care must also be taken that the other sensory qualities of the fruit and the agronomic qualities of the plants selected in this way are not seriously deteriorated. Here, modification of plants using genetic engineering might make possible specific action to limit browning, but this approach does not seem to have been undertaken yet.

III. MODIFICATIONS OF SENSORY PROPERTIES OF FRUITS AND FRUIT PRODUCTS DURING PROCESSING

As discussed in Chapter 4, phenolic compounds contribute to the sensory properties associated with fruit quality. Furthermore, modifications occur when the fruit is detached from the mother plant and subjected to various types of processing. Depending on these, modifications may consist of diminished quality, e.g., by degradation of color, or in contrast by an improvement when phenolics appear to be too bitter or too astringent. These combinations of modifications frequently govern the final quality of fruit-based products, and consumers' requirements may vary according to eating habits and the nature of the product.

A. STABILITY AND INSTABILITY OF ANTHOCYANIN COLOR

It has been mentioned above that considerable color modification can occur when anthocyanins are removed from their natural environment either by dilution in the vacuole contents of cells which contain no anthocyanins, or by the use of solvents for extraction. One of the most immediate consequences is the shift from blue to red in the making of fruit juice; the causes of this are discussed above. However, modifications are not limited to these observations. Indeed, anthocyanins appear to have low stability in all products manufactured from fruits,[212] which thus limits use as food colorants[145,153,213,358] or as consituents of pharmaceutical formulations.[38] As has been well explained by Markakis,[212] two main groups of factors are responsible for the stability or instability of anthocyanin color in fruits during processing: first, factors associated with the initial composition of the fruit with regard to anthocyanins and other constituents, including certain enzymatic systems, and second, external factors as a whole and, in particular, temperature, light, and the physicochemical environment of the fruit during processing.

We intentionally limit ourselves here to a few examples which show the action of these principal factors on color stability. The question of wines is not covered here but is examined in more detail in a later section.

The special interest that can be paid to the initial anthocyanin composition of a fruit lies essentially in the fact that there is differential stability related to molecular structure. For example, among the 3,5-diglucosides of muscadine grape juice, the one most affected by heat is malvidin 3,5-diglucoside (53% destruction at 72°C in 2 h).[95] Diglucosides are generally more heat stable than monoglucosides;[385] diglucoside stability increases with increasing methoxylation, but decreases with increasing hydroxylation of the aglycone.[146] The nature of the sugar may also be involved. In cranberry juice, galactosidic anthocyanins are more stable than the arabinosidic type during storage.[350]

Anthocyanins whose molecules possess two or more acyl groups are much more stable.[46] They are abundant in certain flowers and can be used as food colorants.[15,16] Their stability has been compared to that of anthocyanins extracted from blackberries or from *Vitis*.[214,359] A commercial anthocyanin from red grape skins is slightly more stable during storage in darkness at room temperature, whereas at 5°C, the flower pigment showed greater stability for at least 10 weeks.[214] The relative stability of the two types of pigment varies with temperature, pH, and the presence or absence of light.

Numerous enzymes can spoil the quality of the initial product during the extraction of fruit juices or the preparation of processed products, leading to browning (see above) and to loss of color by enzymatic degradation of anthocyanins. The two phenomena may vary in importance in different types of plant material. Thus, deterioration of the attractive red color during the storage of strawberry preserves is mainly due to a rapid rate of browning related to high procyanidin and catechin contents. In contrast, loss of anthocyanins would occur without marked deterioration of color.[1]

Several enzymatic systems have been found by authors to be involved in this degradation, and in particular phenolases, peroxidases, and β-glucosidases. A succinct review of previous works can be found in Markakis.[211,212] Anthocyanins do not appear to be very good substrates for phenolases, but they may be strongly degraded by coupled oxidation phenomena studied previously (Figure 105).

Ascorbic acid has a protective effect with regard to anthocyanins (Figure 109) since it reduces the *o*-quinones formed before their polymerization.[270,337] In Concord grape, anthocyanin pigment degradation by a crude phenolase extract is dependent on fruit maturity:[55] the most rapid degradation of pigment occurs when crude enzyme from ripe grapes is added. In contrast, oxidative enzymes do not appear to be involved in the loss of color during preparation of strawberry purees, at least for 48 h at 30°C.[349]

Among the numerous approaches which can be envisaged to block phenolase activity (Chapter 4), treatments at high temperature are often used, especially in blanching treatments prior to processing or preservation of fruits. Thus, steam-blanching of red tart cherries for 45 to 60s before freezing[322] blocks the degradation of anthocyanins when the fruits are subsequently thawed and blended in air (Figure 110).

Independently of the favorable action of high temperatures on the blockage of enzymatic activities, numerous studies covering both food products and *in vitro* conditions have shown that anthocyanins are readily destroyed by heat during the processing and storage of foods.[211,212] Loss of color in fruit juices is generally accelerated by raising storage temperature (Figure 111).[202,324,337] In blackcurrant juice, all anthocyanins disappear in 9 weeks at 37°C and pH 2.0 whereas 60% remain at 4°C. Variation related to duration of storage correspond in this case to a second degree nonlinear regression curve.[324] Other authors have formulated logarithmic relationships between anthocyanin destruction and temperature[54,174,202] or time at a given temperature.[212] The half-life of pelargonidin 3-glucoside, the major strawberry preserve anthocyanin, was 1 h at 100°C, 10 d at 38°C, 54 d at 20°C, and 11 months at 0°C.

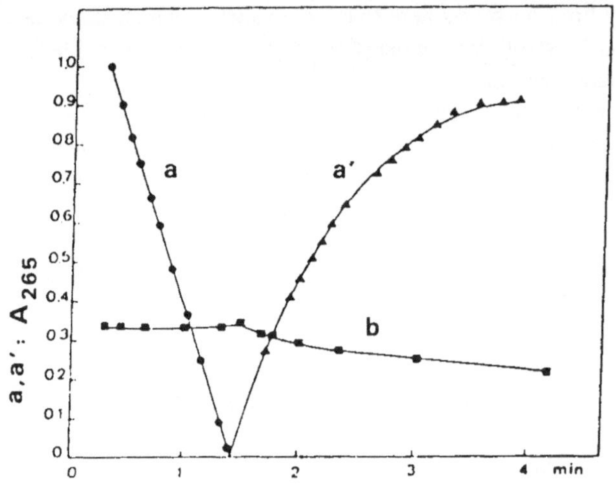

FIGURE 109. Protective action of ascorbic acid on the enzymatic degradation of cyanidin-3-glucoside in the presence of 0.31 m*M* chlorogenic acid: ascorbic acid destruction (a); chlorogenic acid oxidation (a'); anthocyanin degradation (b). (From Pifferi, P. G. and Cultrera, R., *J. Food Sci.*, 39, 786, 1974. With permission.)

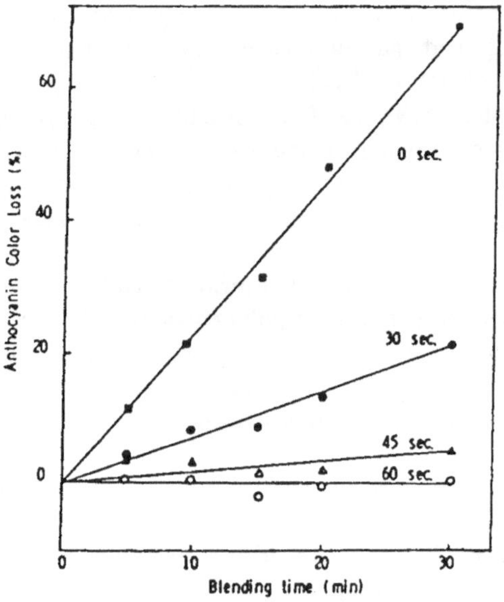

FIGURE 110. Anthocyanin color loss in red tart cherry steam blanched for 0, 30, 45 and 60 s, frozen and thawed in a Waring blendor under air (From Siegel, A., Markakis, P., and Bedford, C. L., *J. Food Sci.*, 36, 962, 1971. With permission.)

Nevertheless, a short-time/high-temperature process was recommended for best pigment retention. Likewise, in the case of red fruit juices, anthocyanin loss appears to be negligible for treatment durations of less than 12 min at 100°C.[212]

During fruit drying, degradation of anthocyanins occurs very rapidly at high temperature. Thus, in prune, only 14.6% of initial cyanidin 3-rutinoside remains after 1 h at 95°C, whereas

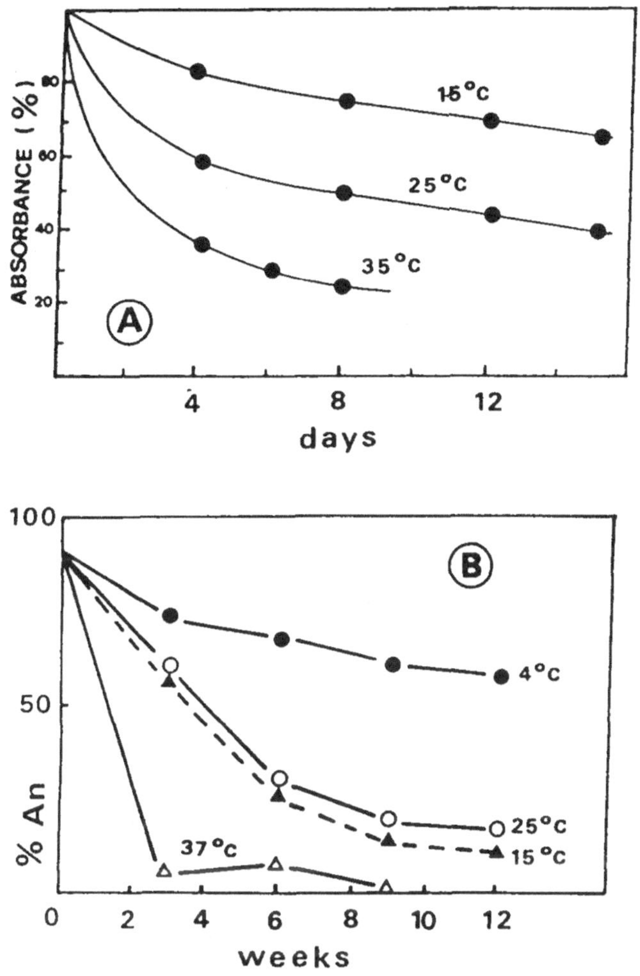

FIGURE 111. Effect of the storage temperature on the color degradation of the orange juice (A) and blackcurrant juice (B). An: Anthocyanin. (Adapted from Maccarone, E., Maccarone, A., and Rapisarda, P., *J. Food Sci.*, 50, 901, 195; and Simard, R. E., Bourzeix, M., and Heredia, N., *Sci. Aliments*, 1, 388, 1981.)

45.6% remains at 55°C (Figure 112). Under the same conditions, degradation of hydroxycinnamic derivatives, and, in particular, caffeoyl 3′-quinic was in contrast slowed down when the temperature was increased from 55 to 95°C, in direct relation with the progressive inactivation of *o*-diphenoloxidases.[282,283] Anthocyanins are probably then degraded by coupled oxidation mechanisms described above in the study of browning.

In fact, anthocyanin degradation mechanisms under the effect of temperature are still not well known,[212,367] but several degradative pathways have been proposed and it would seem that there is first an opening of the central heterocycle leading to a chalcone. Further degradation leads to brown products, especially in the presence of oxygen.[212] In the case of 3,5-diglucosides of anthocyanidins, a 3,5-diglucosyl-7-hydroxycoumarin whose formation may be strongly accelerated by heat but unaffected by O_2 is thought to be a common degradative intermediary.[144]

Balance between the different forms of anthocyanins (see Chapter 4) is shifted by heating towards the chalcone;[46,367] return to the flavylium form is slow, requiring 6 to 12 h at 25°C.[46] This may result in errors in assessment of the red form of anthocyanins when measurement is carried out too soon after heat treatment.[212]

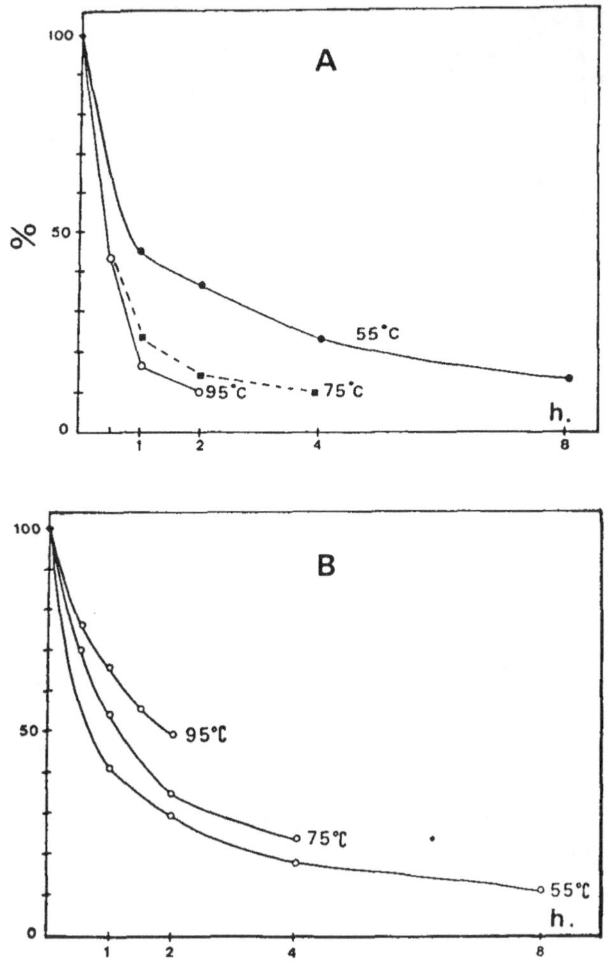

FIGURE 112. Changes in anthocyanin (A) and neochlorogenic acid (B) of plum during drying at different temperatures. (Adapted from Raynal, J., Thèse Doct., Université de Montpellier, France, 1987.)

Light is another factor in the accelerated degradation of anthocyanins. Indeed, anthocyanins possess poor photostability, which generally leads to keeping fruit juices and fruit-based beverages in the dark during storage.[252] The acylated forms are generally the most stable,[385] and copigmentation with polyhydroxylated flavones, isoflavones, and aurone sulfonates[354] also make anthocyanins light-stable.

The pH of the medium can affect both the color of anthocyanins in an aqueous medium (see Chapter 4) and their stability. The latter is particularly affected by interactions between pH and oxygen. Thus, in strawberry juice at 45°C, degradation of pelargonidin 3-glucoside is very weak from pH 2.0 to 4.5 in the absence of O_2 whereas it increases considerably at pH 4.0 in the presence of O_2 (see Reference 212). The role of oxygen in loss of color of red fruits and in the formation of brown substances was shown a long time ago in numerous investigations.[212] It is generally advised that it should be eliminated during processing and storage of juice or for the use of anthocyanins as a food colorant.[64] In cranberry juice, elimination of oxygen tripled the half-life of color[173] and, in the presence of Al^{3+} or Fe^{3+} ions, prevents the appearance of a blue hue. Oxygen may act directly on oxidation of anthocyanins with or without a phenolase enzyme as an intermediary or indirectly by coupled oxidation between ascorbic acid and anthocyanins. Indeed, although — as reminded above

— ascorbic acid can protect anthocyanins from enzymatic oxidation, degradation of juice color is reported fairly frequently when ascorbic acid is added[336,350] although no significant difference was observed in certain cases.[173] Anthocyanins and ascorbic acid may evolve in a parallel manner under the action of different factors. Their degradation is accelerated by copper ions and slowed down by certain flavonols (quercetin, quercitrin) in cranberries.[321] These data suggest possible interaction between the two compounds.[212] In the presence of O_2 and metal ions, such as Cu^{2+}, H_2O_2 formed by degradation of ascorbic acid caused discoloration of anthocyanins. In contrast, other mechanisms may be involved in the absence of oxygen, and in particular products of condensation — unstable and becoming progressively colorless — between ascorbic acid and anthocyanins.[167,212] Other antioxidants, such as cystein (50 to 100 mg/100 ml) or glutathione (100 mg/100 ml) have also been used, for example, in juice of Concord grape[337] and of blood orange[202] to protect anthocyanins from degradation and to obtain better stabilization of red color. However, these effects are very dependent on environmental conditions, and in particular on temperature and pH.

Various other organic chemicals may cause modifications in the color of anthocyanins in products derived from fruits, for example, tartaric acid used to lower the pH of juice of blood orange[202] or sugars, especially fructose, which accelerate degradation, the effect being particularly marked in the presence of O_2.[212] In connection with amino acids and ascorbic acid, ketose sugars contribute to the discoloration of grape juice heated to 75°C by increasing nonenzymatic browning.[336]

The importance that copigmentation phenomena may have in the stabilization of anthocyanin color has been discussed at length in Chapter 4. Rutin and caffeic acid have been used for this at a concentration of 50 mg/ml to stabilize the color of blood orange juice.[202] Half-lives of anthocyanins were then 75 and 90 d, respectively, in comparison with 40 d for the control. However, the use of flavonols as copigments can be limited by their solubility in aqueous media. Thus, in cherry juice with added quercetin or quercitrin, flavonols crystallized out immediately after pasteurization and no increase in anthocyanin stability was observed,[321] whereas flavonols had a positive effect in model systems with cranberry anthocyanins.[345] Copigmentation with dimeric flavans can occur[345] in red grape juice.

The influence of acetaldehyde on color enhancement and condensation reactions of anthocyanins with themselves or with other organic compounds (in particular with tannins) are two factors which intervene to a considerable extent in the aging of wine.[286] This point is discussed later.

As mentioned above in Chapter 4, metal ions may play an important role in the expression of anthocyanin color, and generally lead to shift of pigment color towards the blue end of the spectrum. There may be three causes of the formation of anthocyanin-metallic ion complexes: (1) the existence of natural metalloanthocyanins which, as already mentioned, have mainly been revealed in flowers, (2) contact during crushing or subsequent technological processes between pigment and ions from the fruit itself, and (3) contamination by metals in the extraction and processing line or in the cans in which fruits or juice are stored. Numerous examples of discoloration of anthocyanin-pigmented fruits by reaction with metal ions have been studied,[212] and included here is only a succinct mention of the contradictory effects that may occur when certain metal ions are added. Whereas Ca^{2+}, Fe^{3+}, Al^{3+}, and Sn^{2+} ions increase the stability of the juices of grape,[336] cranberry,[351] crowberry,[173] (Figure 113), tart cherry, grape, and strawberry,[311] the overall effect of complex formation may be negative[98] insofar as supplementation may give the juice a more bluish hue and brown discoloration may occur after the formation of metal-tannin complexes. In addition, red pigmentation may occur in certain fruits that are not colored initially, such as pear.[57] This appears to be caused by the formation of an insoluble complex combining Sn^{2+} and the anthocyanidins which result from the degradation of the abundant proanthocyanidins which occur naturally in the fruit.

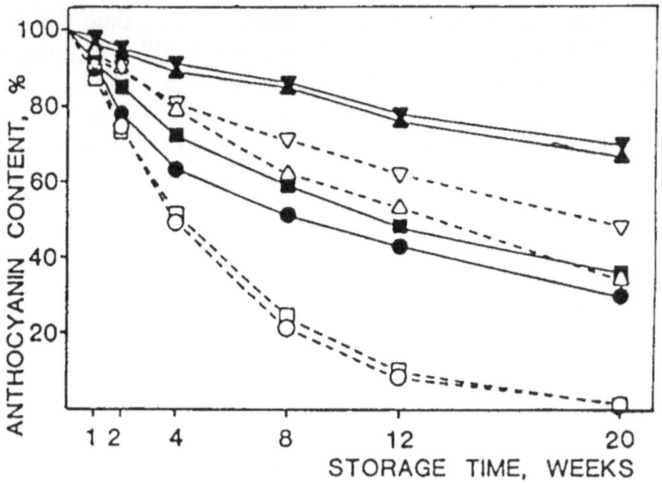

FIGURE 113. Influence of oxygen and Al^{3+} supplementations on the stability of crowberry anthocyanins. ○, R-O_2; ● R-N_2; □, Al01-O_2; ■, Al01-N_2; △, Al25-O_2; ▲, Al25-N_2; ▽, Al50-O_2; ▼, Al50-N_2. (From Kallio, H., Pallasaho, S., Kärppä, J., and Linko, R. R., in *The Shelf Life of Food and Beverages*, Elsevier, Amsterdam, 1986, 285. With permission.)

Sulfur dioxide has been widely used in fruit processing[212] and in the bleaching of anthocyanins applied to their quantitative measurement.[286] According to Markakis[212] and Glories,[106,107] sulfur dioxide can have an effect at four different levels:

- It enhances the extraction of anthocyanins from fruit tissues.[252,368]
- At concentrations as low as 30 ppm, it inhibits the enzymatic degradation of anthocyanins.
- At moderate concentrations of the order of 500 to 2000 ppm, it forms a reversible, colorless SO_2-anthocyanin complex whose formation and stability have been studied extensively.[367]
- At high concentrations (0.8 to 1.5%), it forms an irreversible complex which occurs, for example, in the bleaching of red cherries during brining.[212]

It would thus appear that sulfur dioxide has a complex mode of action, and that contrasting results can be obtained depending on the moment of application and the amount used. SO_2 can also be used to halt the development of microorganisms during the storage of grapes. However, it must be released very slowly in relatively small amounts in order not to discolor the anthocyanins.[256]

B. REMOVAL OF BITTERNESS

Grapefruit juice could form an attractive base for blended fruit juice beverages because of its relatively low cost. However, excessive bitterness prevents it from being used widely and today there is no truly acceptable industrial method of reducing the amounts of bitter compounds in citrus juices.[317] Different complementary approaches have nevertheless been retained,[37] including the following in particular:

- Choice or genetic selection of cultivars with low bitterness; naringin contents do in fact vary considerably between varieties[5] (Chapter 1).
- Choice of ripe fruits in which the naringin content is much lower.[6,166]
- The presence of the smallest possible amount of pulp in the juice; indeed, naringin is found essentially in the albedo, the segment membranes, and the pith,[166] and the juice extraction method can lead to products with very variable bitterness.[74]
- Use of naringinase leading either to nonbitter naringenin by total hydrolysis or to

TABLE 55
Naringin as a Common Starting Material for the Synthesis of Various Flavonoid Neohesperidosides with Useful Taste Properties

Bitterness suppressors		Bittering agents		Sweeteners
Rhoifolin	←	NARINGIN	→	Naringin dihydrochalcone
		↓		
Neodiosmin	←	Neohesperidin	→	Neohesperidin dihydrochalcone

Data from Reference 210.

prunin (naringenin 7-β-D-glucoside), a compound with low bitterness. The enzyme can be used in immobilized form in reactors.[115,246]

- The fixing of phenolic compounds by adsorbants. Among these, particular attention has been paid to cyclodextrin which in monomeric form reduces bitterness when dissolved in grapefruit juice.[181,182] β-Cyclodextrin polymer has also been used to reduce the naringin level in cloudy or clarified grapefruit juice using either batch or continuous flow columns;[316-318]
- The use of compounds which can remove bitterness and in particular the tasteless flavone analogs of the bitter neohesperidosides.[210]

The bitter flavanones in citrus can be converted by simple hydrogenation into the corresponding intensely sweet dihydrochalcones.[143] In the search for new noncaloric sweeteners,[139] these compounds seem an attractive class insofar as the intensity of the sensation of sweetness can be much greater than that of sucrose at the same concentration. Nevertheless, use of dihydrochalcones derived directly from naringin and neohesperidin appears to be limited because of the slow appearance of sweetness; this has led to testing the effectiveness of synthetic dihydrochalcones, glycosylated or not.[78,180]

The taste properties of naringin and its derivatives make them commercially potentially useful compounds, either as a bittering agent in tonic beverages or as starting material for the production of other useful neohesperidosides (Table 55).[210] Grapefruit peel could form a natural source.

Olive technology is still very traditional[87] and the various processes currently used to render these fruits edible are frequently of empirical origin. However, a qualitative and quantitative approach to the removal of bitterness has been developed over the past 20 years and has led to a set of results which should enable the rationalization and modernization of the process. They concern either the fruit itself[11,396] or the water used to remove excess lye.[23,397] The industrial processes used in the preparation of green table olive comprise three main phases:[11] removal of bitterness, water washing, and storage with or without fermentation, including sterilization or not. The removal of bitterness is the only aspect examined here insofar as it leads to complete or almost complete degradation of oleuropein and the other bitter constituents. It is generally carried out by dipping olives in a dilute alkaline solution; dipping time depends on fruit variety, degree of maturity, sodium hydroxide concentration, and temperature. However, in most cases, treatment conditions are decided in an approximate manner, although duration and intensity should be adapted to the characteristics of the plant material to be processed and in particular to the oleuropein content of the fruit at harvesting. A mathematical modeling trial was proposed recently,[11] but it must be refined and generalized. It takes into account the degradation of oleuropein and the progressive appearance of dihydroxyphenylethanol during the process. The evolution of the ratio of the two compounds, measured by HPLC, might provide a good criterion for monitoring the evolution of the technological process (Figure 114) on the condition that one specifies how it reflects the sensory characteristics of the fruit. In addition, this ratio can

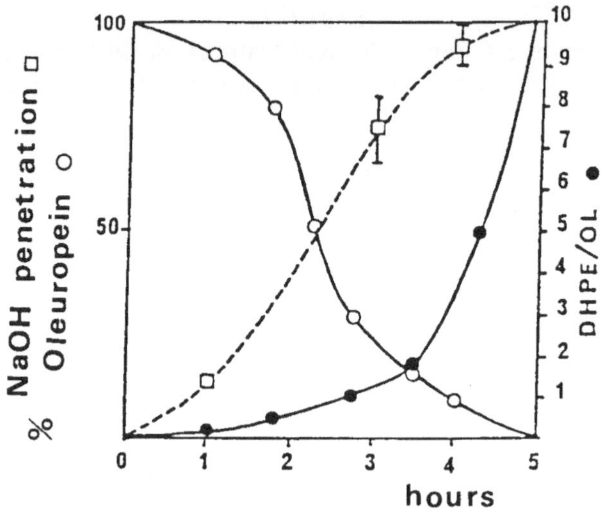

FIGURE 114. Changes in NaOH penetration, oleuropein content, and dihydroxyphenylethanol/oleuropein ratio (DHPE/OL) during alkaline treatment of olive fruit (Adapted from Amiot, M. J., Thesis, University of Montpellier, France, 1986.)

be involved directly in the fermentation which can affect olive during technological processing, since oleuropein degradation products have more marked antimicrobial activities that the original molecule.[93, 412]

C. DEASTRINGENCY OF FRUITS AND FRUIT PRODUCTS

Among the mechanisms proposed to account for the natural decrease in astringency (Chapter 3), the progressive immobilization of tannins by reaction with acetaldehyde is the basis of numerous treatments which have been applied to persimmon fruit for a very long time[110] to reduce astringency and thus render the fruit edible. All these treatments lead to progressive accumulation of acetaldehyde in the flesh of the fruit[220] at a rate depending on the degree of fruit maturity.[356] The principles are summarized briefly here according to Ito.[152]

- Treatment with water at 40°C for 14 to 24 h
- Treatment with ethanol (This is effective but may lead to browning of fruit flesh.)
- Ionizing treatments
- Freezing to −25°C
- Treatment with CO_2 or N_2, which was described by Gore in 1911[110]

The most effective methods have been described by numerous authors,[104,220,222,260,261] with mention in particular of the importance of the duration of application of CO_2 (Figure 115). A combination of CO_2 (20°C/24 h or 30°C/12 h or 40°C/6 h) and subsequent storage in air at 30°C for 3 d results in nonastringent fruits of excellent quality without abnormal softening.[220,222] Whereas soluble tannins decrease rapidly after a fairly short phase, ethanol and acetaldehyde accumulate in the fruit with different kinetics (Figure 116). The combined effect of three factors (CO_2, temperature, duration) reveals two distinct processes in the disappearance of astringency:[104,220] first, an induction process in which the fruit must be exposed to the gas and during which begin the reactions related to deastringency, and then the deastringency process itself during which presence of the gas is no longer necessary. During the first phase, a relatively small amount of acetaldehyde is probably formed by glycolysis. In the second, the soluble tannins in vacuoles are probably converted in nonastringent insoluble forms through coagulation (Chapter 3).

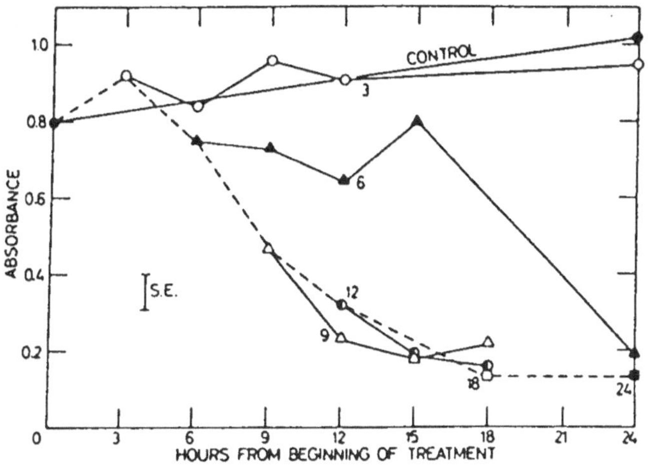

FIGURE 115. The effect of CO_2 treatment (3, 6, 9, 12, 18, and 24 h in CO_2 atmosphere) on the absorption at 277 nm of methanolic extracts from "Triumph" persimmon fruit. (----) Immediately after removal from CO_2; (—) during shelf-life. (From Gazit, S. and Adato, I., *J. Food Sci.*, 37, 815, 1972. With permission.)

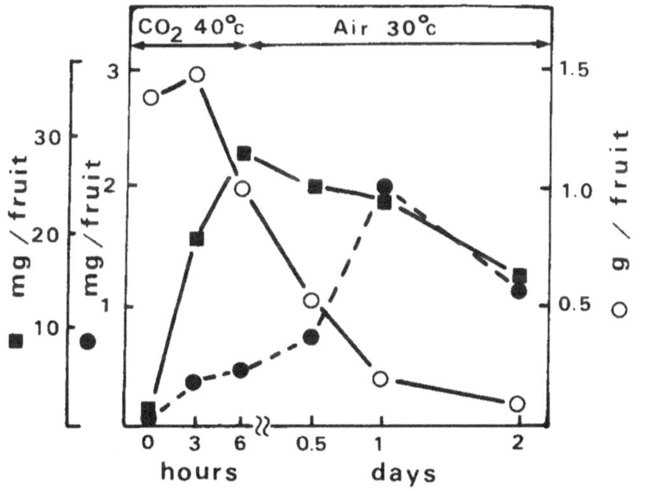

FIGURE 116. Changes in soluble tannin(s) (○), ethanol (■) and acetaldehyde (●) content during the disappearance of astringency in persimmon fruit. (Adapted from Matsuo, T. and Itoo, S., *Plant Cell Physiol.*, 18, 17, 1977.)

The comparative effectiveness of anaerobiosis in the presence of CO_2 or N_2 or under partial vacuum has been re-examined recently.[260-262] Deastringency is always faster under CO_2 (Figure 117) which enhances the relative accumulation of acetaldehyde in contrast with ethanol. However, flesh browning can occur under CO_2 and under vacuum with resulting premature spoilage of the fruit.[262] The higher levels of acetaldehyde produced by the fruit in a CO_2 atmosphere can be explained through fixation of atmospheric CO_2 into malate in which almost all the $^{14}CO_2$ fixed by persimmon accumulates.[261] Malate and pyruvate (produced from malate by action of malic enzyme) are indeed very active in the increase of acetaldehyde production in persimmon. In addition, an enzyme which limits the rate of

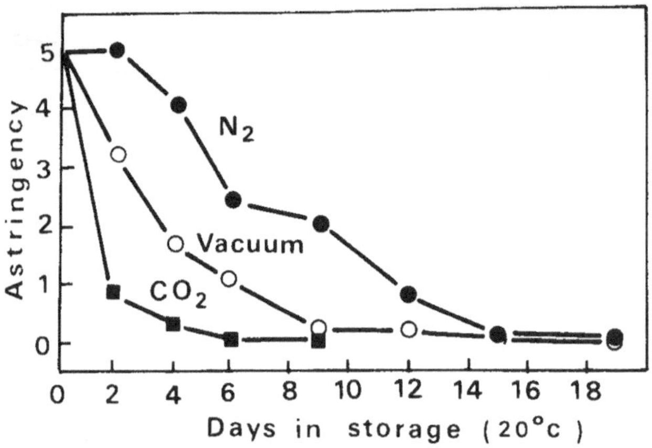

FIGURE 117. Effect of vacuum, 100% CO_2 and 100% N_2 in sealed polyethylene bags containing astringent persimmon fruits on removal of astringency at 20°C. (From Pesis, E., Levi, A., and Ben-Arie, R., *J. Food Sci.*, 53, 153, 1988. With permission.)

acetaldehyde production would be pyruvate decarboxylase, whose activity is stimulated under CO_2.[58]

All these studies were carried out on persimmon, and although natural deastringency mechanisms can differ from one fruit to another (Chapter 3), it would be interesting to study more closely the effect of partial anaerobiosis and of acetaldehyde on the evolution of tannins in the numerous fruits which are kept in cold storage under CO_2-enriched, low-O_2 atmospheres. In addition, the links between polyphenols and acetaldehyde may also be involved in the evolution of the color and astringency of fermented drinks, such as wine (see below) or other beverages derived from fruits.

D. PHENOLS AS NATURAL ANTIOXIDANTS: THE CASE OF OLIVE OIL

Virgin olive oil is the only edible oil which is used directly without being refined. It is also the only one to contain appreciable quantities of natural polyphenols since these compounds disappear during the various stages of refining (neutralization, bleaching, deodorization).[395] The polyphenol content is thus a characteristic of virgin olive oils and can be used to distinguish them from refined oils.

Phenolic compounds separated by gas chromatography and HPLC have been identified as cinnamic (caffeic, *p*-coumaric, *o*-coumaric) and benzoic (vanillic, protocatechuic, *p*-hydrobenzoic, gentisic, syringic) acids, flavones (luteolin and apigenin), and above all tyrosol and 3,4-dihydroxyphenylethanol (or hydroxytyrosol).[112-114,155,232,233,338]

The polyphenols contents of olive oils depend on extraction techniques,[339,340] the varieties used, and above all on the physiologic condition of the fruits. The point discussed in the context of fruits (Chapter 3) arises again here for oil: there is a high polyphenols content in oils extracted from very green fruits (536 ppm) and a much lower content (54 ppm) or even traces in those extracted from ripe fruits.[113]

The consequences of the presence of phenolic compounds in virgin olive oil have been examined with regard to criteria of quality, stability, and sensory characteristics of these oils.[341,395] There is no correlation between the phenolics content and the chemical characteristics of quality of the oil (acidity, peroxide index) and only a fairly small correlation with flavor. However, there is a very good correlation between the stability of olive oils and their phenolics content ($r = 0.8$). This relation is explained by the inhibitory role of auto-oxidation played by these compounds and, in particular, by 3,4-dihydroxyphenylethanol. This compound, which results from hydrolysis of oleuropein, is the only one which oxidizes spontaneously in contact with air because of its *o*-diphenolic nature. It is also the only one to give the typical reactions of commercial antioxygens, such as the 2,6-dichloroquinonimide. Other phenolic compounds present in the oil may, nevertheless, play an additional antioxidant role. Indeed, many of the flavonoids and related compounds have strong antioxidant characteristics in lipid-aqueous and lipid-food systems such as flavones, flavonols, flavanones, flavanonols, and cinnamic acid derivatives.[231,280] In a more general manner, the fruits may be a good source of natural antioxidants for use in foods.

IV. PHENOLS AND BEVERAGES MADE FROM FRUITS

A. PHENOLICS IN WINES

The quality of a wine depends on numerous parameters and phenolic compounds are important among these.[328] The sensory contributions of phenols mainly concern color and flavor (odor and taste). These aspects have been reviewed previously[308,325,329] and information will also be found in preceding paragraphs (Chapter 4) on the role played by phenols in the color, astringency, bitterness, and aroma of fruits. The results obtained also concern fermented beverages, such as wine and cider where the presence of ethanol may, nevertheless, partially modify the physiological response. Study of simplified models which are similar to wine but do not have as complex a structure has often led to improved comprehension of the mechanisms involved.[14,26,60] In addition, one of the difficulties is to establish correlations between the objective data given by analysis of phenolic compounds in wines and those given by a tasting panel.[61,154,344,369,372]

In addition to their direct role, phenolic compounds may also contribute to the sensory and chemical qualities of wine because of their interaction with other molecular types, e.g., proteins, polysaccharides or other polyphenols. These aspects have been reviewed as a whole on several occasions[286,287,325] and only discussed here are the main lines of the evolutions which lead from the initial phenolic composition of grape to that of wine.

Analysis of phenolic compounds in wine concerns both the qualitative and quantitative aspects; the proportions of the various classes present depend in particular on the age of the product. An example of this is given in Table 56.[325] In comparison with the original grapes, qualitative modifications concern first of all the appearance of certain simple phenolic compounds,[328] e.g., tyrosol, free phenolic acids (*p*-hydroxybenzoic, gallic, *p*-coumaric, caffeic, etc.), and certain flaven isomers.[430] In the case of hydroxycinnamic derivatives, one-third are found unchanged in wine, one-third are oxidized and linked with sulfhydryl compounds, and one-third are found in complexes with nucleic acids.[238,325] Volatile phenols also appear (Chapter 4) and contribute to the final aroma of the wine.[81,84,308,319,328,329] In addition, one of the most remarkable modifications consists of the appearance of complexes between tannins and various macromolecules, such as proteins and polysaccharides[105,135,136,375] in the progressive disappearance of free anthocyanins[286] and in the appearance of complexes between tannins and anthocyanins.[135,136,286,325,342,369]

One of the origins of new phenolic compounds in wine should be sought at the level of the transformation (enzymatic or not) of the substances previously present in grapes. The enzymes of the plant material itself and the microorganisms responsible for fermentation may lead for example to oxidation and hydrolysis of native compounds.[240,328] Yeasts increase phenol contents by converting nonphenolic substances and influencing extraction and solubility through the alcohol produced.[329] For example, tyrosol is produced in significant amounts from nonphenolic precursors by yeast fermentation. Fermenting lactic bacteria may also generate or modify the phenols in a wine.[325]

The immediate environment of must and wine may also account for new compounds (Table 56), in particular contact with barrel wood. In this case, small molecules (simple phenols either volatile or not, certain flavonoids) and also hydrolyzable tannins can be

TABLE 56
Gross Phenol Composition Estimated in Gallic Acid Equivalent 1^{-1} for Typical Table Wines from *Vitis vinifera* Grapes

Phenol class	Source[a]	White wine Young	White wine Aged	Red wine Young	Red wine Aged
Nonflavonoids, total		175	160—260	235	240—500
Cinnamates, derivatives[b]	G, D	154	130	165	150
Low volatility benzene deriv.	D, M, G, E	10	15	50	60
Tyrosol	M	10	10	15	15
Volatile phenols	M, D, E	1	5	5	15
Hydrolyzable tannins, etc	E	0	0—100	0	0—260
Macromolecular complexes					
Protein-tannin	G, D, E	10	5	5	10
Flavonoids total		30	25	1060	705
Catechins	G	25	15	200	150
Flavanols	G, D	tr	tr	50	10
Anthocyanins	G	0	0	200	20
Soluble tannins, derivatives	G, D	5	10	550	450
Other flavonoids, derivatives	G, D, E, M	?	?	60?	75?
Total phenols		215	190—290	1300	955—1216

[a] D = Degradation Product, E = Environment, Cooperage, G = Grapes; M = Microbes, Yeast
[b] Values for hydroxycinnamates are probably too high.

Data from Reference 325.

progressively eluted from wood and modify the initial composition of wine[161,273,325] and its organoleptic qualities.[209]

The final phenolic composition of a wine depends principally on three sets of parameters: first, the phenolics in the grapes from which the wine is made, then the extraction and winemaking techniques as a whole, and, finally, the numerous chemical reactions which occur during aging. The result is a product which acquires the gustative qualities sought by the enologist and the discerning consumer. These aspects are examined in turn.

1. Influence of the Raw Material

The characteristics of the grapes used determine to a great extent the quality of the end-product. Several aspects are to be taken into account:

1. The grape variety or combination of varieties used. Numerous examples have been given (Chapter 1) which show the important differences in the phenolic make-up of different cultivars of *Vitis vinifera* and other species of *Vitis* (see Table 10 for hydroxycinnamic derivatives for example). The first major difference concerns the presence or absence of anthocyanins. The color of red wines is connected directly with the anthocyanic composition of the grapes[24,388] although numerous other factors are involved in determination of the final color. Cultivar variations are, nevertheless, much greater than seasonal and site variations.[22] The anthocyanin composition of wines from *Vitis vinifera* is sufficiently close to that of the original grapes for it to be possible to distinguish them from hybrid wines.[286]
2. The degree of maturity of the grapes.[285,289] This question is covered at length in Chapter 3 and shows that important variations can occur during the final weeks before picking, and in particular increase in anthocyanin and some tannin contents and fall in hydroxycinnamic derivative contents.
3. Cultivation conditions, nature of the soil, and local environmental conditions. These

may lead to considerable variations in tannins, anthocyanins, and hydroxycinnamic derivatives.[39,286,332] However, each variety conserves a characteristic phenolic pattern which is independent of the production zone,[36,288] and it has been impossible to date to distinguish between harvests from different Bordeaux vineyards by examining the phenolic compounds alone.[287]

4. Variability of phenolics content according to the year. For example, over a 5-year period the anthocyanin content of berry skins of Cabernet-Sauvignon berries varied between 0.22 and 0.92 g/200 berries and the tannin content varied from 0.36 to 0.76 g/200 berries.[287]
5. The physiological condition and state of health of the grapes at picking.[289,329] The rate of contamination by *Botrytis cinerea* can be related to the subsequent behavior of must and wine.[67] In addition, in certain abnormal grapes, the phenolic fraction and the browning capacity of white wines are lower and the wines display poor organoleptic qualities.[434]
6. The relative contribution of the various constituents of bunches of grapes to the phenolic content of musts and wines. Stalk phenols are partly different — qualitatively and quantitatively — to those of the different parts of the berry. Proanthocyanidins are very unevenly distributed: 22% in stalks, 56% in seeds, 21% in skins, and only 1% in berry pulp.[41] In addition, their properties may vary according to the tissue that they are from. Depending on their solubility and location in the various tissues involved, juices at fermentation contain different proportions of flavan-3-ols, anthocyanins, dimeric procyanidins, and oligomeric proanthocyanidins of different size and constitution.[135,416] Duration of pomace contact of white grapes prior to pressing modifies the phenolic composition of the wines obtained by increasing astringency, caftaric acid content, and sensitivity to browning.[61,334] The high tannin content of stalks may result in wines that are too astringent and so destemming is now a common practice in the making of red Bordeaux wines.[287]

2. Influence of Vinification Techniques

The vinification process includes both physical phenomena, such as the diffusion of phenolic compounds from cells into must and chemical phenomena involving both the reactions triggered by the decompartmentation of enzymes and substrates in berry cells and the reactions associated with fermentative mechanisms.

The duration of vinification has a great influence on the total phenolics and anthocyanidin content of the must[249] (Figure 118), and several hypotheses have been put forward to account for the variations in color during winemaking.[287,342] They take into account the increase in the relative proportions of colored polymers and anthocyanins in relation to maceration time. A large proportion of the polymers was formed during skin fermentation in young port wines.[21]

In conjunction with time, several other parameters strongly influence the extraction of compounds from the whole berry or from seeds,[17,248,249,287,290,329] and, in particular, the crushing of grapes, addition of bisulfite, temperature, ethanol content, homogenization of must, etc. Optimum combination of these factors in the winemaking techniques used results in wines which are more or less rich in anthocyanins and tannins with an average of 1100 to 1800 mg GAE per liter (GAE = gallic acid equivalent) for red wines, 400 to 800 for white wines, and 500 for rosés.[325] In all cases, the figures are lower than those which could potentially be obtained by complete extraction, and they are adapted to the purpose of the wine. Those to be drunk young will have a low tannin content whereas the vinification of a wine intended for long aging requires a higher tannin concentration.[287] Thus, in Bordeaux wines, 4 or 5 d of skin contact is sufficient for a colored, fruity, low tannin wine for early drinking, whereas a wine for aging requires a longer period of maceration which increases

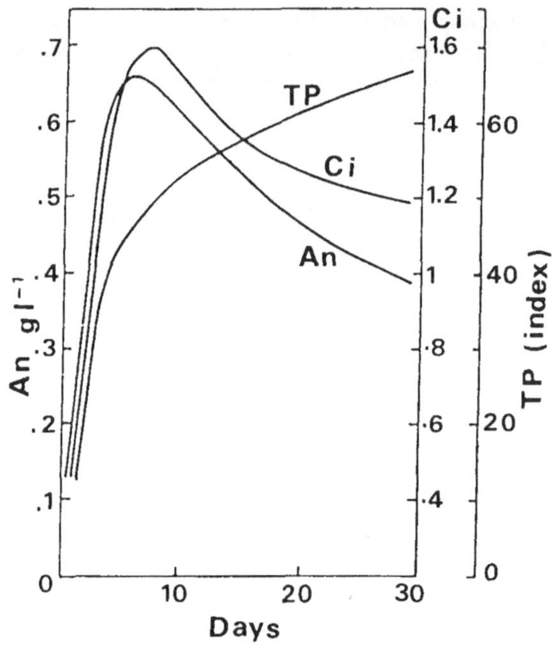

FIGURE 118. Effect of maceration on the evolution of anthocyanins (An), color intensity (Ci), and total phenolics (TP) during vinification. (Adapted from Ribereau-Gayon, P. and Glories, Y., *Proc. 6th Austr. Wine Ind. Tech. Conf.*, Australian Industrial Publisher, Adelaide, 1986, 247.)

the tannin content even if the wine is less colored initially.[287] Vigorous crushing enhances the extraction and diffusion of tannins and anthocyanins, but the corresponding wines are frequently too astringent and bitter. In addition, enzymatic oxidation of phenolic compounds is enhanced, which can lead to browning of the must and a distinct modification in organoleptic qualities. The effects of different treatments to control oxidation of grape have been investigated using the oxygen uptake rate.[10] Low temperature, pectinase, SO_2, inorganic chloride, and free *p*-coumaric acid have been found to be effective in limiting O_2 uptake. In the case of hydroxycinnamoyltartaric derivatives, oxidation in the absence of sulfites results in a considerable fall in the caffeoyltartaric acid content and simultaneous increase in glutathionyl caftaric acid.[327,330,335] In fact, oxygen consumption is directly related to the glutathione/hydroxycinnamic acid ratio rather than to the substrate content alone.[291] Artificial aeration of musts also results in a substantial fall in the hydroxycinnamic derivatives content.[119] Hyperoxidation of white grape musts prior to fermentation induced loss of all phenolic constituents and increased to resistance of the wine to subsequent browning.[61] Oxidative degradation phenomena are probably accompanied by an active role of yeasts in the loss and conversion of hydroxycinnamic esters.[240] However, the direct organoleptic consequences of the partial disappearance of these compounds is still a subject of discussion[61,244,347,400] although it has been shown that oxidation of must significantly increased perceived bitterness in certain wines.[239]

In comparison with traditional vinification, certain techniques use the natural fermentation which can take place in berries under anaerobiosis independently of the presence of microorganisms. In winemaking in which carbonic maceration is used, on the subject of which an update has been published recently,[92] the anaerobic metabolism of the berries is triggered after initial enrichment with exogenous CO_2. During this initial fermentation phase which characterizes this type of winemaking, phenolic compounds diffuse steadily from the

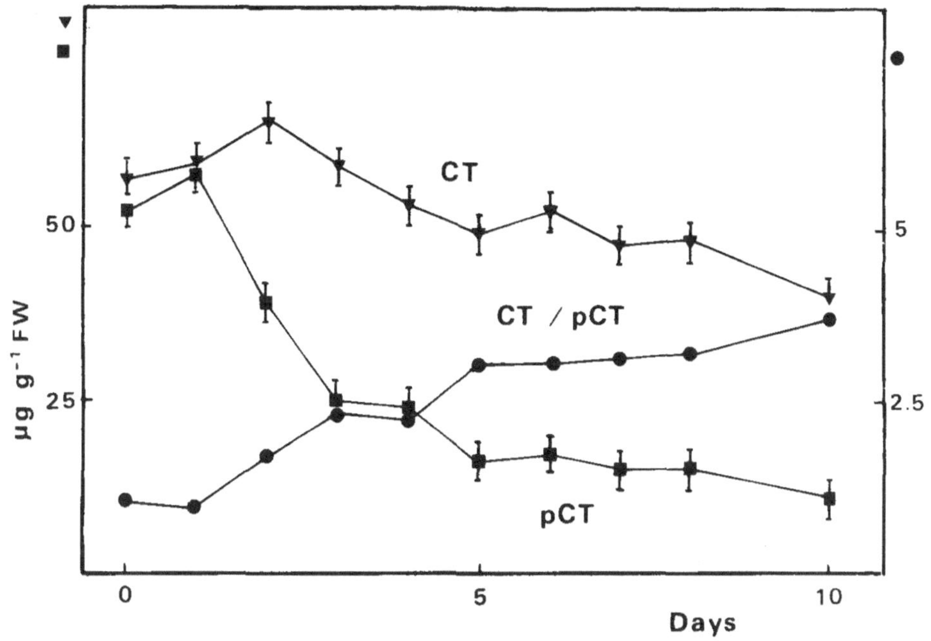

FIGURE 119. Changes in hydroxycinnamoyltartaric esters in cv. Carignane grape under anaerobiosis (CO_2 atmosphere). (CT) caffeoyltartaric, (*p*CT) *p*-coumaroyltartaric acids. (From Ramos, T. and Macheix, J. J., unpublished data).

skin into the pulp, with the phenomenon being accelerated by the temperature.[40] It is observed, as in conventional yeast fermentation, that there is a fall in anthocyanin content and an increase in the polymerized forms which are the product of condensation between tannins and anthocyanins. There is also a rapid decrease in hydroxycinnamic derivatives at the beginning of fermentation (Figure 119), but the degradation rate of *p*-coumaroyltartaric acid is greater than that of caffeoyltartaric acid in all cases. It is probable that hydroxycinnamoyltartaric esters are hydrolyzed, but free *p*-coumaric and caffeic acids do not accumulate and may account for part of the volatile phenols which subsequently appear.[308] An interesting feature of wines produced by carbonic maceration is their special aroma; this is harmonious and sometimes more intense than that of traditional wines.[92] Several volatile phenols (vinyl and ethyl 4 phenols, vinyl and ethyl 4-gaiacol, methyl and ethyl vanillates, ethyl cinnamate, etc.) may form during either the initial fermentation phase or subsequent yeast or lactic fermentations.[80-82,84,91,92]

The polyphenol composition of carbonic maceration wines is closely linked to the combination of temperature and duration of fermentation itself: 10 d at 30 to 32°C to obtain wines rich in tannins, 6 d at 32°C or 12 d at 25°C to obtain a wine that is more colored than tannic.[92]

3. Evolution of Phenolic Compounds during the Aging of Wine

This evolution mainly involves two parameters which characterize red wines: color and astringency. Tannins and anthocyanins are the two groups of compounds which are most directly involved. It has been pointed out (Chapter 4) that the color of anthocyanins is the result of a complex set of factors among which pH plays a particularly important role. In a young wine, there is already an appreciable amount of combined forms in addition to free forms which may be colored or not.[286] During aging, the proportion of free anthocyanins falls and then disappears completely after a period of time which varies with the nature of the wine and the storage conditions.[240,286,325] Degradation is not always linear with time.[228]

It is accompanied by a remarkable change in color, which moves from violet-red to a somewhat brownish-red color which characterizes aged wines and is a criterion of quality. Aging is also accompanied by a decrease in astringency which reaches a level appreciated by the consumer.

There are two successive phases in aging:[287] first, maturation in barrels or tanks where oxygen is present, and a second phase in which the wine is in bottles and protected from air to a considerable extent. The chemical and sensorial evolution of a wine thus depends on the possible presence of oxygen and also on the nature and concentration of phenolic compounds, temperature, bisulfite concentration, and on all the initial physicochemical characteristics (pH, polysaccharide content, proteins, metal ions, etc.). Numerous studies have been carried out on the browning of wines and on white wines in particular.[77,259,303,325,327,417] It has now been shown clearly that there is a good correlation between browning capacity and the flavonoid and hydroxycinnamic ester content when wine is exposed to air.[18,325,327] Oxidation kinetics differ according to the constituents in a complex mixture of phenols,[247] probably because of their variable oxido-reduction potentials and the coupled oxidation phenomena which may occur.[60] In comparison with hydroxycinnamic derivatives, (+)-catechin added to wine results in maximum browning probably because of the possible condensation with tannins and anthocyanins.

The autocatalytic effect that occurs during browning is explained by the regenerative polymerization reactions of the phenols, leading to the reforming of the substrates available for further oxidation and subsequently visibly brown products.[327] A reaction sequence including enzymatic oxidation, coupled oxidation, and the formation of caftaric and catechin copolymers have been proposed (Figure 120).[60,326,327] Phenolic compounds thus have a buffer effect by using oxygen and protecting the wine from oxidative spoilage. This effect is nevertheless limited by polymerization itself and precipitation of oxidation products.[325]

During aging in the presence of air, the absorbance maximum near 520 nm decreases and absorbance near 420 nm rises. These variations correspond to the formation of wine pigments and a decrease in free anthocyanins.[287] The polymers obtained are more stable than the initial anthocyanins in relation to the pH; this results first in an increase in color dependent on aeration (Figure 121) until the condensed forms begin to precipitate. In port wines in the presence of air,[20] the losses of total pigments (measured as absorbance at 520 nm in 1 *M* HCl) and total anthocyanins (obtained as the sum of individual peaks separated by HPLC) were logarithmic with time (Figure 122). At the same time, absorbance due to polymers increased during the first 4 or 5 weeks, reached a plateau, and decreased slowly after 20 weeks (Figure 123). The percentage of color due to the polymeric material increased regularly during aging,[19] whereas the rate of degradation varied in relation to anthocyanin structure: acylated forms disappeared more rapidly.[19,240]

Logarithmic loss of malvidin 3-glucoside in the presence of catechin with or without acetaldehyde was also observed in model wine systems without air.[26] Nevertheless, increases in aldehyde contents increased the rate of loss of anthocyanins, whereas the presence of sulfur dioxide had a slowing effect.[22,287] A moderate addition of acetaldehyde diminishes anthocyanins but increases their condensation and hence their color (Figure 124). Tannins vary little but their condensation increases. If more acetaldehyde is added, polymerization of tannins is considerable and they precipitate. Tannin level and color intensity then decrease.[287]

The mechanisms involved in variations of color and astringency during the aging of red wines have been the subject of numerous studies since the work of Somers.[342] They involve simultaneous variations of anthocyanins and proanthocyanidins, and several condensation mechanisms have been proposed.[106,107,135-137,286-288,290,325,328,342,343]

According to Haslam,[135-137] proanthocyanidins can lead to two types of reaction in a slightly acidic medium: first, by the breaking of interflavan bonds, leading to hydrolytic

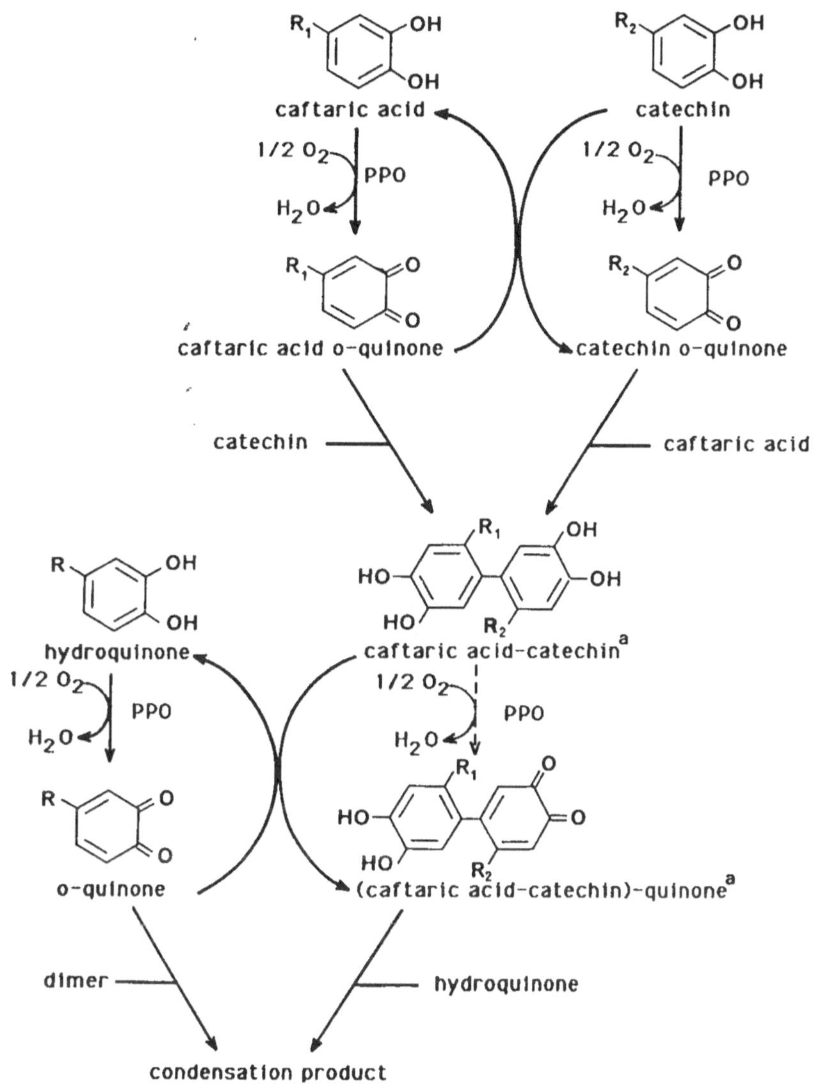

FIGURE 120. Proposed reaction sequence for the enzymatic oxidation of catechin and caffeoyl tartaric mixtures, the formation of copolymers and their subsequent oxidation. R_1 and R_2 are remaining part of caftaric acid and catechin molecules. (From Cheynier, V., Osse, C., and Rigaud, J., *J. Food Sci.*, 53, 1729, 1988. With permission.)

degradation which releases carbocations, and also by electrophilic substitution of the phloroglucinol A-ring of the flavan-3-ol units, leading to condensation products either with aldehydes or with the flavan carbocation. There are mildly acidic conditions (pH 3 to 4) in red wines in which the two carbon-carbon bond-breaking and bond-making processes occur simultaneously.[136] Equilibrium between the different forms of tannins moves away from the soluble oligomeric forms, leading to a fall in solubility, precipitation of the condensed forms and decreased astringency.

An electrophilic condensation reaction also occurs in the presence of acetaldehyde.[136] It involves oxidative phenomena and their control during the aging of wine,[286] with the formation of acetaldehyde thus deriving from oxidation of ethanol. Self-oxidation of the *o*-diphenol groups in wine leads to intermediate formation of H_2O_2 and a coupled oxidation of ethanol to acetaldehyde.[287,421] Cross-linking reactions of separate oligomeric flavan-3-ol

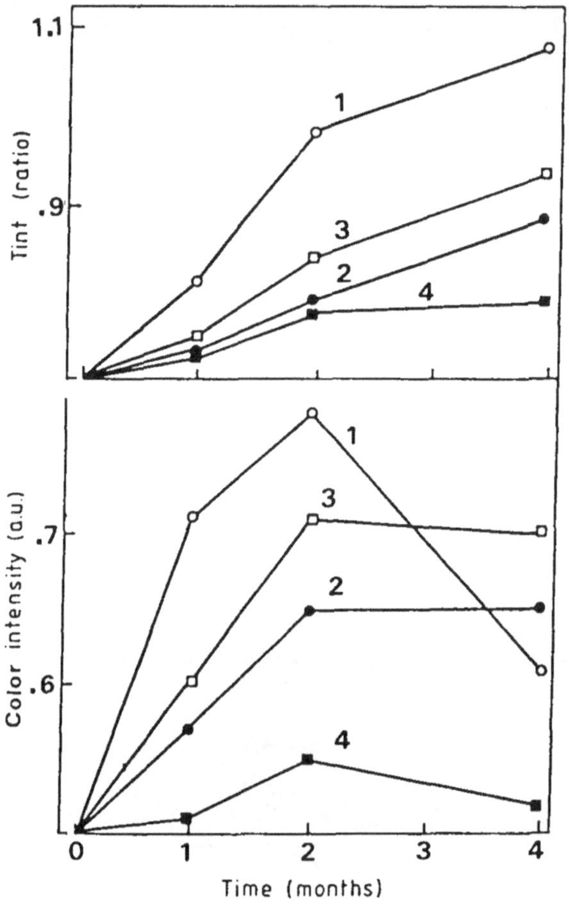

FIGURE 121. Influence of aeration on the color of wine. (1) without aeration; (2) one aeration every 4 weeks; (3) one aeration every 2 weeks; (4) one aeration every week. (Adapted from Ribereau-Gayon, P. and Glories, Y., *Proc. 6th Austr. Wine Ind. Tech. Conf.*, 1986, 247.)

molecules by acetaldehyde (Figure 125) are thought to lead to precipitation from solution, causing both a decrease in astringency and an evolution of color. Phenomena of the same type may account for the natural deastringency of certain fruits as they ripen (Chapter 4). In addition, a similar mechanism involving acetaldehyde has been involved in condensation between proanthocyanidins and anthocyanidins in wine[287,370,371] where it is thought to accentuate the development of color.

Another mechanism is also thought to cause modifications in wine color. This involves direct condensation of anthocyanins with certain phenolic substrates, such as tannins.[136,286,342] According to Somers,[342] anthocyanins are sensitive to attack by nucleophilic reagents including phenolic substrates, such as (+)-catechin in a sequence of reactions which has now been clearly described (Figure 126). The condensed anthocyanin species obtained are chemically much more stable but can react with additional oligomeric proanthocyanidins to give pigmented polymerized forms which precipitate.

All the processes described above (acid-catalyzed breakdown and polymerization of flavan-3-ols, electrophilic substitution and polymerization involving anthocyanins or acetaldehyde) or others involving the direct degradation of anthocyanins by mechanisms which remain to be elucidated[286] intervene simultaneously in the aging of wine. It is, in fact,

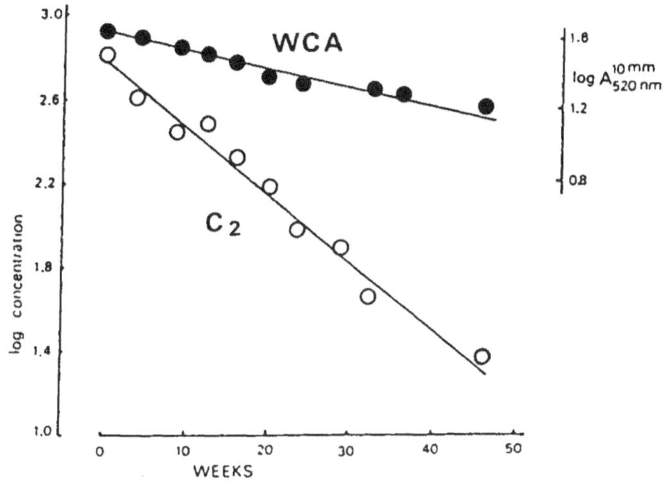

FIGURE 122. Logarithmic changes of total pigment color (WCA as $A_{520\,nm}^{10\,mm}$) and total anthocyanins determined by HPLC (C_2 as mg/l) of a port wine. The scales are such that log concentration = log 18.9 $A_{520\,nm}^{10\,mm}$. (Adapted from Bakker, J., Preston, N. W., and Timberlake, C. F., *Am. J. Enol. Vitic.*, 37, 121, 1986.)

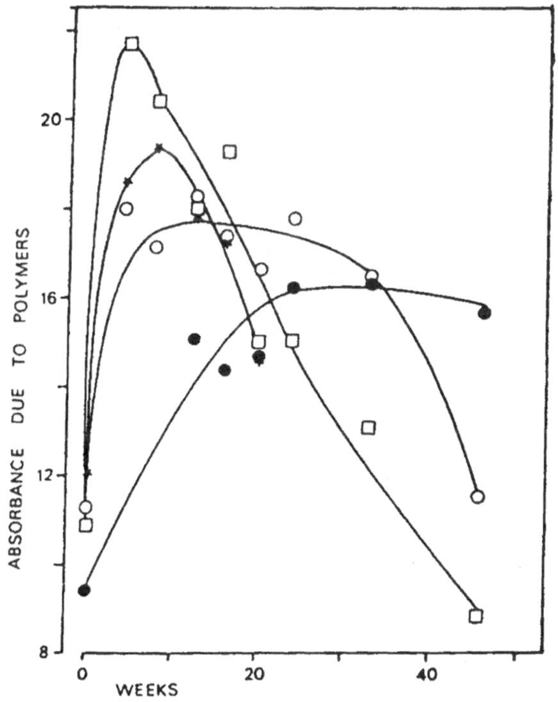

FIGURE 123. Absorbance due to polymers during aging of port wines. (○) control with a normal content of aldehyde; (★) very low aldelyde level; (□) with spirit and 200 $mg \cdot l^{-1}$ acetaldehyde; (●) with spirit and 100 $mg \cdot l^{-1}$ SO_2. (From Bakker, J., *Vitis*, 25, 203, 1986. With permission.)

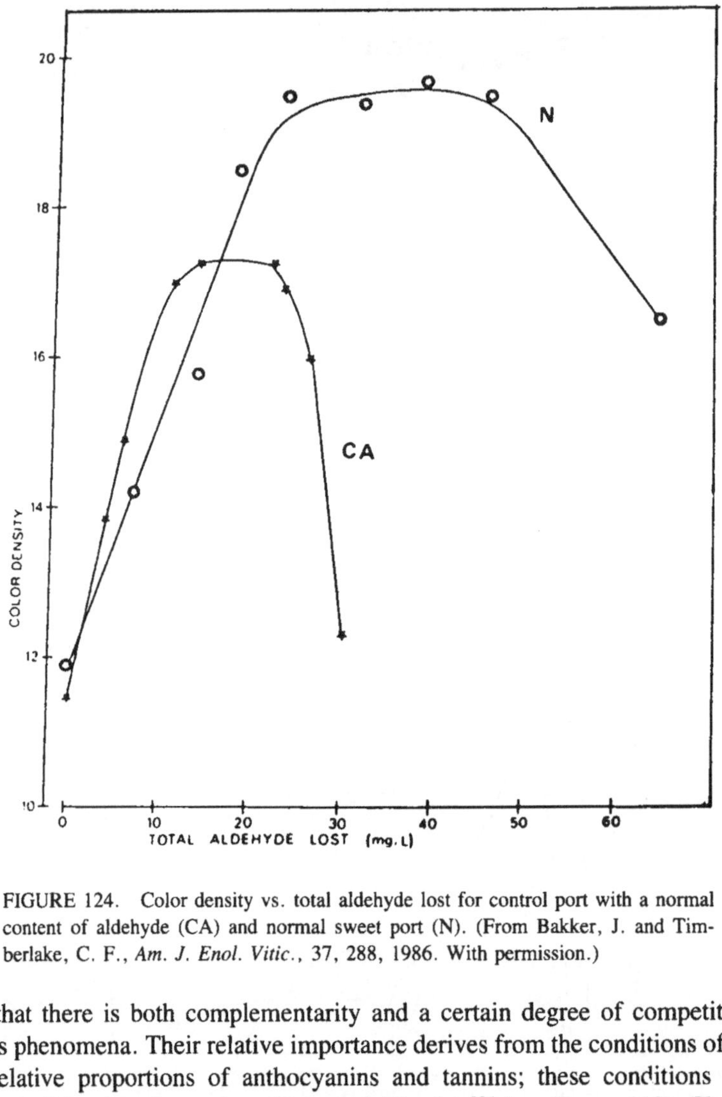

FIGURE 124. Color density vs. total aldehyde lost for control port with a normal content of aldehyde (CA) and normal sweet port (N). (From Bakker, J. and Timberlake, C. F., *Am. J. Enol. Vitic.*, 37, 288, 1986. With permission.)

probable that there is both complementarity and a certain degree of competition between the various phenomena. Their relative importance derives from the conditions of the medium and the relative proportions of anthocyanins and tannins; these conditions are partly a consequence of the duration and method of vinification[286] (see Figure 118). The availability of oxygen during the various stages leading from grape to aged wine is also an important factor which controls biochemical changes.[287,346]

A great deal of progress has been made in recent years in obtaining better objective knowledge of the modifications which occur during winemaking. The aim of knowing the individual variations of the phenolic compounds in wine that was put forward by Singleton and Esau[328] has now been partially reached, at least for the little-polymerized molecules. Nevertheless, the chemical nature of condensed forms is still imprecise, and correlations which are as fine and as complete as possible remain to be established between biochemical data and the sensory qualities of wine.

B. CIDERS AND APPLE JUICE

As in wine, the phenolic composition of apple juice and ciders depends first on the initial composition of the fruit and then on the extraction and solubilization of the phenols during pressing. Finally, modifications occur during the process with a major one leading to the appearance of the characteristic brown color caused by enzymatic oxidation of the phenolic compounds.

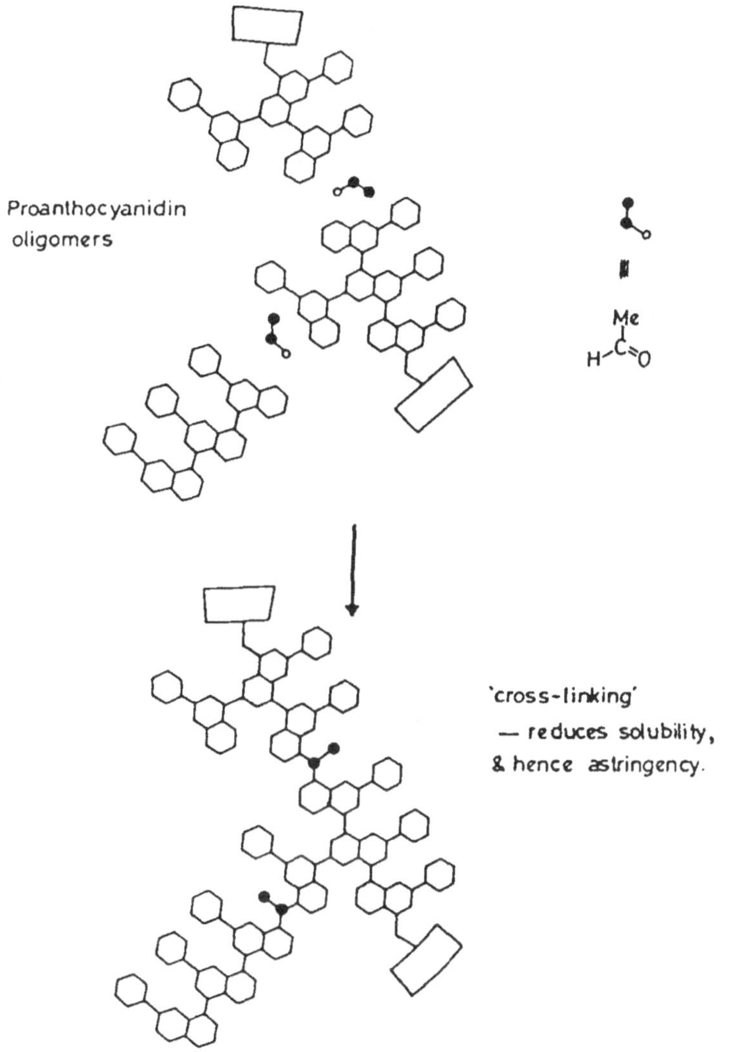

FIGURE 125. Cross-linking reactions of oligomeric proanthocyanidins and acetaldehyde. (From Haslam, E. and Lilley, T. H., *CRC Crit. Rev. Food Sci. Nutr.*, 27, 1, 1988. With permission.)

We have already shown at length (Chapters 1 and 2) that the main phenolic constituents of apple are phenolic acids, catechins. procyanidins, and flavonol glycosides. In addition, phloridzin is present in seeds and peel. The concentrations and relative proportions vary from one cultivar to another and in relation to the degree of fruit maturity, and the compounds are involved in enzymatic browning and haze formation during juice manufacture. They also contribute to the bitterness and astringency of both juices and ciders and appear to be one of the determining parameters in the classification of the different apple cultivars in homogeneous technological groups.[70]

Column chromatography has been widely used to separate the different phenolic compounds in ciders and apple juices, either for analysis of monomeric phenolics[420] or for analysis of procyanidins.[187,193] In addition, HPLC techniques enable good separation of both monomeric and polymeric phenolics in apple juice[188,422] and monitoring, in combination with a pH shift technique, of their evolution during enzymatic browning.[189]

The monomeric phenolics found in ciders vary quantitatively with the nature of the

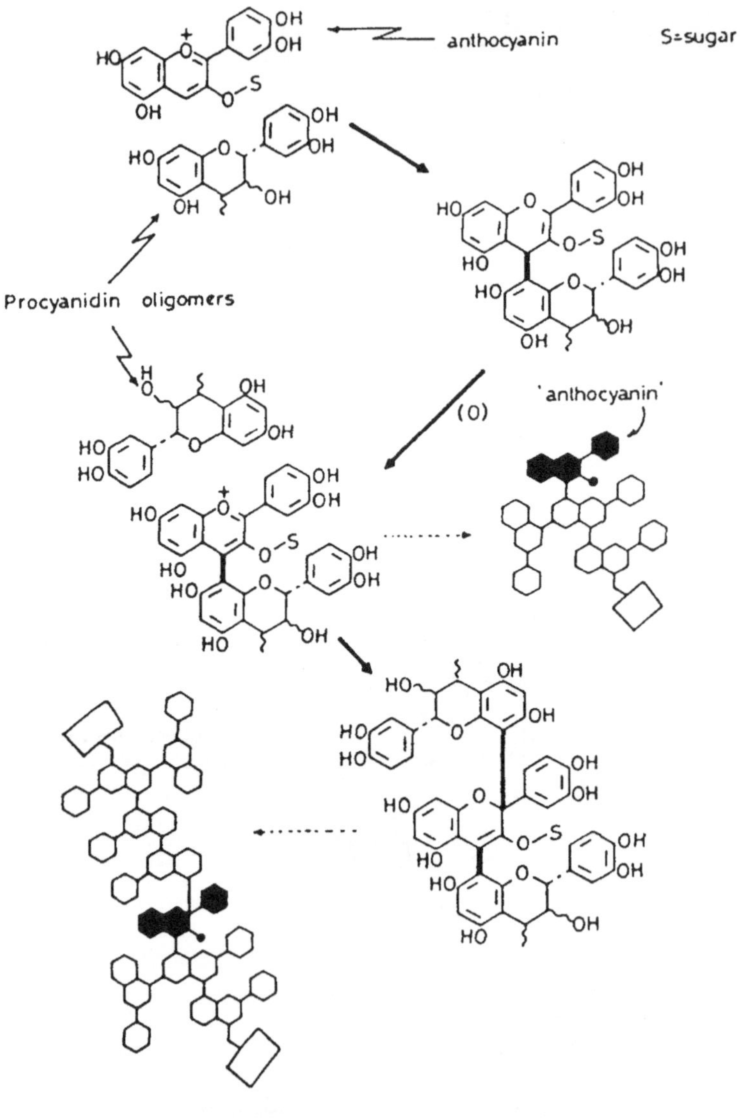

FIGURE 126. Condensation of oligomeric proanthocyanidins and anthocyanins in red wines. (From Haslam, E. and Lilley, T. H., *CRC Crit. Rev. Food Sci. Nutr.*, 27, 1, 1988. With permission.)

apples used (Table 57). The chlorogenic acid content of cider apples of various provenances lies between 0.42 and 0.157% (w/v)[420] and appears to be distinctly higher than that of mature dessert apples or juice made from the latter.[204,236,386] Free hydroxycinnamic acids and caffeic acid, in particular, are only found as traces in freshly made juice but are abundant in certain ciders; this is probably caused by hydrolysis of the combined forms during fermentation by microorganisms.[420] Phloridzin and phloretin xyloside concentrations are fairly high in certain cultivars (Table 57) and may contribute to the bitterness of ciders.

As in the apples from which they are made, various oligomeric and polymeric procyanidins have also been characterized in ciders, and in particular dimers B-1, B-2, B-5, trimer C-1, and more polymerized forms going as far as the heptamer.[187,188,193] The main series is based on (−)-epicatechin, but mixed series of catechin-epicatechin also exist. Procyanidins are found at high levels (2 to 3 g/l) in ciders prepared from bittersweet apples. They are responsible for a large proportion of the bitterness and astringency of the beverage.

TABLE 57
Monomeric Phenolics of Ciders from a Range of Cultivars (% w/v)

Cultivar	Chlorogenic acid	*p*-Coumaroylquinic acid	Catechins	Phloridzin	Phloretin xyloglucoside
Frederick	0.42	0.007	Not detected	0.007	0.010
Dabinett[a]	0.40—0.057	0.003	0.03—0.05	0.015—0.016	0.004—0.012
Michelin	0.051	0.006	0.025	0.007	0.004
Vilberie	0.073	0.043	0.04	0.003	0.003
Yarlington Mill[a]	0.068—0.15	0.062—0.079	0.06	0.003	0.002
Kingston Black	0.025 0.034 (caffeic acid)	0.003	0.010	0.005	0.006

[a] Values for different years.

Data from Reference 420.

As already discussed (Chapter 4), bitterness is associated with oligomeric procyanidins whereas astringency is characteristic of procyanidins with higher molecular weights.[191]

Apart from these varietal aspects, two principal parameters can affect the phenolic composition of ciders: cultivation conditions and processing conditions. Very large variations in tannins (more or less condensed procyanidins), running from 2 to 10 g/l, are found among apples classified as bittersweet. Cultivars with high tannin contents might be useful for blending with low-tannin varieties. However, the criterion of a high-tannin content is insufficiently taken into account to the advantage of orchard performance.[192] In intensive cider orchards, the high levels of nitrogen fertilizer that are applied to increase productivity lead to a distinct decrease in the tannin content of fruits and, more generally, in all phenolic fractions. As a result, there is a decrease in the quality of the ciders, aggravated by modification of the microbial stability of the product.[192]

Factors affecting yield of phenolics during cider processing have been studied.[194] Three main possibilities of loss may be identified: oxidation during milling and before pressing, incomplete extraction of apple tissue, and clarification. The longer the juice remains in contact with air and pomace, the greater the loss of polymeric procyanidins. In contrast, pectinolytic enzyme treatment or warm incubation of apple pulp before juice extraction increased the procyanidin level by 50%. Furthermore, use of a diffuser extractor doubled the phloridzin and polymeric procyanidins contents, whereas subsequent clarification of ciders using gelatine led to significant (up to 20%) loss of both tetrameric and polymeric procyanidins. In all cases, a decrease in procyanidins led to lessened astringency and bitterness judged as being negative by a tasting panel.[194]

A fairly complete study carried out on Golden Delicious made it possible to estimate the influence of different treatments on the phenolic composition of apple juices and their sensory consequences.[387] Two main parameters were studied: first, the effect of oxidation and second, different extraction conditions. Total phenolics content ranged from 25 mg/100 ml in juices from well-oxidized pulps to 150 mg/100 ml in juices where enzymatic oxidation was minimal. The different groups of phenolic compounds studied displayed differences in sensitivity to oxidation (Table 58). Flavonol glycosides were not very sensitive, but their content in finished juices was very largely dependent on the length of time the pulp was held before pressing. These results are caused by the special tissue compartmentation of these compounds (Chapter 1); they are mainly located in epidermal tissue where the cells suffer much less destruction than pulp cells during the pulping operation. Thus, only very small amounts of flavonol glycosides leave unbruised epidermal tissue if extraction is of short duration. In contrast to flavonols, the other flavonoids (essentially monomeric and

TABLE 58
Phenolic Compounds in Apple Juice

Juice procedure	Flavonol glycosides[a]	Other flavonoids[b]	Chlorogenic acid	Other nonflavonoids[c]
Conventional	0.4	20	14	17
Rapidly pasteurized	0.2	71	23	8
NaCl-treated pulp	0.3	73	26	10
Oxidized pulp, 1 h	0.7	9	<1	15
Oxidized pulp, 3 h	6.0	1.0	<1	21
Held and nonoxidized pulp	8.0	101	32	20
Oxidized and enzyme-treated pulp	6.5	12	3	32
PVP and enzyme-treated pulp	8.5	12	<1	32

Note: All values in mg/100 ml.

[a] Expressed as rutin.
[b] Expressed as catechin.
[c] Expressed as chlorogenic acid.

Data from Reference 387.

polymeric flavans) and chlorogenic acid are very sensitive to oxidation, whether this is enzymatic or not (Table 58). It would thus appear that in juice the relative contents of chlorogenic acid, flavonol glycosides, and other flavonoids may be modulated by process conditions. The presence or absence of phenolics in the final apple juice depends to a great extent on factors which include temperature, the presence of SO_2, and the duration of oxidation in the presence of pulp.[190] If the pulp is oxidized before pressing, many of the phenols will be adsorbed on the apple cell debris as brown polymerized derivatives. The juice obtained will be pale in color. On the other hand, if oxidation is carried out on the juice itself in the absence of pulp, oxidized phenols will give the juice a brown color and it will haze simultaneously.[163,387]

Process conditions may, therefore, be determinant for the stability of apple juices during subsequent storage. Indeed, rapidly pasteurized juices darkened most during storage at 25°C whereas juices from oxidized pulp displayed little browning.[387] Flavonoids, and flavanes in particular, are the major compounds which intervene during storage. The samples with the highest initial flavonoid content developed the darkest color. Furthermore, severe darkening occurred when catechin was added.[387]

C. AUTHENTICITY OF FRUIT PRODUCTS AND DETECTION OF ADULTERATION

The adulteration of fruit products and particularly of fruit juices and concentrates is frequently observed in all countries where these are prepared and marketed. The phenomenon covers different aspects, including among others the addition or removal of certain constituents which characterize the normal product and, as a result, make it appear better than it is.[353] For example, orange juice (from *Citrus sinensis*) has historically commanded a higher price than grapefruit juice (from *C. paradisi*). There is, therefore, a temptation to add the lower-priced juice to the more expensive one. However, this does not agree with the definition of orange juice in most countries. Detection and control of adulteration obligatorily requires good chemical knowledge of the initial fruit and the product derived from it. Among the different groups of compounds which may serve as unequivocal markers in a fruit product, phenolics are extremely good for the various reasons discussed at length in preceding chapters and which can be summarized as follows.[178]

Ubiquity — All the fruits and fruit juices which have been studied contain a large number of phenolic compounds in easily measurable quantities.

Specificity — The distribution of phenolic compounds is to a great extent under genetic control which has enabled them to be used as taxonomic markers for the classification of cultivars.

Multiplicity — As can be seen in Chapter 2, a particular fruit may contain several tens of phenolic compounds, which generally makes it possible to associate a true "fingerprint profile" with a certain fruit juice. This "fingerprint" should be considered both qualitatively (presence or absence) and quantitatively, for example, by establishing ratios between the levels of different phenolic constituents.

The evolution of analytical techniques has enabled great progress to be made in the detection of adulterated juices, a logical approach being that of developing a chemical profile of both adulterant and product. Overall approaches based, for example, on the spectral characteristics of orange or apple juices and other edible products[263] make it possible to identify fraudulent juices when they are very marked. A matrix method of analysis[43,45] is a compromise between overall methods or those based on a single test and the very sophisticated techniques using a computerized pattern recognition technique. It is based on a matrix of three to six components analyzed; these include anthocyanins in certain small red fruits, naringin in citrus, and chlorogenic acid in apple juice. These tests will identify fraudulent but not poorly processed apple juice.[44]

Polyphenolic profiles of most fruit juices, adulterated or not, have been obtained mainly through HPLC separation techniques.[178,428] Although most of the components of the profile are not known, they can be used for statistical processing. For example, a method based on the prior separation of carotenoids, polymethoxylated flavones, and flavonol glycosides has been perfected for orange juice.[258] The three groups of compounds are analyzed separately by HPLC and the data obtained are processed with computer program package for multivariate data.[251] Such pattern recognition analysis makes it possible to specify the authentic or adulterated category to which an orange juice belongs. 79 out of 80 authentic orange juices and 18 out of 19 adulterated juices were thus classified correctly.

The methods which combine HPLC separations and statistical processing of data are particularly interesting because of their rapidity and their reproducibility. Nevertheless, interpretation should take two complementary parameters into account:

- First, the variability which has already been discussed at length (Chapters 1 and 3) and which is related to variable growing conditions or the physiological stage of the fruit. However, we have already shown in the case of grape for example (see Table 9) that the proportions of the different constituents are much less variable than the level of each of them.
- Second, the changes associated with the processing itself and which cause considerable modifications to the most fragile molecules or those which are good substrates for *o*-diphenoloxidases. In particular, apple and pear juices display loss of chlorogenic acid, epicatechin, and procyanidin B-2 when oxidation and browning occur.[190]

On the condition that these reservations are taken into account, analysis of polyphenols appears to be a rapid and reliable way of assessing fruit juice authenticity and a good indicator of adulteration. A classic example is the utilization of grape anthocyanins to provide a method for classifying red wines:[284,286] the presence of anthocyanin diglucosides proves the hybrid nature of a grape or a wine since *Vitis vinifera* contains only monoglucosides. Physical and chemical characteristics have frequently been explored for wine authentication.[307] and anthocyanins and flavonoids have recently been used to distinguish between red wines of different varieties and from different regions.[85] Using 4 anthocyanins and catechin

TABLE 59
Caracterization of Pulp Wash Addition in Orange Juice

% Pulp wash	Narirutin/hesperidin ratio
0	0.399
5	0.480
10	0.553
20	0.721

Data from Reference 178.

selected by stepwise discriminant analysis, 96% of the 34 samples analyzed were correctly classified by grape variety. Many other examples could be mentioned, although only reported are a few of the most recent ones.

Conflicting data have been reported on the absence[116] or the presence[76] of naringin and other flavanone glycosides in orange juice. The problem was re-examined recently[298] with comparison of narirutin, naringin, hesperidin, and neohesperidin concentrations in juices from 52 citrus cultivars of different species. Naringin cannot be used as the sole indicator of the presence of grapefruit in orange juice as it has also been found in sour orange and in a grapefruit-like tangelo (K-early). Nevertheless, the concentration profile of naringin and neohesperidin could be used to distinguish between juices that may be legally added to orange juices according to USFDA and those that may not.[298]

In addition, the composition of polymethoxylated flavones (PMF) in citrus juices has been proposed for the measurement of the authenticity of juice.[175] Indeed, high concentrations of PMF are found in peel and low amounts in juice in some species of citrus,[399] and both quantitative and qualitative differences have been found between citrus varieties.[373] Liquid chromatography coupled with mass spectrometry have been used in the determination of PMF in orange peel oil and juice[121] and taxonomic significance of the levels of six PME and the sum of the flavones was shown by factorial discriminant analysis for the differentiation of *Citrus reticulata* and *C. sinensis*.[103]

The addition of pulp wash to orange juice can be detected by quantitative measurement of relative peak areas of HPLC chromatograms.[178] Pulp wash has the same components as orange juice, but the relative amounts vary. Narirutin is the predominant flavonoid in pulp wash serum, and the narirutin/hesperidin peak area ratio was used to track pulp wash addition (Table 59).

The attractive color and flavor of red raspberries have led to increased demand for red raspberry juice and juice concentrates. In order to prevent adulteration, characterization is carried out using different criteria including anthocyanins separated by HPLC.[348] In comparison with other red fruits (blackberry, blackcurrant, strawberry, sour cherry), the anthocyanin profile of raspberry is qualitatively diffferent with two different predominant patterns, of which cyanidin 3-sophoroside is always the most important. Most anthocyanin-containing fruits have qualitatively and quantitatively different anthocyanin profiles and should be detected in the case of blended juices.[428,429] As raspberry contains only cyanidin and pelargonidin (max. 5%), analysis of anthocyanidins would also be useful in detecting adulteration with colorants containing malvidin, delphinidin, peonidin, or petunidin.

Analysis of cranberry anthocyanins by HPLC is also a criterion used to determine the authenticity and the percentage of cranberry juice in cranberry juice cocktail and concentrates.[99,141,142] Good, fast separation of the four major anthocyanins can be carried out in less than 15 min.[66] In any case, the anthocyanin profile of cranberry is relatively simple (six major pigments) in comparison with the anthocyanin composition of grapes (Chapter

2). The presence of oenocyanin-type colorants in cranberry juice cocktail can thus be detected since the oenocyanin pigments produced from grape skins reflect the anthocyanin composition of the grapes even though they are highly polymerized.[213] A simple method based on a color and pigment profile enables rapid detection of adulteration in all cases where at least 50% of the cranberry juice has been replaced by a solution of oenocyanin and citric acid.[99]

Flavonol glycosides are sometimes used as proof of adulteration in food analysis. Thus, a quercetin triglycoside is a suitable indicator for the detection of adulteration of blackcurrant products with redcurrants.[323] Besides the flavonoids, hydroxycinnamic derivatives could be used as coindicators for proof of adulteration in fruits and vegetables.[424]

D. AROMA AND OFF-FLAVORS

During the crushing of fruits and subsequent processing, new compounds can appear from percursors which may or may not be phenolic, in particular under the effect of oxidation enzymes.[75] The phenolic composition of the aroma of the processed product thus depends on these chemical modifications and also on the extraction of aromatic substances from wooden recipients. Microbial origin of certain volatile substances is also probable during fermentation.[308,329]

Numerous volatile phenols are present is small quantities in wines, where they may be responsible for characteristic odors[79] and very full reviews of this subject have been published.[208,308,328,329] In the case of a wine manufactured and stored without any contact with wood, 12 volatile phenols have been identified[84] and in particular 2-ethylphenol and propiovanillone which had only been reported previously in oak-aged wines or distilled spirits. Concentrations were small in all cases, ranging from 8×10^{-4} ppm for 4-vinylphenol to 4×10^{-1} ppm for 4-ethylphenol. Wines made using carbonic maceration generally have a special aroma (Section A above).

When wine is aged in wooden barrels, vanillin and the related coniferaldehyde, syringaldehyde, sinapaldehyde, vanillic acid, and ethyl vanillate are very probably extracted from the wood by alcohol, a phenomenon which is accentuated during the aging of brandies.[281,329] However, some of these compounds may originate in the initial constituents of the wine, in particular, through the degradation of hydroxycinnamic esters.[7,92,205,308] Thus, *p*-coumaric acid, which is abundant in grape skins in the form of *p*-coumaroyltartaric acid or in acylated anthocyanins (Chapter 1) can be decarboxylated into 4-vinylphenol by lactic bacteria. A similar mechanism leads to the formation of 4-vinylguaiacol from ferulic acid which is abundant in barrel wood (Figure 127). If the double bond in the cinnamic acid chain is reduced prior to decarboxylation, 4-ethylguaiacol is then obtained from ferulic acid and 4-ethylphenol from *p*-coumaric acid. All these compounds are responsible for special odors (smoky, leather, cloves) which may contribute to an off-flavor if their concentrations are too high. In brandies, hydroxycinnamic acids lead by oxidation to aldehydes and in particular to vanillone and related compounds.[329]

Identical evolutions have been reported in ciders where *Lactobacilli* can convert chlorogenic, caffeic, and *p*-coumaric acids into 4-ethylcatechol and 4-ethylphenol where they are considered to be part of the characteristic odor of English ciders.[329] Likewise, ferulic acid is a potential precursor of off-flavors in stored orange juice;[241] under mildly acidic conditions, ferulic acid is decarboxylated nonenzymatically to give *p*-vinyl guaiacol (PVG). In orange, ferulic acid is present in bound form (feruloyl putrescine, feruloylglucose) (Chapter 2). However, minor amounts of free ferulic acid (200 μg/l), which certainly originates in the breakdown of combined forms in fresh juice, are probably sufficient as a precursor for the production of PVG at levels above the sensory threshold.[241] Pasteurization of the juice increased the release of free ferulic acid from bound forms and increased PVG formation during subsequent storage of the juice. Furthermore, orange juice incubated with added free ferulic acid displayed an accumulation of PVG during storage (Table 60) and a less acceptable

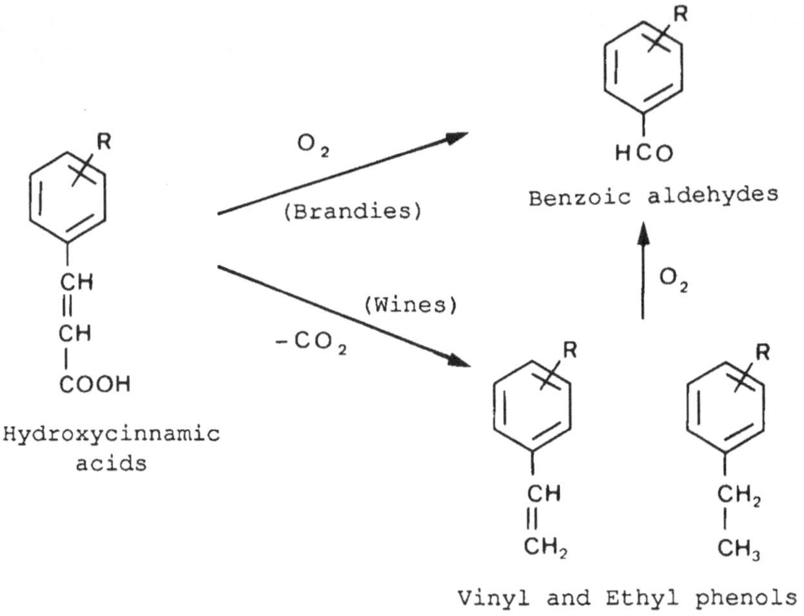

FIGURE 127. Formation of simple volatile phenols from hydroxycinnamic acids (Adapted from Fidler, W., Parker, W. E., Wasserman, A. E., and Doerr, R. C., *J. Agric. Food Chem.*, 15, 757, 1967; and Maga, J. A., *CRC Crit. Rev. Food Nutr.*, 10, 323, 1978.)

TABLE 60
***p*-Vinylguaiacol Content (μg/l) in Orange Juices Incubated 14 or 28 Days with (+ F) and without (−F) Addition of 10 mg/l Ferulic Acid**

	Control	14 d		28 d	
	4°C	35°C	50°C	35°C	50°C
Incubation					
−F	200	300	640	420	680
+F	200	580	1320	640	1800

Data from Reference 241.

aroma to a laboratory test panel. These experiments suggest that the level of free ferulic acid in orange juice provides a good index for the possible appearance of off-flavor during storage.[241]

E. POLYPHENOLS AND FORMATION OF HAZES

Hazes and deposits can occur during the storage of natural beverages, such as wine and fruit juices. These aspects have already been mentioned in the study of wines and ciders where the important role that can be played by proanthocyanidins in the bonds with proteins or other compounds has been shown. A good understanding of the interactions between tannins and proteins is of considerable practical importance for effective control and removal of hazes. The mechanisms of these interactions in plants and plant products have been the subject of in-depth studies.[355] In addition, *o*-diphenoloxidases and peroxidases can interfere with these bonds, either indirectly through the *o*-quinones produced by oxidation of low molecular weight phenolic compounds or directly insofar as the tyrosyl groups of certain proteins can serve as substrates for both enzymes.[218]

During vinification, part of the soluble grape proteins are precipitated through interaction with proanthocyanidins. Removal of protein from wines by immobilized tannic acid[415] or grape proanthocyanidins[276] has been proposed. The tannins extracted from Japanese persimmon or an immobilized form of Chinese gallotannin are also used to remove soluble proteins from sake (Japanese rice wine).[63,221]

Cloud in orange juice was caused by the crystallization of soluble hesperidin with soluble protein.[29] Hesperidin is normally present in the albedo tissue within the cell vacuole. At extraction, it comes into contact with soluble pectin which is also extracted into the medium from the cell wall. Hesperidin then begins to crystallize and to form colloidal particles which contribute to the turbidity of the albedo extract. Recognition of pectin by hesperidin was through the sugar moiety, the aglycone hesperedin giving no turbidity with pectin[31] and the effect of hesperinidase preventing the formation of cloud in the solution. A similar phenomenon of crystallization of hesperidin on pectin can occur directly in the middle lamella of the cell wall when cell compartmentation is destroyed by exposing orange to freezing conditions.[29]

V. BIOTECHNOLOGICAL ASPECTS

Plant biotechnology has developed considerably in two decades both in the field of vegetative propagation and plant improvement and in that of bioproduction and biotransformation of secondary metabolites. One of the most important fundamental aspects, nevertheless, remains that of the extraordinary interest of plant cell cultures for study of the secondary metabolism itself[65,235,419] and in particular the phenolic metabolism. It is, nevertheless, surprising to note that there has been little use of fruits in *in vitro* culture, even through they are suitable for the production of phenolic metabolites. This should probably be associated, on the one hand, with a difficulty in developing *in vitro* cultures from fruits and, on the other, with the fact that when these cultures do succeed they frequently lead to cells whose new biochemical differentiation does not necessarily reflect that of the original fruit. In the case of the genus *Vitis,* for example, anthocyanin-producing cell suspensions can be obtained either from the fruits of *Vitis vinifera*[357] or directly from vines of a *Vitis* hybrid.[431,432]

The phenolic pattern of *in vitro* cultures frequently displays distinct differences from that of the fruit from which they are derived. Thus, the flavonoid pattern of callus and callus-free juice vesicles from lemon fruits was completely different.[393] In apple, *p*-coumaroylglucose was abundant but disappeared almost completely in calli, while new compounds, such as feruloylquinic acid appeared.[265] Likewise, cyanidin 3,5-diglucoside is the major pigment in calli produced from apple cotyledons whereas fruit skin contains mainly cyanidin 3-gludoside.[149] However, in certain other cases, the qualitative phenolic pattern in fruit and *in vitro* cultures is very similar; this applies to polymethoxylated flavones in citrus[48] or anthocyanins in *Vitis.*[200] Quantitatively, phenolic contents are generally lower in *in vitro* cultures than in fruits. This comment applies to polymethoxylated flavones in citrus (65.5 and 880 μg/g DW, respectively, in callus tissue and mature orange flavedo),[48] naringin in callus initiated from albedo explants of grapefruit,[27] and to hydroxycinnamic derivatives in apple.[265] Nevertheless, these quantitative aspects can be improved by research on the conditions for optimization of biosynthesis and accumulation.[148,423] As has been reported for numerous other types of plant material, light and the sucrose, nitrate, and phosphate contents of the medium can make it possible to modulate anthocyanin production by isolated grape berry cells[357] or more generally in *Vitis.*[431,432]

In vitro cultures derived from fruits have mainly been used as a tool for studying the phenolic metabolism. They have made it possible to specify the properties of certain enzymes in phenolic biosynthesis in fruits, and, in particular, phenylalanine ammonia-lyase in citrus[366] and the *O*-methyltransferase system in apple fruit suspensions.[205] The late role of *O*-methylation and glucosylation in the biosynthesis of hydroxycinnamic esters in apple was also revealed in cell suspensions,[183,207] as was the importance of CoA ligases in the formation of various forms of acylated or nonacylated anthocyanins in *Vitis vinifera*[200] and the biotransformation of flavanones by grapefruit.[198] Regulation of capsaicin biosynthesis from phenylalanine has been studied comparatively in whole fruit and in cell suspensions and immobilized cell cultures from *Capsicum fructescens*.[124,125,199]

Calli and cell suspensions derived from fruits are also used for the study of certain enzymes which contribute to the degradation or polymerization of phenolics, and, in particular, catechol oxidases and peroxidases of apple and citrus.[9,34,264,402] These works made it possible to show that there are striking differences during the growth cycle, related to growth phases and biochemical differentiation.

An alternative to synthetic colorants could be the production of pigments by *in vitro* culture of fruits[150] either in bioreactors or in the form of immobilized cells. However, in the case of anthocyanins for which the optimization conditions are now known well, the commercial profitability of the operation still remains a subject for discussion in relation to anthocyanins currently obtained from fruit and flower sources.[150] In addition, although immobilized cells form a promising approach,[199,433] one of the difficulties encountered in the case of anthocyanic pigments is their siting in the vacuole from which they cannot be excreted into the medium. However, here again, optimization of the system is certainly envisageable as for capsaicin which, although normally accumulated in the vacuole,[100] is predominantly released into the medium and not catabolized in immobilized cells from *C. fructescens*.[199] This property makes it possible to envisage industrial production in the near future.[184]

Development perspectives of *in vitro* cultures derived from fruits thus remain limited when the purpose is to achieve commercially profitable production of a phenolic or of a group of phenolics. Nevertheless, particular attention should be paid to fruits which naturally produce original phenolic molecules of pharmacological interest; optimization of *in vitro* production can be envisaged in this case at least on an experimental basis. In addition, the enzymatic capacities of the fruit, which are frequently different to those of the rest of the plant, could be exploited *in vitro* for original biotransformations which would use natural or synthetic substrates and lead to new molecules. Finally, genetic engineering, which is currently developing strongly for plants, might possibly lead ultimately to the production of fruits in which one gene or a group of genes associated with the phenolic metabolism is manipulated to obtain useful properties in fields as varied at phytopathology (disease resistance), sensory properties (improvement of quality), technology (limited browning, etc.), and pharmacology.

REFERENCES

1. **Abers, J. E. and Wrolstad, R. E.,** Causative factors of color deterioration in strawberry preserves during processing and storage, *J. Food Sci.*, 44, 75, 1979.
2. **Abou Aziz, A. B., Abdel-Wahab, F. K., and El-Ghandour, M. A.,** Effect of different storage temperatures on phenolic compounds in banana and mango fruits, *Sci. Hortic.*, 4, 309, 1976.
3. **Adams, J. B. and Blundstone, H. A. W.,** Canned fruits other than citrus, in *The Biochemistry of Fruits and Their Products*, Vol 2, Hulme, A. C., Ed., Academic Press, London, 1971, 507.
4. **Adrian, J.,** La réaction de Maillard vue sous l'angle nutritionnel. I. Domaine de la réaction de Maillard, *Ind. Aliment. Agric.*, 10, 1281, 1972.
5. **Albach, R. F., Redman, G. H., Cruse, R. R., and Petersen, H. D.,** Seasonal variation of bitterness components, pulp and Vitamin C in Texan commercial citrus juices, *J. Agric. Food Chem.*, 29, 805, 1981.

6. **Albach, R. F., Redman, G. H., and Cruse, R. R.,** Annual and seasonal changes in naringenin concentration of Ruby Red grapefruit juice, *J. Agric. Food Chem.*, 29, 808, 1981.
7. **Albagnac, G.,** La décarboxylation des acides cinnamiques substitués par les levures, *Ann. Technol. Agric.*, 24, 133, 1975.
8. **Allan, A. C. and Walker, J. R. L.,** The selective inhibition of catechol oxidases by salicylhydroxamic acid, *Phytochemistry*, 27, 3075, 1988.
9. **Altman, A., Gulsen, Y., and Goren, R.,** Growth and metabolic activity of lemon juice vesicle explants *in vitro*, *Plant Physiol.*, 69, 1, 1982.
10. **Amano, Y., Kubota, M., and Kagami, M.,** Oxygen uptake rate of Koshu grape must and its control, *J. Ferm. Technol.*, 57, 92, 1979.
11. **Amiot, M. J.,** Degradation de l'oleuropeine de l'olive (*Olea europaea* L.) aspects physiologiques, chimiques et technologiques, Thèse Doct., University of Montpellier, France, 1986.
12. **Angel, S. and Ben-Shalom, N.,** Research on olive quality in Israel for processing. I. Prevention of brown spotting in green olives for pickling, *Confructa*, 24, 79, 1979.
13. **Angleton, E. L. and Flurkey, W. H.,** Activation and alteration of plant and fungal polyphenoloxidase isoenzymes in sodium dodecylsulfate electrophoresis, *Phytochemistry*, 23, 2723, 1984.
14. **Arnold, R. A. and Noble, A. C.,** Bitterness and astringency of grape seed phenolics in a model wine solution, *Am. J. Enol. Vitic.*, 29, 150, 1978.
15. **Asen, S., Stewart, R. N., and Norris, K. H.,** Anthocyanin and pH involved in the colour of "Heavenly Blue" morning glory, *Phytochemistry*, 16, 1118, 1977.
16. **Asen, S. K., Stewart, R. N., and Norris, K. H.,** U.S. Patent 4,172,902, 1979.
17. **Aubert, S. and Poux, C.,** Extraction des composés phénoliques du raisin. II. Taux de passage dans les vins, *Ann. Technol. Agric.*, 18, 111, 1969.
18. **Bachmann, O.,** Verbreitung von Phenolcarbonsauren und Flavonoiden bei Vitaceen, *Vitis*, 17, 234, 1979.
19. **Bakker, J.,** HPLC of anthocyanins in port wines: determination of ageing rates, *Vitis*, 25, 203, 1986.
20. **Bakker, J. and Timberlake, C. F.,** The mechanism of color changes in aging port wines, *Am. J. Enol. Vitic.*, 37, 288, 1986.
21. **Bakker, J., Preston, N. W., and Timberlake, C. F.,** The determination of anthocyanins in aging red wines: comparison of HPLC and spectral methods, *Am. J. Enol. Vitic.*, 37, 121, 1986.
22. **Bakker, J., Bridle, P., Timberlake, C. F., and Arnold, G. M.,** The colours, pigment and phenol contents of young port wines: effects of cultivar, season and site, *Vitis*, 25, 40, 1986.
23. **Balice, V. and Cera, O.,** Acidic phenolic fraction of the olive vegetation water determined by a gas chromatographic method, *Grasas Aceites*, 35, 178, 1984.
24. **Ballinger, W. E., Maness, E. P., Makus, A. J., and Caroll Jr, D. E.,** A comparison of anthocyanins and wine color quality in black grapes of 39 clones of *Vitis rotundifolia* Michx, *J. Am. Soc. Hortic. Sci.*, 99, 338, 1974.
25. **Bangerth, F.,** Calcium-related physiological disorders of plants, *Annu. Rev. Phytopath.*, 17, 97, 1979.
26. **Baranowski, E. S. and Nagel, C. W.,** Kinetics of malvidin-3-glucoside condensation in wine model systems, *J. Food Sci.*, 48, 419, 1983.
27. **Barthe, G. A., Jourdan, P. S., Mc Intosh, C. A. and Mansell, R. L.,** Naringin and limonin production in callus cultures and regenerated shoots from *Citrus* sp., *J. Plant Physiol.*, 127, 55, 1987.
28. **Bell, E. A.,** The physiological role(s) of secondary (natural) products, in *The Biochemistry of Plants*, Vol. 7, *Secondary Plant Products*, Conn, E. E., Ed., Academic Press, New York, 1981, 1.
29. **Ben Shalom, N. and Pinto, R.,** The role of hesperidin in the turbidity of orange albedo aqueous extract, *Lebensm. Wiss. Technol.*, 19, 158, 1986.
30. **Ben Shalom, N., Harel, E., and Mayer, A. M.,** Enzymic browning in green olives and its prevention, *J. Sci. Food Agric.*, 29, 398, 1978.
31. **Ben Shalom, N., Shomer, I., and Kanner, J.,** Pectin-Hesperidin interaction in a citrus cloud model system: the effect of pectin degradation, *Lebensm. Wiss. Technol.*, 17, 125, 1984.
32. **Ben Shalom, N., Kahn, V., Harel, E., and Mayer, A.,** Catechol oxidase from green olives: properties and partial purification, *Phytochemistry*, 16, 1153, 1977.
33. **Benjamin, N. D. and Montgomery, M. W.,** Polyphenol oxidase of Royal Ann cherries. Purification and characterization, *J. Food Sci.* 38, 799, 1973.
34. **Berger, R. G., Drawert, F., Kinzkofer, A., Kuntz, C., and Radoulab, B. J.,** Proteins and peroxidase in callus and suspension cultures of apple, *Plant Physiol.*, 77, 211, 1985.
35. **Beveridge, T., and Harrison, J. E.,** Storage browning of pear juice concentrate, *Can. Inst. Food Sci. Technol. J.*, 20, 170, 1987.
36. **Bisson, J. and Ribereau-Gayon, P.,** Influence du cépage et du milieu sur la composition phénolique de cinq raisins noirs, *Ann. Technol. Agric.*, 27, 827, 1978.
37. **Blundstone, H. A. W., Woodman, J. S., and Adams, J. B.,** Canned citrus products, in *The Biochemistry of Fruits and Their Products*, Vol. 2, Hulme, A. C., Ed., Academic Press, London, 1971, 543.
38. **Bonati, A. and Crippa, F.,** Stability of anthocyanosides from *Vaccinium myrtillus* L., in pharmaceutical formulations, *Fitoterapia*, 49, 10, 1978.

39. **Boursiquot, J. M.,** Contribution à l'étude des esters hydroxycinnamoytartriques chez le genre Vitis. Recherche d'application taxonomique, Thèse Doct. Ing., ENSA, Montpellier, 1987.
40. **Bourzeix, M.,** La diffusion des composés phénoliques au cours de la fermentation intracellulaire, *C.R. J. Macération Carbonique,* February 10 to 11, 47, 1971.
41. **Bourzeix, M., Weyland, D., and Heredia, N.,** Etude des catechines et des procyanidols de la grappe de raisin, du vin et d'autres dérivés de la vigne, *Bull. O.I.V.,* 669—670, 1171, 1986.
42. **Bower, J. P. and Van Lelyveld, L. J.,** The effect of stress history and container ventilation on avocado fruit polyphenol oxidase activity, *J. Hort. Sci.,* 60, 545, 1985.
43. **Brause, A. R.,** The matrix method to detect juice adulteration, Fruit Adulteration Workshop, Herndon, VA, July 20 to 21, 1988.
44. **Brause, A. R. and Raterman, J. M.,** Verification of authenticity of apple juice, *J. Assoc. Off. Anal. Chem.,* 65, 846, 1982.
45. **Brause, A. R. Raterman, J. M., Petrus, D. R., and Doner, L. W.,** Verification of authenticity of orange juice, *J. Assoc. Off. Anal. Chem.,* 67, 535, 1984.
46. **Brouillard, R.,** Chemical structure of anthocyanins, in *Anthocyanins as Food Colors,* Markakis P. Ed., Academic Press, New York, 1982, 1.
47. **Bruemmer, J. H. and Roe, B.,** Enzymic browning of simple diphenols and flavonols by orange juice extracts browning, *J. Food Sci.,* 35, 116, 1970.
48. **Brunet, G. and Ibrahim, R. K.,** Tissue culture of citrus peel and its potential for flavonoid synthesis, *Z. Pflanzenphysiol.,* 69, 152, 1973.
49. **Buchanan, D. W., Hall, C. B., Biggs, R. H. and Knapp, F. W.,** Influence of Alar, ethrel and gibberellic acid on browning of peaches, *HortScience,* 4, 302, 1969.
50. **Bureau, D., Macheix, J. J., and Rouet-Mayer, M. A.,** Relations entre la teneur en o-diphénols, l'activité polyphénoloxydasique et l'aptitude au brunissement de quelques variétés de pêches, *Lebensm. Wiss. Technol.,* 10, 211, 1977.
51. **Burnette, F. S.,** Peroxidase and its relationship to food flavor and quality: a review, *J. Food. Sci.,* 42, 1, 1977.
52. **Butt, V. S.,** Oxygenation and oxidation in the metabolism of aromatic compounds, in *Proc. Phytochem. Soc. Europe,* Vol. 25, Van Sumere, C. F., and Lea, P. J., Eds., Clarendon Press, Oxford, 1985, 349.
53. **Butt, V. S. and Lamb, C. J.,** Oxygenases and the metabolism of plants products, in *The Biochemistry of Plants,* Vol. 7, Conn, E. E., Ed., Academic Press, New York, 1981, 627.
54. **Calvi, J. P. and Francis, F. J.,** Stability of concord grape (*V. labrusca*) anthocyanins in model systems, *J. Food. Sci.,* 43, 1948, 1978.
55. **Cash, J. N., Sistrunk, N. A., and Stutte, C. A.,** Characteristics of concord grape polyphenoloxidase involved in juice color loss, *J. Food Sci.,* 41, 1398, 1976.
56. **Chan, J. H. T. and Yang, H. Y.,** Identification and characterization of some oxidizing enzymes of the McFarlin cranberry, *J. Food. Sci.,* 36, 169, 1971.
57. **Chandler, B. V. and Clegg, K. M.,** Pink discoloration in canned pears. I. Role of tin in pigment formation, *J. Sci. Food Agric.,* 21, 315, 1970.
58. **Chang, L. A., Hammett, L. K., and Pharr, D. M.,** Carbon dioxide effects on ethanol production, pyruvate decarboxylase and alcohol dehydrogenase activities in anaerobic sweet Potato roots, *Plant Physiol.,* 71, 59, 1983.
59. **Cheynier, V. F., and Van Hulst, M. W. J.,** Oxidation of trans-caftaric acid and 2-*S*-glutathionylcaftaric acid in model solutions, *J. Agric. Food. Chem.,* 36, 10, 1988.
60. **Cheynier, V., Osse, C., and Rigaud, J.,** Oxidation of grape juice phenolic compounds in model solutions, *J. Food, Sci.,* 53, 1729, 1988.
61. **Cheynier, V., Rigaud, J., Souquet, J. M., Barillière, J. M., and Moutounet, M.,** Effect of pomace contact and hyperoxidation on the phenolic composition and quality of Grenache and Chardonnay wines, *Am. J. Enol. Vitic.,* in press.
62. **Cheynier, V., Trousdale, E., Singleton, V. L., Salgues, M., and Wylde, R.,** Characterization of 2-*S*-glutathionyl caftaric acid and its hydrolysis in relation to Grape wines, *J. Agric. Food Chem.,* 34, 217, 1986.
63. **Chibata, I., Tosa, T., Mori, T., Watanabe, T., and Sakata, N.,** Immobilized tannins. A novel adsorbent for protein and metal ion, *Enzyme Microbiol. Technol.,* 8, 130, 1986.
64. **Clydesdale, F. M., Main, J. H., Francis, F. J., and Damon, R. A.,** Concord grape pigments as colorants for beverages and gelatin desserts, *J. Food Sci.,* 43, 1687, 1978.
65. **Constabel, F. and Vasil, I. K.,** *Cell Culture in Phytochemistry,* Academic Press, San Diego, 1987.
66. **Coppola, E. D.,** Determination of anthenticity and percent juice of cranberry products, Fruit Adulteration Workshop, Herndon, VA, July 20 to 21, 1988.
67. **Cordonnier, R., Hurtrel, J., and Biron, C.,** Relation entre le taux de pourriture de la vendange, l'activité laccase et l'aptitude à la casse oxydasique des mouts et des vins, *Conn. Vigne Vin,* 14, 19, 1980.
68. **Coseteng, M. Y. and Lee, C. Y.,** Changes in apple polyphenoloxidase and polyphenol concentrations in relation to degree of browning, *J. Food Sci.,* 52, 985, 1987.

69. **Cutting, J. G. M., Bower, J. P., and Wolstenholme, B. N.,** Effects of harvest data and applied ABA on polyphenol oxidase levels in avocado (*Persea americana* Mill.) fruit, *J. Hortic. Sci.*, 63, 509, 1988.
70. **Dapena, E., Mangas, J. J., Gonzalez-San José, M. L., and Diez, C.,** Les phenols de la pomme à cidre en rapport avec quelques groupes technologiques, *Bull. Liaison Groupe Polyphénols,* 14, 334, 1988.
71. **De Poix, A., Rouet-Mayer, M. A., and Philippon, J.,** Action combinée des chlorures et de l'acide ascorbique sur l'inhibition des brunissements enzymatiques d'un broyat de pommes, *Lebensm. Wiss. Technol.*, 14, 105, 1980.
72. **Deacon, W. and Marsh, H. V.,** Properties of an enzyme from Bananas (*Musa sapietum*) which hydroxylates tyramine to dopamine, *Phytochemistry,* 10, 2915, 1971.
73. **Dieudonné, G.,** Les dérivés hydroxycinnamiques de l'ananas comosus variété Cayenne Lisse (Cote d'Ivoire) et leurs relations avec le brunissement interne, Thèse Doct. Ingénieur, Université Paris VI, 1977.
74. **Dougherty, M. H., Ting, S. V., Attaway, J. A., and Moore, E. C.,** Grapefruit juice quality improvement studies. Introduction and scope of the study and the effect of processing variables, temperature and duration of storage on the quality of grapefruit juice, *Proc. Flav. State Hortic. Soc.*, 90, 165, 1977.
75. **Drawert, F.,** Formation des aromes à différents stages de l'évolution du fruit, enzymes intervenant dans cette formation, in Facteurs et Régulation de la Maturation des Fruits, *Coll. Int. CNRS,* Paris, 238, 309, 1975.
76. **Drawert, F., Leupold, G., and Pivernetz, H.,** Quantitative determination of rutin, hesperidin and naringin in orange juice by gas chromatography, *Chem. Mikrobiol. Technol. Lebensm.*, 6, 189, 1980.
77. **Du Plessis, C. S.,** Browning of white wines, *Die Wynboer,* 499, 11, 1973.
78. **Dubois, G. E., Crosby, G. A., and Saffron, P.,** Non-nutritive sweeteners: taste-structure relationships for some new simple dihydrochalcones, *Science,* 195, 397, 1977.
79. **Dubois, P., Boulé, G., Illic, M., and Dekimpe, J.,** Etude des phénols volatile de deux vins rouges, *Ann. Technol. Agric.*, 20, 131, 1971.
80. **Dubois, P., Etievant, P., Dekimpe, J., Buret, M., Chambroy, Y., and Flanzy, C.,** Etude sur l'arôme des vins de macération carbonique, *C.R. Acad. Agric.*, 63, 1183, 1977.
81. **Ducruet, V.,** Comparison of the headspace volatiles of carbonic maceration and traditional wines, *Lebensm. Wiss. Technol.*, 17, 217, 1984.
82. **Ducruet, V., Flanzy, C., Bourzeix, M., and Chambroy, Y.,** Les constituants volatiles des vins jeunes de macération carbonique, *Sci. Aliment.*, 3, 413, 1983.
83. **Esterballer, H., Schwarzl, E., Hayn, M.,** A rapid assay for catechol oxidase and laccase using 2-nitro-5-thiobenzoic acid, *Anal. Biochem.*, 77, 486, 1977.
84. **Etievant, P. X.,** Volatile phenol determination in wine, *J. Agric. Food Chem.*, 29, 65, 1981.
85. **Etievant, P., Schlich, P., Bertrand, A., Symonds, P., and Bouvier, J. C.,** Varietal and geographic classification of French red wines in terms of pigments and flavonoid compounds, *J. Sci. Food Agric.*, 42, 39, 1988.
86. **Everington, D.,** Thawing of frozen foodstuffs, *Chem. Ind.*, 35, 973, 1971.
87. **Fernandez Diez, M. J.,** The olive, in *The Biochemstry of Fruits and Their Products,* Vol. 2, Hulme, A. C., Ed., Academic Press, London, 1971, 255.
88. **Fidler, W., Parker, W. E., Wasserman, A. E., and Doerr, R. C.,** Thermal decomposition of ferulic acid, *J. Agric. Food Chem.*, 15, 757, 1967.
89. **Finkle, B. J.,** Freezing preservation, in *The Biochemistry of Fruits and Their Products,* Vol. 2, Hulme, A. C. Ed., Academic Press, London, 1971, 653.
90. **Finkle, B. J. and Nelson, R. F.,** Enzyme reactions with phenolic compounds: effect of *O*-methyltransferase on a natural substrate of fruit polyphenol oxidase, *Nature (London),* 197, 902, 1963.
91. **Flanzy, C., Buret, M., and Chambroy, Y.,** L'acide shikimique au cours du métabolisme anaérobie de la baie de raisin, *Sci. Aliment.*, 1, 283, 1981.
92. **Flanzy, C., Flanzy, M., and Benard, P.,** *La Vinification par Macération Carbonique,* INRA, Paris, 1987.
93. **Fleming, H. P., Walter, W. M., Jr., and Etchells, J. L.,** Antimicrobial properties of oleuropein and products of its hydrolysis from green olives, *Appl. Microbiol.*, 26, 777, 1973.
94. **Flick, G. J., Jr., Ory, R. L., and Stangelo, A. J.,** Comparison of nutrient composition and of enzyme activity in purple, green and white eggplants, *J. Agric. Food Chem.*, 25, 117, 1977.
95. **Flora, L. F.,** Influence of leaf cultivars and maturity on the anthocyanidin-3,5-diglucosides of muscadine grapes, *J. Food Sci.*, 43, 1819, 1978.
96. **Flurkey, W. H. and Jen, J. J.,** 1978, Peroxidase and polyphenoloxidase activities in developing peaches, *J. Food Sci.*, 43, 1828, 1978.
97. **Fordebras, I. and Macheix, J. J.,** Evolution de la teneur en *o*-diphénols et en acide chlorogénique des poires Passe-Crassane et de leurs potentialités de brunissement au cours de la conservationo, *C.R. Acad. Agric.* France, 65, 502, 1979.
98. **Francis, F. J.,** Anthocyanins, in *Current Aspects of Food Colorants,* Furia, T. E., Ed., CRC Press, Boca Raton, FL, 1977, 19.

99. **Francis, F. J.**, Detection of enocyanin in cranberry juice cocktail by color and pigment profile, *J. Food Sci.*, 50, 1640, 1985.
100. **Fujiwake, H., Suzuki, T., and Iwai, K.**, Intracellular localization of capsaicin and its analogues in Capsicum fruit. II. The vacuole as the intracellular accumulation site of capsaicinoid in the protoplast of Capsicum fruit, *Plant Cell Physiol.*, 21, 1023, 1980.
101. **Gajzago, I., Nadudvari-Markus, V., Vamos-Vigyazo, L., and Babos-Szebenyi, E.**, Influence of location and year of harvest on enzymic browning and the polyphenol oxidase-polyphenol complex of apricot cultivars, *Proc. 19th Hung. Annu. Meet. Biochem.*, 221, 1979.
102. **Galeazzi, M. A. M., Sgarbieri, V. C., and Constantinides, S. M.**, Isolation, purification and physicochemical characterization of polyphenoloxidases (PPO) from a dwarf variety of Banana (*Musa cavendishii*, L), *J. Food Sci.*, 46, 150, 1981.
103. **Gaydou, E. M., Bianchini, J. P., and Randriamiharisoa, R. P.**, Orange and mandarin peel oils differentiation using polymethoxylated flavone composition, *J. Agric. Food Chem.*, 35, 525, 1987.
106. **Gazit, S. and Adato, I.**, Effect of carbon dioxide atmosphere on the course of astringency disappearance of persimmon (*Dyospyros kaki* Linn.) fruits, *J. Food Sci.*, 37, 815, 1972.
105. **Glories, Y.**, Recherches sur la matière colorante des vins rouges, Thèse Doct. Etat, Université Bordeaux II, France, 1978.
106. **Glories, Y.**, La couleur des vins rouges. I. Les équilibres des anthocyanes et des tannins. *Conn. Vigne Vin*, 18, 195, 1984.
107. **Glories, Y.**, La couleur des vins rouges, II. Mesure, origine et interprétation. *Conn. Vigne Vin*, 18, 253, 1984.
108. **Golan, A., Kahn, V., and Sadovski, A. Y.**, Relationship between polyphenols and browning in avocado mesocarp. Comparison between the Fuerte and Lerman cultivars. *J. Agric. Food Chem.*, 25, 1253, 1977.
109. **Goodenough, P. W., Kessel, S., Lea, A. G. H., and Loeffler, T.**, Mono and diphenolase activity from fruit of *Malus pumila*, *Phytochemistry*, 22, 359, 1983.
110. **Gore, H. C. and Fairchild, D.**, Experiments on the processing of persimmons to render them nonastringent, *U.S. Dept. Agric. Bur. Chem. Bull*, 141, 1911.
111. **Goupy, P. and Fleuriet, A.**, Caracteristiques du brunissement de l'olive, *Bull. Liaison Groupe Polyphénols*, 13, 455, 1986.
112. **Graciani Constante, E. and Vazquez-Roncero, A.**, Estudio de los componentes del aceite de oliva por cromatogafia liquida de alta eficacia (HPLC). II. Cromatografia en fase inversa, *Grasas Aceites Seville*, 31, 237, 1980.
113. **Graciani Constante, E. and Vasquez-Roncero, A.**, Estudio de los componentes polares del aceite de oliva por cromatografia liquida de alta eficacia (HPLC). III. Aplicacion a diversos tipos de aceites virgenes, *Grasas Aceites Seville*, 32, 365, 1981.
114. **Graciani Constante, E., Colchero Vela, C., and Vasquez Roncero, A.**, Estudio de los componentes polares del aceite de oliva por cromatografia liquida de alta eficacia (HPLC). I. Cromatografia de adsorcion, *Grasas Aceites Seville*, 31, 85, 1980.
115. **Gray, G. M. and Olson, A. C.**, Hydrolysis of high levels of naringin in grapefruit juice using a hollow fiber naringenase reactor, *J. Agric. Food Chem.*, 31, 1002, 1981.
116. **Greiner, G. and Wallrauch, S.**, Naringin as evidence for blending grapefruit juice with orange and tangerine juices, *Fluess. Obst.*, 12, 626, 1984.
117. **Griffin, B. and Stonier, T.**, Studies on auxin protectors. XII. Auxin protectors and polyphenol oxidase in developing coffee fruit, *Physiol. Plant.*, 33, 157, 1975.
118. **Grncarevic, M., and Hawker, J. S.**, Browning of Sultana grape berries during drying, *J. Sci. Food Agric.*, 22, 270, 1971.
119. **Gunata, Y. Z., Salgues, M., and Moutounet, M.**, Evolution des dérivés hydroxycinnamiques au cours de l'élaboration du moût de raisin et du stockage des concentrés, *Sci. Aliment.*, 6, 579, 1986.
120. **Gunata, Y. Z., Sapis, J. C., and Moutounet, M.**, Substrates and aromatic carboxylic acid inhibitors of grape phenoloxidases, *Phytochemistry*, 26, 1573, 1987.
121. **Hadj-Mahammed, M. and Meklati, B. Y.**, Qualitative determination of polymethoxylated flavones in Valencia orange peel oil and juice by LC-UV/VIS and LC-MS techniques, *Lebensm. Wiss. Technol.*, 20, 111, 1987.
122. **Haisman, D. R.**, The effect of sulphur dioxide on oxidising enzyme systems in plant tissues, *J. Sci. Food Agric.*, 25, 803, 1974.
123. **Halim, D. H. and Montgomery, M. W.**, Polyphenol oxidase of d'Anjou pears (*Pyrus communis* L.), *J. Food Sci.*, 43, 603, 1978.
124. **Hall, R. D., Holden, M. A., and Yeoman, M. M.**, Studies on the regulation of capsaicin biosynthesis in immobilized cell cultures and the developing fruit of the Chili pepper, *Capsicum fructescens*, in *Secondary Metabolism in Plant Cell Cultures*, Morris, P., Scragg, A. H., Stafford, A., Fowler, M. W., Eds., Cambridge University Press, Cambridge, 1986, 121.

125. **Hall, R. D., Holden, M. A., and Yeoman, M. M.,** The accumulation of phenylpropanoid and capsaicinoid compounds in cell cultures and whole fruit of the Chili pepper, *Capsicum fructescens*, *Plant Cell, Tissue Organ Culture*, 8, 163, 1987.
126. **Harel, E. and Mayer, M.,** Interconversion of sub-units of catechol oxidase from apple chloroplasts, *Phytochemistry*, 7, 199, 1968.
127. **Harel, E. and Mayer, A. M.,** Partial purification and properties of catechol oxidases in grapes, *Phytochemistry*, 10, 17, 1971.
128. **Harel, E., Mayer, A. M., and Lehman, E.,** Multiple forms of *Vitis vinifera* catechol oxidase, *Phytochemistry*,. 12, 2649, 1973.
129. **Harel, E., Mayer, A. M., and Lerner, H. R.,** Changes in the levels of catechol oxidase and laccase activity in developing peaches, *J. Sci. Food Agric.*, 21, 542, 1970.
130. **Harel, E., Mayer, A. M., and Shain, Y.,** Phenolases from apples, their properties subcellar location and inhibition, *Physiol. Plant.*, 17, 921, 1964.
131. **Harel, E., Mayer, A. M., and Shain, Y.,** Purification and multiplicity of catechol oxidase from apple choroplasts, *Phytochemistry*, 4, 783, 1965.
132. **Harel, E., Mayer, A. M., and Shain, Y.,** Catechol oxidases, endogenous substrates and browning in developing apples, *J. Sci. Food Agric.*, 17, 389, 1966.
133. **Harper, K. A., Morton, A. D., and Rolfe, E. J.,** The phenolic compounds of blackcurrant juice and their protective effect on ascorbic acid. III. The mechanism of ascorbic acid oxidation and its inhibition by flavonoids, *J. Food Technol.*, 4, 255, 1969.
134. **Hasegawa, S. and Maier, V. P.,** Polyphenol oxidase of Dates, *J. Agric. Food Chem.*, 28, 891, 1980.
135. **Haslam, E.,** In vino veritas: oligomeric procyanidins and the ageing of red wines, *Phytochemistry*, 19, 2577, 1980.
136. **Haslam, E. and Lilley, T. H.,** Natural astringency in foodstuffs. A molecular interpretation, *CRC Crit. Rev. Food Sci. Nutr.*, 27, 1, 1988.
137. **Haslam, E., Lilley, T. H., and Ozawa, T.,** Polyphenol complexation. Astringency in fruits and beverages, *Bull. Liaison Groupe Polyphenols*, 13, 352, 1986.
138. **Herrmann, K.,** Flavonols and flavones in food plants: a review, *J. Food Technol.*, 11, 433, 1976.
139. **Higginbotham, J. D.,** A la recherche de nouveaux édulcorants, *Fruits*, 35, 709, 1980.
140. **Hobson, G. E.,** Phenolase activity in tomato fruit in relation to growth and to various ripening disorders, *J. Sci. Food Agric.*, 18, 523, 1967.
141. **Hong, V. and Wrolstad, R. E.,** Cranberry juice composition, *J. Assoc. Off. Anal. Chem.*, 69, 199, 1986.
142. **Hong, V. and Wrolstad, R. E.,** Detection of adulteration in commercial cranberry juice drinks and concentrates, *J. Assoc. Off. Anal. Chem.*, 69, 208, 1986.
143. **Horowitz, R. M. and Gentili, B.,** Taste and structure in phenolic glycosides, *J. Sc. Food Agric.*, 17, 696, 1969.
144. **Hrazdina, G.,** Réactions of the anthocyanidin-3,5-diglucosides: formation of 3, 5-di (*O*-β-D-glucosyl)-7-hydroxycoumarin, *Phytochemistry*, 10, 1125, 1971.
145. **Hrazdina, G.,** Anthocyanins and their role in food products, *Lebensm. Wiss. Technol.*, 14, 283, 1981.
146. **Hrazdina, G., Borzell, A. J., and Robinson, W. B.,** Studies on the stability of the anthocyanidin 3-5-diglucoside, *Am. J. Enol. Vitic.*, 21, 201, 1970.
147. **Hulme, A. C. and Edney, K. L.,** Phenolic substances in the peel of Cox's orange pippin apples with reference to infection by *Gleosporium perennans*, in *Phenolics in Plants in Health and Disease*, Pridham, J. B., Ed., Pergamon Press, Oxford, 1960, 87.
148. **Ibrahim, R. K.,** Regulation of synthesis of phenolics, in *Cell Culture and Somatic Cell Genetics of Plants*, Vol. 4, Constabel, F. and Vasil, I. K., Eds., Academic Press, San Diego, 1987, 77.
149. **Ibrahim, R. K., Thakur, M. L., and Permanand, B.,** Formation of anthocyanins in callus tissue-cultures, *Lloydia*, 34, 175, 1971.
150. **Ilker, R.,** *In vitro* pigment production: an alternative to color synthesis, *Food Technol.*, 70, 1987.
151. **Interesse, F. S., Alloggio, V., Lamparelli, F., and Davella, G.,** Characterization of the oxidative enzymatic system of the phenolic compounds from muscat grapes, *Lebensm. Wiss. Technol.*, 17, 5, 1984.
152. **Itoo, S.,** Persimmon, in *Handbook of Fruit Set and Development*, Monselise, S. P., Ed., CRC Press, Boca Raton, FL, 1986, 355.
153. **Jackman, R. L., Yada, R. Y., Tung, M. A., and Speers, R. A.,** Anthocyanins as food colorants — a review, *J. Food Biochem.*, 11, 201, 1987.
154. **Jackson, M. G., Timberlake, C. F., Bridle, P.,, and Vallis, L.,** Red wine quality: correlations between colour, aroma and flavour and pigment and other parameters of young Beaujolais, *J. Sci. Food Agric.*, 29, 715, 1978.
155. **Janer del Valle, C. and Vazquez-Roncero, A.,** Estudio de los componentes polares del aceite de oliva por chromatografia gaseosa, *Grasas Aceites Seville*, 31, 309, 1980.
156. **Jayaraman, K. S. and Ramanuja, M. N.,** Changes in polyphenoloxidase and other endogenous factors during ripening in some banana varieties, *J. Food Sci. Technol.*, 24, 67, 1987.

157. **Jayaraman, K. S. and Ramanuja, M. N.,** Studies on multiple forms of polyphenoloxidase from some banana cultivars differing in browning rate, *Lebensm. Wiss. Technol.*, 20, 16, 1987.
158. **Jayaraman, K. S., Ramanuja, M. N., Dhakne, Y. S., and Vijayaraghavan, P. K.,** Enzymatic browning in some banana varieties as related to polyphenoloxidase activity and other endogenous fractors, *J. Food Sci. Technol.*, 19, 181, 1982.
159. **Jayaraman, K. S., Ramanuja, M. N., Vijayaraghavan, P. K., and Vaidyanathan, C. S.,** Studies on the purification of banana polyphenoloxidase, *Food Chem.*, 24, 203, 1987.
160. **Jen, J. J. and Kahler, K. R.,** Characterization of polyphenol oxidase in peaches grown in the Southeast, *HortScience*, 9, 590, 1974.
161. **Jindra, J. A., and Gallander, J. F.,** Effect of American and French oak barrels on the phenolic composition and sensory quality of Slyval blanc wines, *Am. J. Enol. Vitic*, 38, 133, 1987.
162. **Joel, D., Harbach, I., and Mayer, A. M.,** Laccase in Anacardiaceae, *Phytochemistry*, 17, 796, 1978.
163. **Johnson, G., Donnelley, B. J., and Johnson, D. K.,** Proanthocyanidins as related to apple juice processing and storage, *Food Technol.*, 23, 1312, 1969.
164. **Joshi, P. R. and Shiralkar, N. D.,** Polyphenolases of a local variety of mango, *J. Food Sci. Technol.*, 14, 77, 1977.
165. **Joslyn, M. A. and Ponting, J. D.,** Enzyme catalysed oxidative browning of fruit products, *Adv. Food Res.*, 3, 1, 1951.
166. **Jourdan, P. S., McIntosh, C. A., and Mansell, R. L.,** Naringin levels in citrus tissues. II. Quantitative distribution of naringin in *Citrus paradisi* Mac Fad., *Plant Physiol.*, 77, 903, 1985.
167. **Jurd, L.,** Some advances in the chemistry of anthocyanin-type plant pigments, in *The Chemistry of Plant Pigments*, Adv. Food Res., Suppl., Chichester C. O., Ed., Academic Press, New York, 1972, 123.
168. **Kagami, M., Amano, Y., and Saito, H.,** On the removal of color from wine. XVII. Studies on the application of ion exchange process to fermentation industries, *J. Ferment. Technol.*, 46, 153, 1968.
169. **Kahn, V.,** Polyphenol oxidase activity and browning of three avocado varieties, *J. Sci. Food Agric.*, 26, 1319, 1975.
170. **Kahn, V.,** Polyphenol oxidase isoenzymes in avocado, *Phytochemistry*, 15, 267, 1976.
171. **Kahn, V.,** Some biochemical properties of polyphenoloxidase from two avocado varieties differing in their browning rate, *J. Food Sci.*, 42, 38, 1977.
172. **Kahn, V. and Pomerantz, S. H.,** Monophenolase activity of avocado polyphenol oxidase, *Phytochemistry*, 19, 379, 1980.
173. **Kallio, H., Pallasaho, S., Kärppä, J., and Linko, R. R.,** Stability of anthocyanins in Crowberry juice, in *The Shelf Life of Food and Beverages*, Charalambous G., Ed., Elsevier, Amsterdam, 1986, 285.
174. **Kearsly, M. W. and Rodriguez, N.,** The stability and use of natural colors in foods: anthocyanins, β-carotene and riboflavin, *J. Food Technol.*, 16, 421, 1981.
175. **Kefford, J. F. and Chandler, B. V.,** *The Chemical Constituents of Citrus Fruits*, Academic Press, New York, 1970.
176. **Kelly, S. H. and Finkle, B. J.,** Action of a ring cleaving oxygenase in preventing oxidative darkening of apple juice, *J. Sci. Food Agric.*, 20, 629, 1969.
177. **Kidron, M., Harel, E., and Mayer, A. M.,** Catechol oxidase activity in grapes and wine, *Am. J. Enol. Vitic*, 29, 30, 1978.
178. **Kirskey, S. T., Schwartz, J. O., and Wade, R. L.,** A new high performance liquid chromatographic procedure for detecting juice adulteration, Fruit Adulteration Workshop, Herndon, VA, July 20—21, 1988.
179. **Knapp, F. W., Hall, C. B., Buchanan, D. W., and Biggs, R. H.,** Reduction of polyphenoloxidase activity in peaches sprayed with alar, ethrel or giberellic acid, *Phytochemistry*, 9, 1453, 1970.
180. **Konichi, F., Esaki, S., and Kamiya, S.,** Synthesis and taste of flavanone and dihydrochalcone glycosides containing 3-*O*-α-L-rhamnopyranosyl-D-glucopyranose or 4-*O*-α-L-rhamnopyranosyl-D-glycopyranose in the sugar moiety, *Agric. Biol. Chem.*, 47, 265, 1983.
181. **Konno, A., Miyawaki, M., and Yasumatsu, K.,** Bitterness reduction of Citrus fruits by beta-cyclodextrin, *Agric. Biol. Chem.*, 45, 2341, 1981.
182. **Konno, A., Misaki, M., Toda, J., Wada, T., and Yasumatsu, K.,** Bitterness reduction of naringin and limonin by beta-cyclodextrin, *Agric. Biol. Chem.*, 46, 2203, 1982.
183. **Koumba-Koumba, D. and Macheix, J. J.,** Biosynthesis of hydroxycinnamic derivatives in apple fruit cell suspension culture, *Physiol. Vég.*, 20, 137, 1982.
184. **Lambie, A. J.,** Commercial aspects of the production of secondary compounds in immobilized plant cells, Int. Symp. Phytochem. Soc. Eur., London, December 19 to 21, 1988.
185. **Lanzarini, G., Pifferi, P. G., and Zamorani, A.,** Specificity of an *o*-diphenol oxidase from *Prunis avium* fruits, *Phytochemistry*, 11, 89, 1972.
186. **Larson, R. A.,** The antioxidants of higher plants, *Phytochemistry*, 27, 969, 1988.
187. **Lea, A. G. H.,** The phenolics of ciders: oligomeric and polymeric procyanidins, *J. Sci. Food Agric.*, 29, 471, 1978.

188. **Lea, A. G. H.,** High performance liquid chromatography of cider procyanidins, *J. Sci. Food Agric.*, 30, 833, 1979.
189. **Lea, A. G. H.,** Reversed-phase high-performance liquid chromatography of procyanidins and other phenolics in fresh and oxidising apple juices using a pH shift technique, *J. Chromatogr.*, 238, 253, 1982.
190. **Lea, A. G. H.,** Oxidation of apple phenols, *Bull. Liaison Groupe Polyphénols*, 12, 462, 1984.
191. **Lea, A. G. H. and Arnold, G. M.,** The phenolics of ciders: bitterness and astringency, *J. Sci. Food Agric.*, 29, 478, 1978.
192. **Lea, A. G. H. and Beech, F. W.,** The phenolics of ciders: effect of cultural conditions, *J. Sci. Food Agric.*, 29, 493, 1978.
193. **Lea, A. G. H. and Timberlake, C. F.,** The phenolics of ciders, 1. Procyanidins, *J. Sci. Food Agric.*, 25, 1537, 1974.
194. **Lea, A. G. H., and Timberlake, C. F.,** The phenolics of ciders: effect of processing conditions, *J. Sci. Food Agric.*, 29, 484, 1978.
195. **Lee, C. Y., Smith, N. L., and Pennesi, A. P.,** Polyphenoloxidase from De Chaunac grapes, *J. Sci. Food Agric.*, 34, 987, 1983.
196. **Lerner, H. R., Mayer, A. M., and Harel, E.,** Evidence for conformational changes in grape catechol oxidase, *Phytochemistry*, 11, 2415, 1972.
197. **Lerner, H. R., Mayer, A. M., and Harel, E.,** Differential effect of phenylhydrazine on the catecholase and cresolase activities of Vitis catechol oxidase, *Phytochemistry*, 13, 397, 1974.
198. **Lewinsohn, E., Berman, E., Mazur, Y., and Gressel, J.,** Glucosylation of exogenous flavanones by grapefruit (*Citrus paradisi*) cell cultures, *Phytochemistry*, 25, 2531, 1986.
199. **Lindsey, K.,** Incorporation of ^{14}C phenylalanine and ^{14}C cinnamic acid into capsaicim in cultured cells of *Capsicum frutescens, Phytochemistry*, 25, 2793, 1986.
200. **Lotfy, S., Fleuriet, A., Ramos, T., and Macheix, J. J.,** Biosynthesis of phenolic compounds in *Vitis vinifera* cv. Gamay Freaux cell suspension cultures: study on hydroxycinnamoyl CoA: ligase, *Plant Cell Rep.*, 8, 93, 1989.
201. **Luh, B. S. and Phithakpol, B.,** Characteristics of polyphenoloxidase related to browning in cling peaches, *J. Food Sci.*, 37, 264, 1972.
202. **Maccarone, E., Maccarrone, A., and Rapisarda, P.,** Stabilization of anthocyanins of blood orange fruit juice, *J. Food Sci.*, 50, 901, 1985.
203. **Macheix, J. J.,** Role de divers facteurs intervenant dans le brunissement enzymatique des pommes pendant la croissance, *Physiol. Vég.*, 8, 585, 1970.
204. **Macheix, J. J.,** Les esters hydroxycinnamiques de la Pomme: identification, variations au cours de la croissance du fruit et métabolisme, Thèse Doct. Etat, Université Pierre et Marie Curie, Paris, 1974.
205. **Macheix, J. J. and Ibrahim, R. K.,** The O-methyltransferase system of apple fruit cell suspension culture, *Biochem. Physiol. Pflanzen*, 179, 659, 1984.
206. **Macheix, J. J., Fleuriet, A., and Quessada, M. P.,** Involvement of phenols and peroxidases in wound healing and grafting, in *Molecular and Physiological Aspects of Plant Peroxidases*, Greppin, H., Penel, C., and Gaspar, Th., Eds., University of Geneva, Switzerland, 1986, 267.
207. **Macheix, J. J., Suen, R., and Ibrahim, R. K.,** Metabolism of phenylpropanoid compounds in apple fruit cell suspension culture, *Biochem. Physiol. Pflanzen*, 176, 195, 1981.
208. **Maga, J. A.,** Simple phenols and phenolic compounds in food flavor, *CRC Crit. Rev. Food Nutr.*, 10, 323, 1978.
209. **Maga, J. A.,** Flavor contribution of wood in alcobolic beverages, in *Progress in Flavour Research*, Adda, J., Ed., Proc. 4th Weurman Flavour Res. Symp., Dourdan, France, 1984, 409.
210. **Maier, V. P.,** Chemistry and significance of selected citrus limonoids and flavonoids, in *Chemistry and Chemical Taxonomy of the Rutales*, Waterman, P. G., and Grundon, M. F., Eds., Academic Press, London, 1983, 319.
211. **Markakis, P.,** Anthocyanins and their stability in foods, *CRC Crit. Rev. Food Technol.*, 4, 437, 1974.
212. **Markakis, P.,** Stability of anthocyanins in Foods, in *Anthocyanins as Food Colors*, Markakis, P., Ed., Academic Press, New York, 1982, 163.
213. **Markakis, P.,** Anthocyanins as food additivies, in *Anthocyanins as Food Colors*, Markakis, P., Ed., Academic Press, New York, 1982, 245.
214. **Markakis, P., Suparmo, and Harp, T.,** Comparative stability of two anthocyanin pigments, in *The Shelf Life of Foods and Beverages*, Charalambous, G., Ed., Elsevier, Amsterdam, 1986. 705.
215. **Martinez-Cayuela, M., Plata, M. C., Faus, M. J., and Gil, A.,** Effect of some phenolic carboxylic acids on Cherimoya (*Annona cherimolia*) polyphenoloxidase activity, *J. Sci. Food Agric.*, 45, 215, 1988.
216. **Martinez-Cayuela, M., Sanchez de Medina, L., Faus, M. J., and Gil, A.,** Cherimoya (*Annona cherimolia* Mill.) polyphenol-oxidase: monophenolase and dihydroxyphenolase activities, *J. Food Sci.*, 53, 1191, 1988.
217. **Matheis, G.,** Enzymatic browning of foods. Quantitative relationships between browning and food constituents, *Z. Lebensm. Unters. Forsch.*, 176, 454, 1983.
218. **Matheis, G. and Whitaker, J. R.,** Modification of proteins by polyphenoloxidase and peroxidase and their products, *J. Food Biochem.*, 8, 137, 1984.

219. **Mathew, A. G., and Parpia, H. A. B.,** Food browning as a polyphenol reaction, *Adv. Food Res.*, 19, 75, 1971.
220. **Matsuo, T. and Itoo, S.,** On mechanisms of removing astringency in persimmon fruit by carbon dioxide treatment. I. Some properties of the two processes in the de-astringency, *Plant Cell Physiol.*, 18, 17, 1977.
221. **Matsuo, T. and Ito, S.,** The chemical structure of Kaki-tannin from immature fruit of the persimmon (*Diospyros Kaki* L.), *Agric. Biol. Chem.*, 42, 1937, 1978.
222. **Matsuo, T., Shinobara, J., and Itoo, S.,** An improvement on removing astringency in Persimmon fruits by carbon dioxide gas, *Agric. Biol. Chem.*, 40, 215, 1976.
223. **Mayer, A. M.,** Polyphenol oxidases in plants. Recent progress, *Phytochemistry*, 26, 11, 1987.
224. **Mayer, A. M. and Harel, E.,** A laccase-like enzyme in peaches, *Phytochemistry*, 7, 1253, 1968.
225. **Mayer, A. M. and Harel, E.,** Polyphenol oxidases in plants, *Phytochemistry*, 18, 193, 1979.
226. **Mayer, A. M. and Harel, E.,** Polyphenol oxidases in fruits. Changes during ripening, in *Recent Advances in the Biochemistry of Fruits and Vegetables*, Friend, J. and Rhodes, M. J. C., Eds., Academic Press, London, 1981, 161.
227. **Mc Bean, D. McG., Joslyn, M. A., and Nury, F. S.,** Dehydrated fruit, in *The Biochemistry of Fruits and Their Products*, Vol. 2, Hulme A. C., Ed., Academic Press, London, 1971, 623.
228. **McCloskey, L. P. and Yengoyan, L. S.,** Analysis of anthocyanins in *Vitis vinifera* wines and red color versus aging by HPLC and spectrophotometry, *Am. J. Enol. Vitic.*, 32, 257, 1981.
229. **Meheriuk, M. and Porrit, S. W.,** Effects of picking date delayed storage, storage temperature, and storage atmosphere on the quality of starking Delicious apples, *Can. J. Plant Sci.*, 53, 593, 1973.
230. **Meheriuk, M., Lau, O. L., and Hall, J. W.,** Effects of some postharvest and storage treatments on the incidence of flesh browning in controlled-atmosphere-stored "Delicious" apples, *J. Am. Soc. Hortic. Sci.*, 109, 290, 1984.
231. **Metche, M.,** Role des flavonoides dans les aliments, *Bull. Liaison Groupe Polyphénols*, 14, 119, 1988.
232. **Montedoro, G.,** I costituenti fenolici presenti negli oli vergini di oliva. Nota I. Identificazione di alcuni acidi fenolici e loro potere antiossidante, *Riv. "S et TA"*, 3, 177, 1972.
233. **Montedoro, G.,** Esame analitico dei constituenti fenolici presenti nell'ollio di oliva in funzione delle caratteristiche della drupa, loro ruolo nella stabilita all'ossidazione e possibile interferenza nella determinazione del numero di perossidi, *Ann. Fac. Agric. Perugia*, 28, 1, 1973.
234. **Montgomery, M. W. and Sgarbieri, V. C.,** Isoenzymes of Banana polyphenol oxidase, *Phytochemistry*, 14, 1245, 1975.
235. **Morris, P., Scragg, A. H., Stafford, A., and Fowler, M. W.,** *Secondary Metabolism in Plant Cell Cultures*, Cambridge University Press, Cambridge, 1986.
236. **Mosel, H. and Herrmann, K.,** Die phenolischen Inhaltsstoffe des Obstes. IV. Die phenolischen Inhaltsstoffe der Brombeeren und Himbeeren und deren Veranderungen wahrend Wachstum und Reife der Fruchte *Z. Lebensm. Unters. Forsch.*, 154, 324, 1974.
237. **Moutounet, M. and Mondies, H.,** La polyphenoloxydase de la Prune d'Ente. Modification de son activité au cours de l'élaboration du pruneau d'Agen, *Ann. Technol. Agric.*, 25, 343, 1976.
238. **Myers, T. E. and Singleton, V. L.,** The non-flavonoid phenolic fraction of wine and its analysis, *Am. J. Enol. Vitic*, 30, 98, 1979.
239. **Nagel, C. W. and Graber, W. R.,** Effect of must oxidation on quality of white wines, *Am. J. Enol. Vitic.*, 39, 1, 1988.
240. **Nagel, C. W., and Wulf, L. W.,** Changes in the anthocyanins, flavonoids and hydroxycinnamic acid esters during fermentation and aging of Merlot and Cabernet Sauvignon, *Am. J. Enol. Vitic.*, 30, 111, 1979.
241. **Naim, M., Striem, B. J., Kanner, J., and Peleg, H.,** Potential of ferulic acid as a precursor to off-flavors in stored orange juice, *J. Food Sci.*, 53, 500, 1988.
242. **Nakamura, K., Amano, Y., and Kagami, M.,** Purification and some properties of a polyphenol oxidase from Koshu Grapes, *Am. J. Enol. Vitic.*, 34, 122, 1983.
243. **Nelson, R. F. and Finkle, B. J.,** Enzyme reactions with phenolic compounds: effects of *o*-methyltransferase and high pH on the polyphenol oxidase substrates in apple, *Phytochemistry*, 3, 321, 1964.
244. **Okamura, S. and Watanabe, M.,** Determination of phenolic cinnamates in white wine and their effect on wine quality, *Agric. Biol. Chem.*, 45, 2063, 1981.
245. **Okuse, A., Okuse, I., and Ryugo, K.,** Effects of certain processing methods, substrate level and polyphenoloxidase on the stability of ascorbic acid in Kiwi fruit, *Hortic. Sci.*, 16, 164, 1981.
246. **Ono, M., Tosa, T., and Chibata, I.,** Preparation and properties of immobilized naringinase using tannin-aminohexyl cellulose, *Agric. Biol. Chem.*, 42, 1847, 1978.
247. **Oszmianski, J., Sapis, J. C., and Macheix, J. J.,** Changes in grape seed phenols as affected by enzymic and chemical oxidation *in vitro*, *J. Food Sci.*, 50, 1505, 1985.
248. **Oszmianski, J., Romeyer, F. M., Sapis, J. C., and Macheix, J. J.,** Grape seed phenolics: extraction as affected by some conditions occurring during wine processing, *Am. J. Enol. Vitic.*, 37, 7, 1986.

249. **Ough, C. S.,** Substances extracted during skin contact with white musts. I. General wine composition and quality changes with contact time, *Am. J. Enol. Vitic.*, 20, 93, 1969.
250. **Padron, M. P., Lozano, J. A., and Gonzalez, A. G.,** Properties of *o*-diphenol: O_2 oxidoreductase from *Musa cavendishii, Phytochemistry,* 14, 1959, 1975.
251. **Page, S. W.,** Computer multivariate analysis of orange juice composition data aids in detection of adulteration, *Food Technol.*, 40, 104, 1986.
252. **Palamidis, N. and Markakis, P.,** Stability of Grape anthocyanin in a carbonated beverage, *J. Food Sci.*, 40, 1047, 1975.
253. **Palmer, J. K.,** The banana, in *The Biochemistry of Fruits and Their Products,* Vol. 2, Hulme, A. C., Ed., Academic Press, London, 1971, 65.
256. **Parish, R. W.,** The intracellular location of phenol oxidases and peroxidase in stems of spinach beet (*Beta vulgaris* L.), *Z. Pflanzenphysiol.*, 66, 176, 1972.
255. **Park, E. Y. and Luh, B. S.,** Polyphenol oxidase of Kiwi fruit, *J. Food Sci.*, 50, 678, 1985.
256. **Paulin, A.,** Procédés pour la protection d'organes végétaux contre les altérations fongiques, Patent no. 1,437,567, France, 1966.
257. **Paulson, A. T., Vanderstoep, J., and Porritt, S. W.,** Enzymatic browning of peaches: effect of gibberellic acid and etephon on phenolic compounds and polyphenol oxidase activity, *J. Food Sci.*, 45, 341, 1980.
258. **Perfetti, G. A., Joe, F. L., Jr., Fazio, T., and Page, S. W.,** Liquid chromatographic methodology for the characterization of orange juice, *J. Assoc. Off. Anal. Chem.*, 71, 469, 1988.
259. **Peri, C., Pompei, C., Montedoro, G., and Cantarelli, C.,** Maderisation of white wines. I. Influence of pressing on the susceptibility of the grapes to oxidative browning, *J. Sci. Food Agric.*, 22, 24, 1971.
260. **Pesis, E. and Ben-Arie, R.,** Involvement of acetaldehyde and ethanol accumulation during induced deastringency of persimmon fruits, *J. Food Sci.*, 49, 896, 1984.
261. **Pesis, E. and Ben-Arie, R.,** Carbon dioxide assimilation during postharvest removal of astringency from Persimmon fruit, *Physiol. Plant,* 67, 644, 1986.
262. **Pesis, E., Levi, A., and Ben-Arie, R.,** Role of acetaldehyde production in the removal of astringency from Persimmon fruits under various modified atmospheres, *J. Food Sci.*, 53, 153, 1988.
263. **Petrus, D. R.,** Spectral characteristics of orange juices and other edible products, Fruit Adulteration Workshop, Herndon, VA, July 20 to 21, 1988.
264. **Phan, C. T.,** Relationship between anatomical structure and metabolism in plants. III. Study of the activities and subunit composition of some enzymes in explants, media and calli of apples in *in vitro* culture, *Rev. Can. Biol. Exp.*, 42, 19, 1983.
265. **Phan, C. T. and Macheix, J. J.,** Relation entre la structure anatomique et le métabolisme des tissus végétaux. I. Differences entre la composition qualitative et quantitative des substances phénoliques d'explantats de pomme et celles de cals et de cellules qui en sont issues par culture, *Rev. Can. Biol.*, 38, 17, 1979.
266. **Philippon, J. and Rouet-Mayer, M. A.,** Décongelation des fruits à noyau. Prévention des brunissements enzymatiques, *Rev. Gén. Froid,* 5, 487, 1973.
267. **Philippon, J. and Rulland, J.,** Qualité de Pêches au sirop appertisées élaborées à partir de fruits congelés entiers-Etude à l'échelon industriel, *Rev. Int. Froid,* 7, 245, 1984.
268. **Philippon, J., Rouet-Mayer, M. A., and Brochier, J. J.,** Conservation par réfrigération de tranches de pommes destinées à l'industrie, *Int. J. Refrig.*, 85, 1979.
269. **Pierpoint, W. S.,** The enzymic oxidation of chlorogenic acid and some reactions of the quinone produced, *Biochem. J.*, 98, 567, 1966.
270. **Pifferi, P. G. and Cultrera, R.,** Enzymatic degradation of anthocyanin: the role of sweet cherry polyphenoloxidase, *J. Food Sci.*, 39, 786, 1974.
271. **Pifferi, P. G., Baldassari, L., and Cultrera, R.,** Inhibition by carboxylic acids of an *o*-diphenol oxidase from *Prunus avium* fruits, *J. Sci. Food Agric.*, 25, 263, 1974.
272. **Pollard, A. and Timberlake, C. F.,** Fruit juices, in the *Biochemistry of Fruits and Their Products,* Vol. 2, Hulme, A. C., Ed., Academic Press, London, 1971, 573.
273. **Pontaillier, P., Salagoity-Auguste, M., and Ribereau-Gayon, P.,** Intervention du bois de chêne dans l'évolution des vins rouges élevés en barriques, *Conn. Vigne Vin,* 16, 45, 1982.
274. **Ponting, J. D., Jackson, R., and Watters, G.,** Refrigerated apple slice: preservative effects of ascorbic acid, calcium and sulfite, *J. Food Sci.*, 37, 434, 1972.
275. **Porritt, S. W.,** The effect of oxygen and low concentration of carbon dioxide on the quality of apples stored in controlled atmosphere, *Can. J. Plant Sci.*, 46, 317, 1965.
276. **Powers, J. R., Nagel, C. W., and Weller, K.,** Protein removal from a wine by immobilized grape proanthocyanidins, *Am. J. Enol. Vitic.*, 39, 117, 1988.
277. **Prabha, T. N. and Patwardhan, M. V.,** Polyphenols of avocado (*Persea ammericana*) and their endogenous oxidation, *J. Food Sci. Technol.*, 17, 215, 1980.
278. **Prabha, T. N. and Patwardhan, M. V.,** A comparison of the polyphenolic patterns in some Indian varieties of apples and their endogenous oxidation. I. *J. Food Sci. Technol.*, 22, 404, 1985.

279. **Prabha, T. N., and Patwardhan, M. V.,** A comparison of the browning potential of some Indian cultivars of ripening apples. II, *J. Food Sci. Technol.*, 22, 431, 1985.
280. **Pratt, D. E.,** Role of flavones and related compounds in retarding lipid-oxidative flavor change in foods, in *Phenolic, Sulfur and Nitrogen Compounds in Food Flavors,* ACS Symposium 26, Charalambous, G. and Katz I., Eds., American Chemical Society, Washington, D.C., 1976, 1.
281. **Puech, J. L.,** Phenolic compounds in oak wood extracts used in the ageing of brandies, *J. Sci. Food Agric.*, 42, 165, 1988.
282. **Raynal, J.,** Modification stucturales et biochimiques de la Prune d'ente au début du séchage. Effets de diverses températures sur la structure cellulaire et ses incidences sur les composés pectiques et phénoliques. Thèse Doct., France, Montpellier, 1987.
283. **Raynal, J., Ginestet, C., Souquet, J. M., and Moutounet, M.,** Evolution des constituants phénoliques de l'épicarpe de la prune d'ente au cours du séchage, *Bull. Liaison Groupe Polyphénols,* 13, 482, 1986.
284. **Ribereau-Gayon, P.,** Les composés phénoliques du raisin et du vin, II. Les flavonoides et les anthocyanosides, *Ann. Physiol. Veg.*, 6, 210, 1964.
285. **Ribereau-Gayon, P.,** Evolution des composés phénoliques au cours de la maturation du raisin, II. Discussion des résultats obtenus en 1969, 1970 et 1971, *Conn. Vigne Vin,* 6, 161, 1972.
286. **Ribereau-Gayon, P.,** The anthocyanins of grapes and wines, in *Anthocyanins as Food Colors,* Markakis, P., Ed., Academic Press, New York, 1982, 209.
287. **Ribereau-Gayon, P. and Glories, Y.,** Phenolics in grape and wines, *Proc. 6th Aust. Wine Ind. Tech. Conf.*, 247, 1986.
288. **Ribereau-Gayon, P., Pontallier, P., and Glories, Y.,** Some interpretations of colour changes in young red wines during their conservation, *J. Sci. Food Agric.*, 34, 505, 1983.
289. **Ribereau-Gavon, J., Peynaud, E., Ribereau-Gayon, P., and Sudraud, P.,** *Traité d'oenologie,* Tome II, Dunod, Paris, 1975.
290. **Ribereau-Gayon, J., Peynaud, E., Ribereau-Gayon, P., and Sudraud, P.,** *Traité d'oenologie,* Sciences et Techniques du Vin, Tome 3, Vinifications, Transformations du Vin, Paris, Dunod, 1976.
291. **Rigaud, J., Moutounet, M., and Cheynier, V.,** Relation entre la consommation d'oxygène et la Consommation en dérivés hydroxycinnamiques de quatre mouts de raisins blancs, *Sci. Aliment.*, 8, 467, 1988.
292. **Rivas, N. and Whitaker, R. J.,** Purification and some properties of two polyphenol oxidases from Bartlett pears, *Plant Physiol.*, 52, 501, 1973.
293. **Romeyer, F. M., Macheix, J. J., and Sapis, J. C.,** Changes and importance of oligomeric procyanidins during maturation of grape seeds, *Phytochemistry,* 25, 219, 1986.
294. **Romeyer, F. M., Sapis, J. C., and Macheix, J. J..,** Hydroxycinnamic esters and browning potential in mature berries of some grape varieties, *J. Sci. Food Agric.*, 36, 728, 1985.
295. **Roudsari, M. H., Signoret, A., and Crouzet, J.,** Eggplant polyphenol oxidase purification, characterization and properties, *Food Chem.*, 7, 227, 1981.
296. **Rouet-Mayer, M. A. and Philippon, J.,** The reversibility of brown discoloration for frozen peaches, *Proc. IV Int. Congress Food Sci. Technol.*, Vol. 1, 33, 1974.
297. **Rouet-Mayer, M. A. and Philippon, J.,** Inhibition of catechol oxidases from apples by sodium chloride, *Phytochemistry,* 25, 2717, 1986.
298. **Rouseff, R. L., Martin, S. F., and Youtsey, C. O.,** Quantitative survey of narirutin, naringin, hesperidin, and neohesperidin in citrus, *J. Agric. Food Chem.*, 35, 1027, 1987.
299. **Sakamura, S., Watanabe, S., and Obata, Y.,** Anthocyanase and anthocyanins occuring in eggplant *Solanum melongena* L. Part III. Oxidative decoloration of the anthocyanin by polyphenol oxidase, *Agric. Biol. Chem.*, 29, 181, 1965.
300. **Sal'kova, E. G., Zuyagintseva, Y. V., Kulien, A. A., and Akhundov, R. M.,** Effect of etherel on biochemical processes during fruit ripening, *Appl. Biochem. Microbiol.*, 13, 71, 1977.
301. **Salgues, M., Cheynier, V., Gunata, Z., and Wylde, R.,** Oxidation of grape juice 2-*S*-glutahionyl caffeoyl tartaric acid by *Botrytis cinerea* laccase and characterization of a new substance: 2,5-di-*S*-glutahionyl caffeoyl tartaric acid, *J. Food Sci.*, 51, 1191, 1986.
302. **Sapers, G. M. and Douglas, F. W., Jr.,** Measurement of enzymatic browning at cut surfaces and in juice of raw apple and pear fruits, *J. Food Sci.*, 52, 1258, 1987.
303. **Sapis, J. C. and Ribereau-Gayon, P.,** Etude du brunissement des vins blancs. II. Influence de certains traitements sur la teneur en composés phénoliques et sur le brunissement. *Conn. Vigne Vin,* 3, 215, 1969.
304. **Sapis, J. C., Macheix, J. J., and Cordonnier, R. E.,** The browning capacity of Grapes. I. Changes in polyphenoloxidase activities during development and maturation of the fruit, *J. Agric. Food Chem.*, 31, 342, 1983.
305. **Sapis, J. C., Macheix, J. J., and Cordonnier, R. E.,** The browning capacity of grapes. II. Browning potential and polyphenol oxidase activities in different mature grape varieties, *Am. J. Enol. Vitic.*, 34, 157, 1983.
306. **Sayavedra-Soto, L. A., and Montgomery, M. W.,** Inhibition of polyphenoloxidase by sulfite, *J. Food Sci.*, 51, 1531, 1986.

307. **Schaeffer, J., Tas, A., Velisek, J., Maarse, H., Ten Noever de Brawn, M., and Slump, P.,** Application of pattern recognition techniques in the differentiation of wines, in *Instrumental Analysis of Food,* Vol. 2, Charalambous, G., Ed., Academic Press, New York, 1983, 335.
308. **Schreier, P.,** Flavor composition of wines: a review, *CRC Crit. Rev. Food Sci. Nutr.,* 12, 59, 1979.
309. **Sciancalepore, V.,** Enzymatic browning in five Olive varieties, *J. Food Sci.,* 50, 1194, 1985.
310. **Sciancalepore, V. and Longone, V.,** Polyphenol oxidase activity and browning in green Olives, *J. Agric. Food Chem.,* 32, 320, 1984.
311. **Segal, B. and Oranescu, E.,** The effect of aluminum (3 +) and tin (2 +) on the juice color of anthocyanins-rich fruits juices, *Bull. Univ. Galati,* 6, 53, 1978.
312. **Sharma, R. C. and Ali, R.,** Isolation and characterization of catechol oxidase from Solanum melongena, *Phytochemistry,* 19, 1597, 1980.
313. **Sharon, O. and Kahn, V.,** The intracellular location of particulate-bound polyphenol oxidase in avocado mesocarp, *Physiol. Plant.,* 45, 227, 1979.
314. **Sharon, O. and Kahn, V.,** Browning potential PPO, catalase and acid phosphatase activities during ripening of non-chilled and chilled avocado, *J. Sci. Food Agric.,* 30, 634, 1979.
315. **Sharon-Raber, O. and Kahn, V.,** Avocado mesocarp: browning potential, carotenoid content, polyphenoloxidase, catalase and peroxidase activities: comparison between six avocado cultivars, *J. Food Sci.,* 48, 1874, 1983.
316. **Shaw, P. E. and Wilson, C. W.,** III, Debittering citrus juices with beta-cyclodextrin polymers, *J. Food Sci.,* 48, 646, 1983.
317. **Shaw, P. E., and Wilson, C. W.,** III, Reduction of bitterness in grapefruit juice with beta-cyclodextrin polymer in a continuous-flow process, *J. Food Sci.,* 50, 1205, 1985.
318. **Shaw, P. E. Tatum, J. H., and Wilson, C. W.,** III, Improved flavor of Navel orange and grapefruit juices by removal of bitter components with beta-cyclodextrin polymer, *J. Agric. Food Chem.,* 32, 832, 1984.
319. **Shimizu, J. and Watanabe, M.,** Formation of C6-compounds in grape berries and musts, *J. Jpn. Soc. Hortic. Sci.,* 50, 393, 1981.
320. **Shomer, I., Ben-Shalom, N., Harel, E., and Mayer, A. M.,** The intracellular location of catechol oxidase in the olive fruit, *Ann. Bot.,* 44, 261, 1979.
321. **Shrikhande, A. J. and Francis, F. J.,** Effect of flavonols on ascorbic acid and anthocyanin stability in model systems, *J. Food Sci.,* 39, 904, 1974.
322. **Siegel, A., Markakis, P., and Bedford, C. L.,** Stabilization of anthocyanins in frozen tart cherries by blanching, *J. Food Sci.,* 36, 962, 1971.
323. **Siewek, F., Galensa, R., and Herrmann, K.,** Nachweis eines Zuzatzes von roten zu schwarzen Johannisbeer-Erzengnissen über die hochdruckflüssig chromatographisc he bestimmung der Flavonolglylkoside, *Z. Lebensm. Unters. Forsch.,* 179, 315, 1984.
324. **Simard, R. E., Bourzeix, M., and Heredia, N.,** Facteurs affectant la degradation de la couleur du jus de cassis (*Ribes Nigrum* L.), *Sci. Aliment.,* 1, 388, 1981.
325. **Singleton, V. L.,** Grape and wine phenolics: background and prospects, *C.R. Coll. Centenaire Paris,* 215, 1980.
326. **Singleton, V. L.,** Grape and wine phenolics; background and prospects, *Proc. Symp. Grape and Wine Centennial,* University of California, Davis, 215, 1982.
327. **Singleton, V. L.,** Oxygen with phenols and related reactions in musts wines, and model systems: observations and practical implications, *Am. J. Enol. Vitic.,* 38, 69, 1987.
328. **Singleton, V. L. and Esau, P.,** *Phenolic Substances in Grapes and Wine and Their Significance,* Academic Press, New York, 1969.
329. **Singleton, V. L. and Noble, A. C.,** Wine flavor and phenolic substances, ACS Symp. Ser., 47, 1976.
330. **Singleton, V. L., Trousdale, E., and Zaya, J.,** Oxidation of wines. I. Young white wines periodically exposed to air, *Am. J. Enol. Vitic.,* 30, 49, 1979.
331. **Singleton, V. L., Trousdale, E., and Zaya, J.,** One reason sun-dried raisins brown so much, *Am. J. Enol. Vitic.,* 36, 111, 1985.
332. **Singleton, V. L., Zaya, J., and Trousdale, E. K.,** Caftaric and coutaric acids in fruit of Vitis, *Phytochemistry,* 25, 2127, 1986.
333. **Singleton, V. L., Salgues, M., Zaya, J., and Trousdale, E.,** Caftaric acid disappearance and conversion to products of enzymic oxidation in grape must and wine, *Am. J. Enol. Vitic.,* 36, 50, 1985.
334. **Singleton, V. L., Sieberhagen, H. A., De Wet, P., and Van Wyk, C. J.,** Composition and sensory qualities of wines prepared from white grapes by fermentation with and without grape solids, *Am. J. Enol. Vitic.,* 26, 62, 1975.
335. **Singleton, V. L., Zaya, J., Trousdale, E., and Salgues, M.,** Caftaric acid in Grapes and conversion to a reaction product during processing, *Vitis* 23, 113, 1984.
336. **Sistrunk, W. A. and Gascoigne, H. L.,** Stability of color in "Concord" grape juice and expression of color, *J. Food Sci.,* 48, 430, 1983.

337. **Skalski, C. and Sistrunk, W. A.,** Factors influencing color degradation in Concord grape juice, *J. Food Sci.*, 38, 1060, 1973.
338. **Solinas, M., Di Giovacchino, L., and Cucurachi, A.,** I polifenoli delle olive e dell'olio di oliva. Nota 1: Variazioni che subiscono alcuni polifenoli delle olive col procedere della maturazione, *Publicazione 18, Ann. Istit. Sperim. Elaiotecn.*, 5, 3, 1975.
339. **Solinas, M., Di Giovacchino, L., and Cucurachi, A.,** I polifenoli delle olive e dell'olio di oliva. Nota 2: Indagine preliminare sull'incidenza delle tecniche operative adottate per l'estrazione dell'olio, *Publicazione 19, Ann. Istit. Sperim. Elaiotecn.*, 5, 3, 1975.
340. **Solinas, M., Di Giovacchino, L., and Mascolo, A.,** I polifenoli delle olive e dell'olio di oliva. Nota 3: Influenza della temperatura e della durata della gramolatura sul contenuto in polifenoli degli oli, *Riv. Ital. Sostanze Grasse*, 55, 19, 1978.
341. **Solinas, M., Di Giovacchino, L., and Mascolo, A.,** I Prolifenoli delle olive e dell'olio di oliva. Nota 4: Loro evoluzione in dipendenza der processi ossidativi primari che intervengono nel corso dela conservazione dell'olio, *Publicazione 43, Ann. Istit. Sperim. Elaiotecn.*, 8, 3, 1978.
342. **Somers, T. C.,** The polymeric nature of wine pigments, *Phytochemistry*, 10, 2175, 1971.
343. **Somers, T. C.,** Interpretations of colour composition in young red wines, *Vitis*, 17, 161, 1978.
344. **Somers, T. C., and Evans, M. E.,** Wine quality: correlations with colour density and anthocyanin equilibre in a group of young red wines, *J. Sci. Food Agric.*, 25, 1369, 1974.
345. **Somers, T. C. and Evans, M. E.,** Grape pigment phenomena: interpretation of major colour losses during vinification, *J. Sci. Food Agric.*, 30, 623, 1979.
346. **Somers, T. C. and Evans, M. E.,** Evolution of red wines. I. Ambient influences on colour composition during early maturation. *Vitis*, 25, 31, 1986.
347. **Somers, T. C., Verette, E., and Pocock, F.,** Hydroxycinnamate esters of *Vitis vinifera:* changes during white vinification and effect of exogenous enzymic hydrolysis, *J. Sci. Food Agric.*, 40, 67, 1987.
348. **Spanos, G. A. and Wrolstad, R. E.,** Anthocyanin pigment nonvolative acid and sugar composition of red Raspberry juice, *J. Assoc. Off. Anal. Chem.*, 70, 1036, 1987.
349. **Spayd, S. E., Morris, J. R., and Robert, C. L.,** Influence of peroxidase and polyphenol oxidase on the color of puree from machine-harvested strawberries, *J. Am. Soc. Hortic. Sci.*, 107, 1070, 1982.
350. **Starr, M. S. and Francis, F. J.,** Effect of ascorbic acid on the relative stability of four anthocyanin pigments in cranberry juice, *Food Technol.*, 22, 1293, 1968.
351. **Starr, M. S. and Francis, F. J.,** Effect of metallic ions on color and pigment content of cranberry juice cocktail, *J. Food Sci.*, 38, 1043, 1973.
352. **Stelzig, D. A., Akhtar, S., and Ribeiro, S.,** Catechol oxidase of Red Delicious apple peel, *Phytochemistry*, 11, 535, 1972.
353. **Stutsman, M. J.,** Fruit juice adulteration. An overview of compliance and regulatory issues, Fruit Adulteration Workshop, Herndon, VA, July 20 to 21, 1988.
354. **Sweeny, J. G., Wilkinson, M. M., and Iacobucci, G. A.,** Effect of flavonoid sulfonates on the photobleaching of anthocyanins in acid solutions, *J. Agric. Food Chem.*, 29, 563, 1981.
355. **Synge, R. L. M.,** Interactions of polyphenols with proteins in plants and plant products, *Qual. Plant. Plant Foods Hum. Nutr.*, 24, 337, 1975.
356. **Taira, S., Kimura, Y., and Watanabe, S.,** Effect of harvest maturity on removal of astringency during sun drying of Japanese persimmon (*Diospyros kaki* Thsnb.) 'Hiratanenashi' fruit after peeling, *J. Jpn. Soc. Food Sci. Technol.*, 35, 528, 1988.
357. **Tan Ham, L., Uhel, C., and Macheix, J. J.,** Anthocyanin accumulation in grape cell suspensions. Poster, Int. Symp. Phytochem. Soc. Europe, London, December 19 to 21, 1988.
358. **Taylor, A. J. and Clydesdale, F. M.,** Potential of oxidised phenolics as food colourants, *Food Chem.*, 24, 301, 1987.
359. **Teh, L. S. and Francis, F. J.,** Stability of anthocyanins from *Zebrina pendula* and *Ipomocea tricolor* in a model beverage, *J. Food Sci.*, 53, 1580, 1988.
360. **Teisson, C.,** Etudes sur le brunissement interne de l'ananas, *Fruits*, 27, 603, 1972.
361. **Teisson, C., Lacoeuilhe, J. J., and Combres, J. C.,** Le brunissement interne de l'ananas. V. Recherches des moyens de lutte, *Fruits*, 34, 399, 1979.
362. **Teisson, C., Martin-Prével, P., and Marchal, J.,** Le brunissement interne de l'ananas. IV. Approche biochimique du phénomène, *Fruits*, 34, 329, 1979.
363. **Teisson, C., Martin-Prével, P., Combres, J. C., and Py, C.,** A propos du brunissement interne de l'ananas accident de la réfrigération, *Fruits*, 33, 48, 1978.
364. **Thomas, P. and Janave, M. T.,** Polyphenol oxydase activity and browning of mango fruits induced by gamma irradiation, *J. Food Sci.*, 38, 1149, 1973.
365. **Thomas, P. and Nair, P. N.,** Effect of gamma irradiation on polyphenol oxidase activity and its relation to skin browning in bananas, *Phytochemistry*, 10, 771, 1971.
366. **Thorpe, T. A., Maier, V. P., and Hasegawa, S.,** Phenylalanine ammonialyase activity in citrus fruit tissue cultured *in vitro*, *Phytochemistry*, 10, 711, 1971.

367. **Timberlake, C. F.**, Anthocyanins — occurrence, extraction and chemistry, *Food Chem.*, 5, 69, 1980.
368. **Timberlake, C. F. and Bridle, P.**, Flavylium salts, anthocyanidins and anthocyanins. II. Reactions with sulphur dioxide, *J. Sci. Food Agric.*, 18, 479, 1967.
369. **Timberlake, C. F. and Bridle, P.**, The effect of processing and other factors on the colour characteristics of some red wines, *Vitis*, 15, 37, 1976.
370. **Timberlake, C. F. and Bridle, P.**, Interactions between anthocyanins phenolic compounds and acetaldehyde and their significance in red wines, *Am. J. Enol. Vitic.*, 27, 97, 1976.
371. **Timberlake, C. F. and Bridle, P.**, Anthocyanins: colour augmentation with catechin and acetaldehyde, *J. Sci. Food Agric.*, 28, 539, 1977.
372. **Timberlake, C. F., Bridle, P., Jackson, M. G., and Vallis, L.**, Correlations between quality and pigment parameters in young Beaujolais red wines, *Ana. Nutr. Aliment*, 32, 1095, 1978.
373. **Ting, S. V., Rouseff, R. L., Dougherty, M. H., and Attaway, J. A.**, Determination of some methoxylated flavones in citrus juices by high performance liquid chromatography, *J. Food Sci.*, 44, 69, 1979.
374. **Torel, J., Cillard, J., and Cillard, P.**, Antioxidant activity of flavonoids and reactivity with peroxy radical, *Phytochemistry*, 25, 383, 1986.
375. **Usseglio-Tomasset, L.**, Acquisitions récentes sur les phénomènes colloidaux dans les moûts et les vins, *Ann. Technol. Agric.*, 27, 261, 1978.
376. **Valero, E., Varon, R., and Garcia-Carmona, F.**, Characterization of polyphenol oxidase from Airen grapes, *J. Food Sci.*, 53, 1482, 1988.
377. **Vámos-Vigyázó, L.**, Polyphenol oxidase and peroxidase in fruits and vegetables, *CRC Crit. Rev. Food Nutr.*, 15, 49, 1981.
378. **Vámos-Vigyaźó, L., and Nádudvari-Márkus, V.**, Enzymatic browning, polyphenol content, polyphenol oxidase and peroxydase activities in pear cultivars, *Acta Alim.*, 11, 157, 1982.
379. **Vámos-Vigyázó, L., and Nádudvari-Márkus, V.**, Inactivation of polyphenoloxidase and depletion of *o*-dihydroxyphenol content during the enzymatic browning reaction of fruit tissues, *Acta Aliment.*, 12, 1, 1983.
380. **Vámos-Vigyázó, L., Gajzágó, I., and Nádudvari-Márkus, V.**, Polyphenol oxidase and peroxidase activities of apricots (*Prunus Armeniaca* L.) as affected by cultivar, location and year of harvest, *Qual. Plant.*, 31, 45, 1981.
381. **Vámos-Vigyázó, L., Gajzágó, I., Mihályi, K., and Nádudvari-Márkus, V.**, Changes in enzymic browning, polyphenol oxidase and polyphenol contents of Jonathan apples during storage, *Acta Aliment. Acad. Sci. Hung.*, 6, 389, 1977.
382. **Vámos-Vigyázó, L., Gajzágó, I., Nádudvari-Márkus, V., and Magdolna-Kerek, M.**, Changes in some enzyme activities and browning rate of apricots during storage, *Qual. Plant. Plant Foods Hum. Nutr.*, 35, 139, 1985.
383. **Vámos-Vigyázó, L., Mihályi, K., Gajzágó, I., and Nádudvari-Márkus, V.**, Studies into the enzymic browning and polyphenol-polyphenol oxidase complex of apple cultivars, *Confructa*, 21, 24, 1976.
384. **Vámos-Vigyázó, L., Schuster-Gajzágó, I., Kiss-Kutz, M., Hoschke, A., Toth, E., Erdos, Z., and Bartucz-Kovacs, O.**, Variations of polyphenol content and polyphenol oxidase activity during ripening of plums, *Bull. Liaison Groupe Polyphenols*, 12, 196, 1984.
385. **Van Buren, J. P., Bertino, J. J., and Robinson, W. B.**, The stability of wine anthocyanins on exposure to heat and light, *Am. J. Enol. Vitic.*, 19, 147, 1968.
368. **Van Buren, J., de Vos, L., and Pilnik, W.**, Measurement of chlorogenic acid and flavonol glycosides in Apple juice by a chromatographic-fluorometric method, *J. Food Sci.*, 38, 656, 1973.
387. **Van Buren, J., de Vos, L., and Pilnik, W.**, Polyphenols in Golden Delicious apple juice in relation to method of preparation, *J. Agric. Food Chem.*, 24, 448, 1976.
388. **Van Buren, J. P., Bertino, J. J., Einset, J., Remaily, G. W., and Robinson, W. B.**, A comparative study of anthocyanin composition in wines derived from hybrid grapes, *Am. J. Enol. Vitic.*, 21, 117, 1970.
389. **Van Lelyveld, L. J. and Bower, J. P.**, Enzyme reactions leading to avocado mesocarp discolouration, *J. Hortic. Sci.*, 59, 257, 1984.
390. **Van Lelyveld, L. J., and De Bruyn, J. A.**, Polyphenols, ascorbic acid and related enzymes activities associated with black heart in Cayenne pineapple fruit, *Agrochemophysica*, 9, 1, 1977.
391. **Van Lelyveld, L. J., Gerrish, C., and Dixon, R. A.**, Enzyme activities and polyphenols related to mesocarp discolouration of Avocado fruit, *Phytochemistry*, 23, 1531, 1984.
392. **Van Sumere, C. F., Albrecht, J., Dedonder, A., De Pooter, H., and Pe, I.**, Plant proteins and phenolics, in *The Chemistry and Biochemistry of Plant Proteins*, Harborne, J. B. and Van Sumere, C. F., Eds., Academic Press, London, 1975, 211.
393. **Vandercook, C. E. and Tisserat, B.**, Flavonoid changes in developing lemons grown *in vivo* and *in vitro*, *Phytochemistry*, 28, 799, 1989.
394. **Varoquaux, P., Souty, M., and Varoquaux, F.**, Le blanchiment d'abricots entiers dans l'eau, *Sci. Aliment.*, 6, 591, 1986.
395. **Vasquez Roncero, A.**, Les polyphenols de l'huile d'olive et leur influence sur les caractéristiques de l'huile, *Rev. Fr. Corps gras*, 1, 21, 1978.

396. **Vasquez Roncero, A. and Janer del Valle, L.,** Evolution de los polifenoles durante el aderezo de aceitunas verdes, I. Estudio cualitativo, *Grasas Aceites,* 28, 421, 1977.
397. **Vasquez Roncero, A., Maestro Duran, R., and Graciani Constante, E.,** Componentes fenolicos de la aceituna. II. Polifenoles del alpechin, *Grasas Aceites,* 25, 341, 1974.
398. **Vaughn, K. C., Lax, A. R., and Duke, S. O.,** Polyphenol oxidase: the chloroplast oxidase with no established funtion, *Physiol. Plant,* 72, 659, 1988.
399. **Veldhuis, M. K., Swift, L. J., and Scott, W. C.,** Fully-methoxylated flavones in Florida orange juices, *J. Agric. Food Chem.,* 18, 590, 1970.
400. **Verette, E., Noble, A. C., and Somers, T. C.,** Hydroxycinnamates of *Vitis vinifera:* sensory assessment in relation to bitterness in white wines, *J. Sci. Food Agric.,* 45, 267, 1988.
401. **Volk, R., Harel, E., Mayer, A. M., and Gan-Zvi, E.,** Catechol oxidase in tissue culture of apple fruit, *J. Exp. Bot.,* 28, 820, 1977.
402. **Volk, R., Harel, E., Mayer, A. M., and Gan-Zvi, E.,** Catechol oxidase in suspension cultures of apple fruit. The effects of growth regulators, *J. Exp. Bot.,* 29, 1099, 1978.
403. **Waite, J. H.,** Calculating extinction coefficients for enzymatically produced *o*-quinones, *Anal. Biochem.,* 75, 211, 1976.
404. **Walker, J. R. L.,** Studies on the enzymic browning of apple fruit. II. Properties of apple polyphenol oxidase, *Aust. J. Biol. Sci.,* 17, 360, 1964.
405. **Walker, J. R. L.,** The polyphenoloxidase of pear fruit, *Aust. J. Biol. Sci.,* 17, 575, 1964.
406. **Walker, J. R. L.,** Inhibition of the apple phenolase system through infection by *Penicillium expansum, Phytochemistry,* 8, 561, 1969.
407. **Walker, J. R. L.,** Enzymic browning in foods, *Enzyme Technol. Dig.,* 3, 89, 1975.
408. **Walker, J. R. L.,** The control of enzymic browning in fruit juices by cinnamic acids, *J. Food Technol.,* 11, 341, 1976.
409. **Walker, J. R. L. and Mc Callion, R. F.,** The selective inhibition of ortho- and para-diphenol oxidases, *Phytochemistry,* 19, 373, 1980.
410. **Walker, J. R. L. and Reddish, C. E. S.,** Note on the use of cysteine to prevent browning in apple products, *J. Sci. Food Agric.,* 15, 904, 1964.
411. **Walker, J. R. L. and Wilson, E. L.,** Studies on the enzymic browning of apples. Inhibition of apple *o*-diphenol oxidase by phenolic acids, *J. Sci. Food Agric.,* 26, 1825, 1975.
412. **Walter, W. M., Fleming, H. P., and Etchells, J. L.,** Preparation of antimicrobial compounds by hydrolysis of oleuropein from green olives, *Appl. Microbiol.,* 26, 733, 1973.
413. **Wang, C. Y. and Mellethin, W. M.,** Relationship of friction discoloration to phenolic compounds in d'Anjou pears, *HortScience,* 8, 321, 1973.
414. **Weaver, C. and Charley, H.,** Enzymatic browning of ripening bananas, *J. Food Sci.,* 39, 1200, 1974.
415. **Weetall, H. H., Zelko, J. T., and Bailey, L. F.,** A new method for the stabilization of white wine, *Am. J. Enol. Vitic.,* 35, 212, 1984.
416. **Weinges, K. and Piretti, M. N.,** Isolierung des $C_{30}H_{26}O_{12}$, Procyanidins B-1 aus Weintrauben, *Liebig's Annalen,* 747, 218, 1971.
417. **Weinges, K. and Piretti, M. V.,** Imbrunimento dei vini bianchi. II. Influenza dei polifenoli, *Ann. Chim. Rome,* 62, 46, 1972.
418. **Weurman, C. and Swain, T.,** Changes in the enzymic browning of Bramley's seedling apples during their development, *J. Sci. Food Agric.,* 6, 186, 1955.
419. **Whitaker, R. J., and Hashimoto, T.,** Production of secondary metabolites, in *Handbook of Plant Cell Culture,* Vol. 4, Evans, D. A., Sharp, W. R. and Ammirato, P. V., Eds., Macmillan, New York, 1986, 264.
420. **Whiting, G. C. and Coggins, R. A.,** Estimation of the monomeric phenolics of ciders, *J. Sci. Food Agric.,* 26, 1833, 1975.
421. **Wildenradt, H. L. and Singleton, V. L.,** Production of aldehydes as a result of oxidation of polyphenolic compounds and its relation to wine aging, *Am. J. Enol. Vitic.,* 25, 119, 1974.
422. **Wilson, E. L.,** High-pressure liquid chromatography of apple juice phenolic compounds, *J. Sci. Food Agric.,* 32, 257, 1981.
423. **Wink, M.,** Physiology of the accumulation of secondary metabolites with special reference to alkaloids, in *Cell Culture and Somatic Cell Genetics of Plants,* Vol. 4, Vasil, I. K., Ed., Academic Press, San Diego, 1987, 17.
424. **Winter, M. and Herrmann, R. K.,** Esters and glucosides of hydroxycinnamic acids in vegetables, *J. Agric. Food Chem.,* 34, 616, 1986.
425. **Wissemann, K. W. and Lee, C. Y.,** Characterization of polyphenoloxidase from Ravat 51 and Niagara grapes, *J. Food Sci.,* 46, 506, 1981.
426. **Wong, E.,** Biosynthesis of flavonoids, in *Chemistry and Biochemistry of Plant Pigments,* Vol. 1, 2nd ed., Goodwin, T. W., Ed., Academic Press, London, 1976, 464.

427. **Wong, T. C., Luh, B. S., and Whitaker, J. R.,** Isolation and characterization of polyphenol oxidase isoenzymes of Clingstone peach, *Plant Physiol.*, 48, 19, 1971.
428. **Wrolstad, R. E.,** Detection of adulteration in red raspberry, blackberry and strawberry products, Fruit Adulteration Workshop, Herndon, VA, July 20 to 21, 1988.
429. **Wrolstad, R. E., Cornwell, C. J., Culbertson, T. D., and Reyes, F. G. R.,** Etablishing criteria for determining the authenticity of fruit juice concentrates, *ACS Symposium Series*, 170, 77, 1981.
430. **Wulf, L. W. and Nagel, C. W.,** Identification and changes of flavonoids in merlot and Cabernet Sauvignon wines, *J. Food Sci.*, 45, 479, 1980.
431. **Yamakawa, T., Ishida, K., Kato, S., Kodama, T., and Minoda, Y.,** Formation and identification of anthocyanins in cultured cells of *Vitis* sp., *Agr. Biol. Chem.*, 47, 997, 1983.
432. **Yamakawa, T., Kato, S., Ishida, K., Kodama, T., and Minoda, Y.,** Production of anthocyanins by *Vitis* cells in suspension culture, *Agric. Biol. Chem.*, 47, 2185, 1983.
433. **Yeoman, M. M.,** Techniques, characteristics, properties and commercial potential of immobilized plant cells, in *Cell Culture and Somatic Cell Genetics of Plants*, Vol. 4, Vasil, I. K., Ed., Academic Press, San Diego, 1987, 197.
434. **Yoneyama, C. and Kushida, T.,** Phenolic fraction and browning capacity of white wine made from abnormal (Ajinashi disease) Koshu grapes, *J. Inst. Enol. Viticult. Yamanashi Univ.*, 16, 9, 1981.

Chapter 6

CONCLUSION: PERSPECTIVES AND PROSPECTS

Knowledge of phenolic compounds in fruits has increased considerably over the past 20 years, particularly with regard to qualitative and quantitative analysis. It has been possible to plot the phenolic pattern of most of the common fruits and to monitor their evolution during growth and above all maturation when very characteristic changes take place. However, in spite of the progress made, gaps still remain and work on analysis in coming years should be concentrated on the following aspects:

- Quantitative studies of each of the phenolic compounds are still too limited, even for the most common fruits which have already been closely studied. They should be extended to a greater variety of cultivars to be able to characterize each of them by a phenolic pattern taking into account both genetic control and the differences related to the biological and physicochemical environment of the fruit.
- Certain common fruits of great economic interest have not been the subject of recent study using modern analytical techniques.
- Some groups of compounds have still received inadequate study, e.g., tannins, where the difficulty lies first of all in very poor solubilization of the most condensed forms and then in chemical analysis itself.

Although fruits have now been recognized as an organ in which the phenolic metabolism is active, very few studies have been devoted to enzymatic regulation. Utilization of *in vitro* culture of fruit material should be more broadly developed in the future to approach this question and in particular the mechanisms of vacuolar accumulation of phenols in fruits. However, the development of *in vitro* culture should not hide the relations between the fruit and the rest of the plant. Does the fruit, which is a very active sink for the products of photosynthesis, also take up phenolic units formed elsewhere or only precursors? This question raises the more general problem of the possible transport of phenolics in the plant, and the fruit/plant model could be a good way of approaching investigation of this.

As for most secondary metabolites, the *raison d'être* of phenolics in fruits is, in part, badly understood. Whereas their role in the interface between the fruit and its environment has now been well established (protection against pathogens, attraction of animals for dissemination), their participation in the physiology of the organ itself (growth, maturation, abscission, etc.) remains a subject for discussion and requires experimental demonstration.

When an overall balance is drawn up taking into account the various domains in which phenolics are involved in a fruit, the conclusions are somewhat contradictory.

- First, high levels of some phenolic compounds are required for protection (pre-existing or induced) against attack by pathogens.
- Second, low levels of phenolic substrates are sought to reduce risk of browning during processing.
- Finally, requirements may be different depending on whether certain phenolic compounds play a direct positive or negative role in the sensory properties of fruits or fruit products (color, bitterness, astringency) or in their chemical reactivity (browning substrates or interaction as antioxidants).

Modification of the phenolic pattern of fruits can be envisaged by means of plant selection either by conventional hybridization techniques and selection or by use in the future of new genetic engineering techniques. However, a reply must first be found to the question whether fruits with high or low phenolic contents should be obtained. All the data that we have reported and discussed in this work show distinctly that there cannot be a single answer to this. The answer must take into account, for a given fruit, both the nature of the different phenolics that are present and their chemical and biological activity and the constraints imposed by the way in which man uses this fruit.

INDEX *

A

* Phenolics of individual fruits are given in abbreviated form in Chapter 2. Those abbreviations are not indexed here and the reader should refer to Chapter 2.

B

C

D

E

G

N

O

P

Q

R

S

T

U

V